PROTEIN METABOLISM
AND NUTRITION

EUROPEAN ASSOCIATION FOR ANIMAL PRODUCTION
PUBLICATION NO.16

Protein Metabolism and Nutrition

Edited by

D.J.A. COLE, PhD
K.N. BOORMAN, PhD
P.J. BUTTERY, PhD
D. LEWIS, PhD, DSc
R.J. NEALE, PhD
H. SWAN, PhD

*University of Nottingham
School of Agriculture*

BUTTERWORTHS
LONDON - BOSTON
Sydney - Wellington - Durban - Toronto

THE BUTTERWORTH GROUP

ENGLAND

Butterworth & Co (Publishers) Ltd
London: 88 Kingsway, WC2B 6AB

AUSTRALIA

Butterworths Pty Ltd
Sydney: 586 Pacific Highway, NSW 2067
Also at Melbourne, Brisbane, Adelaide and Perth

SOUTH AFRICA

Butterworth & Co (South Africa) (Pty) Ltd
Durban: 152-154 Gale Street

NEW ZEALAND

Butterworths of New Zealand Ltd
Wellington: 26-28 Waring Taylor Street

CANADA

Butterworth & Co (Canada) Ltd
Toronto: 2265 Midland Avenue,
Scarborough, Ontario, M1P 4S1

USA

Butterworth (Publishers) Inc
Boston: 161 Ash Street,
Reading, Mass. 01867

All rights reserved. No part of this publication may be reproduced or transmitted in any form or by any means, including photocopying and recording, without the written permission of the copyright holder, application for which should be addressed to the publisher. Such written permission must also be obtained before any part of this publication is stored in a retrieval system of any nature.

This book is sold subject to the Standard Conditions of Net Books and may not be re-sold in the UK below the net price given by the Publishers in their current price list.

First published 1976

ISBN 0 408 70669 4

© The several contributors named in the list of contents, 1976

Printed in Great Britain by Fletcher & Son Ltd, Norwich
and bound by Richard Clay (The Chaucer Press), Ltd., Bungay, Suffolk

PREFACE

The European Association of Animal Production (EAAP) has for some time organised symposia on energy metabolism in farm animals. A similar symposium on trace elements was arranged some years ago. It therefore seemed desirable to have comparable symposia dealing with protein metabolism and nutrition. The first of these was held at the University of Nottingham in July 1974 and was organised by the University of Nottingham and EAAP. Two groups were chosen to carry out the arrangements: a sub-committee of the Animal Nutrition Study Commission of EAAP and a local committee at the University of Nottingham delegated to organise the details of the event. The membership of these groups was as follows:

International Committee

Professor A. Schürch (Chairman)
Dr. K.O. von Selle (Secretary)
Professor A.M. Frens
Professor T. Homb
Professor J. Kielanowski
Professor D. Lewis
Professor A. Poppe
Dr. A. Rerat
Professor A. vaz Portugal

Local Organising Committee

Professor D. Lewis (Chairman)
Dr. D.J.A. Cole (Secretary)
Professor E.F. Annison
Dr. K.N. Boorman
Dr. P.J. Buttery
Dr. R.J. Neale
Dr. H. Swan

It is intended that the symposia will be continued at three yearly intervals and that, as with the other symposia, the proceedings will be published in the form of a book. The next symposium will be held in 1977 in the Netherlands.

It was decided to arrange the programme on the basis of several groups of review papers, each group representing a theme. Those invited to give these papers were encouraged to present a review and then, where appropriate, to comment on more detailed aspects which were part of their current research programmes. Thus, it was hoped that the published proceedings would constitute a cohesive text concerning protein metabolism and nutrition.

The major themes were protein synthesis and turnover, digestion and availability, nitrogen-energy relationships, measurement of protein adequacy, protein nutrition of non-ruminants and protein nutrition of ruminants. It is possible that succeeding symposia will select different themes or perhaps a different approach in regard to invited review or short research papers.

ACKNOWLEDGEMENTS

The organisation of the symposium was made possible by the generous support of:

J. Bibby Agriculture Ltd.,
Boots Company Ltd.,
The British Nutrition Foundation,
Colborn Group Ltd.,
Crosfields Farm Foods Ltd.,
Eastern Counties Farmers Ltd.,
ICI Ltd.,
Milk Marketing Board,
Nitrovit Ltd.,
Pauls and Whites Foods Ltd.,
Pedigree Petfoods Ltd.,
Pfizer Ltd.,
RHM Research Ltd.,
The Rumenco Group,
Spillers Ltd.,
Trouw - Great Britain Ltd.,
The Wellcome Trust.

The Vice-Chancellor of Nottingham University (Dr W.J.H. Butterfield) is thanked for opening the symposium, welcoming visitors to the University and acting as chairman of the first session.

Our thanks are due to Mrs Y. Cole and Mrs J. Buttery for organising the Associate Members programme.

Finally the Local Organising Committee would like to thank all those who have helped in so many different ways and to mention particularly Miss K. Robson for her valuable assistance.

CONTENTS

I	Protein Synthesis and Turnover	1
1	**EUKARYOTE PROTEIN SYNTHESIS AND ITS CONTROL** H.N. Munro, *Massachusetts Institute of Technology, Cambridge, Massachusetts, USA*	3
2	**AMINO ACID SUPPLY AND PROTEIN SYNTHESIS IN ANIMAL CELLS** M.J. Clemens, *National Institute for Medical Research, Mill Hill, London* Virginia M. Pain, *London School of Hygiene and Tropical Medicine, London, W.C.1.*	19
3	**HORMONAL CONTROL OF PROTEIN METABOLISM** K.L. Manchester, *Department of Biochemistry, University of the West Indies, Kingston, Jamaica* now, *Professor of Biochemistry, University of Witwatersrand, Johannesburg, South Africa*	35
4	**PROTEIN TURNOVER** D.J. Millward, P.J. Garlick, W.P.T. James, P. Sender and J.C. Waterlow, *London School of Hygiene and Tropical Medicine, London, W.C.1*	49
II	Digestion and Availability	71
5	**AMINO ACID TRANSPORT** B.G. Munck, *Institute of Medical Physiology, University of Copenhagen, Copenhagen, Denmark*	73
6	**PROTEIN DIGESTION AND ABSORPTION** A. Rerat, T. Corring and J.P. Laplace *Centre National de Recherches Zootechniques, Jouy-en-Josas, France*	97
7	**AMINO ACID AVAILABILITY** H. Erbersdobler, *Institute für Tierphysiologie der Universität Munchen, West Germany*	139
8	**NITROGEN PASSAGE THROUGH THE WALL OF THE RUMINANT DIGESTIVE TRACT** M.I. Chalmers, I. Grant and F. White, *Rowett Research Institute, Bucksburn, Aberdeen*	159

III Nitrogen-Energy Relationships — 181

9 AMINO ACIDS AS SOURCES OF ENERGY — 183
D.B. Lindsay, *Institute of Animal Physiology, Babraham, Cambridge*

10 THE ENERGETIC EFFICIENCY OF AMINO ACID METABOLISM — 197
P.J. Buttery and K.N. Boorman, *Department of Applied Biochemistry and Nutrition, University of Nottingham*

11 ENERGY COST OF PROTEIN DEPOSITION — 207
J. Kielanowski, *Institute of Animal Physiology and Nutrition, Jablonna, Poland*

12 NITROGEN-ENERGY INTERACTIONS IN RUMEN FERMENTATION — 217
N.P. McMeniman, D. Ben-Ghedalia and D.G. Armstrong, *Department of Agricultural Biochemistry, University of Newcastle-upon-Tyne*

IV Measurement of Protein Adequacy — 231

13 BIOLOGICAL CRITERIA OF PROTEIN EVALUATION — 233
J. Delort-Laval, *Centre National de Recherches Zootechniques, Jouy-en-Josas, France*

14 INDIRECT MEASURES OF PROTEIN ADEQUACY — 249
B.O. Eggum, *Department of Animal Physiology and Chemistry, Rolighedsveg 25, Copenhagen, Denmark*

15 DIETARY EFFECTS AND AMINO ACIDS IN TISSUES — 259
R. Pion, *Centre de Recherches Zootechniques, Beaumont, France*

16 THE NUTRITIONAL AND METABOLIC EFFECTS OF AMINO ACID IMBALANCES — 279
Q.R. Rogers, *University of California, California, USA*

V Protein Nutrition of Non-Ruminants — 303

17 AMINO ACID REQUIREMENTS OF MEAT PRODUCING POULTRY — 305
C. Calet, *Institut National de la Recherche Agronomique, Nouzilly, France*

18 **PROTEIN IN THE DIETS OF THE PULLET AND LAYING BIRD** 323
C. Fisher, *Poultry Research Centre, Edinburgh, Scotland*
Present address: *Unilever Ltd., Colworth House, Sharnbrook, Bedford*

19 **PROTEIN NUTRITION OF THE BREEDING SOW** 353
F.W.H. Elsley, *School of Agriculture, University of Edinburgh, Scotland*

20 **THE AMINO ACID REQUIREMENTS OF GROWING PIGS (1)** 369
S. Poppe, *Animal Husbandry Section, University of Rostock, East Germany*

21 **THE AMINO ACID REQUIREMENTS OF GROWING PIGS (2)** 383
T. Homb, *Agricultural College of Norway*

VI Protein Nutrition of Ruminants 395

22 **UTILISATION OF NITROGEN IN RUMINANTS** 397
J.R. Mercer and E.F. Annison, *Unilever Research Laboratory, Sharnbrook, Bedford*
Present address: *Department of Animal Husbandry, University of Sydney, Australia*

23 **AMINO ACID REQUIREMENTS OF RUMINANTS** 417
D. Lewis and R.M. Mitchell, *University of Nottingham*

24 **FACTORS INFLUENCING THE SUPPLY OF NITROGEN AND AMINO ACIDS TO THE INTESTINE OF DAIRY COWS** 425
H. Hagemeister, W. Kaufmann and E. Pfeffer, *Institute für Milcherzeugung, Kiel, West Germany*

25 **PROTEIN REQUIREMENTS IN RELATION TO THE LACTATION CYCLE** 441
A.J.H. van Es and H.A. Boekholt, *Agricultural University, Wageningen, Netherlands*

26 **FACTORS INFLUENCING PROTEIN AND NON-PROTEIN NITROGEN UTILISATION IN YOUNG RUMINANTS** 457
E.R. Ørskov, *Rowett Research Institute, Bucksburn, Aberdeen*

27 **FUTURE ROLE OF COMPUTER SIMULATION IN RESEARCH AND ITS APPLICATION TO RUMINANT PROTEIN NUTRITION** 477
J.L. Black, G.J. Faichney and N. McGraham, *C.S.I.R.O. Division of Animal Physiology, Prospect, NSW, Australia*

List of Delegates 493

Index 509

I

PROTEIN SYNTHESIS AND TURNOVER

1
EUKARYOTE PROTEIN SYNTHESIS AND ITS CONTROL

H.N. MUNRO
Massachusetts Institute of Technology, Cambridge, Massachusetts, U.S.A.

Introduction

In the context of a symposium on protein metabolism and nutrition, it is appropriate to start with the mechanism of protein synthesis and its regulation, since this represents a major route of disposal of amino acids. For example, the amount of protein synthesised daily in the body of a healthy adult man is probably about 300 g, a figure which can be compared with the customary Western dietary intake of about 100 g of protein daily (Munro, 1973). Apart from emphasising the intensity of protein synthesis within the body, these observations also imply considerable re-utilisation of amino acids as a result of recycling within the body.

This introductory paper provides a survey of the main features of the mechanism of protein synthesis in animal cells, and attempts to identify how protein synthesis may be regulated by hormonal and metabolic factors. The first section of the review deals with transcription of nuclear DNA, resulting in the synthesis of various forms of RNA, including messenger RNA. This is followed by a section on the mechanism of translation of this message in the cytoplasm with formation of peptide chains of proteins. The third section discusses post-translational modifications in the peptide chain, including alterations of individual amino acid residues and the formation of secreted proteins and its control. Synthesis of mitochondrial proteins is also included in this section. Finally, an attempt is made to integrate some of the short-term adaptations occurring in the liver in response to a change in amino acid supply, including synthesis of proteins on polyribosomes, alterations in RNA synthesis and breakdown, and changes in mitochondrial nucleic acid metabolism.

Eukaryote Transcription and its Control

Since DNA is the inherited genetic material, it is necessary to have a mechanism which turns this stored information into a form that can be read by successive generations of cells. The first step in making the information available is to transcribe the DNA into RNA. Our current understanding of this mechanism in animal cells is summarised in *Figure 1.1*.

4 Protein synthesis

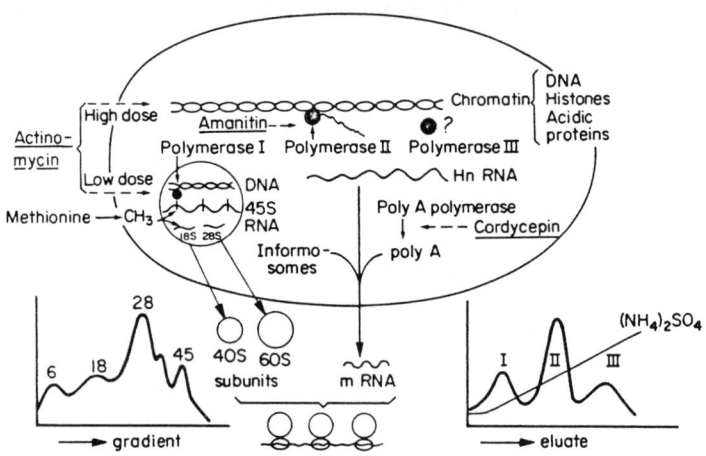

Figure 1.1 Mechanisms of RNA synthesis in the nucleus. The left lower diagram represents resolution of nuclear RNA on a sucrose gradient. The right lower diagram represents DEAE-Sephadex separation of DNA-dependent RNA-polymerases from mammalian nuclei

In eukaryote organisms, the DNA of the nucleus is associated with histones and acidic proteins which make up the chromosomes. In contrast, the DNA of the mitochondria, like that of bacterial cells, consists of a simple loop unassociated with proteins. It is generally accepted that the proteins of nuclear chromosomes play a part in regulating which pieces of the encoded information will be made available, though the detailed mechanism of this regulatory function remains uncertain (Stein, Spelsberg and Kleinsmith, 1974).

The nucleus also contains several forms of RNA separable on sucrose gradients (*Figure 1.1*). The nucleolus is the site of formation of ribosomal RNA. This is first made as a single strand of 45S RNA, which then undergoes fission into shorter lengths, finally resulting in the 28S and 18S RNA found in the two ribosome subunits. This process of selective enzyme cleavage of the 45S RNA (maturation) is regulated by methylation of the ribose residues of certain strategic nucleotides (Greenberg and Penman, 1966). These block enzymic degradation of the nucleotide chain and account for the occurrence of a few methylated riboses in the final ribosomal RNA. In the nuclear sap (nucleoplasm), RNA of large size (heterogeneous nuclear RNA, HnRNA) is transcribed from the chromosomal DNA and from this giant RNA, messenger RNAs with average molecular weights one-fifth that of HnRNA are excised, probably by a mechanism also involving methylation of some parts of the nucleotide chain that is to be conserved as mRNA (Perry and Kelley, 1974). In order to transfer the mRNA from the nucleus to the cytoplasm, it first becomes linked to polyadenylic acid formed by polyadenylic acid polymerase in the nucleus (Adesnik *et al.*, 1972), although the messenger for histone synthesis seems to be an exception in not bearing polyadenylic acid. Finally, the mRNA becomes associated with protein to give an informosome, in which form it leaves the nucleus for the cytoplasm (Schumm and Webb, 1972).

Inhibitors of these processes are important tools for exploring control of protein synthesis. Ribosomal RNA synthesis is inhibited by low doses of actinomycin D, while inhibition of HnRNA (and thus mRNA) synthesis requires a larger amount of actinomycin D, an antibiotic that binds to DNA. Formation of mRNA is also specifically inhibited by amanitin, which, as discussed below, binds to the polymerase rather than the DNA template. Polyadenylic acid polymerase is inhibited by cordycepin (Darnell *et al.*, 1971).

The formation of RNA on a template of eukaryote DNA has long been known to involve the action of DNA-dependent RNA polymerase, and some five years ago three laboratories (Roeder and Rutter, 1969; Kedinger *et al.*, 1970; Jacob, Sajdel and Munro, 1970) demonstrated that the RNA polymerase of the nucleolus has different properties from the enzyme of the nucleoplasm. The RNA polymerase activity of whole nuclei can be resolved on DEAE-Sephadex into several species, of which three major activities, polymerases I, II and III, are shown in the inset of such a column separation at the lower right of *Figure 1.1*. Polymerase I is present in the nucleolus where it is responsible for 45S formation. Polymerase II occurs in the nucleoplasm and is responsible for transcription of HnRNA and thus of mRNA; this polymerase is specifically inhibited by amanitin, and in consequence amanitin can be used to block induction of proteins requiring additional mRNA, such as induction of microsomal enzymes by phenobarbital (Jacob, Scharf and Vesell, 1974). The function of the less abundant nucleoplasmic enzyme, polymerase III, has been identified with synthesis of tRNA and 5s RNA.

The response of the RNA-synthesising system to hormones and to other stimuli (including amino acid supply) is an important means for altering protein synthesis and there are now adequate techniques for exploring how such changes are brought about. For example, it has been well established that liver RNA content increases following administration of corticosteroid hormones to animals (e.g. Goodlad and Munro, 1959). Most of the RNA in cells is ribosomal, so that such an increment in amount represents mainly increased synthesis of rRNA. It has been shown that a single injection of hydrocortisone into rats is indeed followed after three and a half hours by an increase in the amount of 45S RNA made in the liver nucleoli and by a more rapid processing of the 45S RNA into 28S and 18S ribosomal forms (Jacob, Sajdel and Munro, 1969). When the nucleoli from such hormone-injected animals were isolated and incubated in a solution containing nucleotide triphosphates, there was a three-fold greater incorporation of nucleotides into RNA than for nucleoli of control animals. This could be due to exposure of more template DNA to nucleolar polymerase by hormone administration, or to an increase in the amount or activity of the nucleolar polymerase. This problem was investigated by extracting the RNA polymerases from liver nuclei following hydrocortisone treatment (Sajdel and Jacob, 1971). When the extract was resolved on

DEAE-Sephadex, a sharp increase in polymerase I (nucleolar enzyme) was observed, whereas polymerases II and III (nucleoplasmic enzymes) were unchanged in activity. Sajdel and Jacob (1971) concluded that the specific activity of the nucleolar polymerase rather than the total amount of enzyme protein had been increased by hormone treatment.

This finding does not exclude changes in synthesis of specific messenger RNAs following corticosteroid administration. Indeed, it is well known that corticosteroid administration transiently increases the levels of tyrosine aminotransferase and tryptophan oxygenase in rat liver by a process that is inhibited by prior administration of actinomycin D (Kenney, 1970). This evidence that new mRNA for each enzyme is involved in the hormone-induced enzyme increase has been directly confirmed by Schutz, Beato and Feigelson (1973) who extracted the mRNA for tryptophan oxygenase from the liver cytoplasm of control and steroid-treated rats and recovered more messenger at the time when more of the enzyme was being made. None of these studies tells us whether the hormonal treatment alters the availability of DNA for transcription of mRNA. It is perfectly possible that, in accordance with the evidence given above, more ribosomal RNA is synthesised through an increase in the amount or specific activity of polymerase I, whereas the amount of a given messenger RNA is increased by making the corresponding DNA template more available to the transcribing polymerase. This mechanism would have more specificity than a general increase in the amount of polymerase II in the nucleoplasm.

The changes produced by hormones involve specific receptors in the cells of the target organ. The mechanism by which binding of the hormone is followed by increased RNA synthesis has been well worked out in the case of female steroid hormones (Jensen and DeSombre, 1973; O'Malley and Means, 1974). The hormone enters the target cell (e.g. in the uterus) and binds to a specific cytoplasmic receptor. The receptor undergoes a transformation and now enters the nucleus to bind to the chromatin at specific acceptor sites. This is followed by activation of transcription at that site on the chromosome, so that new messenger and other RNA species are made from the underlying DNA. The mRNA's are then transported into the cytoplasm where they cause the synthesis of new proteins in the cell, or increase the amount of existing proteins (as in the case of tryptophan oxygenase).

The detailed response of the nuclear apparatus to hormone stimulation is more complex than this picture implies. After estrogen administration, there is a very rapid and transient increase in RNA formation in the uterus. Means and O'Malley (1972) speculate that this may represent formation of messenger RNA for cytoplasmic synthesis of more acidic chromosomal proteins which then pass to the chromosomes and modify the availability of their DNA template. The requirements for such an intermediate step involving protein synthesis in order to achieve the final affect on mRNA synthesis was also postulated by Begg and Munro (1965) in their study of the action of thyrotropin on thyroid cell RNA synthesis. In these early studies we suggested that more polymerase had first to be made. In this connection, there is evidence that the formation of ribosomal RNA in the nucleolus may be regulated through a mRNA species made by the extranucleolar polymerase II. This is shown by the finding (Jacob et al., 1970) that amanitin, which inhibits only the extranucleolar polymerase II *in vitro* (Jacob, Sajdel and Munro, 1970), nevertheless inhibits synthesis of rRNA when given to the intact animal. Thus we can conceive of an integrated system (*Figure 1.2*) in which a

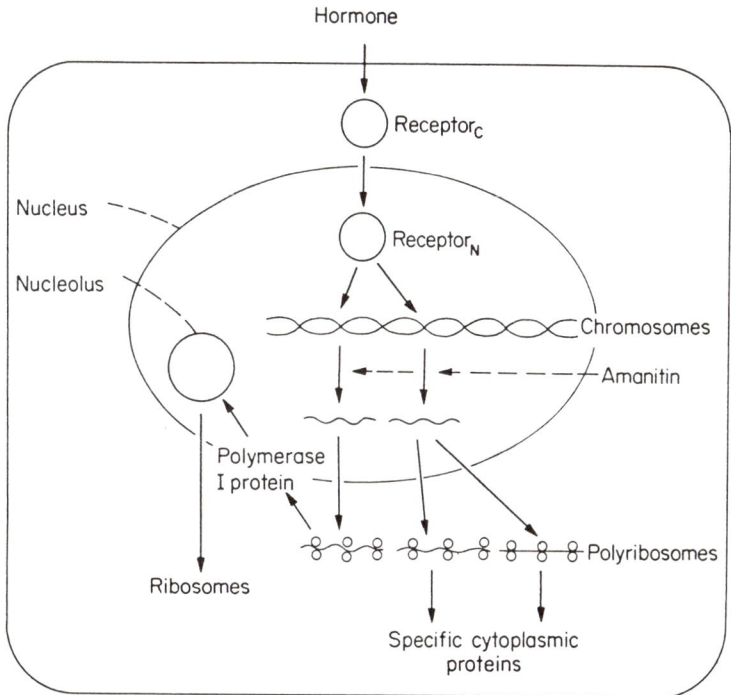

Figure 1.2 Speculative scheme demonstrating the control of selected messenger RNA's and also nucleolar RNA-polymerase following binding of a receptor-associated hormone to chromosomes. Note that the cytoplasmic form of the hormone receptor (receptor$_c$) is transformed to a nuclear form (receptor$_n$) which then binds to specific sites on the chromosomes, unmasking the DNA. This causes the DNA at these sites to be transcribed as messenger RNA's for various proteins, including one regulating the activity of polymerase I which is responsible for formation of ribosomal RNA. This accounts for the in vivo inhibition of ribosomal RNA synthesis by amanitin, although polymerase I is insensitive to amanitin added directly to the enzyme in vitro *(Jacob et al., 1970)*

hormone (attached to its receptor) binds to the chromatin and regulates the synthesis of a series of mRNAs, one of which affects formation of polymerase I or a protein factor required for the activity of polymerase I.

Eukaryote Translation and its Control

The synthesis of peptide chains on ribosomes can be divided into three phases. Firstly, there is initiation, in which ribosome subunits, messenger RNA, initiation factors and initiator methionyl-tRNA are required. In eukaryote cells, three protein initiation factors, M_1, M_2 and M_3, have been identified (Pritchard et al., 1970; Pritchard et al., 1971). Recently, Reichman and Penman (1973) have obtained evidence of a rapid turnover RNA specifically associated with initiation. Secondly, initiation is followed by addition of successive amino acids (peptide chain elongation). Some 60 aminoacyl-tRNA species charged with the 20 amino acids of proteins proceed to form complexes with elongation factor 1 (EF 1) and guanosine triphosphate and then attach to the ribosome to insert the correct amino acid indicated by the codons of the messenger RNA. This binding of aminoacyl-tRNA is followed by translocation of the growing peptide chain across the ribosome surface under the influence of elongation factor 2 (EF 2) and GTP. During this process of peptide chain elongation, the ribosome undergoes conformational changes as the charged tRNA and elongation factors bind and are released again, as we (Steinert, Baliga and Munro, 1974) have recently demonstrated using available sulphydryl groups on the ribosome as the criterion of conformational change (*Figure 1.3*). Finally, termination of the peptide chain requires two additional protein factors. The ribosome then separates from the messenger RNA (run-off ribosome) and dissociates

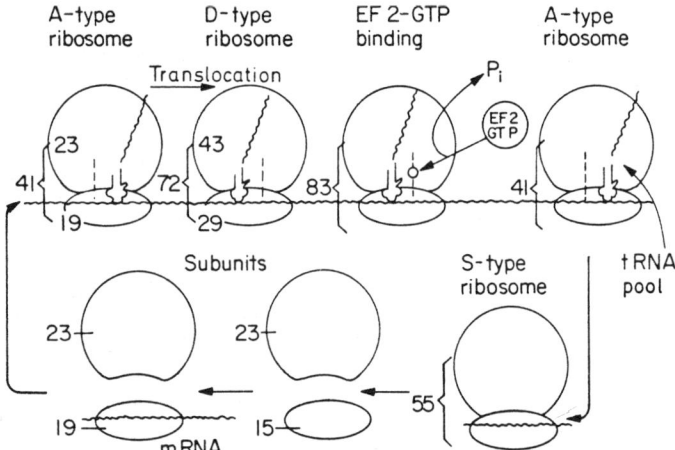

Figure 1.3 Diagram illustrating changes in ribosome conformation during peptide chain elongation. The numbers represent available sulphydryl groups on the intact ribosomes and their two subunits (Steinert, Baliga and Munro, 1974). Ribosome structure opens out as the peptidyl-tRNA translocates from the acceptor (A) position to the donor (D) position and then further from the energy released on binding of EF 2 and GTP. The more compact form is restored by the incoming aminoacyl-tRNA which accepts the peptide chain (A-type again)

into its two subunits; this dissociation requires binding of a protein (dissociation factor) to the smaller subunit of the ribosome to keep the subunits apart.

The sequence described above represents the ribosome cycle (*Figure 1.4*) in which the ribosomes of the cell successively take the form of polyribosomes with a growing peptide chain, then run-off ribosomes after leaving the messenger, followed by subunits, in which the 60S and 40S components are separate and can be used for initiation and another round of the cycle. The proportion of the total ribosome

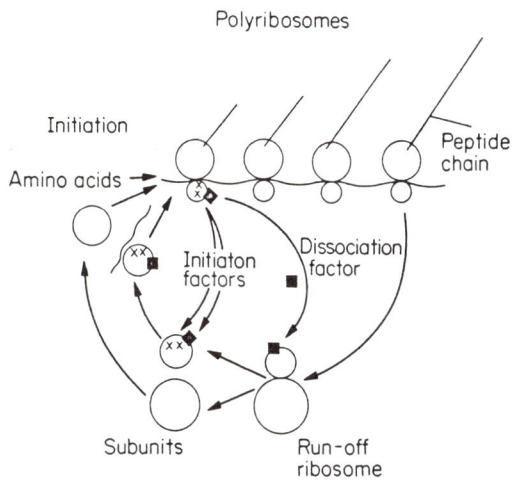

Figure 1.4 The ribosome cycle, illustrating initiation, chain elongation and chain termination (Munro, Hubert and Baliga, 1974)

population in the form of polyribosomes and in the forms of run-off (inactive) ribosomes and as subunits is determined by the balance between initiation, chain elongation and the availability of the ribosome dissociation factor. The main changes observed in rate of translation have been traced to effects on rate of chain initiation or rate of chain elongation. If initiation is reduced, then the ribosomes pass along the messenger RNA strand and accumulate as run-off ribosomes and subunits, the proportion of these latter two forms being dependent on the amount of dissociation factor. The effects of a reduced rate of chain elongation depend on the cause. The antibiotic cycloheximide affects translocation by interfering with binding of EF 2 and guanosine triphosphate (Baliga and Munro, 1971), and in consequence the polysome pattern is not disrupted, although movement in the ribosome cycle slows down. On the other hand, if a single amino acid is not available in adequate amounts, the rate of chain elongation is retarded only at those points where that amino acid has to be inserted in the growing peptide chain, and beyond that point elongation proceeds at a normal rate. This gives rise to various changes in polyribosome distribution on the messenger, as illustrated by the effects of tryptophan deficiency on globin synthesis in reticulocytes (Hori, Fisher and Rabinovitz, 1967).

The influence of amino acid supply on protein synthesis has been extensively investigated using intact animals, perfused organs, tissue slices and cells in culture. The findings of these studies have been reviewed elsewhere (Munro, Hubert and Baliga, 1974). Fasting rats fed protein or amino acid mixtures undergo a rapid aggregation of liver ribosomes into polyribosomes. This is shown in *Figure 1.5*, where the response of rat liver polysomes to a meal containing protein is illustrated. The response of liver polysomes to feeding an

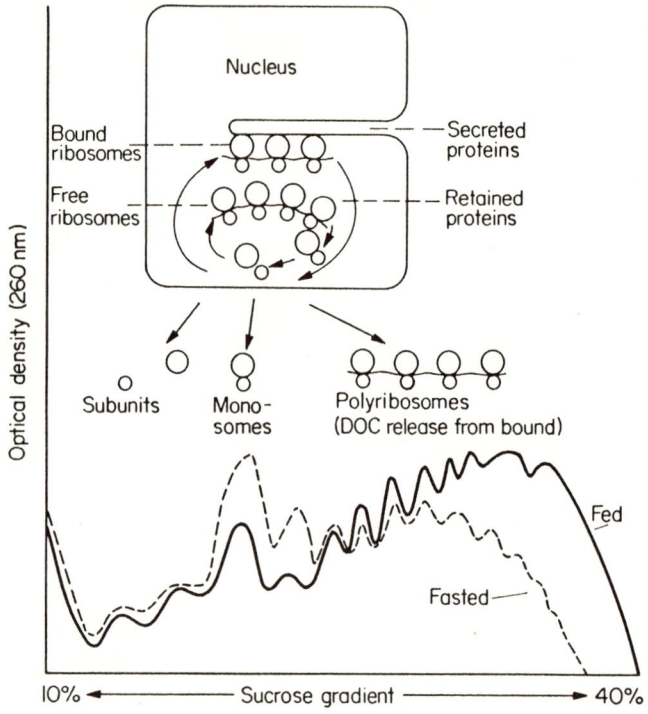

Figure 1.5 Ribosome profiles from the livers of fed and fasted rats showing the intracellular origin of the different ribosome classes (Munro, Hubert and Baliga, 1974)

amino acid mixture is impaired if tryptophan is omitted from the mixture (Fleck, Shepherd and Munro, 1965; Wunner, Bell and Munro, 1966), an effect attributed to tryptophan being the most limiting free amino acid in the liver. This thesis has received some confirmation. Firstly, it has been shown by Allen, Raines and Regen (1969) that administration of a tryptophan-free amino acid mixture to fasting rats causes a drastic reduction in the charging of liver tRNA with tryptophan. Secondly, rats were fed for some days on an amino acid mixture, imbalanced in such a way that threonine or isoleucine levels in the liver were severely reduced. These animals now showed liver polysome responses specifically to threonine or isoleucine administration (Pronczuk, Rogers and Munro, 1970; Ip and Harper, 1973). Finally, it has been established (Fleck, Shepherd and Munro, 1965) that the effect of amino acid supply on liver polyribosomes is not influenced by prior administration of actinomycin D. This suggests a cytoplasmic control not involving a change in the amount of messenger RNA.

This general picture has been confirmed by studies on animal cells in tissue culture (Munro, Hubert and Baliga, 1974; Clemens and Pain, 1974). In particular, Vaughan, Pawlowski and Forchhammer (1971) exposed HeLa cells to media lacking valine, histidine or methionine and observed rapid disaggregation of polysomes accompanied by a considerable reduction in the rate of protein synthesis due both to reduced initiation and prolonged time for chain elongation. The polysome disaggregation was attributed to a greater effect of amino acid deficiency on initiation than on elongation. More recent studies by Vaughan and Hansen (1973) suggest that accumulation of any uncharged tRNA exerts an inhibitory action upon initiation.

These studies emphasise the role of charging of tRNA in determining the rate of protein synthesis. This is dependent not only on the availability of amino acids but also on the amount of intact tRNA. Defects in tRNA charging capacity can arise from loss of the acceptor trinucleotide ApCpC which has been observed by us to occur in samples of placental tRNA (Hubert et al., 1974b) and in myopathic hamster muscle by Bester and Gevers (1973). In addition, aminoacyl-tRNA and elongation factor one mutually stabilise one another (Hubert et al., 1974a) thus providing another control point affected by charging of tRNA with amino acids.

Thus we conclude that charging of tRNA may be the common factor regulating and co-ordinating several aspects of translation. Firstly, the amount of charged tRNA in the cell may determine the stability and thus the quantity of EF 1. Secondly, uncharged tRNA is known to inhibit initiation non-specifically. Thirdly, the rate of chain elongation is affected by the availability of each species of aminoacyl-tRNA.

Post-translational Events in Eukaryote Protein Synthesis

Following formation of the peptide chain, many proteins undergo further change before becoming active structural or enzymic cell components. Modification of amino acids often begins before the peptide chain is complete, and takes many forms, such as hydroxylation of proline in the case of collagen (Miller and Udenfriend, 1970). Such post-translational modifications can provide useful methods of identifying the fate of the peptide chain. Thus for actin and myosin, in which some of the histidine residues become methylated after peptide chain formation, we have shown that the 3-methylhistidine so formed is released during intracellular turnover of these proteins and is not further metabolised, but is excreted quantitatively in the urine (Young et al., 1971). The amount excreted may thus serve as an index of muscle protein turnover.

Some proteins are secreted from cells, many of them being modified by the addition of carbohydrates after translation (glycoproteins). The secretory process has been well documented (Munro and Steinert, 1975). Secreted proteins are made on ribosomes attached to the membranes of the endoplasmic reticulum, which acts as a conducting channel for their secretion from the cell surface. During passage down the channels of the endoplasmic reticulum, sugars are added to transform the secreted proteins into glycoproteins. Other changes in structure can also occur. For example, we showed many years ago that prothrombin undergoes a post-transcriptional change within the endoplasmic reticulum to become the active molecule secreted into the plasma (Goswami and Munro, 1962). It has since been found that vitamin K, which is required for the formation of active prothrombin, acts at a post-transcriptional step. In its absence, inactive prothrombin is secreted into the blood, the coagulation defect being related to an inability of the prothrombin to bind calcium (Stenflo and Ganrot, 1972). Recently, it has been found that vitamin K catalyses the addition of a calcium-binding carboxyl group to the completed prothrombin chain (Johnson, 1974). The finding that the prothrombin peptide chain is made and secreted in normal quantities even in the absence of vitamin K indicates that post-translational modifications do not necessarily regulate the rate of peptide chain formation. It is, however, evident that the levels of many of the plasma proteins are kept within certain limits, sometimes quite narrow limits, and that some form of feedback of information must occur at the site of synthesis.

Finally, it should be pointed out that the mitochondrion is capable of independent protein synthesis, though few proteins are made within that organelle, most being formed on cytoplasmic ribosomes and taken into the mitochondrion. There appears to be a feedback mechanism from the mitochondrion for informing the nucleus when enough mitochondrial protein has accumulated and thus shutting off formation of the appropriate mRNA's in the nucleus (Barath and Kuntzel, 1972). However, some mitochondrial proteins are made on messengers transcribed from mitochondrial DNA. These messengers have recently been shown to be attached to polyadenylic acid (Perlman, Abelson and Penman, 1973); and Jacob and Schindler (1972) have described a mitochondrial poly-A-polymerase in rat liver. In a few cases, some subunits of mitochondrial proteins are made by the mitochondria and some on cytoplasmic ribosomes (Munro and Steinert, 1975), suggesting an elaborate control mechanism for co-ordinating the quantities of each subunit synthesised. In this connection, we (Konijn, Baliga and Munro, 1973) have observed that liver ferritin subunits of different sizes are made on endoplasmic reticulum attached and free ribosomes, and that the synthesis of each species responds differently to iron administration. The membrane-bound ribosomes making ferritin probably belong to the loosely-bound class that direct their nascent peptide chains into the cytosol and not into the cavity of the endoplasmic reticulum (Tanaka and Ogata, 1972).

Integration of Control Mechanisms in Eukaryote Protein Synthesis

Although the evidence reviewed in the preceding section indicates that the immediate response of the cytoplasmic mechanism of mRNA translation to changes in amino acid supply does not necessarily depend upon alterations in RNA synthesis, there is nevertheless good evidence that RNA metabolism (both synthesis and breakdown) does undergo changes in response to amino acid supply.

Evidence from studies on many tissues shows that RNA synthesis responds to changes in amino acid intake (Munro, Hubert and Baliga, 1974). This evidence shows that there are amino acid-dependent responses in RNA metabolism, with special emphasis on ribosomal RNA synthesis and processing by the nucleolus. Thus crude RNA polymerase preparations made from liver nuclei show increased activity one hour after tryptophan has been administered to fasting animals (Henderson, 1970). There is a considerable body of evidence that the synthesis of ribosomal components by cells in culture is sensitive to amino acid availability (Franze-Fernandez and Pogo, 1971) and ribosomal RNA synthesis and maturation (Vaughan, 1972) are also affected by a change in amino acid supply. In addition, the formation of the protein components of the ribosome are restricted by limitation of amino acid supply (Pawlowski and Vaughan, 1972) and the transfer of the subunits to the cytoplasm (Maden, 1969) is also retarded by amino acid starvation. Thus variation in amino acid supply has a co-ordinated effect on ribosome synthesis and transport that is much more rapid and extensive than any changes noted in mRNA synthesis. All of these effects could be secondary to changes in the rate of synthesis on polyribosomes of enzymes and other proteins (e.g. ribosomal proteins) with a rapid rate of turnover.

In addition to amino acid-dependent changes in nuclear RNA synthesis, we have recently observed rapid response of liver mitochondrial RNA synthesis and poly-A-polymerase activity when rats are fed complete amino acid mixtures (Jacob and Munro, 1974). These mitochondrial activities are reduced to about half their normal levels by fasting rats overnight, but increase extensively within three hours after giving a complete mixture of amino acids.

Finally, breakdown of ribosomal RNA in the liver is influenced by amino acid supply, and results in changes in nucleotide pools and in nucleotide biosynthesis (Clifford *et al.,* 1972). The underlying metabolic relationships are shown in *Figure 1.6.* When a rat is fasted or given a protein-deficient diet, the polysomes disaggregate and there is an accumulation of subunits in which activation of latent ribonuclease results in RNA breakdown. Consequently, a reduction in amino acid supply causes RNA breakdown to accelerate and conversely it diminishes when amino acid supply improves. These changes in RNA breakdown rate affect the nucleotide pools of the liver which expand during the period of accelerated RNA breakdown. In turn, the increased levels of purine nucleotides cause feedback inhibition of the

14 *Protein synthesis*

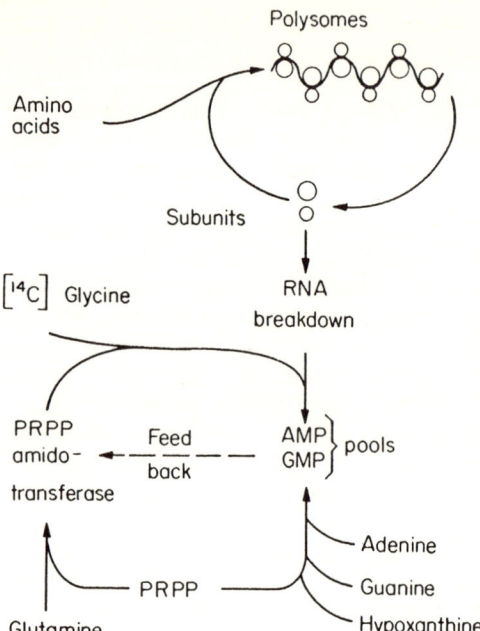

Figure 1.6 Diagrammatic scheme for interrelationships between amino acid supply and liver nucleotide metabolism. PRPP = phosphoribosyl pyrophosphate (Clifford et al., *1972)*

first enzyme in the pathway for biosynthesis of nucleotide bases. In consequence (*Figure 1.7*) the feeding pattern of the rat results in diurnal rhythms in liver polysome abundance, RNA content, purine nucleotide pools (represented here by guanine) and in purine (guanine) synthesis by the *de novo* pathway.

Figure 1.8 is an attempt to provide an integrated picture of the events occurring in the liver cell in response to changes in amino acid supply. Presumably other cells show these changes to a lesser degree, since their amino acid supply undergoes less extreme changes than those of the liver, which receives dietary amino acids directly following absorption.

(1) When amino acids or protein are administered to the fasting animal, the liver responds with a rapid but transient increase in protein synthesis, as evidenced by aggregation of ribosomes.

(2) This response seems to be determined by the free amino acid in the liver that is least abundant in relation to needs.

(3) There is some evidence that the impact of amino acid supply on the rate of protein synthesis may be determined through the degree of charging of tRNA which can affect both initiation and chain elongation.

(4) In addition, the completeness of charging of tRNA may co-ordinate the levels of soluble factors involved in protein synthesis, notably EF 1, through stabilisation dependent upon the presence of charged aminoacyl-tRNA.

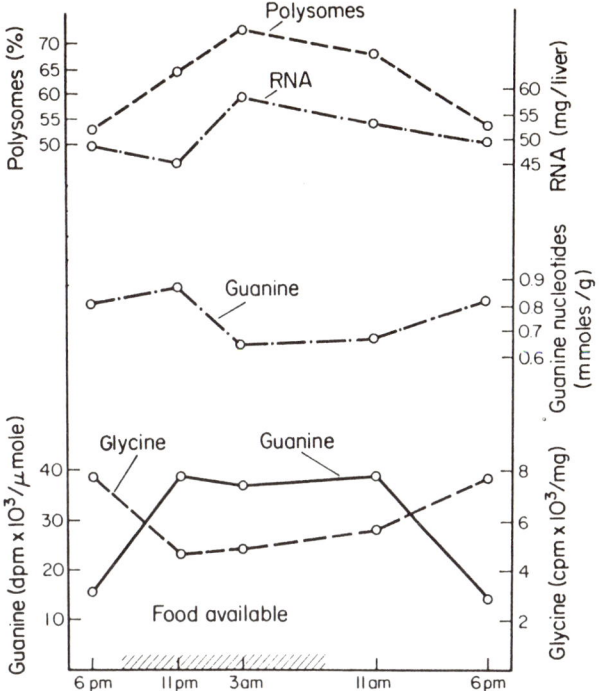

Figure 1.7 Diurnal rhythms in liver polysome abundance and nucleic acid metabolism in rats fed for 12 h per day. From above downwards the lines represent successively the percentage of polysomes in the liver ribosome fraction, the total RNA content per liver, the total free guanine nucleotides. The lower two lines indicate the uptake by free glycine, and free guanine nucleotides of ^{14}C from a pulse dose of [^{14}C]glycine administered at different times of day. Each point is the mean value obtained from four animals (Munro, Hubert and Baliga, 1974)

(5) Also, the amount of RNA in the liver and its turnover are affected by amino acid supply. Breakdown of RNA accelerates when amino acid supply is withdrawn, and results in a transient expansion of the free nucleotide pools and a reduction in the rate of *de novo* biosynthesis of purine nucleotides. These effects are reversed by amino acid administration, and appear to be determined by the state of aggregation or disaggregation of the liver ribosomes in response to amino acid availability.

(6) In addition, synthesis of RNA (notably ribosomal RNA) and maturation of ribosomes is influenced by amino acid supply, possibly through changes in the rate of formation in the cytoplasm of the polymerase and ribosomal proteins.

(7) The effect of amino acid supply on synthesis of mitochondrial RNA polymerase and its poly-A-polymerase is also likely to be secondary to the increased cytoplasmic synthesis of key proteins.

None of these changes in the rate of RNA production and processing in response to amino acid supply seems to be required for the increased synthesis of protein occurring in the cytoplasm, since

16 Protein synthesis

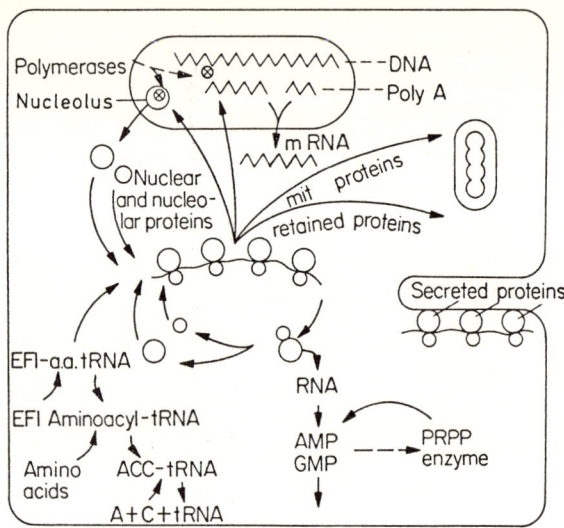

Figure 1.8 Mechanisms in the cell responding to changes in amino acid supply (Munro, Hubert and Baliga, 1974)

the latter is not diminished by pre-treatment with actinomycin. On the contrary, the observed changes in RNA metabolism are probably the consequence of the accelerated rate of protein synthesis. We do not know the significance of such nuclear and mitochondrial responses; they probably meet a requirement for more ribosomal subunits if the increased amino acid supply persists and thus co-ordinate the response of the cell to the availability of this substrate, potentially on a long-term basis. From this picture, it is tempting to conclude that the primary factor in all these events is the charging of tRNA. This would be the mammalian counterpart of ppGpp in the stringent bacterial cell, a nucleotide that forms when tRNA is not charged and which probably regulates many features of bacterial cell metabolism (Cashel and Gallant, 1969).

Finally, in examining these short-term responses of the protein synthesis mechanism, it has to be remembered that there must also be long-term adaptive changes expressed by the fact that the growing organism, such as man, accumulates protein until the adult state is reached and then undergoes a gradual loss of body protein during the rest of adult life (Forbes and Reina, 1970). This represents a programmed unfolding of regulatory mechanisms that operate from conception onwards and are expressed through alterations in protein synthesis, such as the perinatal change in some of the proteins synthesised by the liver (e.g. Linder and Munro, 1973). Although we know little of the underlying control mechanisms involved, it is known that malnutrition of the pregnant rat results in offspring with a reduced growth rate and which form less growth hormone in their pituitary glands (Shrader and Zeman, 1973). The discussion of such changes in control of hormone secretion due to early malnutrition is beyond the scope of this review.

References

ADESNIK, M., SALDITT, M., THOMAS, W. and DARNELL, J.E. (1972). *J. molec. Biol.*, **71**, 21

ALLEN, R.E., RAINES, P.L. and REGEN, D.M. (1969). *Biochim. biophys. Acta*, **190**, 323

BALIGA, B.S. and MUNRO, H.N. (1971). *Nature*, **233**, 257

BARATH, Z. and KUNTZEL, H. (1972). *Nature, (New Biol.)*, **240**, 195

BEGG, D.J. and MUNRO, H.N. (1965). *Nature*, **207**, 483

BESTER, A.J. and GEVERS, W. (1973). *Biochem. J.* **132**, 193

CASHEL, M. and GALLANT, J. (1969). *Nature*, **221**, 838

CLEMENS, M.J. and PAIN, V. (1974). these proceedings

CLIFFORD, A.J., RIUMALLO, J.A., BALIGA, B.S., MUNRO, H.N. and BROWN, P.R. (1972). *Biochim biophys. Acta*, **277**, 443

DARNELL, J.E., PHILIPSON, L., WALL, R. and ADESNIK, M. (1971). *Science*, **174**, 507

FLECK, A., SHEPHERD, J. and MUNRO, H.N. (1965). *Science*, **150**, 628

FORBES, G.B. and REINA, J.C. (1970). *Metabolism*, **19**, 653

FRANZE-FERNANDEZ, M.T. and POGO, A.O. (1971). *Proc. natn. Acad. Sci. U.S.A.*, **68**, 3040

GOODLAD, G.A.J. and MUNRO, H.N. (1959). *Biochem. J.* **73**, 343

GOSWAMI, P. and MUNRO, H.N. (1962). *Biochim biophys. Acta*, **55**, 410

GREENBERG, H. and PENMAN, S. (1966). *J. molec. Biol.*, **21**, 527

HENDERSON, A.R. (1970). *Biochem. J.*, **120**, 205

HORI, M., FISHER, J.M. and RABINOVITZ, M. (1967). *Science*, **155**, 83

HUBERT, C., BALIGA, B.S., VILLEE, C.A. and MUNRO, H.N. (1974a). *Biochim biophys. Acta*, **374**, 359

HUBERT, C., LAGA, E.M., BALIGA, B.S., MURPHY, A., VILLEE, C.A. and MUNRO, H.N. (1974b). unpublished data

IP, C.C.Y. and HARPER, A.E. (1973). *Biochim biophys. Acta*, **331**, 251

JACOB, S.T. and MUNRO, H.N. (1974). unpublished data

JACOB, S.T., SAJDEL, E.M. and MUNRO, H.N. (1969). *Europ. J. Biochem.*, **7**, 449

JACOB, S.T., SAJDEL, E.M. and MUNRO, H.N. (1970). *Biochem. biophys. Res. Commun.*, **38**, 765

JACOB, S.T., MUECKE, W., SAJDEL, E.M. and MUNRO, H.N. (1970). *Biochem. biophys. Res. Commun.*, **40**, 334

JACOB, S.T. and SCHINDLER, D.G. (1972). *Biochem. biophys. Res. Commun.*, **48**, 126

JACOB, S.T., SCHARF, M.B. and VESELL, E.S. (1974). *Proc. natn. Acad. Sci. U.S.A.*, **71**, 704

JENSEN, E.V. and DESOMBRE, E.R. (1973). *Science*, **182**, 126

JOHNSON, B.C. (1974). private communication

KEDINGER, C., GNIAZDOWSKI, M., MANDEL, J.L., GISSINGER, F. and CHAMBON, P. (1970). *Biochem. biophys. Res. Commun.*, **38**, 165

KENNEY, F.T. (1970). in *Mammalian Protein Metabolism*, Vol. **4**, p.131. Ed. H.N. Munro, Academic Press, New York

KONIJN, A.M., BALIGA, B.S. and MUNRO, H.N. (1973). *FEBS Lett.* **37**, 249

LINDER, M.C. and MUNRO, H.N. (1973). *Enzyme*, **15**, 111
MADEN, B.E.H. (1969). *Nature*, **224**, 1203
MEANS, A.R. and O'MALLEY, B.W. (1972). *Metabolism*, **21**, 357
MILLER, R.L. and UDENFRIEND, S. (1970). *Archs Biochem. Biophys.* **139**, 104
MUNRO, H.N. (1973). in *Aromatic Amino Acids in the Brain.* (Ciba Foundation Symposium **22**.) Elsevier, Amsterdam
MUNRO, H.N. and STEINERT, P.M. (1975). in *International Review of Science (Biochemistry Series).* **Vol. 7**, Ed. H.R.V. Arnstein, Butterworths, London
MUNRO, H.N., HUBERT, C. and BALIGA, B.S. (1974). in *Alcohol and Abnormal Protein Synthesis.* Eds M.A. Rothschild, M. Oratz and S.S. Schreiber, Pergamon Press, Oxford
O'MALLEY, B.W. and MEANS, A.R. (1974). *Science*, **183**, 610
PAWLOWSKI, P.J. and VAUGHAN, M.H. (1972). *J. Cell Biol.*, **52**, 409
PERLMAN, S.M., ABELSON, H.T. and PENMAN, S. (1973). *Proc. natn. Acad. Sci. U.S.A.*, **70**, 350
PERRY, R.P. and KELLEY, D.E. (1974). *Cell*, **1**, 37
PRICHARD, P.M., GILBERT, J.M., SHAFRITZ, D.A. and ANDERSON, W.F. (1970). *Nature*, **226**, 511
PRICHARD, P.M., PICCIANO, D.J., LAYCOCK, D.G. and ANDERSON, W.F. (1971). *Proc. natn. Acad. Sci. U.S.A.* **68**, 2752
PRONCZUK, A., ROGERS, Q.R. and MUNRO, H.N. (1970). *J. Nutr.*, **100**, 1249
REICHMAN, M. and PENMAN, S. (1973). *Proc. natn. Acad. Sci. U.S.A.*, **70**, 2678
ROEDER, R.G. and RUTTER, W.J. (1969). *Nature*, **224**, 234
SAJDEL, E.M. and JACOB, S.T. (1971). *Biochem. biophys. Res. Commun.*, **45**, 707
SCHUMM, D.E. and WEBB, T.E. (1972). *Biochem. biophys. Res. Commun.*, **48**, 1259
SCHUTZ, G., BEATO, M. and FEIGELSON, P. (1973). *Proc. natn. Acad. Sci. U.S.A.*, **70**, 1218
SHRADER, R.E. and ZEMAN, F.J. (1973). *J. Nutr.*, **103**, 1012
STEIN, G.S., SPELSBERG, T.C. and KLEINSMITH, L.J. (1974). *Science*, **183**, 817
STEINERT, P.M., BALIGA, B.S. and MUNRO, H.N. (1974). *J. molec. Biol.*, **88**, 895
STENFLO, J. and GANROT, P.O. (1972). *J. biol. Chem.*, **247**, 8160
TANAKA, T. and OGATA, K. (1972). *Biochim. biophys. Res. Commun.*, **49**, 1069
VAUGHAN, M.H. (1972). *Expl Cell. Res.*, **75**, 23
VAUGHAN, M.H., PAWLOWSKI, P.J. and FORSCHHAMMER, J. (1971). *Proc. natn. Acad. Sci. U.S.A.*, **68**, 2047
VAUGHAN, M.H. and HANSEN, B.S. (1973). *J. biol. Chem.*, **248**, 7087
WUNNER, W.H., BELL, J. and MUNRO, H.N. (1966). *Biochem. J.*, **101**, 417
YOUNG, V.R., ALEXIS, S.D., BALIGA, B.S. and MUNRO, H.N. (1971). *J. biol. Chem.*, **247**, 3592

2

AMINO ACID SUPPLY AND PROTEIN SYNTHESIS IN ANIMAL CELLS

M.J. CLEMENS
National Institute for Medical Research, Mill Hill, London

VIRGINIA M. PAIN
London School of Hygiene and Tropical Medicine, London, WC1

Introduction

In this paper we have tried to outline the principal aspects of the control of protein synthesis by amino acids in animal cells. The subject matter is divided into sections dealing with acute and chronic effects, since the mechanisms underlying the two are not necessarily the same. Emphasis has been laid on the regulation of polysome formation as an example of translational control at the level of polypeptide chain initiation, and some recent experiments are described which were designed to investigate the roles of various anabolic hormones in the response of rat liver polysomes to amino acid feeding. In order to stress that regulation by amino acids can occur by mechanisms other than purely translational control we also describe some examples of specific enzyme induction which require new RNA synthesis.

In consideration of the long-term effects of amino acid deprivation we have made comparisons with the effects of acute variations in nutritional status and have discussed the role played by changes in ribosome content of tissues in affecting protein synthetic activity. The relationship between protein synthesis and ribosomal RNA synthesis and processing is briefly considered.

Finally, we have raised the question of whether, by analogy with the stringent response in *E.coli*, the diverse effects of changes in amino acid supply may all be exerted through some common mechanism within the cell.

Acute Effects of Amino Acids on Protein Synthesis

GENERAL PROTEIN SYNTHESIS

The overall rate of protein synthesis in many types of cell is sensitive to amino acid supply (Munro, 1970; Munro, Hubert and Baliga, 1974). This may not seem surprising since amino acids are the substrates for protein synthesis. However, as will be seen, the regulatory effects of amino acids are not compatible solely with a mechanism involving simple substrate limitation.

Polypeptide chain initiation

Protein synthesis occurs on polyribosomes (polysomes), which may have any number of ribosomes per mRNA molecule, from one to fifty or more. The precise number of ribosomes per polysome is a function of:
(1) the length of mRNA strand,
(2) the relative rates of ribosome attachment to and detachment from the mRNA.

It is probable that, *in vivo*, mRNA molecules never carry the maximum number of ribosomes possible, except under specific conditions such as after cycloheximide treatment (Stanners, 1966). This implies that normally the rate of ribosome attachment (initiation) is slower than the rates of movement along the message (elongation) and detachment from it (termination). When polysomes are isolated from cells and their size distribution is monitored by sucrose density gradient centrifugation, the profile observed reflects the balance between the rates of initiation and elongation *in vivo*. A corollary of this is that polysome profiles alone do not tell us about absolute rates of protein synthesis, although a correlation between average polysome size and the rate of amino acid incorporation in a tissue is often apparent.

It is well documented that an increased supply of amino acids to many tissues and cell types results in an increase in the average polysome size. This has been observed in rodent liver, both *in vivo* (Fleck, Shepherd and Munro, 1965; Wunner, Bell and Munro, 1966; Sidransky *et al.*, 1968) and during perfusion (Jefferson and Korner, 1969), and must reflect an enhanced rate of initiation relative to the other steps in protein synthesis. Similar effects of amino acid depletion and replacement have been reported for liver tumour cells in culture (Eliasson, Bauer and Hultin, 1967), L cells (Chen, Hersh and Kitos, 1968), ascites cells (Hogan and Korner, 1968; van Venrooij, Henshaw and Hirsch, 1972), HeLa cells (Vaughan, Pawlowski and Forchhammer, 1971) and perfused heart (Morgan *et al.*, 1971). Many of these reports concern isolated cells or tissues which are very suitable for the study of acute nutritional effects since the absolute rate of protein synthesis at any moment can be measured more reliably than in whole animals. In such systems the polysomal responses to amino acid supply have been shown to correlate very well with the effects on rates of protein synthesis.

In rats and mice, tryptophan is the normal rate limiting amino acid for polysome formation in the liver (Fleck, Shepherd and Munro, 1965; Wunner, Bell and Munro, 1966; Sidransky *et al.*, 1968; Fishman, Wurtman and Munro, 1969; Munro, 1970) but other essential amino acids can be made limiting by feeding imbalanced diets to the animals (Pronczuk, Rogers and Munro, 1970; Ip and Harper, 1973). Similarly, isolated tissues or cells in culture respond to variations in the levels of several amino acids in addition to tryptophan (Jefferson and Korner, 1969; Vaughan, Pawlowski and Forchhammer, 1971; van Venrooij, Henshaw and Hirsch, 1972; McGown et al., 1973). The amino acid requirements for stimulation of protein synthesis appear to be different from those necessary for suppression of protein breakdown, at least in liver (Woodside and Mortimore, 1972).

One of the most striking features of the regulation of polysome formation by amino acids is its insensitivity to inhibition of RNA synthesis (Fleck, Shepherd and Munro, 1965; Sidransky et al., 1968; Chen, Hersh and Kitos, 1968; Hogan and Korner, 1968; Vaughan, Pawlowski and Forchhammer, 1971), which suggests that the control mechanism operates at the level of translation rather than transcription. Other than this, we know very little about the mechanism by which amino acids act. There is a correlation between the rate of initiation and the level of tRNA charging in both rat liver (Allen, Raines and Regan, 1969) and HeLa cells (Vaughan and Hansen, 1973). However, it is not clear how the level of aminoacylation of tRNA can affect initiation in preference to elongation, especially in view of the fact that aminoacyl tRNA is a substrate for the latter process. There have been suggestions that the formation of functional ribosome-mRNA-aminoacyl tRNA 'initiation complexes' may be impaired by de-acylated tRNA (Wettenhall and Wool, 1972; Kyner, Zabos and Levin, 1973; Zasloff, 1973) but greater knowledge of the mechanism of initiation is needed before the significance of these findings may be assessed.

In the whole animal the influence of acute changes in amino acid supply on protein synthesis is complicated by the possible involvement of anabolic hormones. The secretion of both growth hormone (Knopf et al., 1965) and insulin (Floyd et al., 1966) is stimulated by amino acids and these hormones are known to affect protein synthesis in several tissues (Korner, 1965; Wool et al., 1968; Manchester, 1970; Pain and Garlick, 1974). We have carried out a study of the hormonal requirements for amino acid stimulation of polysome formation in rat liver *in vivo* (Clemens and Pain, 1974) and have found that hepatic polysomes in either adrenalectomised or hypophysectomised animals respond to amino acids as in intact rats (*Figure 2.1, Table 2.1*). In hypophysectomised animals growth hormone also promotes polysome formation but does not enhance the amino acid effect (*Figure 2.1*). However, in rats made diabetic by administration of streptozotocin, hepatic polysomes respond very poorly to amino acid force-feeding unless insulin is also given. As can be seen in *Figure 2.2*, insulin or amino acids alone have much less effect than the two together. Essentially the same results are obtained when amino acid incorporation into protein is measured with isolated polysomes *in vitro* (*Table 2.2*; *Figure 2.3*) under conditions in which all the incorporation is due to extension of pre-existing nascent polypeptide chains.

The above results suggest that hormones of the anterior pituitary and adrenal glands are not required for amino acid stimulation of hepatic polysome formation and that amino acid supply is probably the rate-limiting factor in normal, adrenalectomised and hypophysectomised rats. However, availability of insulin is the limiting element in diabetic animals.

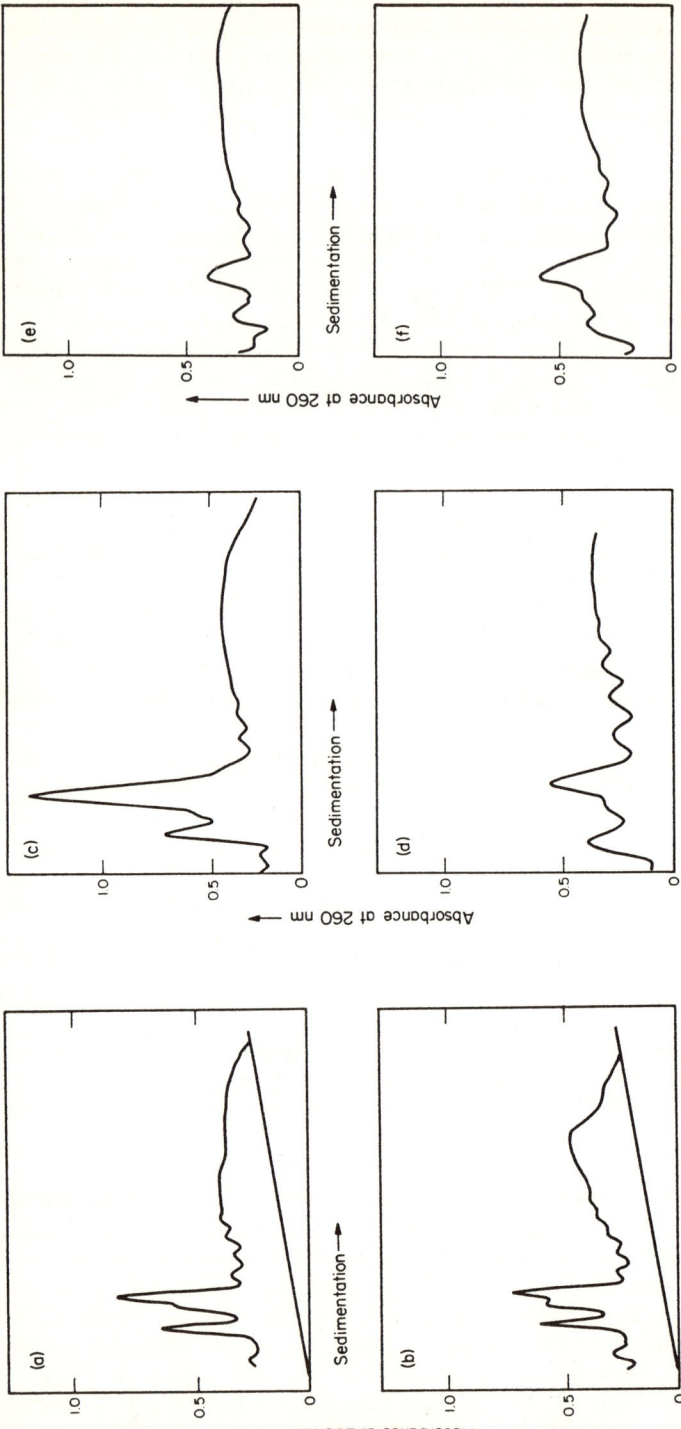

Figure 2.1 Polysome profiles of hepatic ribosomes isolated from normal and hypophysectomised rats 4 h after amino acid force-feeding. Some 80-100 g male rats (hypophysectomised, where indicated, at least 2 weeks previously) were fed ad lib until 14 h before the experiment, at which time each animal was given 7 g of food. This ensured that stomachs were capable of receiving the force-fed mixture subsequently. Groups of rats received either 2 ml of 0.9 per cent NaCl or 2 ml of a mixture of amino acids in the same proportions as used by Wunner, Bell and Munro (1966) but in 1/3 the quantity. Some groups also received intraperitoneally 250 µg of growth hormone at the time of feeding. Ribosomes were prepared from postmitochondrial supernatants by Mg^{2+} precipitation (Clemens and Pain, 1974) and the polysome profiles analysed on 10-30 per cent sucrose gradients (Clemens and Pain, 1974). (a) Normal rats, saline fed; (b) Normal rats, amino acid fed; (c) Hypophysectomised rats, saline fed; (d) Hypophysectomised rats, amino acid fed; (e) Growth hormone treated hypophysectomised rats, saline fed; (f) Growth hormone treated hypophysectomised rats, amino acid fed

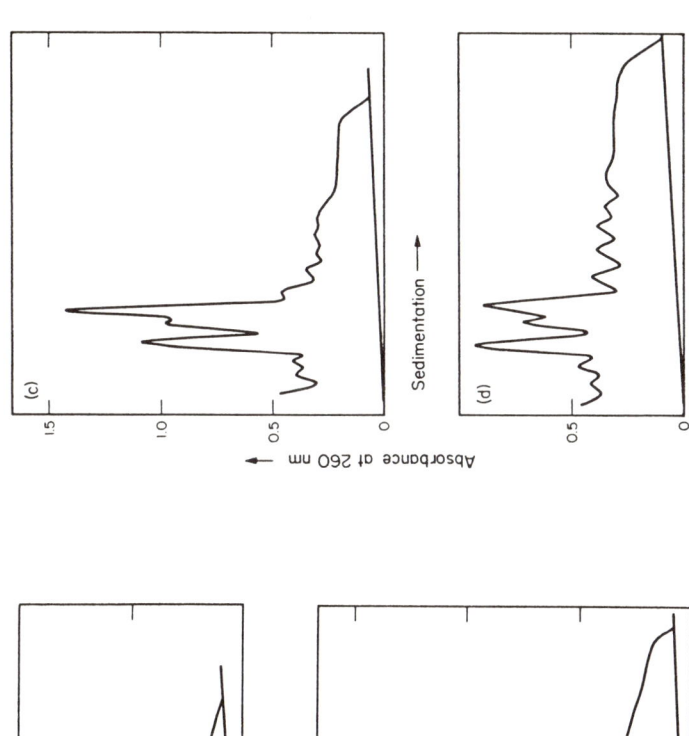
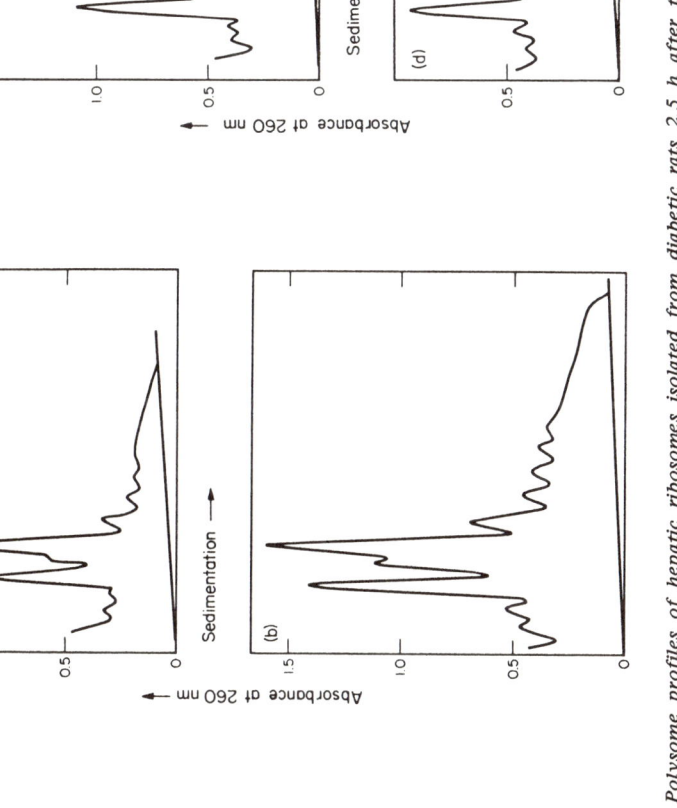

Figure 2.2 Polysome profiles of hepatic ribosomes isolated from diabetic rats 2.5 h after treatment with amino acids and/or insulin in vivo. Some 90-100 g male rats were made diabetic with streptozotocin (100 mg/kg) five days before use. They were fed ad lib until 14 h before the experiment, when each animal was given 7 g of food. Five rats in each group were force-fed either 2 ml of 0.9 per cent NaCl (a and c) or 2 ml of amino acid mixture (b and d) as described in Figure 2.1. Groups c and d also received two units of insulin i.p. at the time of feeding. Ribosomes were prepared and the polysome profiles analysed as described in Figure 2.1

Table 2.1 The proportion of ribosomes in polysomes of livers from rats in various endocrine states

Hormonal status	Ribosomes in polysomes (%)	
	Saline fed	Amino acid fed
Hypophysectomised	45	53
Hypophysectomised, growth hormone treated	69	52
Diabetic	30	42
Diabetic, insulin treated	42	52

The values given refer to liver from rats allowed access to 7 g of food for 14 h before force-feeding with amino acids or saline. The percentage of ribosomes in polysomes are calculated by estimating the percentage of the area of the polysome profile which lies under peaks of trimers and larger species

A similar conclusion can be drawn from the results of Ekren, Jervell and Seglen, (1971) and van den Borre and Webb (1972) who studied isolated livers perfused with media containing various concentrations of amino acids, with or without insulin. Although amino acids have been reported to increase insulin secretion *in vivo*, we have found that a single force-feeding of a dose which stimulated polysome formation for at least four hours did not have any significant effect on the plasma level of this hormone an hour

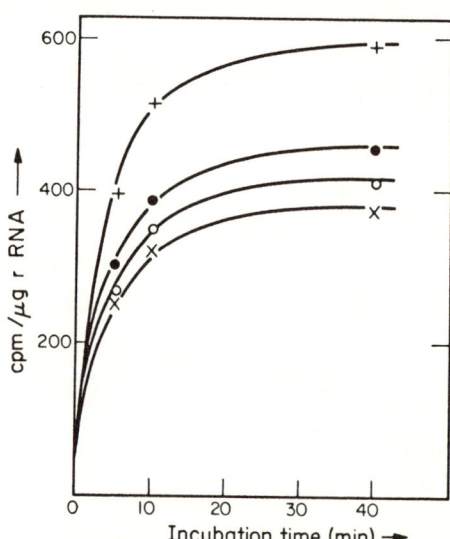

Figure 2.3 Amino acid incorporation into growing peptide chains on polysomes isolated from diabetic rats treated with or without amino acids and insulin in vivo. The same ribosome preparations as illustrated in Figure 2.2 were used. Amino acid incorporation into hot acid insoluble material was assayed by methods described elsewhere (Clemens and Pain, 1974). The points plotted represent values obtained with preparations from the pooled livers of five rats per group. (X———X), saline fed; (●———●), amino acid fed; (o———o), insulin treated; (+———+), amino acid fed and insulin treated

Table 2.2 Amino acid incorporation by polysomes isolated from livers of rats subjected to various endocrine and nutritional treatments

Hormonal status and in vivo treatment of animals	Incubation time (min)	Incorporation (CPM/μg RNA)	Stimulation by amino acid feeding (%)
Adrenalectomised (fasted)	5	22.4, 24.6[1]	
	10	32.1, 41.7	
	40	47.1, 65.9	
Adrenalectomised (fasted) - amino acid fed	5	34.3, 33.2[1]	44
	10	48.3, 48.3	31
	40	72.9, 73.3	29
Adrenalectomised	5	29.8 ± 1.5[2]	
	10	42.0 ± 0.5	
	40	75.7 ± 1.4	
Adrenalectomised - amino acid fed	5	33.1 ± 1.1	11
	10	46.7 ± 1.3	11
	40	80.2 ± 1.8	6
Hypophysectomised	8	15.6 ± 1.0	
	15	24.7 ± 0.2	
	40	41.1 ± 0.8	
Hypophysectomised - amino acid fed	8	17.0 ± 0.7	9
	15	28.7 ± 0.5	16
	40	48.8 ± 2.0	19
Hypophysectomised, growth hormone treated	8	19.1 ± 0.7	
	15	33.1 ± 1.0	
	40	54.1 ± 2.3	
Hypophysectomised, growth hormone treated - amino acid fed	8	$19,5 \pm 2.2$	2
	15	32.1 ± 2.6	0
	40	50.3 ± 2.0	0
Diabetic	5	29 ± 2	
	10	50 ± 3	
	40	83 ± 3	
Diabetic - amino acid fed	5	33 ± 1	14
	10	54 ± 2	8
	40	86 ± 5	4
Diabetic, insulin treated	5	33 ± 1	
	10	61 ± 1	
	40	91 ± 2	
Diabetic, insulin treated - amino acid fed	5	38 ± 1	31
	10	68 ± 0	36
	40	106 ± 3	28

All animals except the first group of adrenalectomised rats, which were fasted, were allowed limited access to food as described in *Table 2.1*. Each rat received either amino acids or saline by stomach tube and was killed 4 h later. Ribosomes were prepared by Mg^{2+} precipitation and their ability to incorporate amino acids into protein *in vitro* was assayed as described by Clemens and Pain (1974)

[1] Individual values quoted due to high mortality rates
[2] Means ± S.E.M. of three animals per group

or more after feeding (Pain et al., unpublished observations). It is more likely that insulin has a permissive role in mediating the response of the liver to amino acids, rather than acting directly. Several possible mechanisms could be involved in such an effect:

(1) Amino acid uptake by the liver may be enhanced in the presence of insulin (Chambers, Georg and Bass, 1965);
(2) The extent or rate of charging of tRNA may be greater when both amino acids and insulin are available. The activity of aminoacyl tRNA synthetases has been shown to be lower than normal in livers of diabetic animals (Germanyuk and Mironenko, 1969);
(3) Some reaction in the sequence of steps involved in initiation, other than that which is affected by amino acid supply, may limit the rate of ribosome attachment to mRNA in the absence of insulin. The reduced proportion of polysomes in livers of diabetic rats (*Table 2.1;* Tragl and Reaven, 1971), together with other evidence (Pain, 1973), strongly suggests that a defect in initiation develops in the absence of insulin. The nature of this defect is, however, unknown.

Polypeptide chain elongation

Since polysomes diminish in size in cells deprived of an essential amino acid, the rate of ribosome movement along the mRNA molecule cannot fall as much as the rate of initiation. Nevertheless, there is evidence that chain elongation is affected by variations in amino acid supply, but again the precise mechanism involved remains obscure.

Measurement of average elongation rates in HeLa cells under various nutritional conditions have shown that it takes up to twice as long as normal to complete a polypeptide chain when an essential amino acid is lacking (Vaughan and Hansen, 1973). Similarly, calculations of rates of amino acid incorporation per polysomal ribosome in ascites cells in culture have shown a reduced efficiency compatible with impaired rates of elongation when essential amino acids are missing (van Venrooij, Henshaw and Hirsch, 1970).

Such findings are not unexpected, since amino acids are the substrates for protein synthesis at the level of chain elongation. Nevertheless, additional regulatory mechanisms may be operative which restrict elongation during amino acid deprivation. For example, Clemens and Korner (1971) found that following a one hour exposure of liver slices to high amino acid levels, the ability of the soluble cell sap fraction to support *in vitro* protein synthesis was enhanced. Clemens (1972) later showed that this elevated activity was maintained *in vitro* in the presence of excess amino acids or aminoacyl tRNA and when inhibitors of initiation were added to cell-free incubations. It can be concluded that exposure of the tissue to a plentiful amino acid supply increases the concentration and/or activity of one or more elongation factors. Care must be taken in interpreting such observations, however, since Smulson and Rideau (1970) have shown that in HeLa cells amino acid deprivation causes association of EF 2 with the large number of monomeric ribosomes that accumulate, leading to an apparent loss of this enzyme activity from the cytoplasm.

SPECIFIC PROTEIN SYNTHESIS
We have described above the regulation of the overall rate of protein synthesis by amino acids, which appears to be independent of RNA synthesis and is exerted primarily at the level of polypeptide chain initiation. There are, however, some examples of amino acid control of synthesis of specific proteins which involve additional mechanisms.

Administration of amino acids to rats by stomach tube induces increased synthesis of several hepatic enzymes including tyrosine aminotransferase (Labrie and Korner, 1968a,b; Rosen and Milholland, 1968) serine dehydratase (Pitot and Peraino, 1964; Jost et al., 1970; Mauron, Mottu and Spohr, 1973) and ornithine decarboxylase (Fausto, 1971). These are all characterised by the fact that actinomycin D blocks their induction by amino acids. The mechanism(s) of induction thus differ from that of the purely cytoplasmic stimulation of polysome formation which similar doses of amino acids produce. In the case of serine dehydratase, cyclic AMP has been implicated as an intermediary in the amino acid effect (Jost et al., 1970). As with polysome formation, the regulation of tyrosine aminotransferase activity is particularly sensitive to tryptophan intake (Wurtman, Shoemaker and Larin, 1968) and this amino acid probably controls the daily rhythmic fluctuations in the level of the enzyme in rat liver. However, in hepatoma cells in culture, tyrosine aminotransferase levels can be regulated by the leucine content of the medium (Lee and Kenney, 1971). In this case the evidence suggests a post-transcriptional action of the amino acid.

As well as causing changes in the rates of synthesis of several enzymes, amino acids also affect the production of serum albumin by the liver. Although this may be considered of primary importance as one of the long-term responses to variations in amino acid intake there is evidence for rapid effects on the synthesis of this protein. Rothschild et al. (1969) were able to stimulate albumin synthesis by perfused liver from fasted rabbits by addition of tryptophan or isoleucine to the perfusion medium. However, livers from fed donors did not respond to these amino acids. Kelman et al. (1972) found that eleven amino acids were necessary to restore albumin synthesis to normal in perfused livers from starved rats.

Chronic Effects of Amino Acids on Protein Synthesis

There are considerable experimental difficulties involved in measuring rates of protein synthesis *in vivo* in animals subjected to acute variations in amino acid intake. However, long-term consequences of low protein diets and their reversal by adequate protein intake or an amino acid meal can be measured more reliably. The results of such *in vivo* studies have produced the somewhat surprising conclusion that the average rate of synthesis of mixed hepatic proteins is not reduced, and may even be increased in animals chronically deprived of protein (Waterlow and Stephen, 1968; Haider and Tarver, 1969; Millward and Garlick, 1972). On the other hand, synthesis of albumin by the liver in protein depleted animals is substantially depressed (Chandrasakharam, Fleck and Munro, 1967; Kirsch *et al.*, 1968; Haider and Tarver, 1969; Morgan and Peters, 1971). This apparently selective effect of protein malnutrition on synthesis of exported proteins may be related to demonstrations of preferential disaggregation of membrane-bound, rather than free, liver polysomes (Gaetani, Massotti and Spadoni, 1969) and extensive disruption of hepatic rough endoplasmic reticulum (Enwonwu, 1972). Polysome profiles from livers of protein deficient animals show disaggregation, and decreased adrenal steroid production has been implicated in this response (Enwonwu and Jacobson, 1973). In contrast, hepatic ribosomes in rats force fed amino acid imbalanced diets show a pronounced shift towards heavier aggregates (Sidransky, Staehelin and Verney, 1964). This phenomenon has been attributed by others to the effects of adrenal steroids secreted in response to the imbalanced diet (Cammarano *et al.*, 1968).

In muscle, a protein deficient diet results in substantial reduction in the protein synthesis rate (Millward, 1970; Young, Stothers and Vilaire, 1971; Millward and Garlick, 1972). Analysis of *in vivo* data indicates that this is due to a rapid decrease in ribosomal activity together with a progressive fall in ribosome content of the tissue (Millward *et al.*, 1973). The decrease in ribosomal activity *in vivo* is reflected in some disaggregation of polysomes (Young and Alexis, 1968) but the change in profile appears insufficient to account for the pronounced decrease seen in the activity of muscle ribosomes in cell-free systems (Young and Alexis, 1968; von der Decken and Omstedt, 1970). Recent studies by Young's group (Alexis, Basta and Young, 1972; Alexis and Young, 1973) have suggested that there is a defect in the activity of the ribosomal salt wash fractions, but the exact mechanism and significance of this change has not been resolved.

A significant part of the response of tissues such as skeletal muscle to chronic amino acid deficiency is the reduction in RNA content which occurs (Young and Alexis, 1968; Howarth, 1972; Millward *et al.*, 1973). Such a decrease reflects loss of ribosomes, since these constitute 80 per cent of total cellular RNA (Young, 1970). Ribosomes have an average half-life of four to five days, at least in rat liver (Loeb, Howell and Tomkins, 1965; Blobel and Potter, 1968) so it is to be expected that fluctuations in tissue RNA concentrations will occur more slowly than changes in polysome size or activity. What we have to consider is whether such changes in RNA concentrations occur as a result of altered protein synthetic rate or whether amino acid intake has some direct influence on either RNA synthesis or breakdown.

Ribosomal RNA synthesis is particularly sensitive to reduced rates of protein synthesis. Thus cycloheximide, an inhibitor of translation, depresses 45S pre-rRNA synthesis and its subsequent processing in

HeLa cells (Willems, Penman and Penman, 1969), rat liver (Muramatsu, Shimada and Higashinakagawa, 1970), yeast (Roth and Dampier, 1972) and many other cell types. Inhibition of protein synthesis by hypertonic conditions has a similar consequence (Pederson and Kumar, 1971). These effects may be due to rapid exhaustion of pools of ribosomal proteins needed for new ribosome synthesis (Cooper and Gibson, 1971) and/or to the rapid turnover of a protein required for activity of the nucleolar RNA polymerase (Yu and Feigelson, 1972; Gross and Pogo, 1974). The effects of amino acids on rRNA synthesis and ribosome maturation in HeLa cells (Maden et al., 1969), ascites cells (Shields and Korner, 1970), LY cells (Tiollais, Galibert and Boiron, 1971), liver cells (Bolcsfoldi and Eliasson, 1972) and fibroblasts (Vaughan, 1972) could therefore be regarded as secondary to changes in the rate of functioning of the protein synthetic machinery. When a lack of an essential amino acid slows the rate of polypeptide initiation there will follow a decline in the levels of rapidly turning over proteins at rates which are a function of their half-lives. Conversely, replacement of a limiting amino acid will restore these proteins equally rapidly (Schimke, 1970). The effects of amino acid feeding on the activity of the nucleolar RNA polymerases of both rat liver (Henderson, 1970) and ascites tumour cells (Franze-Fernandez and Fontanive-Sengüesa, 1973) may be explained by this mechanism, although it may not be the enzyme protein itself which is the rapidly turning over component.

Conclusions

We have considered amino acid supply in relation to polysome size and acute changes in the rate of protein synthesis, induction of enzymes and albumin synthesis and the long-term changes in cellular functions including variations in tissue RNA concentration. The question arises whether all these diverse effects of amino acids are mediated by a common intracellular mechanism. The obvious analogy is with the stringent response of bacteria which is brought about by changes in nutritional environment, including alterations of amino acid supply. In this case the absence of a required amino acid or an impairment of charging of cytoplasmic tRNA leads to the ribosome directed synthesis of two guanosine polyphosphates, ppGpp and pppGpp (Cashel and Gallant, 1969; Haseltine et al., 1972; Pedersen, Lund and Kjeldgaard, 1973). These compounds exert multiple effects on the cells which produce them, including:
    (1) inhibition of rRNA synthesis (Travers, Kamen and Cashel, 1970);
    (2) inhibition of protein synthesis initiation (Yoshida, Travers and Clark, 1972);
    (3) inhibition of elongation reactions of protein synthesis (Legault, Jeantet and Gros, 1972).

This pattern of effects is strikingly similar to that which we have described above for the responses of eukaryotic cells to variations in amino acid supply. However, neither of the guanosine compounds have been detected in higher cells so far, although, of course, some analogous agent may fulfil the same functions (Hershko et al., 1971).

However, the complexity of animal cells may mean that more subtle mechanisms can operate to produce the same effects as in the stringent response. One of these may be simply the differential stability of various cellular components. We have seen how the rate of protein synthesis may influence the rate of rRNA synthesis very quickly, and, conversely, how the rate of new ribosome production may set limits on the total capacity of the cell for protein synthesis. It becomes important to examine the mechanisms underlying the earliest effects of changes in amino acid availability, as these may be the key to understanding all later events. We therefore suggest that of fundamental importance to our knowledge of the regulation of both short and long-term changes brought about by amino acids is the mechanism of regulation of initiation of protein synthesis and polysome formation. Current and future work will add to what is already known concerning this vital aspect of animal cell biology.

References

ALEXIS, S.D., BASTA, S. and YOUNG, V.R. (1972). *Biochem. J.*, **128**, 521

ALEXIS, S.D. and YOUNG, V.R. (1973). *Biochem. J.*, **136**, 773

ALLEN, R.E., RAINES, P.L. and REGEN, D.M. (1969). *Biochim. biophys. Acta*, **190**, 323

BLOBEL, G. and POTTER, V.R. (1968). *Biochim. biophys. Acta*, **166**, 48

BOLCSFOLDI, G. and ELIASSON, E. (1972). *Biochim. biophys. Acta*, **272**, 67

CAMMARANO, P., CHINALI, G., GAETANI, S. and SPADONI, M.A. (1968). *Biochim. biophys. Acta*, **155**, 302

CASHEL, M. and GALLANT, J. (1969). *Nature*, **221**, 838

CHAMBERS, J.W., GEORG, R.H. and BASS, A.D. (1965). *Molec. Pharmac.*, **1**, 66

CHANDRASAKHARAM, N., FLECK, A. and MUNRO, H.N. (1967). *J. Nutr.*, **92**, 497

CHEN, H.W., HERSH, R.T. and KITOS, P.A. (1968). *Expl Cell Res.*, **52**, 490

CLEMENS, M.J. (1972). *Proc. Nutr. Soc.*, **31**, 273

CLEMENS, M.J. and KORNER, A. (1971). *Nature, (New Biol.)*, **232**, 252

CLEMENS, M.J. and PAIN, V.M. (1974). *Biochim. biophys. Acta*, **361**, 345

COOPER, H.L. and GIBSON, E.M. (1971). *J. biol. Chem.*, **246**, 5059

EKREN, T., JERVELL, K.F. and SEGLEN, P.O. (1971). *Nature, (New Biol.)*, **229**, 244

ELIASSON, E., BAUER, G.E. and HULTIN, T. (1967). *J. Cell. Biol.*, **33**, 287

ENWONWU, C.O. (1972). *Lab. Invest.*, **26**, 626

ENWONWU, C.O. and JACOBSON, K. (1973). *J. Nutr.*, **103**, 290

FAUSTO, N. (1971). *Biochim. biophys. Acta*, **238**, 116

FISHMAN, B., WURTMAN, R.J. and MUNRO, H.N. (1969). *Proc. natn. Acad. Sci. U.S.A.*, **64**, 677

FLECK, A., SHEPHERD, J. and MUNRO, H.N. (1965). *Science*, **150**, 628

FLOYD, J.C., FAJANS, S.S., CONN, J.W., KNOPF, R.F. and RULL, J. (1966). *J. clin. Invest.*, **45**, 1487

FRANZE-FERNANDEZ, M.T. and FONTANIVE-SENGÜESA, A.V. (1973). *Biochim. biophys. Acta*, **331**, 71

GAETANI, S., MASSOTTI, D. and SPADONI, M.A. (1969). *J. Nutr.*, **99**, 307

GERMANYUK, Y.L. and MIRONENKO, V.I. (1969). *Nature*, **222**, 486

GROSS, K.J. and POGO, A.O. (1974). *J. biol. Chem.*, **249**, 568

HAIDER, M. and TARVER, H. (1969). *J. Nutr.*, **99**, 433

HASELTINE, W.A., BLOCK, R., GILBERT, W. and WEBER, K. (1972). *Nature*, **238**, 381

HENDERSON, A.R. (1970). *Biochem. J.*, **120**, 205

HERSHKO, A., MAMONT, P., SHIELDS, R. and TOMKINS, G.M. (1971). *Nature, (New Biol.)*, **232**, 206

HOGAN, B.L.M. and KORNER, A. (1968). *Biochim. biophys. Acta*, **169**, 129

HOWARTH, R.E. (1972). *Can. J. Physiol. Pharmac.*, **50**, 59

IP, C.C.Y. and HARPER, A.E. (1973). *Biochim. biophys. Acta*, **331**, 251

JEFFERSON, L.S. and KORNER, A. (1969). *Biochem. J.*, **111**, 703

JOST, J.P., HSIE, A., HUGHES, S.D. and RYAN, L. (1970). *J. biol. Chem.*, **245**, 351

KELMAN, L., SAUNDERS, S.J., WICHT, S., FRITH, L., CORRIGALL, A., KIRSCH, R.E. and TERBLANCHE, J. (1972). *Biochem. J.*, **129**, 805

KIRSCH, R., FRITH, L., BLACK, E. and HOFFENBERG, R. (1968). *Nature*, **217**, 578

KNOPF, R.F., CONN, J.W., FAJANS, S.S., FLOYD, J.C., GUNTSCHE, E.M. and RULL, J.A. (1965). *J. clin. Endocrin.*, **25**, 1140

KORNER, A. (1965). *Recent Prog. Horm. Res.*, **21**, 205

KYNER, D., ZABOS, P. and LEVIN, D.H. (1973). *Biochim. biophys. Acta*, **324**, 386

LABRIE, F. and KORNER, A. (1968a). *J. biol. Chem.*, **243**, 1116

LABRIE, F. and KORNER, A. (1968b). *J. biol. Chem.*, **243**, 1120

LEE, K.L. and KENNEY, F.T. (1971). *J. biol. Chem.*, **246**, 7595

LEGAULT, L., JEANTET, C. and GROS, F. (1972). *FEBS Lett.*, **27**, 71

LOEB, J.R., HOWELL, R.R. and TOMKINS, G.M. (1965). *Science*, **149**, 1093

MADEN, B.E.H., VAUGHAN, M.H., WARNER, J.R. and DARNELL, J.E. (1969). *J. molec. Biol.*, **45**, 265

MANCHESTER, K.L. (1970). in *Mammalian Protein Metabolism*, **Vol.4**, p.229. Ed. H.N. Munro. Academic Press, New York

MAURON, J., MOTTU, F. and SPOHR, G. (1973). *Europ. J. Biochem.*, **32**, 331

MCGOWN, E., RICHARDSON, A.G., HENDERSON, L.M. and SWAN, P.B. (1973). *J. Nutr.*, **103**, 109

MILLWARD, D.J. (1970). *Clin. Sci.*, **39**, 591

MILLWARD, D.J. and GARLICK, P.J. (1972). *Proc. Nutr. Soc.*, **31**, 257

MILLWARD, D.J., GARLICK, P.J., JAMES, W.P.T., NNANYELUGO, D.O. and RYATT, J.S. (1973). *Nature*, **241**, 204

MORGAN, E.H. and PETERS, T. (1971). *J. biol. Chem.*, **246**, 3500

MORGAN, H.E., JEFFERSON, L.S., WOLPERT, E.B. and RANNELS, D.E. (1971). *J. biol. Chem.*, **246**, 2163

MUNRO, H.N. (1970). in *Mammalian Protein Metabolism*, **Vol.4** Ed. H.N. Munro, Academic Press, New York

MUNRO, H.N., HUBERT, C. and BALIGA, B.S. (1975). in *Alcohol and Abnormal Protein Biosynthesis*, p.33. Eds M.A. Rothschild, M. Oratz and S.S. Schreiber. Pergamon Press, Oxford and New York

MURAMATSU, M., SHIMADA, N. and HIGASHINAKAGAWA, T. (1970). *J. molec. Biol.*, **53**, 91

PAIN, V.M. (1973). *FEBS Lett.*, **35**, 169

PAIN, V.M. and GARLICK, P.J. (1974). *J. biol. Chem.*, **249**, 4510

PEDERSEN, F.S., LUND, E. and KJELDGAARD, N.O. (1973). *Nature, (New Biol.)*, **243**, 13

PEDERSON, T. and KUMAR, A. (1971). *J. molec. Biol.*, **61**, 655

PITOT, H.C. and PERAINO, C. (1964). *J. biol. Chem.*, **239**, 1783

PRONCZUK, A.W., ROGERS, Q.R. and MUNRO, H.N. (1970). *J. Nutr.*, **100**, 1249

ROSEN, F. and MILHOLLAND, R.J. (1968). *J. biol. Chem.*, **243**, 1900

ROTH, R.M. and DAMPIER, C. (1972). *J. Bacteriol.*, **109**, 773

ROTHSCHILD, M.A., ORATZ, M., MONGELLI, J., FISHMAN, L. and SCHREIBER, S.S. (1969). *J. Nutr.*, **98**, 395

SCHIMKE, R.T. (1970). in *Mammalian Protein Metabolism*, **Vol.4**. p.177. Ed. H.N. Munro. Academic Press, New York

SHIELDS, R. and KORNER, A. (1970). *Biochim. biophys. Acta*, **204**, 521

SIDRANSKY, H., SARMA, D.S.R., BONGIORNO, M. and VERNEY, E. (1968). *J. biol. Chem.*, **243**, 1123

SIDRANSKY, H., STAEHELIN, T. and VERNEY, E. (1964). *Science*, **146**, 766

SMULSON, M.E. and RIDEAU, C. (1970). *J. biol. Chem.*, **245**, 5350

STANNERS, C.P. (1966). *Biochem. biophys. Res. Commun.*, **24**, 758

TIOLLAIS, P., GALIBERT, F. and BOIRON, M. (1971). *Europ. J. Biochem.*, **18**, 35

TRAGL, K.H. and REAVEN, G.M. (1971). *Diabetes*, **20**, 27

TRAVERS, A., KAMEN, R. and CASHEL, M. (1970). *Cold Spring Harbor Symposium on Quantitative Biology*, **35**, 417

VAN DEN BORRE, M. and WEBB, T.E. (1972). *Life Sci.*, **11**, 347

VAN VENROOIJ, W.J., HENSHAW, E.C. and HIRSCH, C.A. (1970). *J. biol. Chem.*, **245**, 5947

VAN VENROOIJ, W.J., HENSHAW, E.C. and HIRSCH, C.A. (1972). *Biochim. biophys. Acta*, **259**, 127

VAUGHAN, M.H. (1972). *Expl Cell Res.*, **75**, 23

VAUGHAN, M.H. and HANSEN, B.S. (1973). *J. biol. Chem.*, **248**, 7087

VAUGHAN, M.H., PAWLOWSKI, P.J. and FORCHHAMMER, J. (1971). *Proc. natn. Acad. Sci. U.S.A.*, **68**, 2057

VON DER DECKEN, A. and OMSTEDT, P.T. (1970). *J. Nutr.*, **100**, 623

WATERLOW, J.C. and STEPHEN, J.M. (1968). *Clin. Sci.*, **35**, 287

WETTENHALL, R.E.H. and WOOL, I.G. (1972). *J. biol. Chem.*, **247**, 7201

WILLEMS, M., PENMAN, M. and PENMAN, S. (1969). *J. Cell. Biol.*, **41**, 177

WOODSIDE, K.H. and MORTIMORE, G.E. (1972). *J. biol. Chem.*, **247**, 6474

WOOL, I.G., STIREWALT, W.S., KURIHARA, K., LOW, R.B., BAILEY, P. and OYER, D. (1968). *Recent Progr. Horm. Res.*, **24**, 139

WUNNER, W.H., BELL, J. and MUNRO, H.N. (1966). *Biochem. J.*, **101**, 417

WURTMAN, R.J., SHOEMAKER, W.J. and LARIN, F. (1968). *Proc. natn. Acad. Sci . U.S.A.*, **59**, 800

YOSHIDA, M., TRAVERS, A. and CLARK, B.F.C. (1972). *FEBS Lett.*, **23**, 163

YOUNG, V.R. (1970). in *Mammalian Protein Metabolism*, **Vol.4**. p.621, Ed. H.N. Munro. Academic Press, New York

YOUNG, V.R. and ALEXIS, S.D. (1968). *J. Nutr.*, **96**, 255

YOUNG, V.R., STOTHERS, S.C. and VILAIRE, G. (1971). *J. Nutr.*, **101**, 1379

YU, F.L. and FEIGELSON, P. (1972). *Proc. natn. Acad. Sci. U.S.A.*, **69** 2833

ZASLOFF, M. (1973). *J. molec. Biol.*, **76**, 445

3

HORMONAL CONTROL OF PROTEIN METABOLISM

K.L. MANCHESTER,
Department of Biochemistry, University of the West Indies, Kingston, Jamaica
Present address: *Professor of Biochemistry, University of Witwatersrand, Johannesburg, South Africa*

Introduction

Hormones play an important role in initiating and supporting the growth of a variety of tissues. Their continual presence is also necessary for the maintenance of many tissues of the adult at mature size. Absence of particular hormones, as well as retarding growth in the young, will lead to shrinkage and loss of weight of otherwise mature organs. These phenomena are well recognised with regard to the influence of the trophic hormones of the anterior pituitary on their target glands and of the sex hormones on the accessory sexual tissues. Likewise lack of insulin, growth hormone and thyroxine leads to general somatic debilitation.

In the last fifteen or so years the detailed steps of how cells build up amino acids into proteins of a precise primary structure have been elucidated to a remarkable degree. Armed with this knowledge study of how hormones may influence the rate of protein formation has likewise made considerable progress.[1]

Possible Points of Control of Protein Synthesis

Protein synthesis involves translation of the information stored in the sequences of bases in the DNA of the cell nucleus, and various forms of RNA are intimately involved in this process (*Figure 3.1*). Messenger-RNA is synthesised in the nucleus with a sequence of bases determined by that of the DNA chains corresponding to genes, and when the mRNA becomes attached to ribosomes in the cytoplasm the sequence of bases contained in the mRNA is interpreted by translation into a sequence of amino acids to form a new polypeptide chain. Amino acids must obviously be available for the synthesis, they are supplied to the ribosomes in the form of aminoacyl-tRNA. tain extensive protein synthesis the amino acids must either be formed within the cell or be brought into the cell from outside.

There are obviously a variety of ways in which hormones may affect protein synthesis. The more RNA the cell contains, the more

---

[1] For more extensive reviews of this subject see Manchester (1970a; 1975).

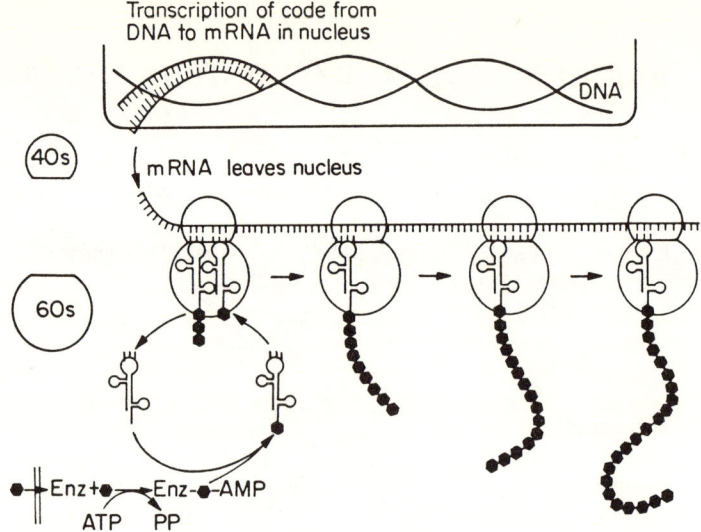

Figure 3.1 A general outline of the mechanism of protein synthesis. Amino acids on entry into the cell undergo activation by ATP and amino acid activating enzymes to form aminoacyl-tRNA. These molecules are then used by the ribosomes to provide the appropriate amino acids for addition into the growing polypeptide chain which is being built up according to the sequence of bases contained in the messenger-RNA. This molecule in turn, together with the RNA of the ribosomes is synthesised in the nucleus. Hormones can make available more ribosomes and more or different mRNA by affecting nuclear transcription. They can enhance the rate of attachment of ribosomes to the mRNA (initiation), and speed up the movement of the ribosomes along the messenger (elongation) by increasing the availability of aminoacyl-tRNA and/or the activity of elongation factors. They can also increase the uptake of amino acids by cells.

potential machinery there is available for protein synthesis. There is no question that hormones exerting growth promoting effects increase cellular RNA concentration, and conversely hormone deficiency results in diminution of RNA levels. A number of hormones promote entry of certain amino acids into cells, but there is no clear evidence that this stimulation of amino acid uptake can operate as a controlling influence on protein synthesis, since many of the amino acids showing the least response are often those already present at the lowest concentrations.

Stimulation of RNA synthesis could be especially of one class, messenger, ribosomal or transfer; or of all forms. In bacteria, messenger-RNA turns over rapidly by comparison with the other forms. Continuing synthesis of protein in bacteria therefore requires continual synthesis of mRNA. A similar situation in cells of higher organisms would offer an obvious point at which hormonal control could operate. Messenger-RNA of animal cells however has a much longer life span than does its bacterial counterpart and is translated many times over (Palmiter, 1973). Present evidence is that the availability of mRNA is unlikely to determine the overall rate of protein synthesis, though what species of mRNA are present may determine which particular proteins are to be formed. Thus, although we can identify situations in which there is an endocrine stimulation of the formation of specific mRNA molecules, leading to enhanced synthesis of specific proteins, increase of RNA in response to hormones is frequently of all forms, including rRNA, and it is the increase of rRNA that is most readily observable.

The mere existence of mRNA and ribosomes together in the cell however does not ensure active protein synthesis. For this to occur there must be initiation; the attachment of the ribosome to the starting point of the sequence of bases in the messenger molecule dictating the peptide sequence to be synthesised; followed once initiation has occurred by elongation. Control of elongation or of initiation is a stage at which hormones might act, and indeed changing rates of initiation in response to hormones seems a frequent occurrence. Measurements of translation rates *in vivo* and in cell-free systems suggest that rates of elongation of the peptide chain once initiated can likewise be controlled. Variation in activity of elongation factors or availability of aminoacylated tRNA may be the mechanisms responsible, but there is a lack of hard information on the subject.

Hormonal Stimulation of Synthesis of RNA

Increased formation of RNA in response to hormones must be mediated at some point in the nucleus and implies that hormones in responsive tissues will act on this organelle. Detailed evidence of this is best illustrated in the case of steroid hormones (*Figure 3.2*).

In the cytoplasm of cells of responsive tissues are proteins to which the steroid binds, e.g. for oestradiol in the uterus and for glucocorticoids in liver. The steroid protein complex so formed undergoes activation in some manner in the cytoplasm, then enters the nucleus. Inside the nucleus the complex forms an association with acidic proteins of the chromatin, and this appears to be one of the critical steps in stimulation of RNA synthesis. The template capacity of chromatin, i.e. the native DNA of the cell nucleus, appears to be dependent on its conformation, which in turn is probably affected by the proteins bound to it. Chromatin associated proteins are of two classes, basic histones and acidic proteins. Histones appear to be relatively non-specific from one tissue or species to another. Acidic proteins on the other hand are more specific for different tissues and species, and it is therefore believed that alterations in the extent of acidic protein binding to DNA or chromatin may be critical in regulating RNA synthesis and transcription of particular RNA sequences (O'Malley and Means, 1974).

There are several forms of RNA polymerase in the nucleus, at least one form being found in the nucleolus and which is believed to be responsible for rRNA synthesis. It is this form of polymerase which usually shows the earliest and most marked response to hormones. This observation however does not mean that it is the nucleolus and its associated polymerase that are the primary sites of hormone action. As has been pointed out by Schimke (1970), if synthesis of a number of proteins is stimulated to a similar proportional extent, those with the more rapid turnover rates will show the larger and more rapid responses. Work with polymerases from rat liver suggests that the nucleolar polymerase has a more rapid rate of

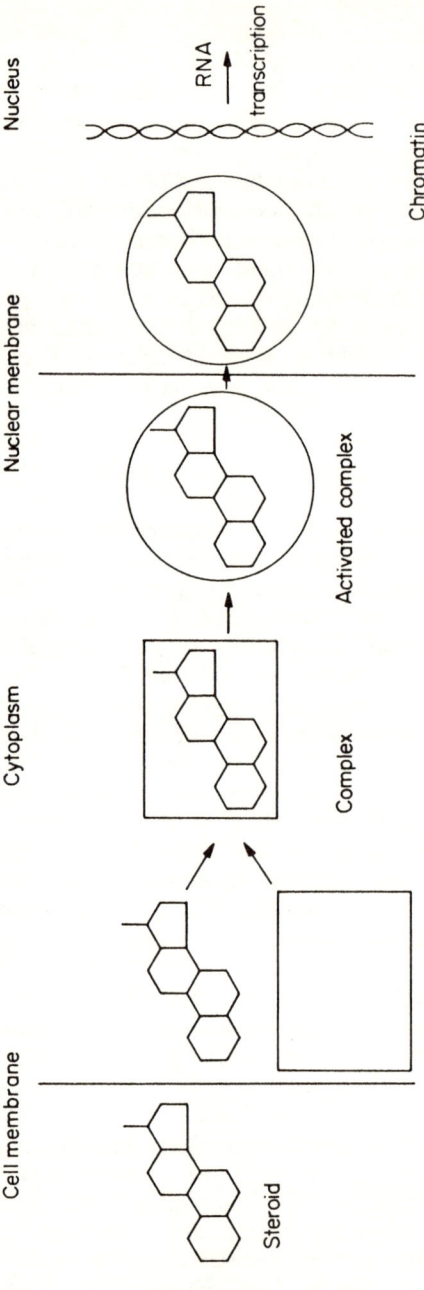

Figure 3.2 Action of steroid hormones. The steroid molecule on entry into a responsive cell is bound by a specific receptor protein present in the cytoplasm. The steroid-receptor complex undergoes some form of activation in the cytoplasm and is then able to enter the nucleus where it becomes attached to proteins associated with the chromatin (DNA). As a result transcription of new or extra RNA occurs

turnover than the major component of the nucleoplasm (Yu and Feigelson, 1972). Thus the more rapid rise in nucleolar polymerase seen after glucocorticoid treatment for example does not necessarily indicate it to result from an action of glucocorticoids specifically directed towards the nucleolus. In fact, results with α-amanitin, an inhibitor of the nucleoplasmic polymerase, suggest that stimulation of at least one form of nucleoplasmic polymerase precedes that of the nucleolar polymerase (Yu and Feigelson, 1973). The general concept therefore is that, in stimulating cellular RNA production, hormones such as oestrogen, testosterone and cortisol bring about first an increase in transcription of the messenger-RNA of one form of nucleoplasmic polymerase which on translation produces a protein which in turn facilitates transcription of messengers for nucleolar and more slowly turning over nucleoplasmic polymerases, which then lead to overall RNA production (*Figure 3.3*).

Hormonal Stimulation of Synthesis of Specific Proteins

Sex steroids in higher animals produce a general increase in protein synthesis without any evidence for stimulated synthesis of any specific protein species. However in a tissue such as the chicken oviduct in which large amounts of a few proteins are secreted, we have a situation in which, in response to a particular hormone, large increases in the production of a particular protein can be seen, e.g. ovalbumin in response to oestrogen and avidin in response to progesterone. In this instance, however, if we start with an immature hen the substantial increase in protein formation in response to oestrogen is preceded by differentiation of a variety of distinct types of epithelial cells from the primitive mucosa before specific protein production begins (O'Malley et al., 1969). Following cell differentiation, as well as there being a large increase in formation of ribosomes and the machinery for protein synthesis, there is also a substantial production of specific messenger-RNA for the particular egg proteins. If after primary stimulation (and differentiation) oestrogen is withdrawn, mRNA for ovalbumin is undetectable, but makes its re-appearance as soon as further oestrogen is given (Palmiter, 1973). Thus we have here a clear example of stimulation of production of a specific messenger-RNA in response to a steroid hormone.

Production of both ovalbumin and avidin by the oviduct requires the presence of both oestrogen and progesterone. Another example of a requirement for several hormones working in concert is in the mammalian equivalent of egg protein production, namely formation of milk proteins in the mammary gland. Development of this gland seems to be dependent on the presence of several hormones, insulin, oestradiol, cortisol and growth hormone, and possibly other factors (Turkington et al., 1973). Increasing concentrations of oestradiol during development and particularly during pregnancy appear to allow insulin to stimulate undifferentiated epithelial cells to divide as a preliminary to their development into secretory alveolar

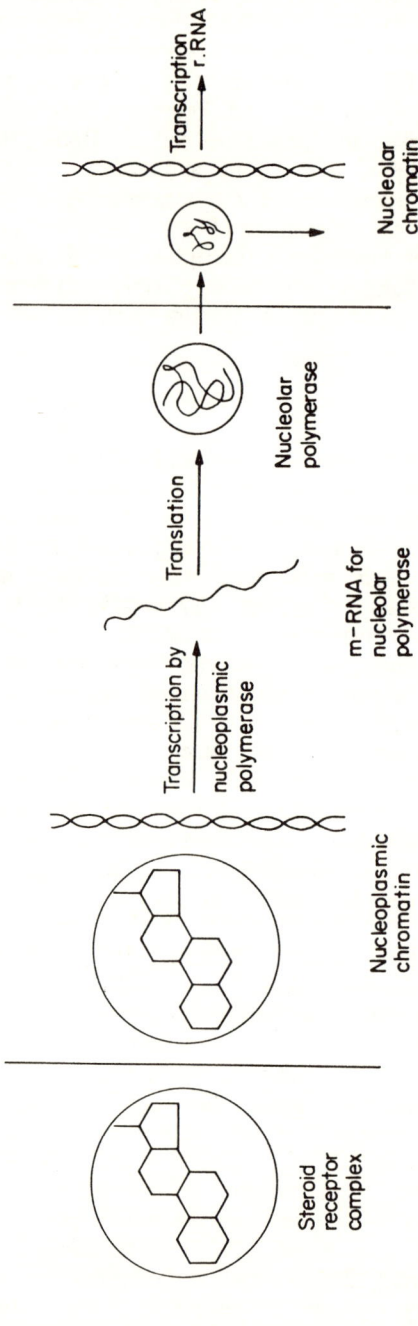

Figure 3.3. Stimulation of synthesis of rRNA in response to glucocorticoids. In the scheme of Yu and Feigelson (1973) the steroid-receptor complex on entry into the nucleus stimulates first transcription of a portion of the nuclear chromatin to produce the messenger-RNA for the nucleolar polymerase (or a rapidly turning over sub-unit of it) responsible for transcribing rRNA. On translation of this messenger more active nucleolar polymerase is available and hence there is more transcription of rRNA

cells. As well as inducing division, insulin stimulates formation of ribosomes and their organisation into polysomes (Turkington and Riddle, 1970). After cell division has occurred, prolactin acts synergistically with insulin to stimulate ribosome formation and in the presence of prolactin the synthesis of specific casein and whey proteins begins. Thus, with both oviduct and mammary gland, cell division tends to occur when the immature organ is stimulated before specific protein production begins, and the hormones stimulating and allowing cell division are not necessarily those responsible for inducing the specific proteins. The mechanisms involved leading to DNA synthesis and cell division in response to hormones have so far received little attention. It is interesting to note that enhanced production of specific mRNA does not appear to involve gene duplication (Harris et al., 1973; Sullivan et al., 1973).

Hormonal Stimulation of Initiation

Uterus, oviduct and mammary gland are all rather specialised tissues. Two hormones that exert a significant controlling influence over major somatic tissues and that have marked effects on nitrogen retention are insulin and growth hormone. As protein molecules, it is unlikely that they enter the cell or are carried to the nucleus, but it seems that their effects are mediated through actions at the membrane of responsive cells. In muscle of diabetic animals the organisation of ribosomes into polysomes is impaired. Treatment with insulin promotes polysome formation, an effect that is not dependent on the synthesis of new RNA (Wool et al., 1968). In isolated tissues when incubated or perfused, there is a disaggregation of polysomes and this disaggregation is prevented in certain cases by the presence of insulin (Morgan et al., 1971; Jefferson, Koehler and Morgan, 1972; Figure 3.4). Thus insulin appears to enhance the formation of polysomes through stimulating the process of initiation. This effect can also be seen with cell free systems obtainable from liver in which initiation is taking place. Addition of polyinosinic acid to bind subunits and single ribosomes that might otherwise re-initiate produces a smaller inhibition of amino acid incorporation when the system is taken from the liver of diabetic rats (Pain, 1973a).

The initiation process, attachment of a ribosome to messenger-RNA and the incorporation of the initiator methionine-tRNA into the complex, involves a number of proteinaceous initiation factors. A possible way in which insulin might control initiation would be if it were to affect a protein kinase system responsible for phosphorylation and dephosphorylation of one of these proteins, as seen with certain enzymes controlling synthetic pathways, but this hypothesis is pure speculation as yet. It is of considerable interest that amino acids seem to be able in certain tissues to enhance polysome aggregation in a manner similar to that seen with insulin (Manchester, 1975; Clemens and Pain, this symposium). It is tempting to suggest that a common mechanism is responsible for these phenomena. Reichman

42 Hormonal control

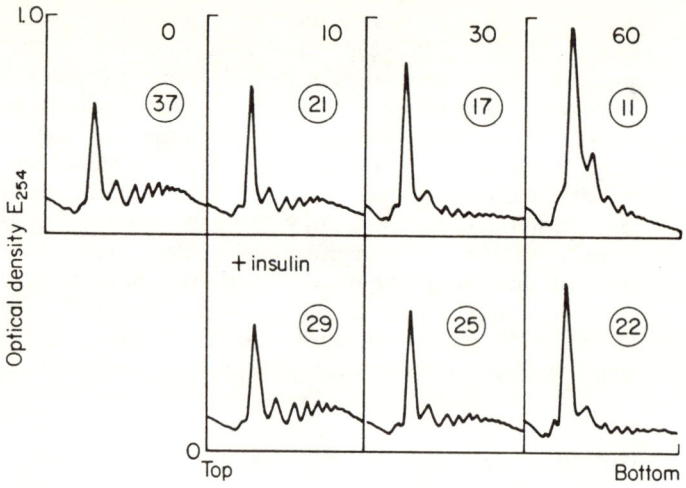

Figure 3.4 Effect of incubation in the presence and absence of insulin on the sucrose density gradient profiles of ribosomes from diaphragm muscle. Diaphragm muscle was incubated for various periods as indicated by the figures (minutes) in the upper panels, then the ribosomes were isolated and their profiles determined. In the upper panels incubation was in the absence of insulin; in the lower panels in its presence. The numbers in circles are the percentage of ribosomes in the preparation carrying nascent peptide chains as determined by labelling with puromycin (Manchester, 1974)

and Penman (1973) have evidence that under some conditions, e.g. amino acid starvation, a new species of RNA that promotes initiation may be formed.

Another possibility is that the extent of aminoacylation of tRNA may control initiation, more specifically than uncharged (de-acylated) tRNA can inhibit the initiation process (Kyner, Zabos and Levin, 1973; Zasloff, 1973). Vaughan and Hansen (1973) have found with HeLa cells that a reduction of only 20 per cent in the level of tRNA aminoacylation may be accompanied by a four-fold decrease in the rate of initiation. Available figures for liver and muscle (*Table 3.1*) show that tRNA in the cell is normally quite fully acylated and it is unlikely that there are large changes in available aminoacyl-tRNA in starvation or diabetes. However, as the figures in *Table 3.1* indicate, only a small change in the extent of tRNA changing could lead to large changes in the amount of unacylated tRNA present. It remains to be seen whether the small fraction of uncharged tRNA in tissues does exert a significant controlling influence.

In addition to effects on initiation it is likely that insulin can also influence elongation (Migliorini and Manchester, 1971; Pain, 1973b).

Table 3.1 Effects of fasting and diabetes on levels of acylated and de-acylated tRNA in liver and muscle

		Units of de-acylated tRNA (cpm/mg)		% acylation on isolation
		Before de-acylation	After de-acylation	
Liver	fed	100	2400	96
	fasted	350	2550	86
Muscle	fed	170	2300	93
	fasted	330	2600	87
Liver	normal	190	2150	91
	diabetic	400	2300	83
Muscle	normal	170	2600	93
	diabetic	160	2470	94

tRNA was isolated from extracts of liver and muscle by treatment with phenol. The purified tRNA was allowed to react with periodate followed by [^{14}C] isonicotinic acid hydrazide. The terminal adenosine of tRNA not bearing an amino acid is subject to attack and takes up ^{14}C. Total amino acid carrying capacity is the amount of ^{14}C taken up by tRNA after removal of amino acids. The amount of de-acylated tRNA in the native material is proportional to the ^{14}C taken up before stripping. The extent of acylation is the difference. (Clarke and Manchester, unpublished observations)

Hormonal Stimulation of Amino Acid Transport

One of the most rapid effects of growth promoting hormones on responsive tissues is often stimulation of amino acid uptake. This action for insulin on muscle is almost immediate and for oestrogen acting upon the uterus precedes increase in protein synthesis. However, a peculiar feature of the hormonal stimulation of amino acid uptake is that it normally shows itself for only a limited group of amino acids (*Table 3.2*), those that are believed to use one of a group of specific transport devices by which different amino acids enter cells (Manchester, 1970b; Riggs and McKirahan, 1973). Strangely it is the amino acids already present in cells in the largest amounts that seem to show the largest responses. That uptake of many amino acids does not respond to the same hormonal stimulus implies that this action is a manifestation rather than a causal enhancement of protein synthesis, and its teleological significance is not clear. Franchi-Gazzola *et al.* (1973) have recently shown for chicken embryo heart cells that at least one component of the transport mechanism used by the responsive amino acids is a protein of short half-life. The rate of breakdown of this protein, which is enhanced by the amino acids it transports, is suppressed by insulin (Guidotti, et al., 1974).

Table 3.2 Influence of hormones on uptake of amino acids by responsive tissues. Based on findings of Manchester (1970b) for effects of insulin on isolated rat diaphragm muscle and of Riggs, Pan and Feng, (1968) for effects of oestrogen administration on uptake by subsequently isolated uteri

Amino acids whose uptake is simulated by	
Insulin	Oestrogen
Alanine	Alanine
Asparagine	Glycine
Glutamine	Lysine
Glycine	Proline
Histidine	Serine
Methionine	
Proline	
Serine	
Uptake not stimulated	
Arginine	Leucine
Aspartate	Phenylalanine
Cystine	Valine
Glutamate	
Isoleucine	
Leucine	
Lysine	
Phenylalanine	
Threonine	
Tryptophan	
Tyrosine	
Valine	

Hormonal Suppression of Protein Catabolism

Suppression of catabolism of the specific 'A mediation' transport protein just referred to is probably not peculiar to that protein, but is one example of a more general ability of certain hormones to inhibit protein breakdown. Although not a subject as yet clearly understood, insulin for example appears to suppress protein catabolism in both perfused liver and heart muscle (Mortimore and Mondon, 1970; Morgan *et al.*, 1972). Goldberg (1969) concluded that cortisol produced a loss of protein from muscle by both inhibition of synthesis and stimulation of breakdown. However more recent work (Millward and Garlick, this symposium) suggests a close correlation in many instances between the rate of synthesis and breakdown of proteins i.e. the rates move up and down together, not inversely.

Hormonal Control of Metabolism Through Specific Protein Formation

As well as with the specific proteins secreted by the oviduct and mammary gland, there are a number of other situations in which hormones bring about selective changes in enzyme activities. Examples are the ability of insulin to promote synthesis of specific enzymes in liver and adipose tissue involved in the metabolic functions of these organs. In these instances, and many others, e.g. the effect of

Table 3.3 Changes in activities of enzymes of glycolysis, gluconeogenesis and fatty acid synthesis induced by hormones and substrates in the perfused rat liver (Wimhurst and Manchester, 1973; Wimhurst and Manchester, 1973; Wimhurst and Harris, 1974; Wimhurst, Manchester and Harris, 1974).

Enzyme	Increase (+) or decrease (−) in enzyme content in livers perfused with						
	Glucose	Fructose	Lactate	Glucagon + lactate	Insulin	Insulin + lactate	Prostaglandin	Prostaglandin + lactate
Glucokinase		−					−	
Phosphofructokinase		−		−	+	−		
Pyruvate kinase			+	+	+	+	+	
Pyruvate carboxylase	−	−						
Phosphoenolpyruvate carboxykinase	−	−		•	−		−	
Fructose 1:6-diphosphatase						+		
Glucose 6-phosphatase		−	−					
Isocitrate dehydrogenase						−	−	−
Malic enzyme							−	
Acetyl-CoA carboxylase							−	

insulin on proteins involved in amino acid transport and of cortisol on the glucose transport mechanism in thymus (Hallahan, Young and Munck, 1973), control of protein synthesis and breakdown results in a directive effect by the hormones on metabolism. One outstanding problem requiring resolution is how the selective influence of the hormones is achieved. Thus, administration of glucocorticoids brings about a marked stimulation of hepatic RNA synthesis, yet enzymes inducible with insulin, which can also enhance RNA formation, are not now induced.

The most marked effects of glucocorticoids in the liver are to enhance the activities of enzymes involved in amino acid degradation and gluconeogenesis, whereas insulin usually tends both to reverse the effects of the steroids and to enhance synthesis of enzymes active in glycolysis and glycogen and fat synthesis. Direct effects of insulin, glucagon and prostaglandin on the activities of a variety of enzymes have recently been shown in perfused rat liver (Wimhurst and Harris, 1974; Wimhurst, Manchester and Harris, 1974). Interestingly, by analogy with the capacity of amino acids to mimic effects of insulin on initiation, substrates such as glucose and lactate are likewise able to induce and suppress activities of a variety of enzymes in the absence of any influence on endocrine secretion (Wimhurst and Manchester, 1973) (*Table 3.3*).

Though we are learning in increasing detail something of the mechanisms by which hormones work, and in particular the way they affect protein synthesis, the pattern that emerges is amazingly complex and leaves a great deal of scope for further investigations.

References

CLEMENS, M.J. and PAIN, V.M., these proceedings
FRANCHI-GAZZOLA, R., GAZZOLA, G.C., RONCHI, P., SAIBENE, V. and GUIDOTTI, G.G. (1973). *Biochim. biophys. Acta*, **291**, 545
GOLDBERG, A.L. (1969). *J. biol. Chem.*, **244**, 3223
GUIDOTTI, G.G., FRANCHI-GAZZOLA, R., GAZZOLA, G.C. and RONCHI, P. (1974). *Biochim. biophys. Acta*, **356**, 219
HALLAHAN, C., YOUNG, D.A. and MUNCK, A. (1973). *J. biol. Chem.*, **248**, 2922
HARRIS, S.E., MEANS, A.R., MITCHELL, W.M. and O'MALLEY, B.W. (1973). *Proc. natn. Acad. Sci . U.S.A.*, **70**, 3776
JEFFERSON, L.S., KOEHLER, J.O. and MORGAN, H.E. (1972). *Proc. natn. Acad. Sci. U.S.A.*, **69**, 816
KYNER, D., ZABOS, P. and LEVIN, D.H. (1973). *Biochim. biophys. Acta*, **324**, 386
MANCHESTER, K.L. (1970a). in *Mammalian Protein Metabolism.* **Vol.4**, 229. Academic Press, New York
MANCHESTER, K.L. (1970b). *Biochem. J.*, **117**, 457
MANCHESTER, K.L. (1974). *Biochemistry*, **13**, 3062
MANCHESTER, K.L. (1975). in *Synthesis of Amino Acids and Proteins.* Ed. H.R.V. Arnstein. *MTP International Review of Science*, p.329-358 Butterworths, London

MANCHESTER, K.L. (1974). *Biochemistry*, **13**, 3062
MIGLIORINI, R.H. and MANCHESTER, K.L. (1971). *FEBS Lett.*, **13**, 140
MILLWARD, D.J. and GARLICK, P.J., these proceedings
MORGAN, H.E., JEFFERSON, L.S., WOLPERT, E.B. and RANNELS, D.E. (1971). *J. biol. Chem.*, **246**, 2163
MORGAN, H.E., RANNELS, D.E., WOLPERT, E.B., GIGER, K.E., ROBERTSON, J.W. and JEFFERSON, L.S. (1972). in *Insulin Action*, p.437. Academic Press, New York
MORTIMORE, G.E. and MONDON, C.E. (1970). *J. biol. Chem.*, **245**, 2375
O'MALLEY, B.W. and MEANS, A.R. (1974). *Science*, **183**, 610
O'MALLEY, B.W., MCGUIRE, W.L., KOHLER, P.O. and KORENMAN, S.G. (1969). *Recent Progr. Hormone Research*, **25**, 105
PAIN, V.M. (1973a). *FEBS Lett.*, **35**, 169
PAIN, V.M. (1973b). *Biochim. biophys. Acta*, **308**, 180
PALMITER, R.D. (1973). *J. biol. Chem.*, **248**, 8260
REICHMAN, M. and PENMAN, S. (1973). *Proc. natn. Acad. Sci., U.S.A.*, **70**, 2678
RIGGS, T.R. and McKIRAHAN, K.J. (1973). *J. biol. Chem.*, **248**, 6450
RIGGS, T.R., PAN, M.W. and FENG, H.W. (1968). *Biochim. biophys. Acta*, **150**, 92
SCHIMKE, R.T. (1970). in *Mammalian Protein Metabolism.* **Vol.4**, 177. Academic Press, New York
SMITH, D.W.E. and McNAMARA, A.L. (1974). *J. biol. Chem.*, **249**, 1330
SULLIVAN, D., PALACIOS, R., STAVNEZER, J., TAYLOR, J.M., FARAS, A.J., KIELY, M.L., SUMMERS, N.M., BISHOP, J.M. and SCHIMKE, R.T. (1973). *J. biol. Chem.*, **248**, 7530
TURKINGTON, R.W., MAJUMDER, G.C., KADOHAMA, N., MacINDOE, J.H. and FRANTZ, W.L. (1973). *Recent Progr. Hormone Res.*, **29**, 417
TURKINGTON, R.W. and RIDDLE, M. (1970). *J. biol. Chem.*, **245**, 5145
VAUGHAN, M.H. and HANSEN, B.S. (1973). *J. biol. Chem.*, **248**, 7087
WIMHURST, J.M. and HARRIS, E.J. (1974). *Europ. J. Biochem.*, **42**, 33
WIMHURST, J.M. and MANCHESTER, K.L. (1973). *Biochem. J.*, **134**, 143
WIMHURST, J.M., MANCHESTER, K.L. and HARRIS, E.J. (1974). *Biochim. biophys. Acta*, **372**, 72
WOOL, I.G., STIREWALT, W.S., KURIHARA, K., LOW, R.B., BAILEY, P. and OYER, D. (1968). *Recent Progr. Hormone Res.*, **24**, 139
YU, F-L. and FEIGELSON, P. (1972). *Proc. natn. Acad. Sci. U.S.A.*, **69**, 2833
YU, F-L. and FEIGELSON, P. (1973). *Biochem. biophys. Res. Commun.*, **53**, 754
ZASLOFF, M. (1973). *J. molec. Biol.*, **76**, 445

4

PROTEIN TURNOVER

D.J. MILLWARD, P.J. GARLICK, W.P.T. JAMES, P.M. SENDER and J.C. WATERLOW
London School of Hygiene and Tropical Medicine

Introduction

We have been interested for some years in the adaptation of protein metabolism to protein deficiency and the development of methods for assessment of rates of protein synthesis and breakdown. This paper describes the methods we have used and the results we have obtained. Most of our results concern protein turnover in liver and muscle since these two tissues demonstrate contrasting responses to inadequate diets.

Measurement of the Rate of Protein Turnover

In order to avoid ambiguity, the rate of protein turnover is best described in terms of the rate of one or other of the two processes of protein synthesis and breakdown. In the steady state, of course, the rate of these two processes is the same. In the non-steady state it is often possible to calculate the rate of one of the processes if the rate of the other is known, since the difference between the two rates can be measured as a net change in protein mass. Indeed a study of the way in which the amount of a protein changes in response to a stimulus can in certain situations itself indicate the rates of synthesis or breakdown of the protein (Schimke, 1970).

In most other cases the measurement of protein turnover rates involves isotopic methods.

Broadly speaking, isotopic methods fall into two groups, those methods based on uptake of isotope into protein which yield information about the rate of protein synthesis and those based on isotope loss from which the rates of both synthesis and breakdown can be measured. These two broad approaches and the results obtained with them are best dealt with separately.

UPTAKE METHODS
These are free from the problems arising from uncertainties about the kinetics of protein degradation. The main difficulties stem from the

need to define and measure accurately the precursor specific activity which is always essential in this type of experiment. Because the actual precursor for protein synthesis, i.e. the transfer RNA-bound amino acid, is technically difficult to isolate and measure, it is usual to compromise and measure the specific activity of the free amino acid in either plasma or the tissue. While there is much evidence to suggest that the amino acid pool is heterogeneous (Hider, Fern and London, 1971), recent work in this laboratory (Fern and Garlick, 1973) suggests that in practice there is little or no difference between the mean intracellular specific activity and that of the precursor in most tissues of the rat.

After the single injection of a labelled amino acid the specific activity of the various free amino acid pools, and presumably that of the precursor, changes very rapidly, rising to a maximum value in minutes or hours, and then falling away (Henriques, Henriques and Neuberger, 1955). To measure the rate of protein synthesis in this situation, the time course of the specific activity of both precursor and protein-bound amino acid must be measured. This can be very laborious and logistically difficult, especially with *in vivo* measurements, since different animals are required for each time point. Nevertheless, when this is done, the rate of incorporation of isotope and hence of protein synthesis can be calculated from the labelling of the protein and the integral of the precursor specific activity-time curve. When properly carried out, these methods do indeed give satisfactory results (e.g. Haider and Tarver, 1969).

One way to avoid such rapid changes in the precursor specific activity after a single injection is to give a large quantity of the labelled amino acid. This results in a more or less constant precursor specific activity for long enough (thirty minutes) to measure the rate of incorporation (Henshaw *et al.*, 1971). The measurements and calculations are simplified by this approach, but it does have the potential drawback of producing metabolic changes in response to the large load of the amino acid, which no longer acts simply as a tracer.

THE CONTINUOUS INFUSION

A more satisfactory way of overcoming the problems of a rapid change in the precursor specific activity is to feed or infuse the amino acid continuously. In this case the labelling of the free amino acid pools of the plasma and tissues reaches a plateau (or more correctly, a quasi-plateau; Aub and Waterlow, 1969). If the way in which the rise to plateau is measured or estimated (which is relatively easy, Garlick, Millward and James, 1973) a single measurement of the protein bound and free amino acid specific activity at the end of the infusion is all that is needed to calculate the rate of protein synthesis.

In addition to measurement of protein synthesis rates in tissues, the continuous infusion method affords the means to measure the turnover of the whole body protein pool by calculations based on the amino acid flux.[1]

THE AMINO ACID FLUX

During a continuous infusion the specific activity of the free amino acid in the plasma reaches a plateau when the rate at which radioactivity is infused is equal to the rate at which it leaves that compartment (Waterlow and Stephen, 1967). Since this is equal to the product of the specific activity and the amount of amino acid leaving in unit time (i.e. the flux rate), the flux is calculated from the expression:

$$ISi = QSp$$

where I = rate of amino acid infusion (μmoles/min);
  Si = specific activity of infused amino acid;
  Sp = specific activity at plateau of the plasma amino acid in the steady state;
and Q = flux rate (μmoles/min) = ISi/Sp

In physiological terms, the flux, Q, is equal to the rate of input into the plasma-free amino acid pool and includes amino acids originating from protein breakdown, *de novo* synthesis and the diet; in the steady state this is equal to the rate at which they leave by way of synthesis into protein, oxidation and other irreversible metabolic pathways. Thus:

$$Q = D + C = E + S$$

where D = intake (diet and *de novo* synthesis)
  C = amino acids entering plasma from protein breakdown
  E = excretion (oxidation and other pathways)
and S = amino acids leaving plasma for protein synthesis.

Strictly speaking, the Q determined from the plasma specific activity will only be a measure of total protein turnover when there is a single homogeneous free amino acid pool. This, of course, is not the case, since in practice dilution of the intracellular pool by protein breakdown results in a specific activity which is less than in plasma (Gan and Jeffay, 1967; Waterlow and Stephen, 1968). The difference will be greatest in the most rapidly turning-over tissues, such as liver, in which the intracellular specific activity is 60 per cent of plasma, compared with muscle, in which it achieves a plateau closer to that of plasma (Garlick, Millward and James, 1973). Thus, because the specific activity of the precursor is lower than in plasma, in the rat the plasma flux underestimates the flux through the protein pool by perhaps 30 per cent (Waterlow and Stephen, 1967).

The first point which emerges from measurements of flux rate by the constant infusion method is that at normal levels of protein intake the amino acid flux is considerably greater than the rate of intake from food. *Table 4.1* shows values obtained by us in both the rat and man. In each case the intake was only one-fifth of the

[1] Flux is here used to indicate what elsewhere has been called production rate or rate of irreversible loss (Shipley and Clark, 1972).

Table 4.1 Flux, intake and excretion

	Intake (I) gN/kg/day	Endogenous excretion (E) gN/kg/day	Flux (F) gN/kg/day	I/F × 100	E/F × 100
MAN	0.155[1]	0.028[2]	0.863[3]	18	3.2
RAT	1.72[4]	0.238[5]	7.83[6]	22	3.0

[1] James et al. (1974)
[2] Scrimshaw et al. (1972) (minus creatinine)
[3] As 1. Includes oxidation
[4] Das and Waterlow (1974). Rats on normal diet
[5] Das and Waterlow (1974). Rats on zero protein diet for two weeks (minus creatinine)
[6] Millward et al. (1974d)

flux. This means that the rest of the flux is synthesised into protein. Since synthesis and breakdown are the same in the steady state, it follows that 80 per cent of the flux is derived from breakdown and re-utilised for protein synthesis. Of course, the 'normal' protein intake cannot be an exact figure since it can and does vary considerably above minimum requirements according to the composition of the diet.

What is perhaps more significant is the relationship between the flux (F) and the endogenous nitrogen excretion (E), a more fundamental physiological parameter representing the basal rate of amino acid oxidation. The ratio (E/F) is remarkably similar in both man and the rat. Does this mean that amino acid requirements are related to whole body protein turnover? Since amino acid requirements represent (at least in part) the rate at which amino acids are channelled into other biosynthetic pathways, and therefore will be related to the turnover of these tissue constituents (such as nucleic acids, porphyrins, creatine, etc.), we can only assume that the relative turnover rates of proteins and other tissue constituents are similar in the rat and man.

Table 4.2 Developmental changes in flux[1]

Age (Days)	Weight (g)	Flux (gN/kg/day)	kg days^{-1}
23	37	14.42	0.517
65	115	8.04	0.288
330	511	4.43	0.144

[1] Millward et al., unpublished observations

One of the difficulties in making these comparisons between species is in the matching of comparable developmental stages. This is especially true since there are profound developmental changes in the flux rate. Table 4.2 shows the flux rate in weanling, adolescent and full-grown rats. The flux rate falls by almost half during the 40 days after weaning, and by a further half in the subsequent nine months. It may be that there are similar changes in the endogenous nitrogen excretion rate, but this is not known. The developmental changes in flux result from changes in body composition and hence in the pattern of protein synthesis between different organs, as well as from changes in the rates of protein synthesis in individual tissues. This will be discussed later.

Changes in the Flux

When a normal dietary intake is replaced by a low protein diet the nitrogen excretion falls; there are two alternative explanations for this. There can either be a fall in the overall flux, or the proportion of the flux which is excreted can fall. The early measurements with ^{14}C lysine showed that the flux did not change appreciably in rats on a low protein diet during the first ten days (Waterlow and Stephen, 1968). Only after five weeks did the flux fall to 50 per cent of its original value (Waterlow and Stephen, 1967).

We know from other experiments that when the protein intake is reduced or eliminated, the fall in nitrogen excretion is more extensive and occurs more rapidly. Indeed, when rats are switched from a 14 per cent to a 5 per cent protein diet the nitrogen excretion is reduced by 75 per cent in only 30 hours (Das and Waterlow, 1974). We believe therefore, that this reduction in nitrogen output which is the primary adjustment to a decreased protein intake cannot be primarily a consequence of a reduction in flux. It results from a shutting down in the urea cycle and hence represents a reduction in the proportion of the flux which is excreted (Das and Waterlow, 1974). *Table 4.3* shows the effect of more severe dietary restriction

Table 4.3 Diet and flux in the rat[1]

Initial flux (100 g rat)		7.83 gN/kg/day
Per cent of initial flux		
Protein-free diet:	Day 1	102
	Day 2	85
	Day 3	81
	Day 21	39
	Day 30	23
Re-feeding high protein	Day 1	38
	Day 3	68
	Day 8	82
	Day 14	90
Starved	Day 1	81
	Day 2	74
	Day 3	56
	Day 4	59

[1]Millward *et al.* (1974d)

on the flux (Millward, Garlick and Nnanyelugo, 1974d). After three weeks of a protein-free diet or three days of total starvation, there was a considerable fall in the flux rate. We do not know at this stage with any certainty whether such a large fall is an adaptative response or a breakdown. We do know that in the rat it is reversible, since on rehabilitation onto a good diet the flux immediately starts to increase, more than doubling in the first three days and reaching near normal levels after eight days *(Table 4.3)*. Also we have shown that when man is given a protein-free low calorie intake, the flux rate is halved *(Table 4.4)*. These results were obtained by constant infusion of ^{14}C tyrosine in obese adults (Sender, James and Garlick, 1974).

Table 4.4 Diet and flux in man[1]

Intake	Flux[2] g protein/day	Oxidation[3] g protein/day	BMR	
			ml O_2/min	kcal/day
2000 kcal 70 g protein	488	88	294	2074
300 kcal 0 g protein	228	34	254	1792

[1] Obese adults (Sender *et al.*, 1974)
[2] Total flux (synthesis + oxidation)
[3] Calculated from the tyrosine oxidation rate and assuming rate of tyrosine oxidation is the same as that of the total amino acid pool

Protein Turnover and Energy Expenditure

The magnitude of whole body protein turnover is such that we expend a considerable fraction of our energy consumption on it, and therefore any fall in whole body protein turnover may be an important mechanism for conserving energy. The exact energy cost of protein synthesis is not known. Theoretically, with four moles of high energy phosphate per mole of peptide bond, the energy cost is about 0.8 kcal/g of protein, if it is assumed that ATP is formed from glucose at a cost of 18 kcal/mole. However, when the energy cost of deposition of protein during growth has been measured in animals (Blaxter, 1967), it appears that about 5 kcal per g are needed, after allowing for the energy content of protein stored. In rapidly growing children recovering from malnutrition it was estimated that the energy cost of protein synthesis was 3 to 4 kcal/g protein (derived from data of Kerr *et al.*, 1973). If total protein turnover in an adult man is about 400 g per day *(Table 4.3)*, these estimates of the energy cost of protein synthesis mean that virtually all the basal energy expenditure would be used up in maintaining protein turnover.

Estimates based on growth measurements must represent a crude figure, which includes the cost of synthesising other cell components associated with protein, such as phospholipid and RNA. The results in *Table 4.4* allow a different approach to the problem. The fall in protein turnover in the obese subjects on a restricted intake was accompanied by a fall in energy expenditure (Sender, James and Garlick, 1974). If we assume that this saving of energy is directly related to the fall in protein turnover, we can calculate the energy cost of protein synthesis as:

$$\frac{\Delta \text{kcal}}{\Delta \text{g protein turned over}} = \frac{282}{206} = 1.41 \text{ kcal/g protein}$$

This is much closer to the theoretical value.

The Synthesis Rate of Tissue Proteins

The synthesis rate of tissue proteins can be calculated from the specific activity of the protein-bound (S_B) and intracellular free (S_i) amino acid specific activity at the end of the infusion (Waterlow and Stephen, 1968; Garlick, Millward and James, 1973). These values are related to the FSR (fractional synthesis rate, k_S) by the expression:

$$\frac{S_B}{S_i} = \frac{\lambda_i}{(\lambda_i - k_S)} \cdot \frac{(1 - e^{-k_S t})}{(1 - e^{-\lambda_i t})} - \frac{k_S}{(\lambda_i - k_S)}$$

The formula contains λ_i which is the rate constant describing the rise to plateau of the free amino acid specific activity. The value of λ_i is difficult to determine directly but we have shown that it can be estimated indirectly with sufficient accuracy.

Table 4.5 *Fractional rate of protein synthesis (FSR), RNA content and 'efficiency' of synthesis in rat tissues*

	FSR (d^1)	RNA/protein ($\times 10^3$)	Efficiency g protein/[g RNA/day]
Liver	0.61	46[1]	13.3[2]
Kidney	0.52	33	15.7
Heart	0.17	16.4	10.4
Muscle	0.15	11.5	13.0
Brain	0.12	22.5	5.3
Whole body	0.26[3]		

[1] Calculated as liver RNA/liver protein + albumin
[2] Assumes albumin synthesis rate is same as endogenous liver protein
[3] Calculated from flux in Table 4.1, assuming rat = 3.03 per cent nitrogen

The synthesis rates of mixed tissue proteins are shown in *Table 4.5*, together with the RNA/protein ratios. If we regard the RNA/protein ratio as an index of capacity for protein synthesis, since the greater part of RNA is ribosomal, then the FSR divided by the capacity is equal to the rate of synthesis per unit RNA, which we can call the efficiency of synthesis (Millward *et al.*, 1973). *Table 4.5* shows that with the exception of brain the efficiency of protein synthesis is very similar in liver, kidney, heart and muscle at 10-16 g protein synthesised/g RNA/day. The low figure for brain may reflect the presence of a larger than average fraction of non-ribosomal RNA, so that the synthesis rate per ribosome may be higher than 5.3 g protein/g RNA/day.

The Sensitivity of Tissue Protein Synthesis to Food Supply

There is much variation amongst the tissues in their sensitivity to food supply (Millward and Garlick, 1972). It had been observed that in malnourished rats the incorporation of labelled amino acids into muscle protein was reduced, whilst in the liver it was maintained or elevated (Waterlow and Stephen, 1966; Munro, 1969). Waterlow and Stephen (1968) confirmed these findings with ^{14}C lysine constant infusion, and showed considerable falls in muscle protein synthesis rates accompanied by elevated fractional rates of liver protein synthesis in malnourished rats. More recently we have extended these studies with measurements by constant infusion of ^{14}C tyrosine.

In order to demonstrate the immediate effects of food intake and subsequent fasting more precisely, we trained rats to consume their daily ration in a four-hour period. Rats were fed for four hours and then fasted. The FSR of muscle and liver protein was measured over consecutive six-hour periods for 54 hours after the meal (Garlick, Millward and James, 1973). In a parallel series of experiments we measured changes in tissue composition (Millward et al., 1974). These experiments showed that in muscle the FSR increased after the meal, reaching a peak at 12-18 hours and fell after this time to a rate which was half the initial rate, whilst in the liver the FSR in general stayed constant. In each tissue the RNA concentrations changed in a parallel way to the FSR.

In the liver the FSR stayed constant while the liver protein mass increased by almost 20 per cent after the meal and fell by 30 per cent during fasting. This must mean that the changes in liver protein mass were initiated by alteration in the breakdown rate, i.e. it fell after the meal and increased during subsequent fasting.

Figure 4.1 shows the results of an experiment which examined the sensitivity of protein synthesis in other tissues to starvation. The brain appears to be insensitive to starvation with no change in RNA content or efficiency of synthesis. In kidney the efficiency of synthesis was maintained but a small loss of RNA induced an 11 per cent fall in the synthesis rate. In both heart and skeletal muscle, however, there were profound falls in the synthesis rate induced by losses of RNA and substantial falls in the efficiency of synthesis.

Figure 4.2 shows the effect of protein deficiency (a protein-free diet) on the synthesis rate in different tissues (Waterlow and Garlick, 1974). The pattern is similar to that of starvation with small changes in brain and kidney, a moderate fall in FSR in the heart (though not so pronounced as in starvation), a very severe reduction in FSR in muscle, whilst in the liver the FSR was increased.

The overall pattern, then, is that in those tissues with essential functions, such as brain and kidney, protein synthesis is reasonably well-protected against dietary insufficiency. With a sufficient energy supply, from ketone bodies for example (Cahill, 1971), protein synthesis can be maintained in these tissues by recycling of amino acids. The protein and RNA content of the brain is maintained, whereas there are only small losses of kidney protein and RNA.

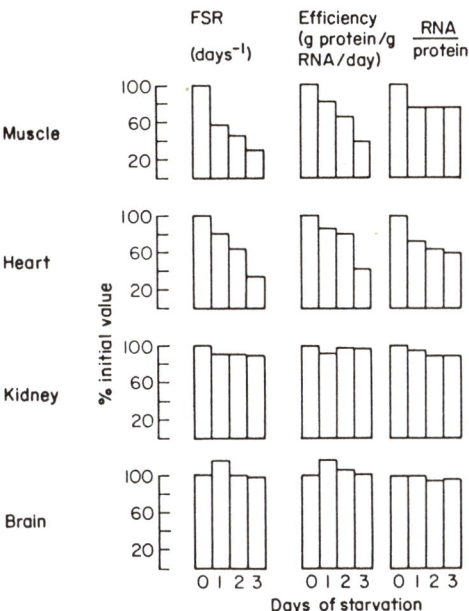

Figure 4.1 Sensitivity of tissue protein synthesis in rats to starvation. The fractional rate of protein synthesis (FSR) was measured by continuous infusion of ^{14}C-tyrosine. The efficiency of synthesis is the FSR divided by the RNA to protein ratio. Results are expressed as per cent of the value obtained in well fed animals (zero days)

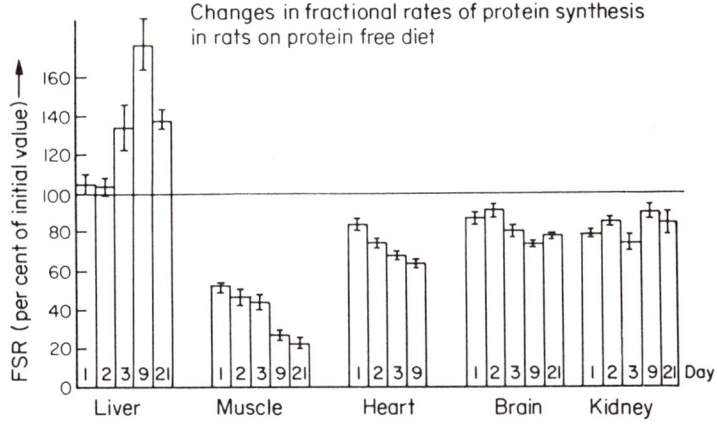

Figure 4.2 Changes in fractional rates of protein synthesis (FSR) in rats on a protein free diet. For details see Figure 4.1

In liver and muscle, however, there are specific responses; in liver the FSR is maintained or elevated, at a time when considerable losses of protein occur, whilst in muscle losses of RNA and reduced efficiency of synthesis result in the FSR falling to one-fifth of normal values. To what extent can we explain these responses.

Adaptation in the Liver

Consider the liver of a 100 g rat on a normal diet. We can calculate (Jeejeebhoy *et al.*, 1972) that 88 mg of albumin are exported each day. This, together with the other exported proteins, may constitute over one-fifth of all liver protein synthesis (*Table 4.6*).

Table 4.6 Endogenous and export protein synthesis in liver[1]

Endogenous[2]

Liver protein = 750 mg
FSR = 0.61 day^{-1} Total synthesis = 458 mg/day

Export[3]

Albumin synthesis 11.2 mg/hr/300 g rat = 88 mg/day
Fibrinogen + transferrin etc. = 10 mg/day (estimate)

Total synthesis (approx) = 98 mg/day

[1] 100 g rat on normal diet
[2] Millward *et al.*, unpublished observations
[3] Data from Jeejeebhoy *et al.* (1972)

Most of this is degraded in extrahepatic tissues such as gastrointestinal tract, kidney and the reticuloendothelial system. These amino acids mix with the various tissue pools, and some may eventually return to the liver, particularly those from the gastrointestinal tract. It is possible, however, that some, such as the branched chain amino acids, do not return to the liver, since they are oxidised primarily in extrahepatic tissues. The liver is therefore dependent to some extent at least on a continuous dietary supply of amino acids, particularly the branched chain amino acids, to balance its accounts.

When the dietary protein supply ceases, the liver has breakdown of its own endogenous protein and proteins of other tissues as a source of amino acids for albumin synthesis. As mentioned above, however, amino acids arriving at the liver from other tissues (such as muscle) are liable to be deficient in the branched chain amino acids and so endogenous liver protein and amino acids returning from albumin degraded in the gastrointestinal tract may be its only source. Now net catabolism of endogenous liver proteins can be achieved by increasing protein breakdown or decreasing synthesis, but we know from our measurements that the FSR in liver does not fall. This being the case it follows that an increase in protein breakdown must occur. This is shown schematically in *Figure 4.3*.

In this figure we have used our own data for the change in liver protein mass and endogenous synthesis rates, and that of Jeejeebhoy

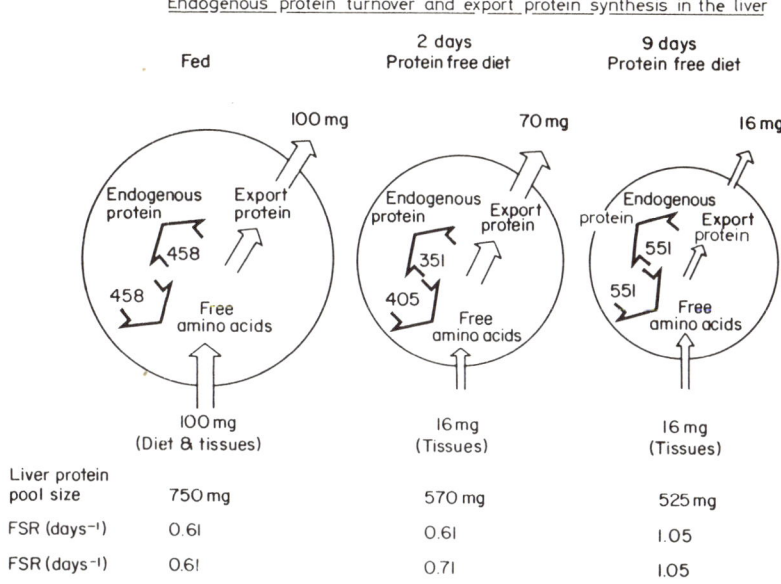

Figure 4.3 Synthesis of endogenous and export protein in rat liver. The data on fractional synthesis rate and pool size of endogenous protein are taken from the study illustrated in Figure 4.2 *and the fractional breakdown rate (FBR) is calculated from them. The number of milligrams of protein synthesised is equal to the FSR multiplied by the liver protein content. Data for export protein synthesis are taken from Jeejeebhoy* et al. *(1973)*

et al. (1973) for the export protein synthesis rate. This latter figure must be approximate since we do not know the rate of synthesis of other export proteins. Furthermore, we have assumed, in line with other arguments presented above, that in the absence of a dietary supply, the peripheral tissues might only be able to supply sufficient amino acids for a limited amount of export protein synthesis in the liver.

The initial fed condition is a more or less steady state condition, with dietary amino acids replacing those exported as proteins. When dietary supply stops we observe a loss of liver protein at an initial rate which is very similar to the rate of export protein synthesis, and indeed part of this loss must result from the export of proteins. After two days even though export protein synthesis has fallen by 30 per cent, it still results in a considerable imbalance so that the liver protein will continue to fall as endogenous protein breakdown is in excess of synthesis. However, after this time the albumin synthesis rate falls off dramatically and so does the observed rate of loss of liver protein. (The catabolic rate of albumin synthesis also falls at this time preventing a major fall in albumin concentration). At nine days export protein synthesis is less than a fifth of the original level, and the liver can presumably extract sufficient amino acids from the plasma to cope with this export rate. As shown in *Figure 4.2* the endogenous FSR is elevated at this time. We do not know at present the mechanism for this change, and whether it results from the suppression of albumin synthesis.

Furthermore, little is known about the initiation of the increase in the rate of protein catabolism. The fall in hepatic amino acid concentrations in response to the poor diet (Wannemacher and Allison, 1968; Adibi, *et al.*, 1973) may be involved, since Mortimore (Mortimore and Mondon, 1970; Woodside and Mortimore, 1972) has shown that in the perfused liver, high amino acid concentrations or the addition of insulin decrease protein breakdown. Thus a fall in amino acid and insulin concentrations may be an important part of the signal, which increases protein breakdown in the liver without prejudicing the FSR of liver protein. However, the literature is somewhat contradictory on the subject of amino acid levels and liver protein synthesis (Jefferson and Korner, 1969; Tavill *et al.*, 1973). In the latter paper it was reported that in a perfused liver preparation from protein-deficient rats, supplementation with amino acids failed to stimulate albumin synthesis, and polysome aggregation was maintained demonstrating that there is no significant translational dysfunction. It is interesting to note that in some studies in which loss of aggregation of liver ribosomes is reported, *in vitro* incorporation of labelled amino acids is maintained or elevated (Enwonwu and Jacobson, 1973). Our studies show that the rate of protein synthesis per unit RNA is either maintained or elevated in the liver of malnourished rats.

The net catabolism of liver protein leads to a fall in liver weight and this is accompanied by a rapid loss of RNA (Millward *et al.*, 1974a). The loss of RNA may result from a failure of synthesis (Wannemacher, Wannemacher and Yatvin, 1971) as well as an increase in breakdown (Enwonwu and Munro, 1970) which may be associated with the fall in ribonuclease inhibitor activity (Enwonwu and Sreebny, 1971).

In summary, the initial response of the liver to a lack of dietary supply is an increase in endogenous protein breakdown to supply substrate for plasma protein synthesis. However, there are selective controls which subsequently reduce the rate of plasma protein synthesis thus obviating the continued loss of liver protein. The rate of endogenous protein synthesis is however well maintained throughout the period of the deficiency.

Adaptation in Muscle

In skeletal muscle the machinery cannot function properly under conditions of starvation or protein deficiency. The rate of protein synthesis falls in response to a variety of insults and this is usually associated with a loss of RNA and a fall in efficiency. The extent of the fall in efficiency of protein synthesis is shown in *Table 4.6*. In protein-depleted rats the efficiency fell to 57 per cent, whilst in starvation it fell to 41 per cent of the control value. As a result of the fall in synthesis rate, net catabolism occurs and muscle protein is mobilised. In the young rat the rate of loss of muscle protein is slow, since visceral tissues seem to bear the brunt of acute dietary insufficiency, but in chronic exposure to a poor diet a great deal of muscle protein is lost (Mendes and Waterlow, 1958). Just as important, perhaps, is the shutting down of muscle protein synthesis so that what little dietary protein is available can be selectively channelled to those tissues with greater survival value.

If we consider the mechanism of these changes in muscle, it is again unlikely that free amino acid concentrations play a direct role in initiating the change. When net catabolism of muscle protein occurs, as in starvation, the concentration of the rate-limiting amino acids rises considerably (Millward et al., 1974).

Insulin is undoubtedly of paramount importance in mediating the response. Muscle ribosomes are dependent on insulin for normal function (Wool, 1972; Jefferson, Koehler and Morgan, 1972), and our measurements show that there are exact parallels between changes in plasma insulin levels and muscle protein synthesis rates in fed and fasted rats (Garlick, Millward and James, 1973; Millward et al., 1974).

Table 4.7 Factors which affect muscle protein synthesis

	FSR (d^{-1})	Efficiency g protein/g RNA/day
High protein[1]	0.134	13.8
Four weeks' protein-free diet[1]	0.0269	7.97
Three days' starvation[2]	0.046	7.48
Streptozotocin[2]	0.048	6.04

[1] Millward et al. (1974c)
[2] Millward et al., unpublished observations

Table 4.7 shows that the effect of streptozotocin-induced diabetes is to induce a more pronounced fall in efficiency than three days' starvation or four weeks' protein deficiency.

Protein synthesis in heart muscle also shows this sensitivity to insulin (Morgan et al., 1971) and there are considerable falls in the FSR in starvation. There is some evidence, however, to suggest that in heart protein synthesis is to some extent protected against insulin deficiency by free fatty acids (Rannels et al., 1970). In other tissues, however, there is little to suggest that insulin is so important for ribosomal function. No doubt endocrine factors such as growth hormone, glucocortoids and glucagon play an important part in mediating these changes in tissue protein synthesis (Manchester, this symposium), but it seems to us that the relative dependence on insulin of the initiation of translation is of over-riding importance in determining the pattern of response of tissue protein synthesis to failure of the food supply.

Protein Breakdown

The direct measurement of protein breakdown is very difficult. The traditional method has been to measure the rate of loss of isotope from proteins pulse labelled with a suitable amino acid. In the absence of significant re-utilisation of label, the rate of loss of isotope is considered to be equal to the rate of protein breakdown. (The rate of fall of specific activity is equal to the rate of synthesis; Koch, 1962.) In practice it is very difficult to avoid the problem of re-utilisation. We know from the magnitude of the flux rate (*Table 4.1*) that most of the amino acids liberated by protein breakdown are re-utilised rather than oxidised and excreted. To minimise this difficulty, labelled amino acids have been used which have a high flux rate through some other pathway, such as the guanidine carbon of arginine (excreted through the urea cycle) or the carboxyl groups of aspartate and glutamate (excreted through the Krebs cycle) (Millward, 1970a). In each case re-utilisation is probably negligible but a price has to be paid in that large (and expensive) amounts of the isotopes must be used to achieve a reasonable level of labelling.

A second problem with this type of approach arises when decay rates of mixtures of protein with different turnover rates are measured. Heterogeneity of turnover results in a decay curve which is difficult to interpret since the early part is disproportionately influenced by the rapidly turning-over protein and will, therefore, over-estimate the average turnover rate. Whilst we have calculated (Garlick and Millward, 1972) that for most mixtures the decay curve measured over a time period equal to three times the average half life gives a reasonably accurate value for weighted mean turnover rate, it is probably best to limit this type of experiment to the measurement of the breakdown rate of specific purified proteins.

Kinetics of Loss of Isotope

When care is taken to avoid re-utilisation of isotopes and the decay curve of a purified protein is measured, an exponential loss of radioactivity is usually observed indicating random, first order breakdown of proteins (Schimke, 1964) and it is generally accepted that most tissue proteins follow these kinetics (Schimke, 1970). There is some (but by no means conclusive) evidence to suggest that this is not the case for the myofibrillar proteins of muscle. Since these proteins constitute up to two-thirds of all muscle proteins, non-random kinetics of breakdown would seriously influence experiments in which the loss of radioactivity from muscle was measured. Dreyfus, Kruh and Schapira (1960) measured myosin turnover with ^{14}C glycine and reported lifetime kinetics (i.e. non-random breakdown) analogous to haemoglobin in red cells. These results have not been widely accepted because of the extensive re-utilisation which is presumed to occur with ^{14}C glycine. Certainly when the decay rate of myofibrillar proteins was measured with carboxyl-labelled aspartate and glutamate an exponential decay was observed (Millward, 1970a). Nevertheless, labelled amino acids have been reported to become initially incorporated into the periphery of the myofibrils (Morkin, 1970) and subsequently become distributed throughout the matrix (Venable, 1969). This in itself implies non-random turnover. Now when protein is lost from muscle as in starvation, it has been reported that the diameter of individual myofibrils is reduced and it is possible that the peripheral myofilaments are selectively degraded. If these proteins had been labelled immediately prior to the starvation then a very rapid loss of label would be observed. We have in fact reported such results but did not interpret them in this way at the time (Millward, 1970b). *Figure 4.4* shows the loss of radioactivity from aspartate and glutamate isolated from the myofibrils of control and

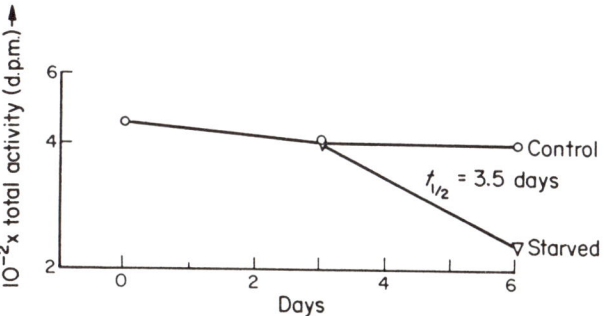

Figure 4.4 Loss of radioactivity from aspartate and glutamate isolated from myofibrillar proteins labelled by ^{14}C-Na_2CO_3 injection

starving rats. After three days protein began to be lost from the muscle and this coincided with a very rapid fall in radioactivity, indicating a breakdown rate of 20 per cent per day or a half life of 3.5 days. Since recent measurements with the constant infusion method indicate that the breakdown rate is not increased in starvation we feel that the results in *Figure 4.4* indicate non-random loss of labelled proteins.

The opposite experimental situation to this is that of catch-up growth (i.e. after protein depletion), in which the myofibrillar diameter rapidly increases. Young, Stothers and Vilaire (1972) have reported that there is no loss of ^{14}C-aspartate from muscle at this time, indicating a cessation of protein breakdown. However, our measurements (*Table 4.7*) suggest that muscle protein breakdown is maintained or even elevated during rapid catch-up growth. It is possible, therefore, that the kinetics of loss of radioactivity from myofibrillar proteins is variable depending on the particular nutritional state of the animal.

Measurements of decay rates can yield, therefore, very valuable information about the mechanism of protein breakdown, particularly in complex protein structures such as the myofibrils. The actual rate is, however, in most cases more easily determined indirectly by way of measuring the synthesis and growth rates.

The Effect of Diet on Protein Breakdown in Muscle

In the liver it seems that the overall control of protein content is exerted through changes in the breakdown rate as discussed previously. Some of our recent results of the effect of diet on muscle protein breakdown are shown in *Table 4.8*. In starvation and protein deficiency the rate of muscle protein breakdown is reduced. The net loss of muscle protein induced by these dietary states results

Table 4.8 Diet and Muscle Protein Breakdown

Diet	Breakdown rate ($days^{-1}$)	$t_{1/2}$ days
Control 100 g[1]	0.103	6.7
Protein free 30 days[1]	0.057	12.1
Starvation four days[2]	0.060	11.5
Re-feeding[1] high protein eight days	0.130	5.3
Weanling control[3]	0.225	3.1
Weanling malnourished[3]	0.082	8.5

[1] Millward *et al.* (1974c).
[2] Millward *et al.*, unpublished observations
[3] Millward *et al.* (1974b).

from a more pronounced fall in synthesis rates. There is indirect evidence that in man, in both starvation (Young, Haverberg and Munro, 1973) and protein-calorie malnutrition (Rao and Nagabhushan, 1973), muscle myofibrillar protein breakdown is reduced, since the excretion of 3-methyl histidine is reduced in each case. Also, 3-Me histidine is found in myofibrillar proteins and since methylation occurs after translation, re-utilisation is not possible and the excretion rate should reflect muscle protein breakdown (Young *et al.*, 1970).

In contrast to these situations, when rats are recovering from protein deficiency and growing rapidly, rates of muscle protein breakdown are elevated (*Table 4.8*; Millward, Garlick and Nnanyelugo, 1974c). Once again there is tentative evidence that similar changes occur in man, since in the infant recovering from malnutrition the excretion of the 3-Me histidine rapidly returns to normal levels during the rapid growth of recovery (Rao and Nagabhushan, 1973).

We have also found rapid breakdown rates in the rapidly growing weanling rat (*Table 4.8*) and Turner and Garlick (1974) have reported increased breakdown rates in the compensatory hypertrophy of the diaphragm following resection of the phrenic nerve.

While there is no conclusive explanation for this phenomenon, the high rates of breakdown may reflect increased turnover associated with the myofibrillar proliferation which occurs during growth, and may be related to the splitting of myofibrils which has been reported by Goldspink (1970).

These results mean that rapid muscle growth which occurs on a good diet is not energetically economical, since total protein synthesis greatly exceeds net synthesis. There is one situation in which the breakdown rate is low during growth of muscle; this occurs in weanling rats which have been marginally congenitally malnourished (*Table 4.8*). The rate of protein synthesis is only half that of well-fed rats, yet growth occurs, since the rate of breakdown is reduced by two-thirds (Millward, Garlick and Nnanyelugo, 1974b). This, therefore, may represent an adaptive fall in protein breakdown.

Protein Turnover and Cellular Environment

One of the remarkable features of protein turnover is its heterogeneity. In the liver, half lives of protein vary from less than one hour to four days (Schimke and Doyle, 1970). Whilst we do not know the extent of heterogeneity of turnover in other tissues, we certainly know that overall turnover rates vary considerably between tissues. One possible reason for this is the presence in some tissues of inherently stable proteins which result in low average turnover rates for the tissues. Since the myofibrils are obvious candidates for such a role we have compared the turnover rate of only those tissue proteins soluble in low ionic strength buffers, thus excluding myofibrillar proteins. *Figure 4.5* shows that the heterogeneity of turnover is still apparent with the average half life of soluble muscle proteins over ten times as long as those of liver.

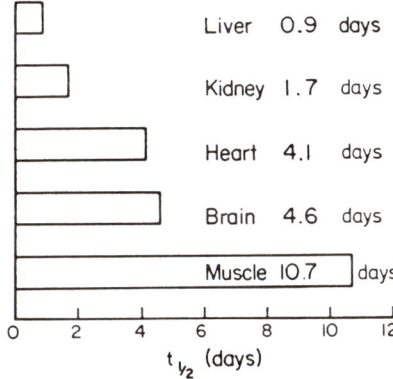

Figure 4.5 Turnover of soluble proteins from tissues. The fractional synthesis rate (FSR) was measured by continuous infusion of ^{14}C-tyrosine into growing rats. The fractional breakdown rate (FBR) was derived by subtracting the growth rate from the FSR. Half-life ($t_{1/2}$) is then given by $t_{1/2} = 0.693/FBR$

When we compare the turnover of organelles in different tissues, the heterogeneity is still apparent. Thus, *Figure 4.6* shows that mitochondria in muscle turn over more slowly than in liver, and myofibrils in muscle turn over more slowly than in heart. These data suggest, therefore, that the variations in turnover rate between tissues

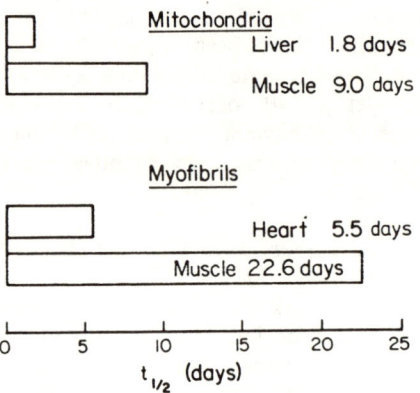

Figure 4.6 Protein turnover in tissue organelles. Half-lives ($t_{1/2}$) were obtained as in Figure 4.5

do not necessarily reflect populations of proteins in the tissues with inherently different turnover rates. Rather, the differences seem to reside in what could be called the 'relative proteolytic environment' of the tissue which may be optimal for the particular role of the tissue in the organism. Thus, in the liver and perhaps kidney, the fluctuating metabolic role requires high turnover rates in order that rapid adjustments of the enzyme content can be made. To achieve this, high concentrations of proteolytic enzymes are required which result in rapid turnover of all the proteins of the tissue. Thus, organelles such as mitochondria may well be forced to turn over quickly, even though their stability and function would allow a longer half life as suggested by their slower turnover in muscle.

As to the nature of the 'proteolytic environment', the proteolytic enzymes of the lysosome are the only clearly characterised group of enzymes found in most tissues (De Duve and Wattiaux, 1966; De Duve, 1969). How these enzymes achieve the specificity which is needed to produce the wide range of turnover rates found in the tissues is a question that remains to be answered.

References

ADIBI, S.A., MODESTO, T.A., MORSE, E.L. and AMIN, P.F. (1973). *Am. J. Physiol.*, **225,** 408
AUB, M. and WATERLOW, J.C. (1970). *J. theor. Biol.*, **26,** 243
BLAXTER, K.L. (1967). in *The Energy Metabolism of Ruminants*, p.263. Hutchinson, London
CAHILL, G.F. Jr. (1971). *Diabetes*, **20,** 785
DAS, T.K. and WATERLOW, J.C. (1974). *Br. J. Nutr.*, (in the press)
DE DUVE, C. and WATTIAUX, R. (1966). *A. Rev. Physiol.*, **28,** 435
DE DUVE, C. (1969). in *Lysosomes in Biology and Pathology*. Eds. J.T. Dingle and H. Fell. **Vol.l,** p.3. North Holland Publishing Co., Amsterdam
DREYFUS, J.C., KRUH, J. and SCHAPIRA, G. (1960). *Biochem. J.*, **75,** 574
ENWONWU, C.O. and JACOBSON, K. (1973). *J. Nutr.*, **103,** 290
ENWONWU, C.O. and MUNRO, H.N. (1970). *Archs. Biochem. Biophys.*, **138,** 532
ENWONWU, C.O. and SREEBNY, L.M. (1971). *J. Nutr.*, **101,** 501
FERN, E.B. and GARLICK, P.J. (1973). *Biochem. J.*, **134,** 1127
GAN, J.C. and JEFFAY, H. (1967). *Biochim. biophys. Acta.*, **148,** 448
GARLICK, P.J. and MILLWARD, D.J. (1972). *Proc. Nutr. Soc.*, **31,** 249
GARLICK, P.J., MILLWARD, D.J. and JAMES, W.P.T. (1973). *Biochem. J.*, **136,** 935
GOLDSPINK, G. (1970a). *Am. J. Physiol.*, **209,** 100
GOLDSPINK, G. (1970b). *J. Cell. Sci.*, **6,** 593
HAIDER, M. and TARVER, H. (1969). *J. Nutr.*, **99,** 433
HENRIQUES, O.B., HENRIQUES, S.B. and NEUBERGER, A. (1955). *Biochem. J.*, **60,** 409
HENSHAW, E.C., HIRSCH, C.A., MORTON, B.E. and HIATT, H.H. (1971). *J. biol. Chem.*, **246,** 436
HIDER, R.C., FERN, E.B. and LONDON, D.R. (1971). *Biochem. J.*, **121,** 817
JAMES, W.P.T., GARLICK, P.J. and SENDER, P.M. (1974). *Clin. Sci. and Mol. Med.*, **46,** 8
JEEJEEBHOY, K.N., BRUCE-ROBERTSON, A., HO, J. and SODTKE, U. (1972). in *Protein Turnover*, CIBA Foundation Symposium No.9, Associated Scientific Publishers, Amsterdam
JEFFERSON, L.S., KOEHLER, J.O. and MORGAN, H.E. (1972). *Proc. natn. Acad. Sci., USA*, **68,** 816
JEFFERSON, L.S. and KORNER, A. (1969). *Biochem. J.*, **111,** 703
KERR, D., ASHWORTH, A., PICOU, D., POULTER, N., SEAKINS, A., SPADY, D. and WHEELER, E. (1973). in *Endocrine Aspects of Malnutrition*, Kroc Foundation Symposia No. 1, Published by K.F. Kroc, Santa Ynez, California
KOCH, A.L. (1962). *J. theor. Biol.*, **3,** 283
MENDES, C.B. and WATERLOW, J.C. (1958). *Br. J. Nutr.*, **12,** 74
MILLWARD, D.J. (1970a). *Clin. Sci.*, **397,** 577
MILLWARD, D.J. (1970b). *Clin. Sci.*, **39,** 591

MILLWARD, D.J. and GARLICK, P.J. (1972). *Proc. Nutr. Soc.*, **31**, 157
MILLWARD, D.J., GARLICK, P.J., JAMES, W.P.T., NNANYELUGO, D.O. and RYATT, J.S. (1973). *Nature*, **241**, 204
MILLWARD, D.J., GARLICK, P.J., JAMES, W.P.T. and NNANYELUGO, D.O. (1974a). *Br. J. Nutr.*, **32**, 127
MILLWARD, D.J., GARLICK, P.J., and NNANYELUGO, D.O. (1974b). *Proc. Nutr. Soc.*, **33**, 55A
MILLWARD, D.J., GARLICK, P.J., and NNANYELUGO, D.O. (1974c). *Proc. Nutr. Soc.*, **33**, 115A
MILLWARD, D.J., GARLICK, P.J. and NNANYELUGO, D.O. (1974d). *Proc. Nutr. Soc.*, **34**, 33A
MORGAN, H.E., JEFFERSON, L.S., WOLPERT, E.B. and RANNELS, D.E. (1971). *J. biol. Chem.*, **246**, 2163
MORKIN, E. (1970). *Science*, **167**, 1499
MORTIMORE, G.E. and MONDON, C.E. (1970). *J. biol. Chem.*, **245**, 2375
MUNRO, H.N. (1969). in *Mammalian Protein Metabolism*, **Vol.III**, p.237. Ed. H.N. Munro. Academic Press, New York and London
PAIN, V.M. and GARLICK, P.J. (1974). *J. biol. Chem.*, **249**, 4510
RANNELS, D.E., JEFFERSON, L.S., HJALMARSON, A.C., WOLPERT, E.B. and MORGAN, H.E. (1970). *BBRC*, **40**, 1110
RAO, B.S. and NAGABHUSHAN, V.S. (1973). *Life Science*, **12**, 205
SCHIMKE, R.T. (1964). *J. biol. Chem.*, **239**, 3808
SCHIMKE, R.T. (1970). in *Mammalian Protein Metabolism*. **Vol.IV**, Ed. H.N. Munro. Academic Press, New York and London
SCHIMKE, R.T. and DOYLE, D. (1970). *A. Rev. Biochem.* p.929
SCRIMSHAW, N.S., HUSSEIN, M.A., MURRAY, E., RAND, W.M. and YOUNG, V.R. (1972). *J. Nutr.*, **102**, 1595
SENDER, P.M., JAMES, W.P.T. and GARLICK, P.J. (1974). *Proceedings of the Second Congress on Energy Regulation in Man.* in *Regulation of Energy Balance in Man*, p.224. Ed. E. Jequier. Editions Médecine et Hygiène, Geneva. 1975
SENDER, P.M., JAMES, W.P.T. and GARLICK, P.J. (1974). *Proceedings of the Second Congress on Energy Regulation in Man* (in the press)
SHIPLEY, R.A. and CLARK, R.E. (1972). in *Tracer Methods for in vivo Kinetics.* Academic Press, New York and London
TAVILL, A.S., EAST, A.G., BLACK, E.G., NADKARNI, D. and HOFFENBERG, R. (1973). in *Protein Turnover*, p.155. CIBA Foundation Symposium No.9. Associated Scientific Publishers, Amsterdam
TURNER, L.V. and GARLICK, P.J. (1974). *Biochim. biophys. Acta.*, **349**, 109
VENABLE, A.T. (1969). *Anat. Rec.*, **163**, 179
WANNEMACHER, R.W. and ALLISON, J.B. (1968). in *Protein Nutrition and Amino Acid Pools.* Ed. J.H. Leatham, p.206. Rutgers University Press, New Brunswick
WANNEMACHER, R.W., WANNEMACHER, C.F. and YATVIN, M.B. (1971). *Biochem. J.*, **124**, 385
WATERLOW, J.C. and GARLICK, P.J. (1974). in *Alcohol, Nutrition and Protein Synthesis.* Eds. Rothschild, Oratz and Schreiber, Pergamon Press
WATERLOW, J.C. and STEPHEN, J.M.L. (1966). *Br. J. Nutr.*, **20**, 461
WATERLOW, J.C. and STEPHEN, J.M.L. (1967). *Clin. Sci.*, **33**, 489
WATERLOW, J.C. and STEPHEN, J.M.L. (1968). *Clin. Sci.*, **35**, 287
WOODSIDE, K.H. and MORTIMORE, G.E. (1972). *J. biol. Chem.*, **247**, 6474

WOOL, I.G. (1972). *Proc. Nutr. Soc.*, **31**, 185
YOUNG, V.R., BALIGA, B.S., ALEXIS, S.D. and MUNRO, H.N. (1970). *Biochim. biophys. Acta*, **199**, 297
YOUNG, V.R., HAVERBERG, L.N. and MUNRO, H.N. (1973). *Metabolism*, **22**, 1429
YOUNG, V.R., STOTHERS, S.C. and VILAIRE, G. (1971). *J. Nutr.*, **101**, 1379

II

DIGESTION AND AVAILABILITY

5

AMINO ACID TRANSPORT

B.G. MUNCK,
Institute of Medical Physiology A, University of Copenhagen, Copenhagen, Denmark

Introduction

The problems with which the present review deals have previously been considered in more comprehensive publications. The monograph by Wilson (1962) and Wiseman's chapter in 'Handbook of Physiology' (1968) should be consulted on carrier specificities of intestinal transport of amino acids. The role of sodium and the energetics of amino acid transport have been discussed in detail by Christensen (1970, 1972); Schultz and Curran (1970) and by Kimmich (1973). The two latter reviews also deal particularly with the interactions between sugars and amino acids. LeFevre (1972) and Heinz (1972) studied sugar and amino acid transport by mammalian cells in general. Christensen (1972) discusses side chain requirements of amino acid transport by Ehrlich cells. Recent work by the author (1972a,b) has covered some of the aspects of the present review. In this situation it seems preferable to bring this recent review up to date and to present a personal evaluation of the current status of some aspects of intestinal absorption, rather than to register carefully all papers published within the last four years.

Methodological Considerations

In vivo, the small intestinal epithelium is cylindrical in structure, the epithelial lining being very complex. As a transporting epithelium, its characteristics differ markedly along the length of the intestine (Baker and George, 1971). Even within a crypt there are major functional differences when compared with the immediate surrounding villi. The epithelial lining of the small intestine is continuously renewed. The cells originate by mitosis at the bottom of the crypt. By a process of continued maturation they migrate up the villi and are shed from the tips of the villi in 50-80 hours. There are reasons to believe (Munck, 1972c) that the functional characteristics of absorption are confined to the villous epithelium, whereas the crypt epithelium is a secretory structure (Hendrix and Bayless, 1970).

In order to understand the function of the absorptive epithelium, it is necessary to describe the transport characteristics of its two bordering membranes, the luminal (brush border) membrane and the contraluminal (basolateral) membrane, and of the shunt provided by the junctional complexes and the intercellular spatia of the epithelium.

Of these structures the luminal membrane and the shunt can be directly studied with respect to influx (Schultz et al., 1967; Frizzell and Schultz, 1972). Efflux across this membrane and fluxes across the contraluminal membrane, however, must be studied more indirectly, either through a three compartment analysis of the epithelium (Ussing and Zerann, 1951; Schultz et al., 1967) including varying fractions of the non-epithelial parts of the gut wall and the media bathing its luminal and serosal surfaces; or by measuring efflux to the serosal side from preloaded mucosal tissues; that is, epithelium, villous cores, and submucosa (Hajjar, Khuri, and Curran, 1972). The procedure of the first method is to measure the unidirectional fluxes J_{ms}, J_{sm}, and J_{mc} together with the steady state epithelial accumulation $[A]_c$ and to calculate J_{cs} and J_{sc} from the following equations (*Figure 5.1*) (Munck and Schultz, 1969a):

(1) $J_{net} = J_{ms} - J_{sm} = J_{mc} - J_{cm} = J_{cs} - J_{sc}$
(2) $J_{ms} = J_{mc} \cdot J_{cs}/J_{cm} + J_{cs}$

The critical point of this approach is that it must be assumed that the non-epithelial tissues of the preparation constitute a well-stirred extension of the fluid bathing the serosal side of it. There can be little doubt that in the end this analysis leads to underestimates of J_{cs}. The second method is subject to the same sources of error. The everted sac preparation is clearly inferior to either of the two methods described above (Munck, 1972a).

The steady state uptake by the epithelium must be measured using the isolated mucosa (Schultz, Fuisz and Curran, 1966). The alternative method of using rings of everted intestine is clearly inferior (Munck, 1972b) in part owing to the fact that the rings are transformed into miniature sacs.

Isolated intestinal epithelial cells are being used more frequently for studies of intestinal transport (Reiser and Christiansen, 1971a; Kimmich, 1970a,b; Momtazi and Herbert, 1973). This method can provide cells predominantly of either villous or crypt origin (Harrison and Webster, 1969; Webster and Harrison, 1969) and thus serve to characterise these two cell populations. It is also the necessary tool for Kimmich's extensive studies of the role of the sodium-potassium gradients in sugar and amino acid transport.

The isolated cell should be well suited for studies of the cellular equipment of transport mechanisms and enzyme systems. However, with regard to its use for studying the dynamics of transport, it has some important weaknesses. Firstly, with this preparation the polarity of the enterocyte is eliminated. It cannot, therefore, be used to separately characterise the two parts of the cell membrane. Secondly, it appears that by its preparation, the functional integrity of the cell as an amino acid accumulating unit suffers markedly. Although its metabolic activities, measured as CO_2 and lactic acid production, proceed at constant rates for at least 60 minutes (Kimmich, 1970a), its accumulating ability declines abruptly after 15 or fewer minutes of incubation (Kimmich, 1970a; Reiser and Christiansen, 1971a,c; Kimmich and Randles, 1973b). In fact a steady-state with respect to accumulation never seems to be achieved. It is, therefore, not even possible to judge how early the functional deterioration affects the time course of accumulation. Thirdly, probably because of its nature as a monocellular suspension, it is a prerequisite for measuring initial rates of uptake, that very short periods of incubation are used. Several experimental studies demonstrate that incubation periods in excess of 0.5 min cannot be used (Reiser and Christiansen, 1971b; 1973b; Kimmich and Randles, 1973b; Tucker and Kimmich, 1973). This also appears from Reiser and Christiansen's (1971c) demonstration of trans-stimulation of lysine uptake by leucine after only one minute of incubation.

Important information about the biochemical basis of the function of the epithelial cell membranes, comes from studies on isolated brush border membranes. Fujita, Parsons and Wojnarowska, (1972); Peters, (1970, 1973), and Donlon and Fottrell, (1972) have studied the distribution of peptidases between this membrane and the epithelial cytoplasma. Fujita *et al.* (1972) have studied the distribution of Na-K-ATPase. Hopfer *et al.* (1973) have demonstrated sodium dependence of sugar uptake by brush border membranes which are isolated as vesicles. In recent years, Stirling has successfully used radioautographic techniques to demonstrate specific binding to intestinal cell membranes and most recently (1972) he has used ouabain to localise Na-K-ATPase.

Specificities of Intestinal Amino Acid Transport

The issue involves a number of questions: (1) Are both the amino and the carboxylic group necessary? (2) Must the amino group be in the a-position to the carboxyl group? (3) What is the effect of changing the side chain? (4) How is transport affected by making the side chain electrically charged?

Studies of these questions have led to the conclusion that intestinal absorption of amino acids is served by two systems for neutral amino acids; that is, one for a-amino-mono-carboxylic acids and one for imino acids; by one system for diamino-monocarboxylic acids, and possibly one for the dicarboxylic acids glutamic and aspartic acid (Wilson, 1962; Wiseman, 1968; Munck, 1972a; Heinz, 1972).

THE TRANSPORT MECHANISM FOR a-AMINO-MONOCARBOXYLIC (NEUTRAL) AMINO ACIDS
Wiseman's original studies indicated that this carrier would accept only amino acids with an unsubstituted amino group in a-position to the carboxyl group except for proline and hydroxyproline. Studies by

Spencer, Bow, and Markulis, (1962); Wilson et al., (1960), and Lin, Hagihira and Wilson, (1962) strongly indicated that it would not accept amino acids with electrically charged side chains and that both the amino and the carboxylic acid group were indispensable. Using inhibition of influx of phenylalanine as an indicator, Hajjar and Curran, (1970) reached the conclusion that either of these groups could be omitted. Using the same technique and the same preparation (rabbit ileum) but measuring influx of the appropriate substances, Schultz, Yu-Tu, and Strecker, (1972) could not confirm the data of Hajjar and Curran. It thus seems reasonable to conclude that these two groups are indispensable for transport by the carrier of neutral amino acids. In this group are included the usual amino acids glycine, alanine, valine, leucine, isoleucine, methionine, cysteine, serine, threonine, phenylalanine, tyrosine, mono-iodotyrosine, tryptophan, histidine, arginine, glutamine, proline and hydroxyproline. From these studies it also appears that, as long as they do not give rise to a charged side chain, a number of different substituents are acceptable in the side chains of both aliphatic and cyclic amino acids (Wilson et al., 1960; Lin, Hagihira and Wilson, 1962; Hajjar and Curran, 1970).

Studies of mutual inhibition for transport by sacs of everted intestine (Wiseman, 1956) and kinetics of uptake by rings of everted intestine (Finch and Hird, 1960) and transport by sacs (Matthews and Laster, 1965), indicated that the concentrations of half maximal rate of transport (K_t) fell with increasing lipophili of the amino acids. This relationship has been confirmed by measurement of the kinetics of influx across the brush border membrane (Hajjar and Curran, 1970; Schultz, Yu-tu and Strecker, 1972). Wiseman's data also indicated that the rate of transport (J_{max}) was higher the less lipophilic the amino acid. Similar results were reported by Matthews and Laster (1965). Later Munck, (1966c) demonstrated that sacs, as normally used, were unsuitable for establishing maximum rates of transepithelial transport. Also more recent studies have established values for J_{max} across the brush border, showing that the relationship suggested by Wiseman does not exist (Schultz, Yu-Tu and Strecker, 1972).

As briefly discussed, the data on inhibition of valine transport (Lin, Hagihira and Wilson, 1962) indicated that a charged side chain is not tolerated by the carrier of neutral amino acids. However, Haghira et al., (1961) found that methionine inhibited uptake of lysine by rings of hamster intestine. Varying, but over-all weak inhibition by neutral amino acids of uptake of basic amino acids by rings, was subsequently observed by Robinson and Felber (1964), and Robinson (1968). Using sacs and rings respectively from rat intestine, Munck (1968) and Robinson (1968) observed a more substantial inhibitory effect of lysine on transport of neutral amino acids. Munck and Schultz (1969a,b) demonstrated that this mutual inhibition took place at the brush border membrane. And, as it further demonstrated that both amino acids were transported by more than one carrier, this study left the question unanswered as to whether lysine was transported by the carrier of neutral amino acids also. This uncertainty seems now to be eliminated by Reiser and Christiansen (1972, 1973a). They found no mutual inhibition between valine and lysine, but observed that lysine could inhibit as much as one third of the total alanine uptake by isolated epithelial cells from rat small intestine. Alanine was also found to be among the neutral amino acids which stimulate uptake of lysine by this preparation. From these data it appears that, whereas lysine does not use the carrier of neutral amino acids, some of these (alanine, leucine, methionine, phenylalanine and probably serine and threonine) are transported also by the carrier of basic amino acids.

The conclusiveness of these data suffer to some extent because of the use of the preparation of isolated cells. More importantly, however, the data do not all represent initial rates of uptake because of too long an incubation period. Some of the rates may have been influenced by transconcentration effects. It is therefore of interest that recent studies (Munck and Rasmussen, 1975) led to the same conclusion as that drawn by Reiser and Christiansen (1972). Studying the interactions between lysine and tryptophan; tryptophan, a very potent inhibitor of transepithelial lysine transport, (Munck, 1966c), was found to be a weak inhibitor of lysine influx across rat jejunal brush border. Furthermore, tryptophan influx across the brush border could not be inhibited by lysine, but was inhibited by methionine according to the separately determined transport kinetics for these amino acids. Such results would be obtained if tryptophan and lysine were exclusively transported by the carrier of neutral amino acids and that of basic amino acids respectively.

The studies by Hagihira, Wilson and Lin, (1962); Daniels, Newey and Smyth, (1969a,b) and Hajjar and Curran, (1970) indicate that except for proline and hydroxyproline, the carrier of neutral amino acids is not used by imino acids.

THE TRANSPORT MECHANISM FOR BASIC AMINO ACIDS

From the preceding section it should be clear that a number of neutral amino acids are transported by the carrier of basic amino acids. The first evidence that basic amino acids were transported by a special mechanism and the proof that they were actively transported was presented by Hagihira et al. (1961). These data, together with those of Robinson and Felber (1964), Robinson (1966, 1968); Munck (1966c), and Chez et al. (1971) suffice to group arginine, cystine, histidine, lysine and ornithine as what could be called the primary users of the carrier of basic amino acids.

The main stimulus of continued interest in intestinal transport of basic amino acids and its relation to that of the neutral amino acids, has probably come from the observed ability of some neutral amino acids to stimulate transepithelial transport of lysine (Munck, 1965, 1966b,c; Munck and Schultz, 1969b); to apparently stimulate steady-state epithelial accumulation of lysine and arginine (Robinson and Felber, 1964, 1965; Robinson, 1968); and finally from the trans-stimulation by leucine of lysine influx across the brush border membrane of rabbit ileum (Munck and Schultz, 1969b). These interactions have subsequently been successfully studied by Reiser and Christiansen with respect to types and specificities of interactions, and with respect to the role of sodium.

These observations constitute a natural prelude to the treatment in following sections of trans effects on transport across the brush border membrane and of the relatively meagre information on the characteristics of transport across the basolateral membrane. It, therefore, seems appropriate to discuss these interactions.

Stimulation of Transepithelial Transport of Lysine

Studies (Munck and Schultz, 1969b) of the effect of 2 mM leucine on the steady-state parameters of lysine transport at 10 mM lysine, showed that a doubling of J_{ms} with no effect on J_{sm} resulted in a 300 per cent stimulation of J_{net}, and that steady-state uptake of lysine was significantly reduced by about 15 per cent, and that J_{mc} was the same after pre-incubation at 10 mM lysine and at 10 mM lysine + 2 mM leucine; demonstrating that influx of lysine across the brush border was unchanged at conditions under which J_{net} was markedly stimulated. Although efflux of lysine across the brush border by necessity was reduced in the presence of leucine, these results clearly demonstrated that the stimulation of J_{ms} was not secondary to events at the brush border membrane, but resulted from events at the basolateral membrane. The dissociation between events at these membranes is furthermore evident from the observations that alanine is the most potent stimulant of lysine uptake by isolated epithelial cells (Reiser and Christiansen, 1971a,b) with maximum effect at 5 mM alanine against 10 mM lysine (Reiser and Christiansen, 1972). Although at 10 mM lysine, 5 mM alanine increased lysine transfer to the serosal side of sacs of everted intestine by only 65 per cent, 5 mM leucine enhanced this transport by 300 per cent (Munck, 1972a).

TRANSCONCENTRATION EFFECTS ON TRANSPORT OF LYSINE ACROSS THE BRUSH BORDER MEMBRANE

When first observed, the leucine stimulation of transepithelial transport of lysine was considered an example of competitive exchange diffusion (Munck, 1965; Stein, 1967). As discussed, this interpretation had to be abandoned. However, the studies on which this was based (Munck and Schultz, 1969b) did demonstrate a transconcentration effect of leucine on lysine influx across the brush border membrane. However, this trans effect differed radically from that of competitive exchange diffusion in being observed as a stimulation of the unidirectional influx. Using isolated epithelial cells, Reiser and Christiansen (1971c) have obtained results which are consistent with those of Munck and Schultz (1969b). For methodological reasons, however, it is not possible to classify the type(s) of trans effects observed by these authors with certainty. These problems are further discussed below. Here, two interesting groups of observations remain to be discussed.

The studies of Reiser and Christiansen have especially established that mutual inhibition exists between basic amino acids and those neutral amino acids which at certain concentrations stimulate different parameters of the transport of basic amino acids. There are, however, a group of neutral amino acids which are rather potent inhibitors of lysine transport, whose own transport is not easily, or may not be at all, inhibited by lysine. This was first observed for tryptophan (Munck, 1966c) and has later been found for isoleucine (Robinson, 1968) and valine (Reiser and Christiansen, 1969, 1972). In detail this type of interaction has been examined only for tryptophan (Munck and Rasmussen, 1975). The results of this study (discussed elsewhere in this review) indicate (*Figure 5.1*) that tryptophan is not transported by, but immobilises, the lysine carrier of the brush border membrane. In

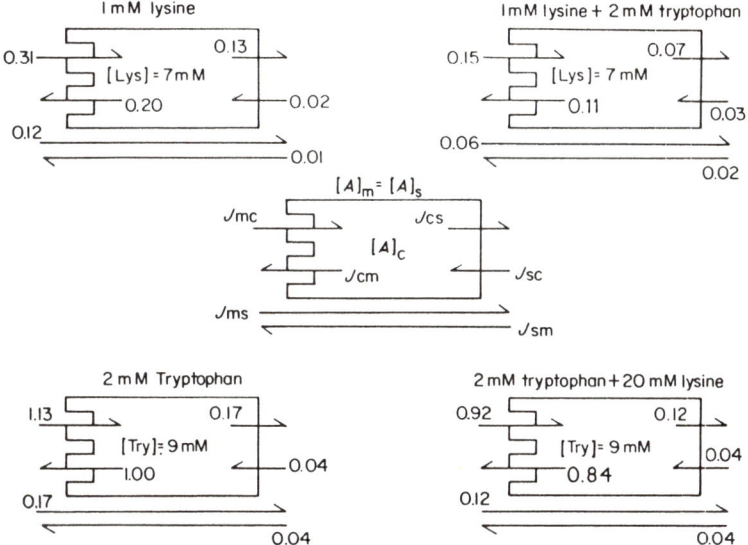

Figure 5.1 Steady state unidirectional fluxes across the epithelium and across the brush border and basolateral membranes. J_{ms}, J_{sm}, and J_{mc} are measured fluxes, the other fluxes are calculated using equations 1 and 2. All fluxes are stated in $\mu moles/cm^2$ h. The conditions of incubation are stated for each of the 'boxes'

addition, lysine and tryptophan appear to have an ordinary competitive relationship with the lysine carrier of the basolateral membrane. It clearly would be of interest to study further the interactions of valine and isoleucine with lysine transport.

The second point of interest is the demonstration by Reiser and Christiansen that, although neither lysine nor alanine transport by isolated epithelial cells are absolutely sodium dependent (1972), lysine inhibition of alanine uptake can only take place with sodium present in the media of incubation (1973a). This observation is the more interesting because the very same relationship has been observed with Ehrlich cells, where sodium in the media is indispensable for the inhibition of lysine uptake by, for example, leucine (Christensen, Handlogten and Thomas, 1969). In this context, it is natural to draw the attention to the suggestion by Christensen et al. (1973) that reversible protonation of one of the cationic groups may allow transport of basic amino acids to be energised by a hydrogen ion gradient in the cell membrane. This proposal is based upon transport studies using synthetic diamino acids which have not so far been used with intestinal preparations.

Table 5.1 K_t values for influx of amino acids across the brush border membrane of rabbit ileum and rat jejunum. The transport mechanisms are indicated as follows:
1 - Neutral amino acids
2 - Basic amino acids
3 - Imino acids

	Rabbit Ileum			Rat Jejunum[4]		
	1a[1]	1b[2]	2[3]	1	2	3
Glycine		20.5				
Alanine	9	10.6				
Serine	12					
Valine	7	4.4				
Leucine	6	4.8		4	5	
Methionine		1.3		3	11	
Phenylalanine	3.5	2.7				
Tyrosine		4.5				
Tryptophan	6	2.2		8	26	
Lysine			10		3	
Histidine	15					
Sarcosine						11
Proline				10		20
Gaba						26

[1] Schultz, Yu-Tu, and Strecker, 1972
[2] Hajjar and Curran, 1970
[3] Munck and Schultz, 1969a
[4] Munck and Rasmussen, 1975 and unpublished observations

The Transport Mechanism for Imino Acids

Hagihira, Wilson and Lin (1962) demonstrated that proline, hydroxyproline, and N-, mono-, di-, and trimethylglycine were transported in hamster small intestine by a separate mechanism. Spencer and Brody (1964) added the cyclic iminoacids azetidine-2-carboxylic acid pipecolic acid. Newey and Smyth (1964) demonstrated that at least two carriers were needed to account for rat intestinal transport of glycine and proline. Munck (1966a) confirmed for rat intestine that a separate transport mechanism existed for proline, hydroxyproline, sarcosine (N-mono-methylglycine), betaine, and alanine. These results were confirmed by Daniels, Newey and Smyth (1969a,b) who furthermore found that β-alanine, and β-, and γ-aminobutyric acid were specific inhibitors of the imino acid carrier, and that this carrier apparently did not prefer 1- for d-isomers. Although Randall and Evered (1964) were unable to demonstrate active transport of ω-amino acids, γ-aminobutyric acid included, the data of Daniels *et al.* (1969) strongly indicated that the imino acid carrier was unspecific with respect to steric configuration and position of amino group. These data, however, had come from experiments with sacs or pieces of the whole gut wall. By these methods it is not possible to separate with certainty transport characteristics provided by the luminal membrane from those of the basolateral membrane. In addition Daniels *et al.* (1969) had not measured transport of these interesting substances. Techniques measuring influx across the brush border of rabbit ileum and jejunum (Alvarez, Goldner and Curran, 1969; Peterson, Goldner and Curran, 1970) did not find evidence of a substantial secondary transport mechanism for glycine or alanine. Very recently, Munck and Rasmussen studied the transport of proline, γ-aminobutyric acid and sarcosine. They were able to separate clearly influx of proline across the brush border into two components, one of which is competitively inhibited by methionine and one by sarcosine. Influx of proline could be inhibited by γ-aminobutyric acid. Influx of γ-aminobutyric acid could be inhibited by sarcosine, but not by methionine. Furthermore the epithelial accumulation and transmural fluxes were measured at 1 and 40 mM γ-aminobutyric acid. At these two concentrations the tissue-medium concentration ratios were 12.2 and 1.8. In the short-circuited state with equal concentrations on both sides, the flux ratios were 4.0 and 1.3 respectively (*Figure 5.2*). These results are highly confirmatory to the conclusions drawn from

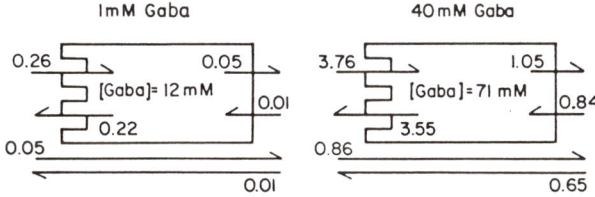

Figure 5.2 Steady state unidirectional fluxes across the epithelium and across the brush border and basolateral membranes. J_{ms}, J_{sm} and J_{mc} are measured fluxes, the other fluxes are calculated using equations (1) and (2). All fluxes are stated in moles/cm^2 h. The conditions of incubation are stated for each of the 'boxes'

the studies of Daniels, Newey and Smith (1969b). On this background, it is surprising that Evered and Hargreaves (1973) report that even at 1 mM, γ-aminobutyric acid transport against a concentration difference could not be demonstrated with sacs of everted rat small intestine. The most likely explanation is that the net transport was insufficient to fill the sub-epithelial tissues and the volume of transporter water insufficient to change the initial concentration of the media. The evidence discussed here indicates that unbranched aliphatic α-aminomonocarboxylic acid with more than one carbon atom in the side chain, cannot use the carrier of imino acids. There is no evidence that this carrier can be used by basic amino acids.

The Transport Mechanism for a-Amino-dicarboxylic Acids

Because of extensive transamination of aspartic and glutamic acids by intestinal tissues (Neame and Wiseman, 1956, 1957; Ramaswamy and Radhakrishnan, 1970) the transport characteristics for these amino acids can only be defined for unidirectional influx across the brush border membrane. So far, only one study of this function has been published (Schultz et al., 1970). Here, values of K_t and J_{max} were established, mutual inhibition according to these constants was demonstrated, and the affinities of sodium were established for the carrier complexes of these amino acids. The question of mutual inhibition between these amino acids and neutral and basic amino acids was not examined. Thus, it is presently not known whether aspartic and glutamic acids are transferred across the brush border membrane by a carrier of their own.

Transconcentration Effects

In the numerous studies on which the discussions of transport specificities are based, observations have never been made which in the end could be shown to represent cis stimulation. That is, stimulation by substance A (amino acid or sugar) present in the solution bathing the brush border membrane, of influx of substance B across this membrane.

For a long time it was an acceptable characteristic of the model of carrier mediated intestinal transport of sugars and amino acids, that carrier and carrier-substrate complexes had equal mobilities in the cell membrane (Wilbrandt and Rosenberg, 1961). That this was not necessarily the case became clear when Goldner, Schultz and Curran (1969) demonstrated that in rabbit ileum the effect of sodium on sugar transport could be accounted for by assuming that the carrier became immobilised by the attachment of sugar and that mobility was restored when the sugar-carrier complex became supplemented by sodium. Additional evidence for variable permeability came from the observation by Caspary, Stevenson and Crane (1969) that although being a sodium-dependent competitive inhibitor of sugar transport by hamster small intestine, 1-fucose itself was not transported. As discussed above, tryptophan is a similar example. It inhibits influx of lysine although it seems not to be transported by the carrier of basic amino acids. The studies on glutamate and aspartate influx across rabbit ileal brush border (Schultz et al., 1970) probably provides another example that, in addition, illustrates a mechanism which, together with the use of more than one mechanism of transport, can account for the large differences between the J_{max} values for individual amino acids. It was reported that J_{max} for glutamate was twice that for aspartate, but that mutual inhibition between them followed predictions based on their K_t values. It is, therefore, more likely that the glutamate-carrier complex has a higher mobility than that there are different carrier concentrations available to the two amino acids.

Caspary, Stevenson and Crane, (1969) also found that fucose in the incubation medium inhibited efflux of a sugar with which the tissues had previously been preloaded. With respect to the cytoplasm, this effect of fucose must be considered an inhibitory trans effect. In the case of intestinal amino acid transport trans effects have only been observed with respect to influx.

Substrates in trans location with respect to a substance of which the transport is being measured, can affect this transport in different ways. Its binding can be a prerequisite for the carrier to be mobile. In this case, exchange diffusion is the only possible mode of transport, the term *'obligatory exchange diffusion'* could be used. By increasing the mobility of a carrier which is mobile also when empty, the substance in trans position can enhance transport. For this effect the terms *'accelerative exchange diffusion'* (Stein, 1967) and *'trans-stimulation'* (Heinz, 1972) have been used; the latter shall be used here. The opposite effect of blocking or slowing the mobility of a carrier will lead to an inhibition of transport. Therefore the term *'trans-inhibition'* (Heinz, 1972) shall be used. These three effects all apply to the true unidirectional flux across the cell membrane. When, however, measurements of influx are extended over periods long enough that the transconcentration suffices to create a significant efflux, then the presence of a competitive inhibitor in trans position can eliminate this efflux and thus create a false impression of stimulation of influx rate. This effect, which has been termed *'competitive exchange diffusion'* (Stein, 1967) is a straightforward consequence of the mobile carrier model. Its demonstration therefore provides an argument for the validity of the carrier concept.

In discussing and interpreting data on intestinal transport, it is extremely important to make clear which of these types of trans effect is at hand or under discussion.

'Obligatory exchange diffusion' has not been observed for intestinal amino acid transport and shall not be discussed any further.

'Competitive exchange diffusion' has been observed between alanine and a number of neutral amino acids for uptake by isolated epithelial cells (Reiser and Christiansen, 1973c). But as long as the experimental technique suffices to eliminate this phenomenon, it has no significance for the evaluation of epithelial transport by means of data on J_{ms}, J_{sm}, J_{mc}, and $[A]_c$. This phenomenon shall therefore not be subject to further consideration.

The two remaining trans effects, 'trans-inhibition' and 'trans-stimulation' do, however, profoundly affect the conditions under which J_{mc} must be measured in order to be used to describe the transport across the basolateral membrane. In case of such trans effects, J_{mc} must be measured after pre-incubation with media, which apart from the presence of radioactive labels, are identical to those from which influx across the brush border is eventually to be measured.

Trans-inhibition of intestinal amino acid transport has never been reported, and trans-stimulation has only been observed for a few neutral amino acids acting on the transport of a basic amino acid. In this connection it is noteworthy that leucine, which stimulated influx of lysine and was transported by the lysine carrier, apparently did not stimulate its own influx (Munck and Schultz, 1969b).

Munck and Rasmussen, (1975) have now observed that tryptophan is a very potent trans-inhibitor of the influx of lysine and tryptophan (*Figure 5.3*). Although tryptophan does not, in the cis position inhibit

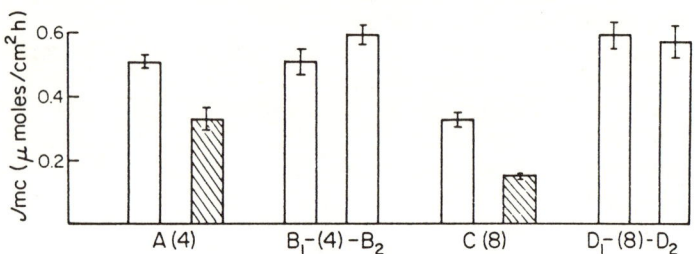

Figure 5.3 Transconcentration effects of tryptophan. A, B, C, D Open bars pre-incubated in Krebs phosphate buffer (KB) for 30 minutes. Cross hatched bars pre-incubated for 30 minutes in Krebs phosphate buffer containing 2 mM tryptophan

A.   Test incubation in KB containing 1 mM galactose
B_1 -   -   -   -   -
B_2 -   -   -   - containing 1 mM galactose + 5 mM tryptophan
C   Test incubation in KB containing 1 mM lysine
D_1 -   -   -   -   -
D_2 -   -   -   - containing 1 mM lysine + 2 mM tryptophan

The s.e. values are indicated for each bar. The number of paired experiments is given in brackets with the letter used to indicate each series

influx of galactose, it is a rather marked trans-inhibitor thereof. This, of course, severely complicates the interpretations of the effects of tryptophan. In the same study, however, methionine (*Figure 5.4*) was found to trans-stimulate influx of lysine, to trans-inhibit influx of tryptophan, but to have neither cis nor trans effects on influx of galactose. On this background trans effects of leucine on leucine were further examined. When leucine is excluded from transport by the lysine carrier through lysine inhibition, previous loading with leucine was seen to inhibit leucine influx. Together, these observations demonstrate that leucine, methionine and tryptophan are trans-inhibitors of the carrier of neutral amino acids, and that leucine is a trans-stimulator of the influx of both leucine and lysine by the carrier of basic amino acids. In the previous experiments (Munck and Schultz, 1969; Munck, 1972d) these two effects have simply cancelled each other.

All the data cited above demonstrate that trans effects, inhibiting as well as stimulating, are elicited at quite low concentrations of amino acids. It is thus conceivable that these phenomena may play a role during digestion and absorption of an ordinary meal and during absorption of tube fed protein hydrolysates or amino acid mixtures.

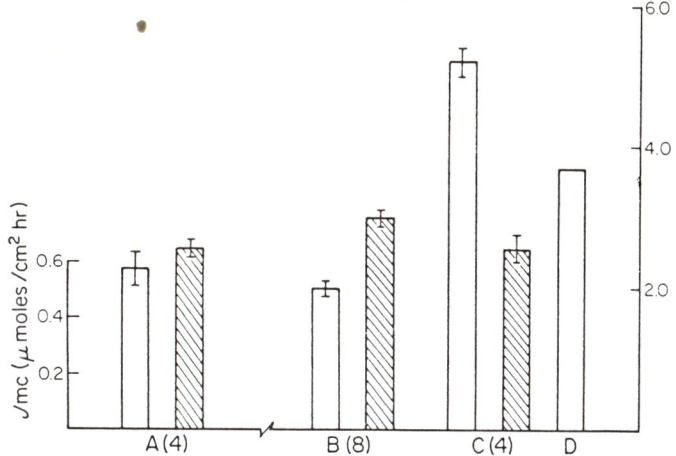

Figure 5.4 Transconcentration effects of methionine, A, B, C, D. Open bars pre-incubated in Krebs phosphate buffer (KB) for 30 minutes. Cross hatched bars pre-incubated for 30 minutes in KB containing 2 mM methionine
A Test incubation in KB containing 1 mM galactose
B -   -   - -   containing 10 mM lysine
C -   -   - -   containing 5 mM tryptophan
D Calculated flux of tryptophan at 5 mM tryptophan + 2 mM methionine using the K_t values stated in Table 5.1 and the control flux of series C
Note that the scale for series A differs from that for series B, C, and D

Transport Across the Basolateral Membrane

This final step in transepithelial transport has not frequently been dealt with.

On the basis of the effects of anoxia upon amino acid transport by sacs of everted rat intestine, Newey and Smyth (1962) discussed a second step in transepithelial transfer of amino acids. Rather than demonstrating a second step, however, their data demonstrated that epithelial accumulation, but not influx across the brush border membrane, is energy dependent.

Using sacs of everted rat intestine, Munck (1965) found that heavy preloading of serosal fluids of sacs with proline inhibited transport of glycine to the serosal fluid. These results were taken as evidence of competition for mediated transport across the basolateral membrane. It was, however, not completely ruled out that this inhibition could have been caused by trapping of proline escaping from the epithelium in the unstirred layer at the outside of the brush border membrane.

The first reasonably unequivocal demonstration of saturation of amino acid transport across the basolateral membrane, is that of lysine transport by rabbit ileum (Munck and Schultz, 1969a). It was based on measurements of J_{ms}, J_{sm}, J_{mc}, and $[lys]_c$. Using the isolated mucosal tissues from the small intestine of a fresh water turtle, Hajjar, Khuri and Curran (1972) used the time course for wash out of previously accumulated alanine, and the time course of development of steady-state unidirectional transmucosal fluxes, to calculate flux rates across the basolateral membrane. The values were correlated with estimates of the epithelial concentrations at which the appropriate measurements had been made. They found that passage of alanine across the basolateral membrane of the turtle small intestine is a saturable sodium independent process.

Using the procedure of Munck and Schultz (1969) and Munck and Rasmussen (to be published) the transport of proline (*see* Curran, 1973) and tryptophan across the basolateral membrane of rat intestine was studied. Both fluxes appear to be saturable. That of tryptophan can be inhibited by lysine. Vice versa, efflux of lysine across the basolateral membrane seems competitively inhibited by tryptophan (*Figure 5.1*). More evidence of mediated transport across the basolateral membrane is that this step in the transepithelial transport of lysine and tryptophan can be stimulated by leucine and methionine respectively (Munck and Schultz, 1969b; Munck and Rasmussen, 1975).

The Sodium Gradient Hypothesis

The role in intestinal transport of sugars and amino acids over this and other electrochemical gradients, must be considered in connection with what is known about the same processes in other mammalian cell systems, especially the Ehrlich ascites tumour cell. The problems and unanswered questions are many and the literature is vast. It is, therefore, necessary to refer to recent illuminating reviews (Schultz and Curran, 1970; Christensen, 1970; Christensen *et al.*, 1973; Heinz, 1972; Kimmich, 1973).

According to the sodium gradient hypothesis, active transepithelial transport of sugars and amino acids is energised by coupling between sodium and non-electrolyte transport across the luminal membrane. Thus, downhill flux of sodium into the cell leads to uphill flux of the non-electrolyte. The low cellular concentration of sodium is maintained by ATP-ATPase driven transport of sodium outwards across the basolateral membrane.

For rabbit ileum, both unidirectional fluxes of alanine and sodium across the brush border have been shown to be qualitatively consistent with the hypothesis (Schultz and Curran, 1970). In accordance with Cranes' (1965) suggestion that the potassium gradient also contributed to sugar transport, potassium has been shown, especially at low sodium concentrations, to inhibit sugar influx across rabbit ileal brush border (Goldner, Schultz and Curran, 1969). Influx of alanine has furthermore been shown to be independent of intracellular sodium and the availability of metabolic energy (Chez *et al.*, 1967). The weak effects of some metabolic inhibitors seem, in the light of effects of the SH-reagent p-chloromercuriphenyl sulfonic acid (Schaeffer, Preston and Curran, 1973), to be explained as resulting from the ability of these inhibitors to react with SH-groups.

It has never been demonstrated that the sodium gradient could account quantitatively for the observed transport of sugars and amino acids. A recent estimate by McD. Armstrong, Byrd and Hamang (1973) indicated that for frog small intestine, the electrochemical gradient for sodium could suffice for galactose transport.

In a series of papers, Kimmich and co-workers have examined some aspects of the effects of sodium and potassium on sugar and amino acid uptake by isolated chicken intestinal cells (Kimmich, 1970a,b; Tucker and Kimmich, 1973; Kimmich and Randles, 1973a,b; Kimmich, 1973). They found that this preparation could transport the non-electrolytes against a concentration difference even when both the sodium and the potassium gradients were reversed. A sodium gradient alone could drive sugar transport, but a potassium gradient could not. Based on these studies a model was proposed in which transfer of sugars and amino acids across the cell membrane is directly energised by a membrane ATPase function.

Clearly the results presented by Kimmich and co-workers are not consistent with the sodium gradient hypothesis. Neither, however, does Kimmich's model in a simple way account for the transport characteristics of rabbit ileum. It is inconsistent with the location of ATPase at the contraluminal membrane (Fujita *et al.*, 1972; Stirling, 1972).

A unifying view of the role of the sodium gradient in intestinal transport immediately raises the question of whether other gradients are involved and/or whether the micro-environment of the microvilli could differ enough from the cytoplasm to explain the results.

If the sodium, co-transported with sugars or amino acids, must leave across the basolateral membrane, a favourable microvillus environment with respect to sodium seems unlikely. But this is not so if a substantial fraction of the sodium returned to the luminal fluid in an electrically neutralised state.

Contrary to the model proposed by Schultz and Zalusky (1964) Barry, Smyth and Wright (1965) found that net sodium transport did not equal the short circuit current (I_{sc}) during rat jejunal sugar transport. Taylor *et al.* (1968) proposed that the reason could be that the sugars induced an NaCl secretion approximately equal to the sugar-coupled sodium absorption. Munck (1972c) found that proline and glucose, in addition to galactose affected I_{sc}, sodium and chloride fluxes in quantitative agreement with the model of Taylor *et al.* (1968). And the idea was dismissed that the increment in J_{sm}^{Na} resulted alone from the simultaneously observed increments in passive transepithelial permeability. Frizzell and Schultz (1972), however, provided evidence for rabbit ileum that J_{sm}^{Na} as suggested by Schultz and Zalusky (1964) represents passage of sodium through a paracellular shunt. Using the technique of Frizzell and Schultz on rat jejunum, in experiments which with respect to glucose concentration, I_{sc}, tissue resistance, and glucose induced change in I_{sc} were identical with those described by Munck (1972d). Munck and Schultz (1974, and unpublished results) found that the sodium conductance of the shunt pathway compared extremely well with J_{sm}^{Na} as measured by Munck (1972d). Thus, 28 mM glucose increased influx of sodium across the luminal cell membrane by

8 µeq/cm² hr and the I_{sc} by 5.7 µeq/cm² hr (n = 10, s.e. = 0.5); combining these numbers (*Figure 5.5*) with those of Munck (1972d) and Munck and Schultz (1974) it appears that of 8 µeq sodium co-transported with glucose, only 0.5 - 1.0 µeq can be accounted for by J_{net}^{Na}, whereas of the remaining seven, five must return to the luminal fluid electrically neutralised by co-transport with some anions. The data for glucose indicate that a substantial fraction can be chloride, and the data for proline and galactose indicate that it can all be chloride.

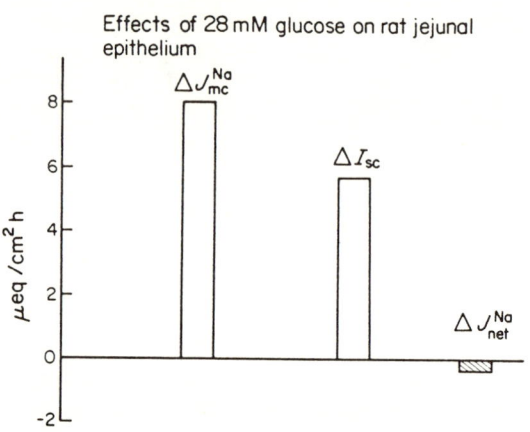

Figure 5.5 Effects of 28 mM of glucose on J_{mc}^{Na} corrected for flux through a diffusional pathway (ΔJ_{mc}^{Na}) short circuit current (ΔI_{sc}) and upon net transmural sodium transport across the short circuited wall of rat jejunum (ΔJ_{net}^{Na}). (Data from Munck and Schultz, 1974 and Munck, 1972d)

Coupled influx of sodium and chloride has been demonstrated across the luminal membrane of rabbit ileum (Nellans, Frizzell and Schultz, 1973). A similarly coupled efflux of NaCl (Nellans, Frizzell and Schultz, 1974) could be the mechanism behind the data discussed for rat jejunum. This being the case it seems likely that in the brush border region of the intracellular space, a microenvironment could exist with a sodium activity well below that registered for the cytoplasma by Lee and Armstrong (1972), and McD. Armstrong, Byrd and Hamang (1973) or the concentrations reported by Kimmich (1970b). As des- described here the maintenance of this micro-environment depends on the presence of chloride. It is, therefore, consistent with these ideas that net transport of proline and proline induced increments in I_{sc} are lower in media where chloride is replaced by sulphate (Munck, 1972d).

Using a modified Cori technique, Shishova (1956) observed that luminal ATP stimulated the absorption of amino acids. More recently Kohn, Newey and Smyth (1970); Reiser and Christiansen (1971); Field (1971); Gerencser and Armstrong (1972) and Hardcastle (1973) have observed different stimulating, as well as inhibitory effects of extra- cellular ATP on transport activities of rat, rabbit, and frog small intes- tine. A common mechanism of action has not been proposed.

Interactions Between Sugars and Amino Acids

The question of the mechanism of this interaction is closely related to the questions about the energetics of epithelial transport. The alternative view (Alvarado, 1966; 1971) that sugars and amino acids compete for a common multifunctional carrier must, for lack of evidence, be abandoned. In all cases where influx across the brush border membrane has been studied as an isolated phenomenon, influx of amino acids has been found not to be influenced by the presence of actively transported sugars (Chez, Schultz and Curran, 1966; Frizzell and Schultz, 1971; Munck, 1972d).

Schultz and Curran (1970) and Semenza (1971) have proposed that the inhibition which appears to result from increased efflux across the brush border membrane is caused by an increase in sodium activity inside the brush border membrane.

Kimmich (1973) has interpreted mutual inhibition between sugars and amino acids in terms of his recent model (Kimmich and Randles, 1973) and proposed that the limiting factor is the supply in the membrane of energised transport-ATPase. This model is not consistent with the influx data cited above. For rat intestine it seems disproved by the inability of preloading at 28 mM galactose to make it possible to inhibit leucine influx (Munck, 1972d). In connection with Kimmich and Randles' results with isolated intestinal cells, it seems of interest that Munck (1972d) observed that galactose markedly increased the amino acid permeability of the basolateral membrane. It seems possible that such an effect may have played a role in the studies of Kimmich and Randles.

Transport of Peptides

Newey and Smyth (1959, 1960) suggested that peptides were taken up by the intestinal epithelium independent of the transport mechanisms for amino acids. Only glycyl-glycine, however, was found to pass the epithelial cell layer unhydrolysed.

Within the last few years, experimental and clinical research carried out especially by Matthews and co-workers and by Milne and his associates, has established beyond doubt that peptide transport by the small intestinal epithelium proceeds by mechanisms different from those responsible for transport of free amino acids.

In vivo as *in vitro* the absorbing capacity for amino acids seems higher when the amino acids are presented as di- and tripeptides than when presented as free amino acids (Matthews *et al.*, 1968; Lis, Crampton and Mathews, 1971; Cheng *et al.*, 1971; Adibi, 1971; Cook, 1972). Exceptions to that rule have been presented (Asatoor *et al.*, 1970; Navab and Asatoor, 1970). *In vivo* more amino-nitrogen is absorbed from pancreatic digests of proteins than from the equivalent amount of complete hydrolysates (Crampton *et al.*, 1971). *In vivo* as *in vitro* the extramucosal hydrolytic capacity is insufficient to account for the degrees of hydrolysis observed (Newey and Smyth, 1960;

Adibi, 1971; Hellier et al., 1972a). *In vitro* the sites of maximal absorption are differently located for free amino acids and for peptides (Crampton, Lis and Matthews, 1973), and there is evidence that the two categories of transport are differently affected by prolonged fasting (Lis, Matthews and Crampton, 1972). Even more convincing evidence has come from studies on patients with malabsorption of neutral amino acids (Hartnup disease) (Navab and Asatoor, 1970; Asatoor et al., 1970a,b) and basic amino acids (cystinuria), (Hellier et al., 1972b). These patients poorly absorb neutral and basic amino acids respectively from solutions of free amino acids, whereas the same amino acids are normally absorbed when presented as dipeptides.

Peptide transport across the brush border membrane of rabbit ileum has been studied by Rubino, Field and Schwachman (1971) using the labelled glycine residue of glycyl-proline as indicator. They demonstrated saturability, inhibition by other peptides, lack of inhibition by amino acids, and sodium dependences. Using rings or sacks of everted hamster small intestine Addison, Burston and Matthews (1972) have studied transport of glycyl-sarcosine. This transport was sodium dependent and could be inhibited by methionyl-methionine, but not by methionine. It was furthermore inhibited by anoxia and metabolic inhibitors.

Since peptides taken up by intestinal preparations are generally not found in the tissues as peptides, but as free amino acids, it must be assumed that the amino acid transport mechanisms of the basolateral membrane are responsible for the final step in the transepithelial transport of peptides. The speculations on the nature of peptide transport can, therefore, be limited to the events at the brush-border membranes and are as follows:

(1) Absorption from the luminal fluid and unstirred layer at the brush border is secondary to hydrolysis by peptidases here and on the surface of the membrane and can be ruled out by the observations discussed above, especially by the lack of competition by initially free amino acids.

(2) Transport by a peptide carrier followed by hydrolysis by peptidases of the cytoplasm cannot be completely ruled out. It is supported by the abundance of peptidases in the cytoplasm (Peters, Donlon and Fottrell, 1972; Donlon and Fottrell, 1973; Peters, 1973; Fujita, Parsons and Wojnarowska, 1972). However, tri- and tetrapeptidases seem predominantly localised to the brush border fraction of the epithelial cell homogenate (Peters, Donlon and Fottrell, 1972; Peters, 1973; Fujita et al., 1972). This indicates that tri- and tetrapeptides, which as far as examined have transport characteristics similar to dipeptides, are hydrolysed in the cell membrane. It is also unexpected, though not inconceivable, that cytoplasmic hydrolysis can maintain cytoplasmic peptide concentrations below measurable levels.

(3) A sequence of hydrolysis in the membrane followed by capture of each liberated amino acid by a carrier which takes care of transport across the inner strata of the membrane, is feasible.

(4) The simplest model, which is consistent with the observations discussed in this section, consists of a unification of the peptidase and carrier function. That is, as a peptide is accepted by a membrane

peptidase, it is transferred to the cytoplasma while split into its constituent amino acids. A similar combined function has been discussed for the intestinal sucrase-isomaltase (Semenza, 1970). The observations on glycylsarcosine transport indicate that the carrier function can proceed independently of the peptidase function. This model is at least consistent with the results of Fujita *et al.* (1972) which indicate that the variety and activities of brush border peptidases may suffice to account for the intestinal protein absorption.

It is necessary to end this short discussion of peptide absorption by referring to recent detailed accounts by Ugolev (1965), Ugolev and De Laey (1973) and Matthews (1972).

References

ADDISON, J.M., BURSTON, D. and MATTHEWS, D.M. (1972). *Clin. Sci.*, **43**, 907

ADIBI, S.A. (1971). *J. clin. Invest.*, **50**, 2266

AKEDO, H. and CHRISTENSEN, H.N. (1962). *J. biol. Chem.*, **237**, 113

ALVARADO, F. (1966). *Science*, **151**, 1010

ALVARADO, F. (1971). in *Intestinal transport of electrolytes, amino acids and sugars.* Ed. McD. Armstrong. p.281. Charles C. Thomas, Springfield, Illinois, USA

ALVAREZ, O., GOLDNER, A.M. and CURRAN, P.F. (1969). *Am. J. Physiol.*, **217**, 946

ANTONIOLI, J.A. and CHRISTENSEN, H.N. (1968). *Am. J. Physiol.*, **215**, 951

ARMSTRONG, McD., BYRD, B.J. and HAMANG, P.M. (1973). *Biochim. biophys. Acta*, **330**, 237

ASATOOR, A.M., BANDOH, J.K., LANT, A.F., MILNE, M.D. and NAVAB, F. (1970a). *GUT*, **11**, 250

ASATOOR, A.M., CHENG, B., EDWARDS, K.D.G., LANT, A.F., MATTHEWS, D.M., MILNE, M.D., NAVAB, F. and RICHARDS, A.J. (1970b), *GUT*, **11**, 380

BAKER, R.D. and GEORGE, M.J. (1971). *Biochim. biophys. Acta*, **225**, 315

BARRY, R.J.C., SMYTH, D.H. and WRIGHT, E.M. (1965). *J. Physiol.*, **181**, 410

BURSTON, D., ADDISON, J.M. and MATTHEWS, D.M. (1972). *Clin. Sci.*, **43**, 823

CASPARY, W.F., STEVENSON, N.R. and CRANE, R.K. (1969). *Biochim. biophys. Acta*, **193**, 168

CHENG, B., NAVAB, F., LIS, M.T., MILLER, T.N. and MATTHEWS, D.M. (1971). *Clin. Sci.*, **40**, 247

CHEZ, R.A., PALMER, R.R., SCHULTZ, S.G. and CURRAN, P.F. (1967). *J. gen. Physiol.*, **50**, 2357

CHEZ, R.A., SCHULTZ, S.G. and CURRAN, P.F. (1966). *Science*, **153**, 1012

CHEZ, R.A., STRECKER, C.K., CURRAN, P.F. and SCHULTZ, S.G. (1971). *Biochim. biophys. Acta*, **233**, 222

CHRISTENSEN, H.N. (1970). in *Membranes and ion transport.* Ed. E.E. Bittar. Vol.1, p.365; Wiley-Interscience, New York

CHRISTENSEN, H.N. (1972). *Bioenergetics*, **4**, 233
CHRISTENSEN, H.N., CESPEDES, C. de, HANDLOGTEN, M.E. and RONQUIST, G. (1973). *Biochim. biophys. Acta*, **300**, 487
CHRISTENSEN, H.N., HANDLOGTEN, M.E. and THOMAS, E.L. (1969). *Proc. Nat. Acad. Sci., U.S.A.*, **63**, 948
COOK, G.C. (1972). *Clin. Sci.*, **43**, 443
CRAFT, I.L., GEDDES, D., HYDE, C.W., WISE, I.J. and MATTHEWS, D.M. (1968). *GUT*, **9**, 425
CRAMPTON, R.F., GANGOLLI, S.D., SIMSON, P. and MATTHEWS, D.M. (1971). *Clin. Sci.*, **41**, 409
CRAMPTON, R.F., LIS, M.T. and MATTHEWS, D.M. (1973). *Clin. Sci.*, **44**, 583
CRANE, R.K. (1965). *Fed. Proc.*, **24**, 1000
CURRAN, P.F. (1973). in *Transport mechanisms in epithelia*. Eds. H.H. Ussing and N.A. Thorn. p.298. Munksgaard, Copenhagen
DANIELS, V.G., DAWSON, A.G., NEWEY, H. and SMYTH, D.H. (1969). *Biochim. biophys. Acta*, **173**, 575
DANIELS, V.G., NEWEY, H. and SMYTH, D.H. (1969a). *J. Physiol.*, **205**, 15
DANIELS, V.G., NEWEY, H. and SMYTH, D.H. (1969b). *Biochim. biophys. Acta*, **183**, 637
DONLON, J. and FOTTRELL, P.F. (1972). *Comp. Biochem. Physiol.*, **41B**, 181
DONLON, J. and FOTTRELL, P.F. (1973). *Biochim. biophys. Acta*, **327**, 425
EVERED, D.F. and HARGREAVES, B.M.C. (1973). *Xenobiotica*, **3**, 753
FIELD, M. (1971). *Am. J. Physiol.*, **221**, 992
FINCH, L.R. and HIRD, F.J.R. (1960). *Biochim. biophys. Acta*, **43**, 278
FRIZZELL, R.A. and SCHULTZ, S.G. (1971). *Biochim. biophys. Acta*, **233**, 485
FRIZZELL, R.A. and SCHULTZ, S.G. (1972). *J. gen. Physiol.*, **59**, 318
FUJITA, M., OHTA, H., KAWAI, K., MATSUI, H. and NAKAO, M. (1972). *Biochim. biophys. Acta*, **274**, 336
FUJITA, M., PARSONS, D.S. and WOJNAROWSKA, F. (1972). *J. Physiol.*, **227**, 377
GERENCSER, G.A. and ARMSTRONG, W.McD. (1972). *Biochim. biophys. Acta*, **255**, 663
GOLDNER, A.M., SCHULTZ, S.G. and CURRAN, P.F. (1969). *J. gen. Physiol.*, **53**, 362
HAGIHIRA, H., LIN, E.C.C., SAMIY, A.H. and WILSON, T.H. (1961). *Biochem. biophys. Res. Commun.*, **4**, 478
HAGIHIRA, H., WILSON, T.H. and LIN, E.C.C. (1962). *Am. J. Physiol.*, **203**, 637
HAJJAR, J.J. and CURRAN, P.F. (1970). *J. gen. Physiol.*, **56**, 673
HAJJAR, J.J., KHURI, R.N. and CURRAN, P.F. (1972). *J. gen. Physiol.*, **60**, 720
HARDCASTLE, P.T. (1973). *Biochim. biophys. Acta*, **332**, 114
HARRISON, D.D. and WEBSTER, H.L. (1969). *Expl Cell Res.*, **55**, 257
HEINZ, E. (1972). in *Metabolic Pathways, vol.6, Metabolic Transport*. Ed. L.E. Hokin, p.455. Academic Press, New York and London

HELLIER, M.D., HOLDSWORTH, C.D., McCOLL, I. and PERRETT, D. (1972a). *Gut*, **13**, 965
HELLIER, M.D., HOLDSWORTH, C.D., PERRETT, D. and THIRUMALAI, C. (1972b). *Clin. Sci.*, **43**, 659
HENDRIX, T.R. and BAYLESS, T.M. (1970). *A. Rev. Physiol.*, **32**, 139
HOPFER, U., NELSON, K., PERROTTO, J. and ISSELBACHER, K.J. (1973). *J. biol. Chem.*, **248**, 25
KIMMICH, G.A. (1970a). *Biochem.*, **9**, 3659
KIMMICH, G.A. (1970b). *Biochem.*, **9**, 3669
KIMMICH, G.A. (1973). *Biochim. biophys. Acta*, **300**, 31
KIMMICH, G.A. and RANDLES, J. (1973a). *J. Membr. Biol.*, **12**, 23
KIMMICH, G.A. and RANDLES, J. (1973b). *J. Membr. Biol.*, **12**, 47
KOHN, P.G., NEWEY, H. and SMYTH, D.H. (1970). *J. Physiol.*, **208**, 203
LARSEN, P.R., ROSS, J.E. and TAPLEY, D.F. (1964). *Biochem. biophys. Acta*, **88**, 570
LEE, C.O. and ARMSTRONG, W.McD. (1972). *Science*, **175**, 1261
LE FEVRE, P.G. (1972). in *Metabolic Pathways, vol.6, Metabolic Transport.* Ed. L.E. Hokin, p.385. Academic Press, New York and London
LIN, E.C.C., HAGIHIRA, H. and WILSON, T.H. (1962). *Am. J. Physiol.*, **202**, 919
LIS, M.T., CRAMPTON, R.F. and MATTHEWS, D.M. (1971). *Biochim. biophys. Acta*, **233**, 453
LIS, M.T., MATTHEWS, D.M. and CRAMPTON, R.F. (1972). *Br. J. Nutr.*, **28**, 443
MATTHEWS, D.M. (1972). in *Peptide transport in bacteria and mammalian gut*, p.71. A Ciba Foundation Symposium. ASP, Amsterdam
MATTHEWS, D.M., CRAFT, I.L., GEDDES, D.M., WISE, I.J. and HYDE, C.W. (1968). *Clin. Sci.*, **35**, 415
MATTHEWS, D.M. and LASTER, L. (1965). *Am. J. Physiol.*, **208**, 593
MOMTAZI, S. and HERBERT, V. (1973). *Am. J. clin. Nutr.*, **26**, 23
MUNCK, B.G. (1965). *Biochim. biophys. Acta*, **109**, 142
MUNCK, B.G. (1966a). *Biochim. biophys. Acta*, **120**, 97
MUNCK, B.G. (1966b). *Biochim. biophys. Acta*, **120**, 282
MUNCK, B.G. (1966c). *Biochim. biophys. Acta*, **126**, 299
MUNCK, B.G. (1968). *Biochim. biophys. Acta*, **156**, 192
MUNCK, B.G. (1972a). in *Glaxo Symposium on Transport across the intestine.* Eds. W.L. Burland and P.D. Samuel, p.169; Churchill Livingstone, Edinburgh and London
MUNCK, B.G. (1972b). in *Glaxo Symposium on Transport across the intestine.* Eds. W.L. Burland and P.D. Samuel, chapter 15, p.187; Churchill Livingstone, Edinburgh and London
MUNCK, B.G. (1972c). *J. Physiol.*, **223**, 699
MUNCK, B.G. (1972d). *Biochim. biophys. Acta*, **266**, 639
MUNCK, B.G. and RASMUSSEN, S.N. (1975). *Biochim. biophys. Acta*, **389**, 261
MUNCK, B.G. and SCHULTZ, S.G. (1969a). *J. gen. Physiol.*, **53**, 157
MUNCK, B.G. and SCHULTZ, S.G. (1969b). *Biochim. biophys. Acta*, **183**, 182
MUNCK, B.G. and SCHULTZ, S.G. (1974). *J. Membr. Biol.*, **16**, 163

NAVAB, F. and ASATOOR, A.M. (1970). *GUT*, **11**, 373
NEAME, K.D. and WISEMAN, G. (1956). *J. Physiol.*, **133**, 39
NEAME, K.D. and WISEMAN, G. (1957). *J. Physiol.*, **135**, 442
NELLANS, H.N., FRIZZELL, R.A. and SCHULTZ, S.G. (1973). *Am. J. Physiol.*, **225**, 467
NELLANS, H.N., FRIZZELL, R.A. and SCHULTZ, S.G. (1974). *Am. J. Physiol.*, **226**, 1131
NEWEY, H. and SMYTH, D.H. (1959). *J. Physiol.*, **145**, 48
NEWEY, H. and SMYTH, D.H. (1960). *J. Physiol.*, **152**, 367
NEWEY, H. and SMYTH, D.H. (1962). *J. Physiol.*, **164**, 527
NEWEY, H. and SMYTH, D.H. (1964). *J. Physiol.*, **170**, 328
PETERS, T.J. (1970). *Biochem. J.*, **120**, 195
PETERS, T.J. (1973). *Clin. Sci. Mol. Med.*, **45**, 803
PETERS, T.J., DONLON, J. and FOTTRELL, P.F. (1972). in *Glaxo Symposium on Transport across the intestine.* Eds. W.L. Burland and P.D. Samuel, p.153. Churchill Livingstone, Edinburgh and London
PETERSON, S.C., GOLDNER, A.M. and CURRAN, P.F. (1970). *Am. J. Physiol.*, **219**, 1027
RAMASWAMY, K. and RADHAKRISHNAN, A.N. (1970). *Indian J. Biochem.*, **7**, 50
RANDALL, H.G. and EVERED, D.F. (1964). *Biochim. biophys. Acta*, **93**, 98
REISER, S. and CHRISTIANSEN, P.A. (1969). *Biochim. biophys. Acta*, **183**, 611
REISER, S. and CHRISTIANSEN, P.A. (1971a). *Biochim. biophys. Acta*, **225**, 123
REISER, S. and CHRISTIANSEN, P.A. (1971b). *Biochim. biophys. Acta*, **233**, 480
REISER, S. and CHRISTIANSEN, P.A. (1971c). *Biochim. biophys. Acta*, **241**, 102
REISER, S. and CHRISTIANSEN, P.A. (1972). *Biochim. biophys. Acta*, **266**, 217
REISER, S. and CHRISTIANSEN, P.A. (1973a). *Biochim. biophys. Acta*, **307**, 212
REISER, S. and CHRISTIANSEN, P.A. (1973b). *Biochim. biophys. Acta*, **307**, 223
REISER, S. and CHRISTIANSEN, P.A. (1973c). *Life Sciences*, **12**, part II, 25
ROBINSON, J.W.L. (1966). in *Certain aspects of intestinal amino-acid absorption.* Verlag der Odenwald, H. Frotscher KG, Darmstadt.
ROBINSON, J.W.L. (1968). *Europ. J. Biochem.*, **7**, 78
ROBINSON, J.W.L. and FELBER, J.P. (1964). *Gastroenterology*, **101**, 330
ROBINSON, J.W.L. and FELBER, J.P. (1965). *Biochem. Z.*, **343**, 1
RUBINO, A., FIELD, M. and SCHWACHMAN, H. (1971). *J. biol. Chem.*, **246**, 3542
SCHAEFFER, J.F., PRESTON, R.L. and CURRAN, P.F. (1973). *J. gen. Physiol.*, **62**, 131
SCHULTZ, S.G. and CURRAN, P.F. (1970). *Physiol. Rev.*, **50**, 637
SCHULTZ, S.G., CURRAN, P.F., CHEZ, R.A. and FUISZ, R.E. (1967). *J. gen. Physiol.*, **50**, 1241

SCHULTZ, S.G., FUISZ, R.E. and CURRAN, P.F. (1966). *J. gen. Physiol.*, **49**, 849

SCHULTZ, S.G., YU-TU, L., ALVAREZ, O.O. and CURRAN, P.F. (1970). *J. gen. Physiol.*, **56**, 621

SCHULTZ, S.G., YU-TU, L. and STRECKER, C.K. (1972). *Biochim. biophys. Acta*, **288**, 367

SCHULTZ, S.G. and ZALUSKY, R. (1964). *J. gen. Physiol.*, **47**, 567

SEMENZA, G. (1970). *FEBS Symposium*, **20**, 117

SEMENZA, G. (1971). *Biochim. biophys. Acta*, **241**, 637

SHISHOVA, O.A. (1956). *Biochemistry*, **21**, 105

SPENCER, R.P., BOW, T.M. and MARKULIS, M.A. (1962). *Am. J. Physiol.*, **202**, 171

SPENCER, R.P. and BRODY, K.R. (1964). *Biochim. biophys. Acta*, **88**, 400

SPENCER, R.P., WEINSTEIN, J., SUSSMAN, A., BOW, T.M. and MARKULIS, M.A. (1962). *Am. J. Physiol.*, **203**, 634

STEIN, W.D. (1967). in *The movement of molecules across cell membranes*, p.156. Academic Press, New York and London

STIRLING, C.E. (1972). *J. Cell Biol.*, **53**, 704

TAYLOR, A.E., WRIGHT, E.M., SCHULTZ, S.G. and CURRAN, P.F. (1968). *Am. J. Physiol.*, **214**, 836

TUCKER, A.M. and KIMMICH, G.A. (1973). *J. Membr. Biol.*, **12**, 1

UGOLEV, A.M. (1965). *Physiol. Rev.*, **45**, 555

UGOLEV, A.M. and De LAEY, P. (1973). *Biochim. biophys. Acta*, **300**, 105

USSING, H.H. and ZERAHN, K. (1951). *Acta Physiol. Scand.*, **23**, 110

WEBSTER, H.L. and HARRISON, D.D. (1969). *Expl Cell Res.*, **56**, 245

WILBRANDT, W. and ROSENBERG, T. (1961). *Pharmacol. Rev.*, **13**, 109

WILSON, T.H. (1962). in *Intestinal Absorption*. W.B. Saunders Company, Philadelphia

WILSON, T.H., LIN, E.C.C., LANDAU, B.R. and JORGENSEN, C.R. (1960). *Fed. Proc.*, **19**, 870

WISEMAN, G. (1956). *J. Physiol.*, **133**, 626

WISEMAN, G. (1968). in *Handbook of Physiology*, Ed. C.F. Code. Section 6, vol.3, p.1277. Am. Physiol. Soc., Washington, D.C.

6

PROTEIN DIGESTION AND ABSORPTION

A. RERAT, T. CORRING and J.P. LAPLACE
Centre National de Recherches Zootechniques, Jouy-en-Josas, France

Introduction

The theme of this report has already been studied in a number of reviews in the last few years (Snook, 1973; Gray and Cooper, 1971; Rogers and Harper, 1966; Crane, 1969; Cuthbertson and Tilstone, 1972; Fisher, 1967a,b; Erbersdobler, 1973; Porter and Rolls, 1971; Nasset, 1972; Salter, 1973; Fauconneau and Michel, 1970). On account of the magnitude of the subject, it is quite out of the question to examine it in detail, but three main points are investigated considering in particular the most recent research data concerning:
(1) The role of gastrointestinal transit in the digestive phenomena.
(2) The sequence of the enzymatic hydrolyses.
(3) The absorption phenomena.

Gastrointestinal Transit

The pattern (duration - volume) of the digestive transit determines the contact of dietary substances with the digestive enzymes and the absorptive area. The role of gastric emptying is of primary importance since this mechanism controls, for the main part, the passage of digesta to the small intestine. This is particularly true regarding the proximal segments, which are important in the absorption of the products of protein digestion. The concept that the main function of the stomach is that of storage is too restrictive, as the mechanism is complex and submitted to many controls.

Studies on gastric emptying in man have been made by various authors (Hunt and Spurrel, 1951; Hopkins, 1966; Brömster, Carlberger and Lundh, 1968; Stordy, Greig and Bogoch, 1969; Coates *et al.*, 1973). According to Johansson *et al.* (1972), stomach emptying of proteins takes place regularly at a rate of 1000 mg/20 min, the amount of protein emptied by the stomach exceeding by 3.5 g the amount present in the test meal (5.0 g). In addition the concentrations of protein are three to four times lower in the jejunum than in the stomach. In the rat, several studies of gastric emptying and transit in whole or part of the intestine have been reported (Reynell and Spray, 1956; Derblom,

Johansson and Nylander, 1966; Francois, Compere and Rondia, 1968; Sikov, Thomas and Malhum, 1969). Similar data, although limited, are available for mice (Dawson, 1972) or guinea pigs (Beccari, 1957). For these two species, however, the physiological aspects were neglected because of the nutritional objective of the experiment, and in particular the experimental conditions (liquid test meals or synthetic diets) which were very different than for normal-feeding animals kept in artificial conditions with suppression of coprophagy. The effects related to fasting or digestive repletion (Wiepkema *et al.*, 1972; Poulakos and Kent, 1973) are rather important. Fasting prior to the experiments induces gastric emptying and accelerates digestive transit. Research on pigs, however, has the advantage that it can be carried out with natural balanced diets and normal feeding rhythms and intake levels, not to be compared with 'test-meals'. In addition, in this species, studies on gastroduodenal motility have been carried out and have provided the best present knowledge on gastric emptying and its control mechanisms.

The systematic study of the gastroduodenal emptying is facilitated by the mathematical expression obtained in growing pigs. Thus, the studies of Auffray (1965) and Auffray, Martinet and Rerat (1967) based on direct observations in fistulated pigs and on the study of the mechanical phenomena, followed by those of Laplace and Tomassone (1970) and Laplace (1971; 1972; 1974a) concerning investigations in fistulated pigs and examination of the electromyographic activities of the gastroduodenal area, led to conclusions that the gastric emptying pattern could be represented as a third degree function (*Figure 6.1*).

$$y = b_0 + b_1 x + b_2 x^2 + b_3 x^3$$

By means of this equation it is possible to study the factors controlling the emptying by variance analysis applied to the polynomial coefficients. In this way, b_0 accounts for the previous state of the digestive system and, in particular, the intestinal emptiness; b_1 indicates the influence of the degree of repletion of the stomach (volume of the meal) and b_2 expresses the intervention of reflex inhibitions of intestinal origin as a response to all mechanical or chemical stimulations. The direct experimental determination of these mathematically calculable mechanisms has been made for most of the factors involved, and is reported in the papers previously mentioned.

The stomach emptying of proteins as well as that of other constituents of the diet can be considerably modified by various factors, such as the volume of the meal (Hunt and McDonald, 1954) osmotic pressure (Hunt and Pathak, 1960; Hunt, 1961) density of the dietary components (Chang, McKenna and Beck, 1968) acidity of endogenous or exogenous origin (Hunt, 1971), pH of the intestinal contents (Borgström *et al.*, 1957; Rhodes, Goodall and Apsimon, 1966; Rune, 1968) and acid-base balance (Rune, 1972). Besides these physical factors, the biochemical factors related to the composition of the diet are essential although sometimes dependent upon a physical influence (osmolarity for example).

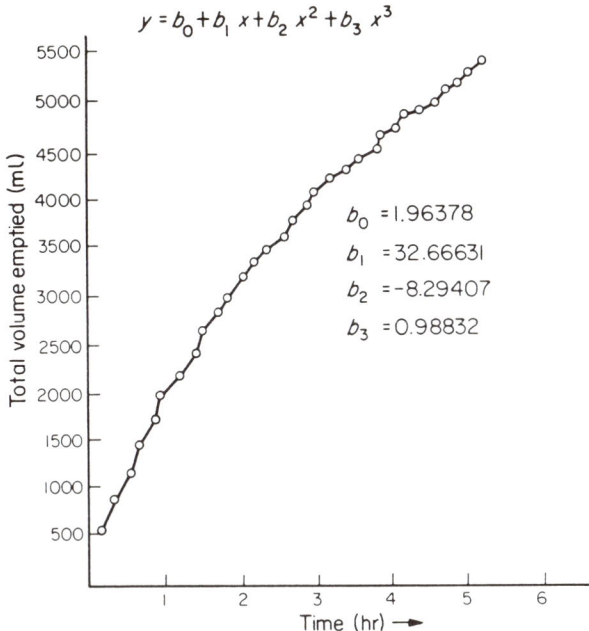

Figure 6.1 Computer simulation of gastric emptying of a natural meal (600 g flour + 1800 ml water) in the pig, according to a third degree polynomial adjustment. (Re-entrant duodenal fistula). This computer-traced graph from the calculated values is almost identical to that traced from the measured values (from Laplace and Tomassone, 1970)

The influence of the dietary protein is debatable, and it seems to be related to the amount supplied. Thus, according to Peraino *et al.* (1959), casein produced an inhibitory effect on gastric emptying, a fact which is not in agreement with the conclusions drawn by Rosenthal and Nasset (1958) and Buraczewski *et al.* (1971), who showed casein had no effect. This contradiction can be explained by differences in the food intake level, the enhancement of which would lead to appearance of the inhibitory effect. In addition, the nature of the proteins affects their gastric emptying (Canzler and Glatzel, 1966a,b; Zebrowska, 1968; 1973a; Porter and Rolls, 1971), but this phenomenon, observed when the food is offered for the first time (Chen, Rogers and Harper, 1962) disappears when the animals have become adapted (Rogers *et al.*, 1960). It must be emphasised that it is very difficult to interpret data collected from analysis of the intestinal contents, on account of the large variability in the digestibility and absorption rate of the proteins considered, as well as their poor palatability.

The influence of technological treatments has also been discussed, this action being considered as positive (Erbersdobler, 1973) or non-existent (Zebrowska, 1968) for moderate heat treatments. Finally, the presence of free amino acids induces different effects according to their concentration. When these substances are used as dietary supplements, the gastric emptying of proteins is not modified (Rogers, Spolter and Harper, 1962; Rolls, Porter and Westgarth, 1972). Meals containing large amounts of amino acids, however, are emptied more slowly (Buraczewski *et al.*, in Rolls, Porter and Westgarth, 1972). When added to a solution of phenol red, at a total concentration of 0.2 M, some amino acids (glycine, β-alanine, 1-leucine, 1-lysine and DL-methionine) significantly delay the gastric emptying. This inhibition is increased as the molar concentration increases (Cooke and Moulang, 1972).

In the case of the pig, Zebrowska and Buraczewska (1972a) conclude that stomach emptying is increased with protein-free diets, whereas Rerat and Lougnon (1963) did not observe any difference between protein and protein-free diets during a period of six hours; after that time, the residual gastric contents were larger in the case of protein-free diets. Furthermore, the nature of the protein (casein, zein) leads to gastric emptying curves, the features of which are always characteristic of the feed (Auffray, 1965).

The inhibitory effect of carbohydrates is well known (Hunt, 1960; Elias *et al.*, 1968; Mehnert and Förster, 1968; Husband, Husband and Mallinson, 1970). In man given almost normal mixed meals, Johansson (1973) and Lagerlöf, Johansson and Ekelund (1973) observed that after addition of glucose, the emptying of protein and fat is reduced during the first phase (important emptying of carbohydrates); during the first 20 minutes, 1.38 g of proteins are emptied versus 4.15 g in the absence of glucose. This inhibitory effect has also been found in the rat by Mehnert and Förster (1968) and Reynell and Spray (1956). It must be noted that the rate of gastric emptying conditions the rate of absorption of a substance, but the latter, in return, may also affect the emptying. This is the case for glucose, since the absorption reduces the level of this substance in the intestine, consequently the level of inhibition is related to the resulting osmotic pressure. Rosenthal and Nasset (1958) have emphasised the fact that the gastric emptying (and absorption) of proteins is very different according to the nature of the dietary carbohydrates. This effect of carbohydrates has also been observed in the pig (Buraczewski *et al.*, 1971) and is illustrated in *Figure 6.2*.

The influence of fats is also very important. Their inhibitory effect increases (Hunt and Knox, 1968) or decreases (Pirk and Skala, 1970) with the varying chain length of the fatty acids. The effect of fats on the transit of proteins varied according to the nature of the fat source (Canzler and Glatzel, 1966a,b). Likewise, in the rat, Harkins, Longenecker and Sarett, (1964) demonstrated overall enhancement of gastric emptying when the fat content increases from 4 to 12 per cent, but the emptying of lactose and proteins is only slightly modified.

Some other original data have been obtained with pigs. In contrast with other species, the pH of the emptied gastric contents only plays a minor part in the control of the emptying (Auffray, 1965; Laplace,

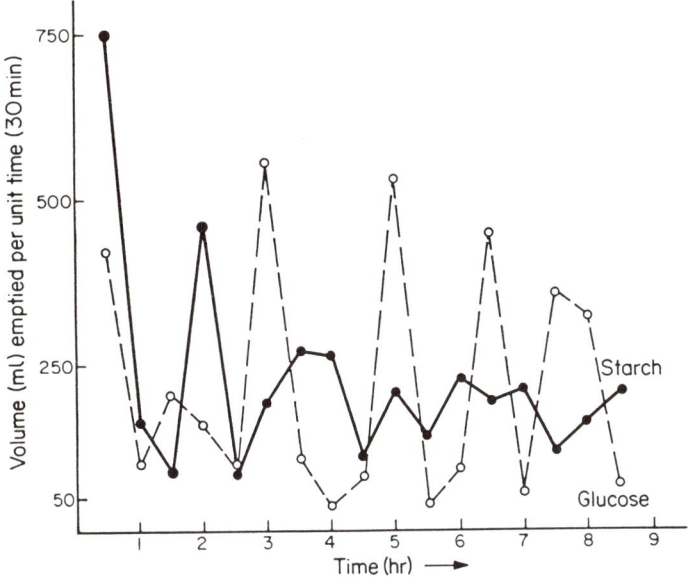

Figure 6.2 Gastric emptying of a meal (12 per cent casein and 65 per cent carbohydrate) according to the carbohydrates supplied: starch or glucose. Digesta collected at the proximal duodenal cannula. The intermittence of gastric emptying of the glucose containing diet reveals the inhibitory effect of glucose (through duodenal osmoreceptors) until it is absorbed (from Auffray, 1965)

1974b). In addition, the influences of duodenal origin seem to proceed from a larger area of small intestine. Thus, at different levels of the jejunum this effect of the intestinal emptiness, versus repletion, arises within the first eight metres of the small intestine and not only from the duodenum (Auffray, 1965). It can be added that the gastric motility which is almost continuous for the first two hours of emptying, is interrupted by rest phases coinciding with the presence in the duodenum of contents, including a large amount of solid material. These phases may be reproduced by the introduction into the duodenum of digesta with a high dry matter content taken from another animal or from the same animal on another occasion. The antagonism of gastric and duodenal motilities is particularly obvious when comparing variable dilutions of the same dietary ration. Thus, a starch easily hydrolysed *in vivo,* such as maize starch, entails a slowing down of the gastric emptying as a result of the osmotic pressure due to too rapid a release of glucose (Laplace, 1971). Finally, with the piglet, gastric emptying and intestinal transit have been described by several authors (Kidder, Manners and McCrea, 1961; Kidder and Manners, 1968; Cuperlovic, Hristic and Cmiljanic, 1972; Hill, Noakes and Lowe, 1969) and in this case the effect of the coagulability of the milk replacer used appears to be particularly important in early weaning. Thus, Seve and Laplace (1974) have shown that one hour after the beginning of a meal, 30 per cent of the nitrogen ingested is recovered in the solid phase of the gastric contents, for a diet in which 48 per cent of the proteins were provided by a concentrate of soluble fish proteins, compared to 70 per cent for a diet based on cow's milk (dense coagulum).

Enzymatic Studies

SEQUENCE OF ENZYMATIC HYDROLYSES

The enzymatic hydrolysis in the stomach is mainly performed by the pepsins, secreted in an inactive form (pepsinogens) and activated by hydrochloric acid. Several pepsins, four to five, have been demonstrated in man and pig (Taylor, 1968; Seigffers, Segal and Miller, 1963) with two pH zones of maximum activity, i.e. 1.6-2.4 and 3.3-4.0 (Taylor, 1959). According to Tang et al. (1959) and Ryle and Porter (1959), there are also other proteinases, the nature of which is different from that of the pepsins. According to the intake level and nature of the protein (Zebrowska, (1973) as well as the nature of the lipids (Ziemlanski, Cieslakowa and Palaszewska, 1972) and carbohydrates Buraczewski et al., 1971) composing the diet; the dietary protein is subjected to an enzymatic hydrolysis, the magnitude of which varies according to the duration of stomach emptying. So, the latter determines the duration of the contact between enzymes and substrates.

The dietary proteins are broken down into soluble proteins and large peptides, the amount of free amino acids being very small (Buraczewski et al., in Porter and Rolls, 1971; Gullberg and Olhagen, 1959; Glass and Ishimori, 1961). This first stage in the digestion of proteins can be considered as a preparatory phase. All the post-gastric enzymatic hydrolyses can be differentiated into intraluminal digestion and membrane and intracellular digestions. The intraluminal digestion is mainly due to the proteolytic enzymes, secreted by the pancreas in an inactive form in the duodenum. The activation of trypsinogen by the intestinal enterokinase produces trypsin which activates the chymotrypsinogens procarboxypeptidases A and B and proelastase, mainly into active enzymes (Gray and Cooper, 1971). The peptidases associated with the outer surface of the epithelial cell of the intestine, as well as those acting inside the cell which are of a different nature (Gray and Cooper, 1971) participate in the membrane and intracellular digestions (Ugolev, 1965; Kim et al., 1971; Heizer, Kerley and Isselbacher, 1972).

After the gastric preparatory stage, the dietary proteins are submitted to a sequence of enzymatic hydrolyses. The nature of the products of this enzymatic digestion is not clearly demonstrated in the different research works and depends, in particular, on the nature of the dietary protein (Zebrowska, 1973). This author found that in the duodenal contents of the pig, the fraction of free amino acids represents about 60 per cent of the amino nitrogen, whereas the peptides constituted 20 per cent. Furthermore a high proportion of soluble proteins have been found in the terminal ileum which could be due to an accumulation of molecules more resistant towards the action of proteolytic enzymes.

In addition, the rapidity of digestion has been emphasised by a great number of authors (Crane and Neuberger, 1960; Goldberg and Guggenheim, 1962). According to Nixon and Mawer (1970b) milk or gelatin are reduced in man within 15 minutes to 30-50 per cent free amino acids and peptides. The notion of rapidity of digestion has to be associated with that of the sites of absorption, i.e. most of the dietary proteins being digested and absorbed in the distal duodenum and proximal jejunum (Zebrowska and Buraczewska, 1972a; Borgström et al., 1957; Schlüssel, 1959; Nixon and Mawer, 1970b).

ROLE OF THE DIFFERENT ENZYMATIC SECRETIONS IN THE DIGESTION OF PROTEIN

Gastric secretion

The observations of Everson (1952), Welbourn and Doggart (1956) and Lundh (1958) have shown in man that nutritional state and growth are generally affected in the gastrectomised subject. In man, the absorption falls to 61.7-73.8 per cent after total gastrectomy, versus 83.1-97.4 per cent in intact subjects (Jeejeebhoy, 1964). In the pig, total gastrectomy produces a lowering of the apparent digestibility of the proteins of 17-18 per cent (Cunningham, 1967).

Pancreatic secretion

Shingleton et al. (1956) and Douglas et al. (1953) have shown that after total pancreatectomy in the dog, nitrogen absorption decreases by 30 per cent one hour after the meal. Pekas, Hays and Thompson, (1964); Anderson and Ash (1971) noticed a decrease in the apparent digestibility of nitrogen of 15-70 per cent after ligature of the pancreatic canal in the pig. Measurement of the α-amino nitrogen level in the portal blood has also been used to point out total enzymatic deficiencies of the pancreas (West, Wilson and Eyles, 1946; Anfanger and Heavenrich, 1949; Christensen and Shwachman, 1949). From the porto-arterial differences in the concentration of α-amino nitrogen and from the measure of the portal blood flow rate, it has been possible to determine the role of pancreatic digestion (Rerat and Corring, unpublished). Three pigs were fitted with pancreatic and duodenal fistulae (Corring, Aumaitre and Rerat, 1972), with an electromagnetic probe to measure the blood flow rate in the portal vein, and two catheters for arterial and venous blood samplings (carotid and portal vein) (Rerat, 1971a). The animals were subjected to several series of assays: re-introduction of pancreatic juice, non-re-introduction of pancreatic juice, re-introduction of a solution of salts ($NaHCO_3$ + NaCl, pH: 8.4) after intake of a protein meal. The α-amino nitrogen absorption shows the specific role of pancreatic enzymes and of the neutralising capacity of the pancreatic juice *(Figures 6.3* and *6.4)*. In the absence of pancreatic secretion 18 hours before the meal, this absorption decreases by about 50 per cent within the post-prandial period. When the intestinal pH, which remains acid because of the deprivation of pancreatic secretion, is neutralised by the solution of salts, the absorption is only 25 per cent about eight hours after the meal, probably because of a reduced action of pepsin in the duodenum.

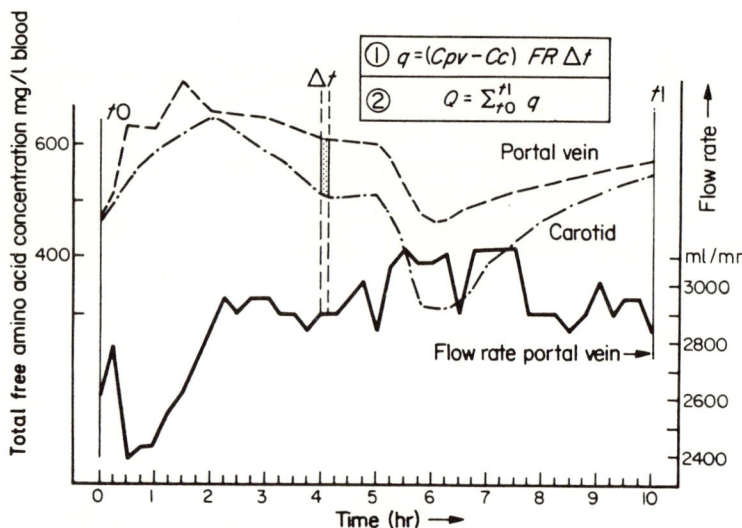

Figure 6.3 Method for calculating the quantity of nutrient absorbed during the postprandial period
q: quantity absorbed during the time Δt
C_{pv}: concentration of the nutrient in the portal vein
C_c: concentration of the nutrient in the carotid artery
FR: flow rate

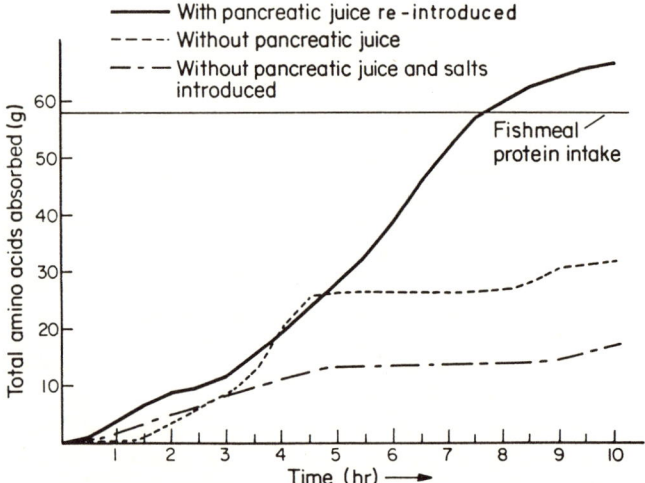

Figure 6.4 Absorption of aminoacids (g) during 10 hours after the ingestion of 500 g of a semisynthetic meal (containing 58 g fish meal protein). Pig fitted with a permanent cannula of the pancreatic duct and of the duodenum: with or without re-introduction of the pancreatic juice (Rerat and Corring, unpublished)

Intestinal secretion

Little data is available about the role of the intestinal secretion (Gardner, Brown and Laster, 1970). Sadikali (1971) emphasises the importance of malabsorption in the case of enzymatic deficiencies resulting in delayed absorption of amino acids. Buraczewska and Zebrowska (1972) have shown in the pig, that in the absence of pepsins and pancreatic enzymes, there is only a small or no release of some amino acids from casein introduced directly into the proximal part of the intestine. A maximum absorption of only 20 per cent has been recorded for arginine and serine.

ENZYMATIC ADAPTATION

A great number of studies have been carried out in order to determine the adaptation of digestive enzymes, mainly those of the exocrine pancreas, to the diet. In the stomach and the intestinal mucosa, the data available are less numerous and it is important to distinguish gastric or intestinal enzymatic variation due to changes depending upon enzyme synthesis or protecting effect of the dietary protein (Snook and Meyer, 1964; Buraczewski, 1966).

Adaptation of gastric secretion

Snook and Meyer (1964) in the rat and Blagivodov, Pomeltsov and Shatalov, (1972) in the dog, have shown an increase in the peptic activity either after enhancement of the protein level of the diet, or after exclusion of the exocrine pancreatic function. Storozuk (1968) has also pointed out a variation in the gastric hydrolysis depending on the protein ingested.

Adaptation of the intestinal secretion

The enzymes of the intestinal mucosa do not seem to be much affected by a modification of the diet (Dargel and Hock, 1971; Rustow and Hock, 1970). Scharrer and Zucker (1967) determined an increase in the absorption of amino acids in the rat, when the casein content of the diet increased from 13 to 88 per cent. However, these few data do not permit a conclusion on enzymatic adaptation, more especially as the pancreatic enzymes have not been considered.

Adaptation of the pancreatic secretion

All studies made in this field show that a diet with a high level of protein produces an enhancement of the proteolytic activity of the exocrine pancreas (Grossman, Greengard and Ivy, 1942, Desnuelle, Reboud and Ben Abdeljlil, 1962; Howard and Yudkin, 1963; Snook

106 *Digestion and absorption*

and Meyer, 1964). Corring and Saucier (1972) demonstrated in fistulated pigs that the activity of chymotrypsin (*Figure 6.5*) varies greatly according to the protein content of the diet (increase of 250 per cent in the case of 10 to 30 per cent change of the protein level in the diet.) The variation of trypsin is not as high (20 per cent). The adaptation is complete after 5-7 days and the chymotryptic variation regularly occurs after only 48 hours.

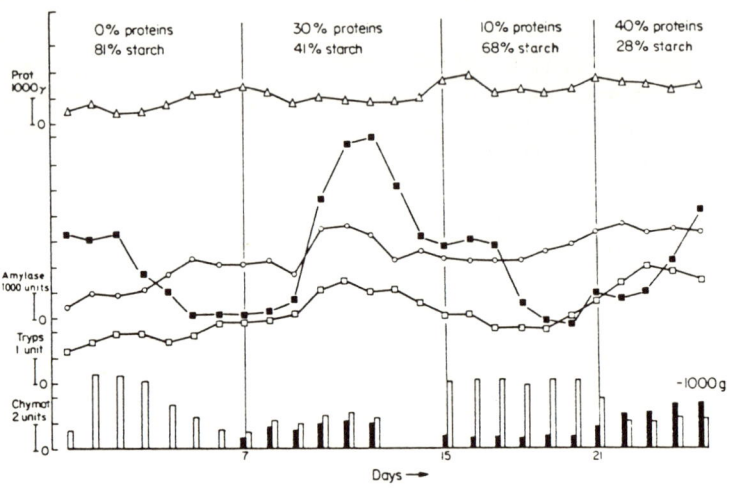

Figure 6.5 *Pancreatic enzyme adaptation to the protein content of the diet in the pig. Mean specific enzymatic activities (per mg protein) and total protein mean concentration of the daily collected pancreatic juice (Corring and Saucier, 1972)*

■———■ chymotrypsin
□———□ trypsin
○———○ amylase
△———△ total protein

ingested starch

ingested protein

The adaptation of pancreatic proteolytic enzymes has also been observed to depend upon the nature of the dietary protein (Snook, 1973; Storozuk, 1970).

The mechanisms of this pancreatic adaptation are still unknown. The works of Green and Lyman (1972) and of Corring (1973) suggest that this phenomenon could be based upon a regulation by feedback control of the proteolytic enzymes on the pancreatic secretion. The dietary proteins could act by binding the enzymes in the intestinal lumen and by blocking this regulatory mechanism, thus entailing an increase in the pancreatic secretion. The larger the amount of dietary protein, the higher the increase in the pancreatic secretion of enzymes.

Absorption of Protein Digestion Products

Two main questions can be asked about absorption:
(1) What is the nature, distribution and origin of the chemical substances present in the small intestine?
(2) Are the products which disappear from the digestive tract during digestion, identical with those which appear in the organs of the animal?

To answer the latter question, it is necessary to know, on the one hand, the nature of the products which disappear and, on the other, the nature of those which appear in the portal vein and the intestinal lymph.

COMPOSITION AND ORIGIN OF DIGESTA

The nitrogen fractions of digesta

After a protein meal, an increase in the amounts of nitrogen present in the small intestine is observed. This accumulation of nitrogen depends on the nature of the protein and is higher for insoluble protein (zein) or for a partly enzyme-resistant protein (gelatin) than for casein or mixtures of amino acids (Chen, Rogers and Harper, 1962; Geiger, 1951); it also depends on the protein level and any increases in the latter (Twombly and Meyer, 1961; Zebrowska and Buraczewska, 1972c).

Examination of the various nitrogen fractions shows that the digestive contents of rats receiving diets containing an easily digestible protein (casein) are analogous with those of animals submitted to fasting or fed a protein-free diet (Buraczewski *et al.*, 1971): the concentration of peptides is low, whereas that of the protein or a-amino fractions is high (Rolls, Porter and Westgarth, 1972). On the other hand, after ingestion of less easily digestible proteins, the levels of peptides recorded are higher (Rolls, Porter and Westgarth, 1972; Chen, Rogers and Harper, 1962; Zebrowska, 1968). These peptides often contain more lysine and glutamic acid (as in the case of heated proteins) than the initial protein (Buraczewski, 1972).

There is a similarity between the amino acid composition of duodenal contents and that of the dietary proteins ingested (Zebrowska, 1973b; Tkachev and Pakhno, 1970; Crompton and Nesheim, 1969) which is all the more marked as the protein level is increased (Zebrowska and Buraczewska, 1972b). On the other hand, the amino acid content of digesta from the terminal ileum corresponds more closely to that when protein-free diets are fed (Zebrowska and Buraczewska, 1972a; Zebrowska, 1973b).

Proportion of endogenous nitrogen

At the different levels of the digestive tract the contents are composed not only of non-absorbed residues from the food ingested, but also of products of endogenous origin (digestive enzymes, mucoproteins, desquamations, urea, amino acids produced by the cellular catabolism, and albumin) which dilute the exogenous residues.

From a nutritional point of view, various problems concern the endogenous nitrogen, i.e. quantity secreted, its composition and consequently eventual utilisation as a complement in the case of an imbalanced supply of exogenous nitrogen, its digestibility, chronology of absorption and recycling. Because of the diversity of methods used (Snook, 1973), with varying accuracy according to species, there are two sorts of opinions about the volume of endogenous nitrogen. For some authors (Nasset and Ju, 1961; Twombly and Meyer, 1961), endogenous nitrogen represents, at certain periods after the meal and in certain sites of the digestive tract, a large supply, diluting up to 7-9 times the residual exogenous nitrogen in the dog and the rat: this nitrogenous matter might temporarily buffer the variations in the composition of the diet. Thus, in the dog, the contents of free amino acids at the level of the duodenum (40 per cent of total nitrogen), are very different from the amino acid composition of the proteins found in the diet. The composition of these free amino acids in the duodenum remains rather constant whatever the protein or amino acid mixture ingested (Nasset, Schwartz and Weiss, 1955; Nasset, 1962). This effect of 'levelling' has also been noticed to a lesser extent in man, (Olmsted, Nasset and Kelly, 1966) and in the rat (Bergen and Purser, 1968; Gent and Creamer, 1972).

In contrast, more recent experiments provide conclusive arguments which demonstrate that a rather large fraction of the protein found in the intestine after a meal is of exogenous origin: 25 per cent in the rat (Ochoa-Solano and Gitler, 1968) about 50 per cent in the duck (Crompton and Nesheim, 1969), 47-87 per cent in man (Nixon and Mawer, 1970b), and 20-80 per cent in the pig (Horszczaruk, 1971a,b). The total free amino acid composition of the duodenal contents is much affected by the diet and the dilution of exogenous nitrogen should therefore not be as important as suggested by Nasset. In any case, the percentages calculated in this manner depend both on the digestibility of the exogenous nitrogen and on the time elapsed after the meal, factors that determine the level of residual exogenous nitrogen. The example given in *Table 6.1* (Rerat and Lougnon, 1963) shows, indeed, that the percentage of exogenous nitrogen calculated varies according to the degree of digestion of the meal.

However that may be, these divergencies are even more obvious when calculating the endogenous supplies. *Table 6.2* indicates, as an example, the estimations made by Nasset (1965) and those of Fauconneau and Michel (1970) concerning the production of endogenous nitrogen per day in a man weighing 70 kg.

These estimations vary from one to five according to the hypotheses used for these calculations. Most of the authors agree, however, on

Table 6.1 Distribution of nitrogenous contents at different levels of the gut. Variation with time (Rerat and Lougnon, 1963) elapsed since the meal (880 g dry matter; 10 per cent fish meal protein, 5 per cent fibre and 3 per cent mineral and vitamin concentrate, the rest being maize starch; the protein-free meal is the same except for fish meal proteins)

Time elapsed since the meal (h)	1	2	3	4	5	7	8
Stomach							
Protein diet	11.6	8.5	7.4	6.2	6.1	2.4	1.5
(n)	(1)	(1)	(6)	(2)	(3)	(1)	(2)
Protein free diet	1.9	-	1.4	1.1	1.0	0.9	1.7
	(2)	(0)	(6)	(3)	(3)	(1)	(2)
Apparent percentage of endogenous nitrogen	16	-	19	18	16	38	>100
Small intestine							
Protein diet	4.9	4.6	7.2	7.5	4.1	2.8	2.5
Protein free diet	5.1	-	4.7	4.7	3.4	2.9	2.1
Apparent percentage of endogenous nitrogen	>100	-	66	62	82	>100	85

n = number of animals

Table 6.2 Estimated amounts of crude protein (g) secreted in the gut (70 kg man)

	Nasset (1965)	Fauconneau and Michel (1970)
Saliva	2-12	3
Gastric juice	8-18	5
Bile	2-3	1
Pancreatic juice	12-30	8
Intestinal juice	40-200	--
Mucosal desquamation	77-91	50

the fact that the amounts are rather low: 70 g for a 70 kg man (Munro, 1966); about 40 g for a 50 kg pig (Zebrowska and Buraczewska, 1972a,b,c). The most important source of divergence concerns the magnitude of the whole of the intestinal secretions. Apart from the intestinal juice, the digestive secretions can easily be collected quantitatively and their nitrogen content can be determined. As an example in a 50 kg pig, the amount of protein produced by pancreatic secretion ranges about a mean value of 18 g, and the resulting supply of amino acids has also been calculated (*Tables 6.3* and *6.4*) (Corring and Jung, 1972; Corring, unpublished data).

The estimations of the loss of nitrogen corresponding to the desquamations are very variable (*Table 6.2*). According to Da Costa (1971), this loss is higher (180 g/day) than that reported in the table.

Table 6.3 Endogenous protein excreted by the pancreatic juice of the pig (45 kg mean live weight) within 24 hours (Corring, unpublished data)

Pig no.	Mean per pig (g)	Variation coefficient (%)
1 (20)[1]	22.8 ± 1.3	25.8
2 (13)	18.0 ± 1.9	37.9
3 (5)	11.8 ± 0.6	11.3
4 (9)	18.9 ± 1.6	25.0
5 (10)	16.8 ± 1.8	35.0
6 (17)	17.6 ± 0.9	21.2
7 (11)	18.5 ± 0.7	12.8
All pigs	General mean (g)	Variation coefficient (%)
(86)	18.6 ± 0.6	30.6

[1] Number of data

Table 6.4 Amounts of amino acids (g) secreted in the pancreatic juice of the pig (mean live weight: 45 kg) within 24 hours (Corring, unpublished data)

Lysine	0.76	Cystine	0.46
Histidine	0.35	Valine	0.89
Aspartic acid	1.55	Methionine	0.16
Threonine	0.81	Isoleucine	0.70
Serine	1.09	Leucine	1.11
Glutamic acid	1.41	Tyrosine	0.76
Proline	0.74	Phenylalanine	0.64
Glycine	0.81	Arginine	0.70
Alanine	0.79		

The proteins secreted and catabolised in the digestive tract represent another source of endogenous nitrogen (Jeffries and Sleisenger, 1968). Less than 10 per cent of the catabolised plasma proteins are broken down in the digestive tract (Freeman, 1964; Waldman et al., 1967).
As the catabolism of albumin is estimated to be 180 mg/kg live weight, it is apparent that the addition of endogenous nitrogen resulting from this catabolism is very small.

These quantities of nitrogen are large whatever the mode of estimation. Furthermore they vary according to feeding; but the effect is difficult to estimate except for special cases (protein-free diets, raw soya bean, etc.).

DISAPPEARANCE OF THE PRODUCTS OF DIGESTION

Total absorption can be estimated by comparison of the amount excreted in the faeces with that of the food ingested. It can also be studied by means of more elaborate techniques *in vivo* (e.g. intubation allowing recovery of the contents, perfusion of intestinal loops between two cannulae) or *in vitro* (e.g. everted intestinal segments, according to the technique of Wilson and Wiseman, 1954; intestinal rings) using different types of meal (proteins, protein hydrolysates or mixtures of synthetic amino acids) and allowing the determination of sites of absorption, the forms of absorption and mechanisms of transport.

Digestibility and availability

Because of the endogenous supply, and also because of the modifications of the undigestible fraction of the feeds occurring during their transit, comparison of nitrogen excreted with nitrogen ingested allows only an approximate estimation of the nature and amount of the dietary nutrients disappearing during the course of the digestive transit. Therefore it is not possible by this method to give an accurate estimation either of the quantity and nature of the nutrients really absorbed or of the chronology of their absorption.

Primarily it is necessary to define the losses of endogenous origin, excluding oxidative catabolism. Since the first studies of Kuiken and Lyman (1948) in the rat and Dammers (1964) in the pig, a great number of investigations have been made and most of them are cited in the reviews of Meade (1972) and Rerat (1971b). Among these studies, let us mention those by scientists who have calculated the digestive endogenous loss of all amino acids in three animal species (rat, pig, chicken) (Meier, Poppe and Weisemuller, 1970a,b,c; Poppe, Meier and Uhlemann, 1971; Uhlemann and Poppe, 1970; Kristen, Poppe and Meier, 1970). These authors have determined the true digestibility-coefficient of the dietary amino acids, thus providing comparative data for the different species (Poppe and Meier, 1971a) as well as a description of a simplified method (Poppe and Meier, 1971b). It can be noted that the differences between the rat and the pig are smaller than those between these two species and the chicken, the pig showing the highest absorption rate. This type of method has been used to measure the availability of lysine (Booth and Carpenter, 1973).

Kinetics (in vivo) *of the disappearance of the products of digestion*

The composition of intestinal contents varies with time and site after a protein-containing or protein-free meal.

The arrival of material coming from the stomach after a protein meal results in a rapid increase (1 hour) in the amount of nitrogen in the small intestine (Twombly and Meyer, 1961; Mettrick, 1970; Zebrowska, 1968; Rogers *et al.*, 1960) followed by a more or less early decrease of variable duration according to the protein ingested and its level in the diet (Twombly and Meyer, 1961). In the piglet, the disappearance of dietary proteins is at a level of about 45 per cent two hours after the meal (Padalikova, 1966). In the rat, a second period of accumulation is observed eight hours after the meal, which is probably of endogenous origin (Twombly and Meyer, 1961).

Moreover, the composition of the intestinal contents are modified as they are pushed towards the distal extremity of the small intestine (Crompton and Nesheim, 1969), and the composition of the contents in each intestinal segment varies with time and according to the protein ingested (Mettrick, 1970; Nixon and Mawer, 1970a,b). The fraction of undigested exogenous proteins is accumulated in the distal jejunum and in the ileum at the same time as most of the endogenous proteins and the tryptic activity are present in the small intestine (Ochoa-Solano and Gitler, 1968; Crompton and Nesheim, 1969; Zebrowska and Buraczewska, 1972a,b,c; Buraczewski, 1970). It is to be noted that a certain number of amino acids (iminoacids, dicarboxylic acids, threonine, serine, glycine) are present as peptides, in a proportion of 90 per cent till their absorption (Nixon and Mawer, 1970a,b).

Mechanisms of absorption

Forms of absorption. It is well known now that the products of digestion leave the intestinal lumen either in the form of free amino acids or in the form of oligopeptides (Wiseman, 1968; Saunders and Isselbacher, 1966; Matthews, 1972).

Free Amino Acids: Most L-amino acids pass through cellular membranes of the intestine against a concentration gradient, a fact which indicates that an active transport takes place (Wiseman, 1951, 1953; Agar, Hird and Sidhu, 1953). Some amino acids of the D form are also transported actively (De La Noue, Newey and Smyth, 1971). Each amino acid has a characteristic absorption rate (Wiseman, 1956) but this rate varies with the presence of other amino acids, proving the existence of competition within the transport mechanisms (Wiseman, 1955; Agar, Hird and Sidhu, 1956; Hagihira *et al.* (1960). Thus, the amino acids may be classified in various groups, according to the system of transport they adopt:

(1) Group of neutral amino acids (mono-amino-mono-carboxylic acids).
(2) Group of iminoacids (proline, hydroxyproline), glycine and sarcosine.
(3) Group of dibasic acids (lysine, arginine, ornithine) and cystine.
(4) Group of dicarboxylic acids (aspartic acid, glutamic acid).

Besides the fact that some amino acids may be transported by several carriers, interactions exist between the members of the different groups, leading to stimulation or inhibition of absorption. These interactions are due to a competition for energy or to configurational changes in an adjacent carrier. These competition phenomena are particularly interesting when studied *in vivo* on complex mixtures of amino acids (Orten, 1959; Delhumeau, Velez Pratt and Gitler, 1962; Anderson and Linkswiler, 1969; Adibi and Gray, 1967; Gitler and Martinez-Rojas, 1964; Scharrer, 1971; Scharrer and Brüggemann, 1971; Bergen, 1969; Lazarov and Ivanov, 1973; Klemina, Sisova and Kasatockin, 1965). According to Adibi, Gray and Menden (1967), the absolute quantity of each amino acid absorbed from 'simulated' hydrolysates mainly depends on its molar concentration and on its characteristic absorption rate. In an equimolar mixture of amino acids, relative absorption sequences of the acids are always the same whatever the concentration. The most rapidly absorbed amino acids are: methionine, isoleucine, leucine and valine; then proline, arginine, alanine and phenylalanine. Lysine and dicarboxylic acids show the slowest absorption. The absorption of lysine is probably inhibited by that of arginine, and that of arginine accelerated by the absorption of other amino acids. According to Gitler (1964), the competition during absorption may induce a modification in the composition of the amino acid mixture offered to the organism from that of the digestive contents. However, the nutritional consequences of such a phenomenon have to be minimised. Thus, Christensen (1963) emphasises that the competition phenomena may be compensated for by the fact that as the absorption takes place all along the digestive tract, the amino acids presenting a high affinity for the systems of transport, can be absorbed in the proximal intestine and those presenting a lower affinity can be absorbed in a more distal part after a very short lapse of time.

It is known that different *oligopeptides* enter the mucosal cells, where they are hydrolysed, before being transported to the blood in the form of amino acids (Wiggans and Johnston, 1959; Newey and Smyth, 1959, 1962). This mode of absorption is more rapid than from equimolar mixtures of free amino acids. This is true not only for oligopeptides formed by the assembling of two, three or four molecules of glycine (Craft and Matthews, 1968; Craft *et al.*, 1968; Cook, 1972), generally not occurring during digestion, but also for mixed peptides formed by the assembling of various neutral, basic and acid amino acids (Matthews, Crampton and Lis, 1968; Matthews *et al.*, 1969; Craft *et al.*, 1969; Hellier *et al.*, 1972; Cheng *et al.*, 1971; Adibi, 1971; Lis, Crampton and Matthews, 1972a). In this case, the competition between amino acids is partially or completely avoided (Matthews *et al.*, 1969). These phenomena are certainly of great importance and can be illustrated by two facts: tryptic hydrolysates of various proteins, composed of small peptides, are absorbed more rapidly than equivalent mixtures of amino acids (Crampton *et al.*, 1971). Furthermore, the sites where the peptides are mostly absorbed are located in the proximal intestine (Crampton, Lis and Matthews, 1973). This fact explains why proteins are digested more rapidly than could be expected if they had to undergo a complete hydrolysis in the intestine (Fisher, 1967a,b). It is possible that the mechanism responsible for the entry of peptides into the mucosa is independent of that of the free amino acids (Asatoor *et al.*, 1970; Rubino, Field and Schwackman, 1971; Lis, Crampton and Matthews, 1972a). It is still not known if they are hydrolysed by the enzymes of the brushborder (Ugolev, 1966) the free amino acids being attached simultaneously to the transport sites, or if their entering into the intestinal mucosal cell is followed by a hydrolysis inside the latter (Newey and Smyth, 1962; Peters and McMahon, 1970).

Nutritional factors: Various nutritional factors are liable to change the absorption of amino acids *in vitro* and *in situ*, i.e. the presence or absence of vitamin B6 (Huang, 1961; Akedo *et al.,* 1960; Asatoor *et al.,* 1972; Munck, 1965), nature of the dietary sugars (Annegers, 1966; Saunders and Isselbacher, 1965; Newey, Sanford and Smyth, 1970; Cook, 1972), protein level (Scharrer and Brüggemann, 1971; Lis, Matthews and Crampton, 1972b; Nakamura, Yasumoto and Mitsuda, 1972; Kirsch, Saunders and Brock, 1968; Neale, 1971).

PRODUCTS OF DIGESTION APPEARING IN THE ORGANISM: NATURE, KINETICS OF APPEARANCE, QUANTITIES

Most of the substances released during digestion (oligopeptides, amino acids) pass through the intestinal wall where they may be modified or disappear for tissue synthesis. The product of these actions appears in the portal or lymphatic intestinal circulation, both of which drain the gut. For estimating the total action of the digestive tract, it is therefore necessary to identify the substances appearing in the circulation during digestion, to describe their kinetics of appearance and to measure the amounts found.

Route of absorption: lymphatic system or portal vein?

Very few studies have been attempted to determine more accurately the role of the lymphatic pathway in the absorption of protein digestion products. Jacobs and Largis (1969) have reported the appearance of labelled amino acids in the lymph immediately after their administration by stomach tube in the rat. According to Dawson and Porter (1962) three hours after a radioactive labelled meal, the total activity of the amino acid fraction of the lymph is the same per unit volume as that found in the portal blood. The incorporation of a radioactive marker into the proteins of the intestinal lymph is also relatively rapid (Morris, 1956); however, on account of the time elapsed after the meal, this might only reflect the changes in the composition of the general circulation.

Consequently, the lymphatic pathway may be used by the products of protein digestion. However, even if an increase in the lymphatic flow rate is recorded during the postprandial period (Simmonds, 1955; Barrowman and Roberts, 1967) this flow rate is very low when compared to the portal one (1/800, according to Dawson and Porter, 1962). The increases in the concentration of amino acids being similar in the two systems, the amount absorbed by the lymph can only be very small.

Forms of absorption

It is not possible to discuss here the absorption of whole proteins, which takes place during a variable length of time after the birth of some animals, in the case of immunity carrying proteins, and even sometimes in adults (Morris, 1968).

What is happening to the oligopeptides which, according to some authors, might enter the mucosal cells of the intestine? It would seem that some of these oligopeptides are able to cross the intestinal wall and to appear in the blood without being modified. Thus Prockop, Keiser and Sjoerdsma (1962) observed an increase in the plasma level of hydroxyproline, free and bound in the form of oligopeptide, after the ingestion of gelatin. The appearance of glycyl-glycine in the portal blood has also been noticed during the absorption of peptides composed of glycine (Peters and McMahon, 1970), a fact which corroborates the findings made *in vitro* by Newey and Smyth (1959). Besides, it is known that molecules analogous with peptides appear in the urine after intake of meals containing heated protein (Bjarnasson and Carpenter, 1969, 1970; Ford and Shorrock, 1971).

However, these facts seem to constitute particular cases and generally peptides only appear as traces during digestion either in the portal circulation, or in the peripheral one (Christensen *et al.*, 1947; Stein and Moore, 1954; Christensen, 1949; Dent and Schilling, 1949; Parshin and Rubel, 1951; Levenson, Roser and Upjohn, 1959; Newey and Smyth, 1959; Dawson and Holdsworth, 1962).

Consequently, most of the products of protein digestion appear in the organism as free amino acids. Despite the opposite opinion of Fisher (1967) this is what has been determined during qualitative and quantitative studies based upon the variations in the portal aminoacidemia.

Passage of amino acids in the organism: qualitative aspects

Effect of the dietary supply of protein on the aminoacidemia: It has been known for a long time that the aminoacidemia increases after a protein meal (Howell, 1906; Folin and Denis, 1912; Van Slyke and Meyer, 1912; Kalmykoff, 1924). The enhancement of the amino acid level is observed both in the systemic circulation (Frame, 1958; Rogers, Spolter and Harper, 1962; Richardson, Hale and Ritchey, 1965; Nordstrom *et al.*, 1970; Stockland *et al.*, 1971) and in the mesenteric or portal circulation (Eggum, 1966; Nielsen, 1966; Lebedeva and Aliev, 1970; Newey and Smyth, 1959; Pion, Fauconneau and Rerat, 1963; 1964a,b, 1966). However, this increase is generally less important in the systemic circulation than in the portal vein (Dent and Schilling, 1949; Denton and Elvehjem, 1954; Peraino and Harper, 1963; Bolton and Wright, 1937; Clark *et al.*, 1963) because of the uptake of amino acids by the liver (Ostrowski, 1969; Typpo *et al.*, 1970; Elwyn, Parikh and Shoemaker, 1968; Bloxam, 1972a,b).

The increase in the level of circulating amino acids is very rapid; it can be seen immediately, following the first 30 minutes after the meal (Pion, Fauconneau and Rerat, 1963; Dawson and Holdsworth, 1962) and can be followed over a period of variable length according to the quality and nature of the protein ingested and the animal species considered. Elwyn, Parikh and Shoemaker (1968) have recorded a duration of absorption of eleven hours in the dog and Noda et al., (1969) have shown that such a phenomenon might last twelve hours in the rat. However, in most of the cases, the fasting blood level is reached less than eight hours after the meal, and the portal-systemic differences are reduced to zero at this time. *Figure 6.6* gives an example of these facts (Pion and Rerat, 1969a,b)

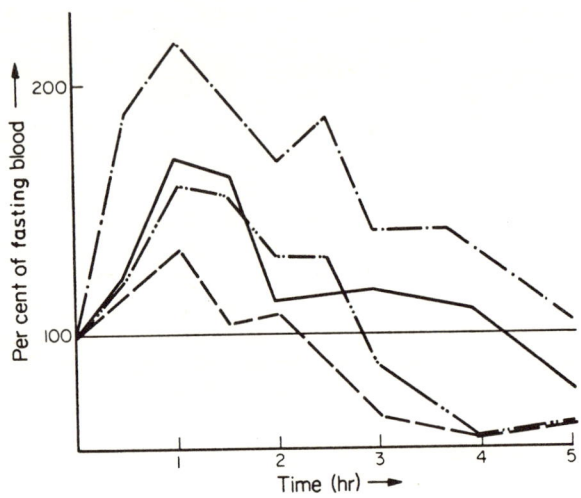

Figure 6.6 Concentration of free lysine in the portal blood during the digestion (per cent of fasting blood)
———•——— Sunflower meal + lysine (2.8 g/kg)
———••——— Sunflower meal
——————— Wheat + lysine (2.29 g/kg)
— — — — — Wheat
Diets: 400 g
Sunflower meal: 12 per cent crude protein in a semisynthetic diet
Wheat: 15.7 per cent crude protein in dry matter
(Pion and Rerat, 1967, 1969)

Some workers (Nielsen, 1966; Denton and Elvehjem, 1954; Hartmann et al., 1956; Rogers, Spolter and Harper, 1962) point out the existence of two or three waves of appearance and not only one peak.

The magnitude and duration of the rise in the blood level of various amino acids depends upon the quantity (Wiss, 1948; McLaughlan et al., 1963; Weller, Margen and Calloway, 1969; Henderson, Schurr and Elvehjem, 1949; Schurr et al., 1950) and composition of the proteins ingested (Chernikov and Usacheva, 1972; Canzler, Glatzel and Zimmermann, 1966; Chiericato and Lanari, 1972; Anderson and Linkswiler, 1969; Denton and Elvehjem, 1954; Peraino and Harper, 1963; Whitaker and Patrick, 1971; Puchal et al., 1962; Krysciak, Ostrowski and Rys, 1966).

It is generally considered that a relationship exists between the composition of the supplementary amino nitrogen found in free form in the blood, and that of the diet (Elwyn, 1970). This fact, more obvious in the portal vein, is illustrated by the example given in *Table 6.5* (Pion, Fauconneau and Rerat, 1966).

Although some parallelism exists between the increase in the blood level of each amino acid and its content in the diet, there are also a number of discrepancies. Thus, in the example considered, the portal blood contents of leucine, isoleucine and phenylalanine, as compared to the composition of the feed, are low in the case of diets based on barley. Threonine and lysine deficiencies of barley and peanut oilmeal seem to be exaggerated in the corresponding portal blood. The levels of diaminoacids appearing in the blood are very low when compared to the amounts supplied by the diet, whereas the glycine and alanine contents increase markedly in comparison with the dietary supply of these two amino acids. These facts are due to phenomena of transamination in the intestinal wall (Neame and Wiseman, 1957, 1958; Peraino and Harper, 1963).

The addition of an essential free amino acid to rebalance a protein results in particular in an enhancement of the blood level of this amino acid during the course of digestion (Pion and Rerat, 1967, 1969; Pion, Prugnaud and Rerat, 1972; Wells, 1968; Anderson and Linkswiler, 1969; Rolls, Porter and Westgarth, 1972; Stockland, Meade and Melliere, 1970).

Variation factors of aminoacidemia: The similarity between the composition of the supplementary free aminoacidemia and the amino acid composition of the diet is, however, contested by some authors (Levenson, Rosen and Upjohn, 1959; Dent and Schilling, 1949; Nasset, Gonapatty and Goldswill, 1963; Goldberg and Guggenheim, 1962; Eggum, 1966; Feigin, Beisel and Wannamacher, 1971). This similarity is closer for some amino acids (Gruhn and Hennig, 1972) and over a short period of time after the meal (Whitaker and Patrick, 1971). The differences are particularly evident for certain non-essential amino acids (Pion, Fauconneau and Rerat, 1966).

These disagreements reveal that the composition of the proteins ingested, although representing one of the important factors of postprandial aminoacidemia is not the only one. Furthermore, there are factors pertaining to digestion and others corresponding to the nutritional state of the subject on which the magnitude of the basal aminoacidemia will depend. A marked tissue protein synthesis will result in a removal of free amino acids from the circulating blood, whereas protein catabolism, on the contrary, may temporarily increase the level of these circulating amino acids (Pion, 1974).

Table 6.5 *Composition (percentage of the sum of total amino acids) of the increase in portal aminoacidemia as compared with ingested feeds (Pion et al., 1966)*

Feeds	Increase in portal blood concentrations as related to fasting blood (% of total amino acids)				Composition of feeds (% of total amino acids)		
	Herring meal		Peanut oil meal with starch	Barley	Herring meal	Peanut oil meal	Barley
	with starch	without starch					
Aspartic acid	2.7 ± 0.3	1.8 ± 0.2	5.1 ± 0.6	3.3 ± 0.5	11.7	14.9	7.1
Serine	4.5 ± 0.3	5.1 ± 0.7	7.8 ± 1.0	6.1 ± 1.0	5.1	6.6	5.7
Glycine	14.0 ± 2.9	13.6 ± 2.1	13.8 ± 2.4	21.3 ± 2.0	6.8	8.1	5.3
Lysine	8.5 ± 0.5	10.4 ± 2.2	3.2 ± 0.7	3.1 ± 0.9	9.8	4.6	4.9
Threonine	6.9 ± 0.3	10.2 ± 1.5	3.0 ± 0.6	2.2 ± 0.6	6.2	3.5	4.5
Alanine	13.6 ± 1.1	6.7 ± 2.1	15.5 ± 0.8	18.1 ± 0.7	8.5	5.2	5.3
Glutamine + glutamic acid	15.2 ± 1.3	11.2 ± 1.3	14.2 ± 1.7	19.6 ± 3.2	16.4	25.2	33.6
Valine	10.6 ± 0.4	13.3 ± 1.1	7.3 ± 0.4	6.4 ± 0.8	8.0	5.9	7.0
Leucine, isoleucine, phenylalanine	20.8 ± 0.8	22.6 ± 1.9	22.8 ± 1.2	15.5 ± 1.8	22.6	20.1	21.9
Tyrosine	3.2 ± 0.4	4.9 ± 0.5	7.2 ± 0.7	4.3 ± 0.7	4.7	5.9	4.9
Number of data	26	8	15	9			

Each value represents the mean increase of aminoacidemia recorded at different time intervals after the meal in four to six pigs. Feeding: either 48 to 64 kg of fish proteins with or without 252-336 g of maize starch; or 64 g of peanut proteins + 336 g maize starch; or 300-500 g of barley (13 per cent crude protein)

Digestive Factors: among these factors let us mention the general ones connected with the dietary sugars (Guggenheim, Halevy and Friedmann, 1960; Buraczewski et al., 1971) and the fats (Swenseid et al., 1967) or with the previous meal of the animal (Scharrer and Zucker, 1967; Ziemlanski et al., 1967, 1970; Hugo, Gallo Torres and Ludorf, 1973). Others are related to the availability of amino acids in the proteins and to their balance, upon which may depend a certain competition at the moment of absorption. The effect of heat treatments and especially incorrect heating on the aminoacidemia has been demonstrated. Erbersdobler (1967), et al. (1968, 1972) show that the damage caused to a protein (soyabean or casein) by such treatments, results in a poorer appearance of some amino acids in the portal vein. This phenomenon is stressed by the fact that the animals get accustomed to high protein diets, leading to greater differences between fasting and postprandial values (Scharrer, Erbersdobler and Zucker, 1967; Erbersdobler, 1967). The changes in the rate of appearance of amino acids in the portal blood plasma depending upon the heat treatment applied to the protein, have already been the subject of other studies drawing the same conclusions (Shimada and Zimmerman, 1973; Goldberg and Guggenheim, 1962; Wheeler and Morgan, 1958).

The fact that the availability of amino acids plays an important part in their further absorption is shown by the lower rapidity of appearance of the amino acids when they proceed from various proteins than when they constitute hydrolysates derived from these proteins. The different authors are in full agreement with respect to this matter (Zimmerman, Nielsen and Schoenheyder, 1962; Crane, 1969; Dent and Schilling, 1949; Peraino and Harper, 1963; Kratzer, 1944).

This has been confirmed when, instead of hydrolysates, the animals are fed mixtures of free amino acids of the same pattern as the proteins (Anderson and Linkswiler, 1969; Adibi, Gray and Menden, 1967; Free and Leonards, 1944; Scharrer and Zucker, 1968). Likewise, the absorption of free amino acids seems to be slightly quicker than that of the amino acids coming from the proteins (Pion, Prugnaud and Rerat, 1972; Rolls, Williams and Porter, 1969; Rolls, Porter and Westgarth, 1972) which is in agreement with the findings concerning the kinetics of disappearance of free amino acids from the digestive tract (Buraczewska, Buraczewski and Raczynski, 1972; Polivoda and Gritsenko, 1973). The form of stereoisomer administered may have some effect on the absorption rate, which is generally slower for D than for L forms (Matthews and Smyth, 1954) except in the case of methionine, which shows no obvious difference between the two forms (Kalafian, 1970).

Lastly, two other factors intervene at the level of the digestive tract, modifying the composition of the amino acid mixtures entering the organism, i.e. production and absorption of amino acids derived from endogenous proteins and competition for absorption sites.

Metabolic Transformation of Amino Acids and Uptake by the Intestinal Wall: After passing through the intestinal wall, the amino acids may have several roles, i.e. they either participate directly in the synthesis of proteins, very active at this time, especially during the period of digestion; or they enter a cycle of chemical reactions resulting in transformations; or pass into the portal vein unchanged.

Direct uptake of amino acids by the intestinal wall during the course of absorption, with a view to renewing tissues and enzymes, is a possibility which should not be completely rejected, especially during periods of denutrition (Levin, 1970). Alpers (1972) has very clearly pointed out this fact by means of labelled leucine (^{3}H or ^{14}C) injected or infused into animals. Under these experimental conditions, the specific activity of the incorporated leucine proceeding from the intestinal lumen is about 100 times higher than that proceeding from the blood. This uptake phenomenon is true for the 10 essential amino acids (Steiner and Gray, 1969). These facts confirm the presumptions of some authors (Ju and Nasset, 1959; Schlüssel, 1953; 1956a,b, 1959; Dawson and Holdsworth, 1962).

Furthermore an uptake of amino acids may occur from the arterial blood. Thus, Peraino and Harper (1962) have shown that the portal blood level of lysine may be inferior to that of the arterial blood after a diet based on hydrolysed zein. This fact is related to the high amino acid requirement of the intestinal mucosa, and is more marked for some amino acids when using imbalanced diets. This high requirement corresponds to the very rapid turnover of intestinal proteins (Alpers, 1972; Arnal, Fauconneau and Pech, 1971, 1972; Hirschfield and Kern, 1969). In addition, the lowering in the level of amino acids in the portal vein as compared with the systemic blood, may correspond to the flux of amino acids towards the intestinal lumen demonstrated *in situ* by Christensen (1963) and Jacobs (1965) and *in vivo* by Starovojtov and Mirosnicenko (1968). Some amino acids are transformed during their passage through the intestinal wall: this is the case for dicarboxylic acids (aspartic and glutamic acids) which are metabolised into alanine (Neame and Wiseman, 1957, 1958; Matthews and Wiseman, 1953; Peraino and Harper, 1962; Pion, Fauconneau and Rerat, 1963; Nielsen, 1966; Eggum, 1966). Finch and Hird (1960) as well as Pion, Fauconneau and Rerat (1963) have also shown the transformation of arginine into ornithine during the passage through the intestinal wall.

Kinetics of appearance of amino acids in the organism: quantitative aspects

It is clear that the processes of digestion and absorption cannot be analysed only upon the basis of portal amino-acidemia.

The elimination of the sources of the differences depends upon the variations in the systemic aminoacidemia and allows a better approach to be made to the study of these processes: thus, the similarity of the dietary amino acid supply with that pattern of the porto-arterial difference, is better than with the composition of the additional aminoacidemia found in the portal vein (Shimada and Zimmerman, 1973). It should be noted that the site of sampling of the systemic blood only causes slight variations, whether it is the aorta, jugular vein or vena cava (Nordstrom et al., 1970; Stockland et al., 1971). The porto-jugular differences must therefore be very similar to the porto-arterial ones. In addition (*Figure 6.3*), measurement of the portal blood flow rate allows the porto-systemic differences to have a quantitative significance, (Rerat, 1971a, 1973; Rerat and Duee, 1973; Elwyn, Parikh and Shoemaker, 1968), but it is not possible to estimate the magnitude of retention by the intestinal wall of amino acids proceeding from the intestinal lumen or from the arterial blood. In spite of this insufficiency, this methodology is the only one, at present, permitting a general survey of digestive phenomena. It allows more accurate data to be obtained concerning questions that have previously been subjected to long discussions, e.g.: endogenous nitrogen, form of absorption, absorption rate of free amino acids, measurement of the metabolic transformation in the intestinal wall.

Amount of endogenous nitrogen absorbed. The above mentioned methodology was applied to pigs receiving a protein-free meal after 18 hours of starvation. *Figure 6.7* shows the cumulative appearance of total and essential amino acids in the portal blood after absorption. *Table 6.6* indicates the apparent quantities of each amino acid appearing in the blood during the postprandial period (7 h) as well as the apparent quantities retained by the intestinal wall.

The absorption of amino acids during a postprandial period of seven hours after a protein-free meal, is rather high. Quantitatively, it corresponds to about a sixth of an amino acid mixture supplied by a 'normal' meal, this percentage being variable for each amino acid. Although the secretion of endogenous nitrogen is probably less important between meals, it is possible to calculate that the daily absorption of endogenous crude protein corresponds to about 50 g. It must be emphasised that the quantity measured in this way only represents the excess of amino acids which have not been retained to reconstitute the tissues of the intestinal wall. Considering the period where the systemic concentration is higher than the portal one, this synthesis is important and probably exceeds the absorption of endogenous nitrogen. However that may be, it appears that the utilisation of endogenous nitrogen has been highly overestimated by Nasset (1965). These data can be compared with estimations of the same kind made by Zebrowska and Buraczewska (1972a) in the pig, and Elwyn, Parikh and Shoemaker (1968) in the dog. These authors, using the same techniques, consider that under 'normal' conditions, the amounts of amino acids derived from the hydrolysis of endogenous proteins must be absolutely the same as the supplies of amino acids necessary to re-synthesise these proteins. This would not be the case, however, during periods of depletion or repletion during which the intestine becomes temporarily a supplier or demander of amino acids (Elwyn, 1970).

Table 6.6 Amount of amino acids (g) appearing in the portal vein (a) or disappearing from the general circulation (b) during a period of 7 h following a protein-free meal (mean from four animals) (Rerat, unpublished data)

	a	b		a	b
Lysine	1.12 ± 0.34	0.94 ± 0.24	Ornithine	0.47 ± 0.07	0.96 ± 0.31
Histidine	0.50 ± 0.16	0.51 ± 0.13	Aspartic acid	0.98 ± 0.35	0.57 ± 0.19
Threonine	0.38 ± 0.12	0.74 ± 0.28	Serine	0.71 ± 0.16	0.84 ± 0.40
Valine	0.72 ± 0.17	0.98 ± 0.31	Asparagine	0.37 ± 0.17	0.17 ± 0.08
Methionine	0.40 ± 0.21	0.40 ± 0.05	Glutamine	0.38 ± 0.18	3.74 ± 1.19
Arginine	0.39 ± 0.18	0.56 ± 0.15	Glutamic acid	2.11 ± 0.57	2.94 ± 1.76
Isoleucine	0.32 ± 0.11	0.35 ± 0.05	Proline	1.19 ± 0.50	3.03 ± 0.22
Leucine	0.82 ± 0.41	0.59 ± 0.10	Glycine	1.16 ± 0.26	1.97 ± 0.73
Phenylalanine	0.47 ± 0.20	0.40 ± 0.12	Alanine	1.53 ± 0.70	0.86 ± 0.29
Total essential amino acids	5.12 ± 1.37	5.48 ± 1.03	Citrulline	1.37 ± 0.22	1.62 ± 0.91
			Cystine	0.61 ± 0.29	0.61 ± 0.21
Total amino acids	16.28 ± 2.19	21.46 ± 4.50	Tyrosine	0.38 ± 0.09	0.44 ± 0.11

(a) $g = (C_p - C_a)D dr$
(b) $g' = (C_a - C_p)D dr$

C_p = portal concentration
C_a = arterial concentration
D = portal flow rate
dr = time elapsed since the meal

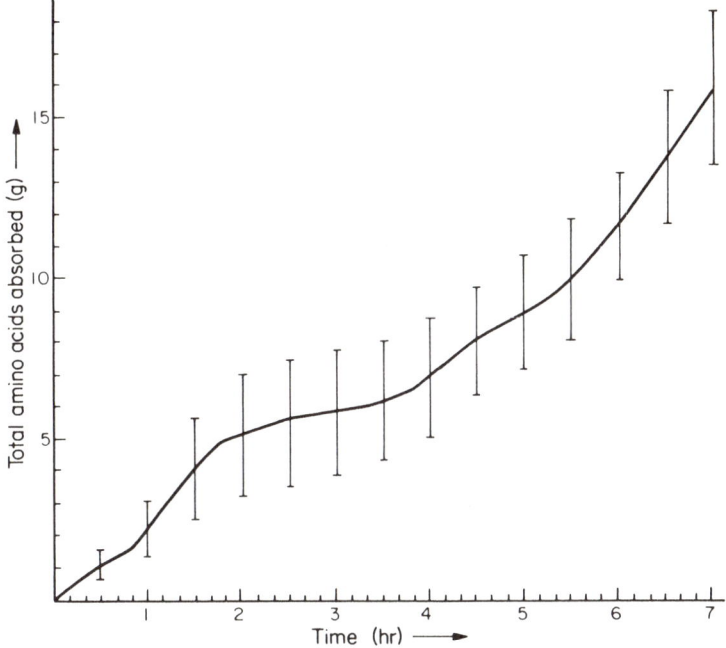

Figure 6.7 Total amino acids appearing in the portal blood after a 400 g protein free diet (maize starch, mineral and vitamin mixture, and fibre). Mean of four pigs. The vertical bars indicate standard deviation of the mean (Rerat, unpublished data)

Form of passage of proteins within the organism. With regard to this question, one must bear in mind the first quantitative estimations based on theoretical data concerning the portal blood flow rate and on measurements of the porto-systemic differences in the concentration of amino acids. Dent and Schilling (1949), Dawson and Porter (1962) have calculated that the addition of amino nitrogen to the blood flow during the postprandial period represents 30-40 per cent of the protein ingested. Fisher (1967a,b) considers that these data are overestimated because of the absorption of endogenous nitrogen and concludes that the main part of the ingested protein does not appear in the blood in the form of amino acids.

In contrast, the quantitative studies of Elwyn, Parikh and Shoemaker, (1968) show that during the 24 hours following the intake of meat in the dog, the amount of amino acids reaching the blood generally exceeds the amount ingested, proving that amino acids derived from the hydrolysis of endogenous proteins contribute to the digestive outputs. The pattern of the mixture of amino acids absorbed is analogous to the composition of the proteins ingested, except for some few modifications. As a matter of fact, the bulk of the proteins ingested appears in the blood in the form of amino acids, rather than in the form of proteins or peptides.

Analogous experiments have been performed with pigs (Rerat and Corring, unpublished). After administration of variable amounts of complex diets based on cereals (*Figure 6.8*) large quantities of amino acids rapidly appear in the portal vein in excess of the amounts ingested, these being small. In these conditions, the amounts of endogenous nitrogen absorbed do not seem to be very large.

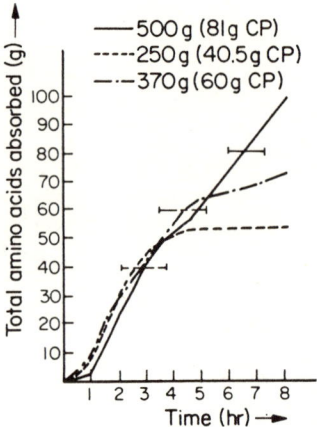

Figure 6.8 Total amino acids (g) absorbed after ingestion of different quantities of 16 per cent protein diets (composition: barley 35 per cent; wheat 15 per cent; maize 15 per cent; wheat bran 10 per cent; soyabean flour 18 per cent; salt and vitamin mixture). The horizontal bars indicate the amount of protein ingested. (Rerat and Corring, unpublished data) CP = crude protein

In the case of semi-synthetic diets containing various levels of fish meal (*Figure 6.9*), the absorption seems to be slower and to last longer, so that the quantities of amino acids passing through the portal vein during the period studied, are not always analogous to those ingested, in particular when the latter are high. Nevertheless, it can be deduced that the main part of the protein appears in the form of free amino acids.

When considering each amino acid separately (*Table 6.7*) their absorption rate is not uniform. The composition of the mixture of essential amino acids absorbed is however fairly similar to that of the feed (*Table 6.8*). In contrast, the modification of the non-essential amino acids is considerable (*Table 6.9*). The absorption rate of diacids is low, in particular in the case of aspartic acid. Contrary to that, the amounts of glycine, alanine, citrulline and ornithine are very large as indicated by other authors (Elwyn, Parikh and Shoemaker, 1968).

Absorption rate of free amino acids. Quantitative measurement of the appearance of labelled methionine (^{35}S) in the blood circulation (*Table 6.10*) shows that free methionine added to a diet presents absorption kinetics different to those of methionine from fish protein.

Table 6.7 Absorption coefficients[1] (and standard deviation of the mean) of essential amino acids coming from fish meal proteins (three pigs ingesting respectively 60 g, 74 g, 134 g protein, 12-16 per cent protein semi-synthetic diet) (Rerat, unpublished data)

Histidine	98.0 ± 6.4
Lysine	74.0 ± 9.4
Phenylalanine	65.3 ± 10.0
Leucine	67.3 ± 8.4
Isoleucine	54.0 ± 4.7
Methionine	79.0 ± 7.6
Valine	73.7 ± 6.3
Threonine	58.0 ± 10.1
Arginine	64.0 ± 12.2
Total essential amino acids	70.0 ± 5.3

[1] $\dfrac{\text{Quantity of amino acids absorbed during 7 h}}{\text{Quantity of ingested amino acids}} \times 100$

Table 6.8 Comparative composition of dietary proteins and of the mixture of essential amino acids (as a percentage of their sum) absorbed during a postprandial period of 7 h. Diet of fish meal + free methionine (mean from three animals) (Rerat, unpublished data)

	Dietary proteins	Absorbed mixture
Histidine	4.6	6.5 ± 0.3
Lysine	15.8	17.0 ± 1.7
Phenylalanine	8.0	7.6 ± 1.8
Leucine	15.6	15.1 ± 0.8
Isoleucine	9.6	7.5 ± 0.7
Methionine	7.2	8.4 ± 1.6
Valine	11.6	12.4 ± 0.8
Threonine	9.0	7.5 ± 0.7
Arginine	10.9	10.1 ± 1.7
Tyrosine	5.8	7.1 ± 0.3

Table 6.9 Comparative composition of dietary proteins and of the mixture of non essential amino acids (as a percentage of their sum) absorbed during a postprandial period of seven hours. Diet of fish meal + free methionine (mean from three animals) (Rerat, unpublished data)

	Dietary proteins	Absorbed mixture
Glutamine + Glutamic acid	26.1	17.1 ± 4.9
Aspartic acid	17.0	3.5 ± 1.6
Serine	7.5	9.3 ± 1.6
Proline	8.0	9.7 ± 4.8
Glycine	11.4	18.8 ± 4.7
Alanine	12.4	18.6 ± 2.8
Citrulline	-	8.5 ± 3.0
Ornithine	-	3.8 ± 0.5

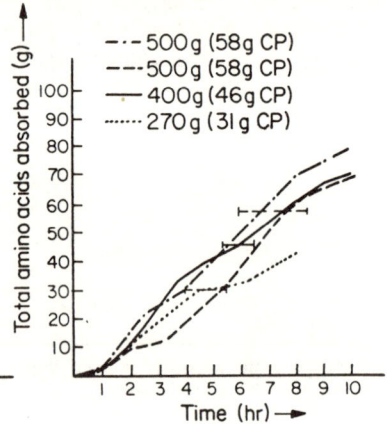

Figure 6.9 Total amino acids absorbed (g) after the ingestion of different quantities of a semisynthetic diet (12 per cent fish meal protein, maize starch, 5 per cent crude fibre, 4 per cent mineral and vitamin mixture). The horizontal bars indicate the amount of protein ingested (Rerat and Corring, unpublished data) CP = crude protein

The absorption of the free amino acid added to the diet is much more rapid than that of fish methionine, during the first hour. However, after 3-4 hours, the absorption percentage is the same whatever the origin of methionine. This kind of study should be repeated with other proteins with a digestive transit different from that of the fish meal contained in a semi-synthetic diet. However that may be, the results of this study correspond to those of previous investigations realised with other industrial amino acids and other protein feeds (Pion and Rerat, 1967; 1969; Pion, Prugnaud and Rerat, 1972; Rolls, Porter and Westgarth, 1972). In the light of present knowledge and except for particular cases, it seems to be accepted that the supplementation cannot play its role because of a too rapid absorption of the additives. This is also the opinion of Longenecker and Hause (1958).

Table 6.10 Absorption coefficient (amount absorbed/amount ingested × 100) of free methionine (F) or methionine proceeding from fish meal protein (P) - Variation according to the time elapsed since the meal - Mean from three pigs (Rerat and Pion, unpublished data)

Origin of the methionine	Amount ingested (g)	Time elapsed since the meal			
		1 h	2 h	3 h	4 h
F	0.8 - 1.7	16.3 ± 2.4	34.7 ± 1.7	43.7 ± 1.8	51.0 ± 3.8
P	1.6 - 3.7	9.3 ± 2.8	20.7 ± 8.2	37.0 ± 5.7	60.7 ± 9.2

Regression equation of absorption coefficient (y) on time (t)
$F : y = (11.3 \pm 1.3)t + 8.2$; $r = 0.94$
$P : y = (17.0 \pm 2.9)t - 10.7$; $r = 0.88$

Summary

It is perhaps best to conclude by summarising the different questions discussed in this report.

The digestive transit represents the first stage of digestion (enzymatic hydrolysis and absorption) since the movement of digesta determines, in time and quantity, the contact of the latter with the digestive secretions and sites of absorption. The transit is principally marked by the pattern of gastroduodenal emptying. The factors controlling gastric emptying are therefore mainly responsible for the variations that can be observed. In spite of the large differences in the methodology (artificial conditions for the test meals and previous fasting in man and rat, mostly normal conditions in the pig), the principal sources of variation are the same for these species, i.e. number and volume of meals, inhibitory effect of sugars or fats and, to a lesser extent, the nature of the proteins. The multinomial mathematical model of gastric emptying in the pig provides a new possibility for testing the magnitude of the effects, expressed by multinomial coefficients of groups of factors (stimulatory or inhibitory). The possible interpretations, according to the effect studied, have all been confirmed in the pig by the study of the mechanical or electrical phenomena in the gastroduodenal region.

The enzymatic digestion of dietary protein is a sequential succession of hydrolyses at different levels of the digestive tract. The gastric hydrolysis can be considered as a preparatory phase prior to the intestinal hydrolysis during which a complete digestion takes place under the action of pancreatic enzymes, mainly in the intestinal lumen, and of peptidases at the level of the epithelial mucosa. Examination of the relative importance of these hydrolyses clearly shows the predominant role of the pancreatic digestion. The studies reported point out the rapidity of digestion and allow a determination of the digestion sites. Finally, the enzymes of the digestive tract adapt themselves to the diet and to its changes. Thus, the gastric and pancreatic enzymes vary at the same time as the protein content of the diet and according to the quality of the dietary protein.

The absorption has been considered from several points of view. The analysis of intestinal contents, taking into account the enzymatic hydrolyses, shows that the endogenous supply is very important quantitatively. From the progression of these contents in time and space, it may be noticed that the digesta enter the intestinal mucosa not only in the form of free amino acids, but also as oligopeptides, a fact that partly explains the high digestion and absorption rate of the proteins. Apart from exceptions, these oligopeptides do not, however, pass through the intestinal mucosa, but are hydrolysed either in the mucosal cell or in the brush border. On the other hand, during the postprandial period, a very large rush of free amino acids is observed in the portal circulation, whereas the amount found in the lymphatic route seems to be negligible. The utilisation of a quantitative method combining the portal-arterial differences and the blood flow rate in the portal vein, shows that except for particular cases (e.g. heat-treatment) where the availability of amino acids is changed; the composition of the mixture of amino acids being added to the initial aminoacidemia, is generally close to that of the dietary supply, especially as regards the essential amino acids. Conversely, a very large modification of non-essential amino acids is noticed due to the transamination in the gut wall and uptake of the mucosal cell. The nutritional significance of the endogenous protein seems therefore not to be important even though a certain amount of free amino acid does appear in the portal blood after a protein-free meal. The addition of free amino acids to the diet is followed by an early appearance of these substances in the blood flow as compared to the free amino acids derived from the dietary proteins; however, the nutritional significance of this phenomenon has not yet been assessed.

References

ADIBI, S.A. (1971). *J. clin. Invest.,* **50** (II), 2266
ADIBI, S.A. and GRAY, S.J. (1967). *Gastroenterology,* **52**, 837
ADIBI, S.A., GRAY, S.J. and MENDEN, E. (1967). *Am. J. clin. Nutr.,* **20**, 24
AGAR, W.T., HIRD, F.J.R. and SIDHU, G.S. (1953). *J. Physiol.,* **121**, 255
AGAR, W.T., HIRD, F.J.R. and SIDHU, G.S. (1956). *Biochim. biophys. Acta,* **22**, 21
AKEDO, H.T., SUGAWA, S., YOSHIKAWA, S. and SUDA, M. (1960). *J. Biochem. Tokyo,* **47**, 124
ALPERS, D.H. (1972). *J. clin. Invest.,* **51**, (I), 167
ANDERSON, D.M. and ASH, R.W. (1971). *Proc. Nutr. Soc.,* **30**, 34A-35A
ANDERSON, H.L. and LINKSWILER, H. (1969). *J. Nutr.,* **99**, 91
ANFANGER, H. and HEAVENRICH, R.M. (1949). *Am. J. Dis. Child.,* **77**, 425
ANNEGERS, J.H. (1966). *Am. J. Physiol.,* **210**, 701
ARNAL, M., FAUCONNEAU, G. and PECH, R. (1971). *Annls Biol. anim. Biochim. Biophys.,* **II** (2), 245
ARNAL, M., FAUCONNEAU, G. and PECH, R. (1972). *Annls. Biol. anim. Biochim. Biophys.,* **12**, 91
ASATOOR, A.M., CHENG, B., EWARDS, K.D.G., LANT, A.F., MATTHEWS, D.M., MILNE, M.D., NAVAB, F., and RICHARDS, A.J. (1970). *Gut,* **II**, 380
ASATOOR, M., AMRIT, K., CHAD, H.A., DAWSON, I.M.P., MILNE, M.D., and PROSSER, V.D.I. (1972). *Br. J. Nutr.,* **28**, 417
AUFFRAY, P. (1965). *Conférence au Comité Scientifique de Pathologie porcine,* I.N.R.A., Paris
AUFFRAY, P., MARTINET, J., and RERAT, A. (1967). *Annls. Biol. anim. Biochim. Biophys.,* **7**, 261
BARROWMAN, J., and ROBERTS, K.B. (1967). *Q. Jl exp. Physiol.,* **52**, 19
BECCARI, E. (1957). *Boll. Soc. ital. Biol. sper.,* **33**, 523
BERGEN, W.G. (1969). *Proc. Soc. exp. Biol. Med.,* **132**, 348
BERGEN, W.G. and PURSER, D.B. (1968). *J. Nutr.,* **95**, 333

BJARNASSON, J. and CARPENTER, K.J. (1969). *Br. J. Nutr.*, **23**, 859
BJARNASSON, J. and CARPENTER, K.J. (1970). *Br. J. Nutr.*, **24**, 313
BLAGIVODOV, D.F., POMELTSOV, A.N. and SHATALOV, V.N. (1972). *Bull. exp. Biol. med.*, **74**, 19
BLOXAM, D.L. (1972a). *Br. J. Nutr.*, **27**, 233
BLOXAM, D.L. (1972b). *Br. J. Nutr.*, **27**, 249
BOLTON, C. and WRIGHT, G.P. (1937). *J. Physiol.*, **89**, 269
BOOTH, V.H. and CARPENTER, K.J. (1973). *Nutr. Abstr. Rev.*, **43**, 423
BORGSTRÖM, B., DAHLQUIST, A., LUNDH, G. and SJÖVALL, J. (1957). *J. clin. Invest.*, **36**, 1521
BRÖMSTER, D., CARLBERGER, G. and LUNDH, G. (1968). *Scand. J. Gastroent.*, **3**, 641
BURACZEWSKA, L., BURACZEWSKI, S. and RACZYNSKI, G. (1972). *Zesz. Probl. Postep. Nauk. Roln.*, **129**, 163
BURACZEWSKA, L., and ZEBROWSKA, T. (1972). *Zesz. Probl. Postep. Nauk. Roln.*, **126**, 129
BURACZEWSKI, S. (1966). *Factors affecting amino acid levels in the blood.* PhD Thesis, University of Reading
BURACZEWSKI, S. (1970). *Roczn. Naukro. In.*, **134D**, 1
BURACZEWSKI, S. (1972). *Zesz. Probl. Postep. Nauk. Roln.*, **126**, 157
BURACZEWSKI, S., PORTER, J.W.G., ROLLS, B.A. and ZEBROWSKA, T. (1971). *Br. J. Nutr.*, **25**, 299
CANZLER, H. and GLATZEL, H. (1966a). *Nutritio et Dieta*, **8**, 49
CANZLER, H. and GLATZEL, H. (1966b). *Nutritio et Dieta*, **8**, 88
CANZLER, H., GLATZEL, H. and ZIMMERMANN, H. (1966). *Nahrung*, **10**, 281
CHANG, C.A., MCKENNA, R.D. and BECK, I.T. (1968). *Gut*, **9**, 420
CHEN, M.L., ROGERS, Q.R. and HARPER, A.E. (1962). *J. Nutr.*, **76**, 235
CHENG, B., NAVAB, F., LIS, M.T., MILLER, T.N. and MATTHEWS, D.M. (1971). *Clin. Sci.*, **40**, 247
CHERNIKOV, M.P., USACHEVA, N.T. (1972). *Vop. Pitan.*, **31**, 3
CHIERICATO, G.M. and LANARI, D. (1972). *Alimentazione Animale*, **16**, 41
CHRISTENSEN, H.N. (1949). *Biochem. J.*, **44**, 333
CHRISTENSEN, H.N. (1963). *Fedn Proc. Fedn Am. Socs exp. Biol.*, **22**, 1110
CHRISTENSEN, H.N., COOPER, P.F. Jr, JOHNSON, R. and LYNCH, E.L. (1947). *J. biol. Chem.*, **168**, 191
CHRISTENSEN, H.N. and SHWACHMAN, H. (1949). *J. clin. Invest.*, **28**, 319
CLARK, A.J., HAYS, V.W., McCALL, J.T. and SPEER, V.C. (1963). *J. Anim. Sci.*, **22**, 1118
COATES, G., GILDAY, D.L., CRADDUCK, T.D. and WOOD, D.E. (1973). *Can. med. Ass. J.*, **108**, 180
COOK, G.C. (1972). *Clin. Sci.*, **43**, 443
COOKE, A.R. and MOULANG, J. (1972). *Gastroenterology*, **62**, 528
CORRING, T. (1973). *Annls. Biol. anim. Biochim. Biophys.*, **13**, 755

CORRING, T., AUMAITRE, A. and RERAT, A. (1972). *Annls Biol. anim. Biochim. Biophys.*, **12**, 109

CORRING, T. and JUNG, J. (1972). *Nutr. Rep. Int.*, **6**, 187

CORRING, T. and SAUCIER, R. (1972). *Annls Biol. anim. Biochim. Biophys.*, **12**, 233

COSTA, L.R. Da (1971). *J. Nutr.*, **101**, 431

CRAFT, I.L., CRAMPTON, R.F., LIS, M.T. and MATTHEWS, D.M. (1969). *J. Physiol.*, **200**, 1112

CRAFT, I.L., GEDDES, D., HYDE, C.W., WISE, I.J. and MATTHEWS, D.M. (1968). *Gut*, **9**, 425

CRAFT, I.L. and MATTHEWS, D.M. (1968). *Br. J. Surg.*, **55**, 158

CRAMPTON, R.F., GANGOLLI, S.D., SIMSON, P. and MATTHEWS, D.M. (1971). *Clin. Sci.*, **41**, 409

CRAMPTON, R.F., LIS, M.T. and MATTHEWS, D.M. (1973). *Clin. Sci.*, **44**, 583

CRANE, C.W. (1969). in *Malabsorption, Pfizer medical monographs*, **4**, 25. Eds. R.H. Girdwood and A.N. Smith, University Press, Edinburgh

CRANE, C.W. and NEUBERGER, A. (1960). *Biochem. J.*, **74**, 313

CROMPTON, D.W.T. and NESHEIM, M.C. (1969). *J. Nutr.*, **99**, 43

CUNNINGHAM, H.M. (1967). *J. Anim. Sci.*, **26**, 500

CUPERLOVIC, M., HRISTIC, V. and CMILJANIC, R. (1972). *Acta. Vet. Beograd*, **22**, 111

CUTHBERTON, D.P. and TILSTONE, W.J. (1972). in *Protein and amino acid functions*, **Vol.II**, 119, Ed. E.J. Bigwood. Pergamon Press, Oxford, New York

DAMMERS, J. (1964). *Digestibility in the pig. Factors influencing the digestion of the components of the feed and the digestibility of the amino acids.* These Drukkery West Friesland, Hoorn, pp.152, Université Van Leuven, Landbouvinstitut

DARGEL, D. and HOCK, A. (1971). *Nahrung*, **15**, 353

DAWSON, N.J. (1972). *Comp. Biochem. Physiol.*, **41A**, 877

DAWSON, R. and HOLDSWORTH, E.J. (1962). *Br. J. Nutr.*, **16**, 13

DAWSON, R. and PORTER, J.W.G. (1962). *Br. J. Nutr.*, **16**, 27

DE LA NOUE, J., NEWEY, H. and SMYTH, D.H. (1971). *J. Physiol.*, **214**, 105

DELHUMEAU, G., VELEZ PRATT, G. and GITLER, C. (1962). *J. Nutr.*, **77**, 52

DENT, C.E. and SCHILLING, D.A. (1949). *Biochem. J.*, **44**, 318

DENTON, A.E. and ELVEHJEM, C.A. (1954). *J. Biochem.*, **206**, 449

DERBLOM, H., JOHANSSON, H. and NYLANDER, G. (1966). *Acta Chir. Scand.*, **132**, 154

DESNUELLE, P., REBOUD, J.P. and BEN ABDELJLIL, A. (1962). in *Ciba Foundation Symposium on the exocrine pancreas*, Churchill (London), p.90

DOUGLAS, G.J. Jr, REINAUER, A.J., BROOKS, W.C. and PRATT, J.H. (1953). *Gastroenterology*, 452

EGGUM, B.O. (1966). *Z. Tierphysiol. Terernähr. Futtermittelk.*, **22**, 32

ELIAS, E., GIBSON, G.J., GREENWOOD, L.F., HUNT, J.N. and TRIPP, J.H. (1968). *J. Physiol.*, **194**, 317

ELWYN, D.H. (1970). in *Mammalian protein metabolism,* **Vol.IV,** 523. Ed. H.N. Munro. Academic Press, New York and London
ELWYN, D.H., PARIKH, H.C. and SHOEMAKER, W.C. (1968). *Am. J. Physiol.,* **215,** 1260
ERBERSDOBLER, H. (1967). *Analytical and Physiological characterisation of damage to amino acids in heat treatment of foods and feedingstuffs.* Thesis, Tierärztl. Fak., Univ. München,
ERBERSDOBLER, H. (1973). in *Proteins in human nutrition,* 453. Ed. J.W.G. Porter and B.A. Rolls. Academic Press, London and New York
ERBERSDOBLER, H., DÜMMER, H. and ZUCKER, H. (1968). *Z. Tierphysiol. Tierernähr. Futtermittelk.,* **24,** 136
ERBERSDOBLER, H., WEBER, G. and GUNSSER, I. (1972). *Z. Tierphysiol., Tierernähr. Futtermittelk.,* **29,** 325
EVERSON, T.C. (1952). *Int. Abstr. Biol. Sci.,* **95,** 209
FAUCONNEAU, G., and MICHEL, M.C. (1970). in *Mammalian protein metabolism,* **4,** 481. Ed. H.N. Munro. Academic Press, New York and London
FEIGIN, R.D., BEISEL, W.R. and WANNAMACHER, R.W. (1971). *Am. J. clin. Nutr.,* **24,** 329
FINCH, L.R. and HIRD, F.J.R. (1960). *Biochem. biophys. Acta,* **43,** 278
FISHER, R.B. (1967a). *Proc. Nutr. Soc.,* **26,** 23
FISHER, R.B. (1967b). *Br. med. Bull.,* **23,** 241
FOLIN, O. and DENIS, W. (1912). *J. biol. Chem.,* **11,** 87
FORD, J.E. and SHORROCK, C. (1971). *Br. J. Nutr.,* **26,** 311
FRAME, E.G. (1958). *J. clin. Invest.,* **37,** 1710
FRANCOIS, E., COMPERE, R. and RONDIA, G. (1968). *Bull. Rech. agron. Gembloux,* **3,** 655
FREE, A.H. and LEONARDS, J.R. (1944). *J. Lab. clin. Med.,* **29,** 963
FREEMAN, T. (1964). in *The role of the gastrointestinal tract in protein metabolism.* p.125. Ed. H.N. Munro. Blackwell, Oxford
GARDNER, J.D., BROWN, M.S. and LASTER, L. (1970). *New Engl. J. Med.,* **283,** 1317
GEIGER, E. (1951). *Fedn Proc. Fedn Am. Socs exp. Biol.,* **10,** 670-675
GENT, A.E. and CREAMER, B. (1972). *Digestion,* 7, 13
GITLER, C. (1964). in *Mammalian protein metabolism,* **Vol.I.,** 35 Eds H.N. Munro and J.B. Allison. Academic Press, New York
GITLER, C. and MARTINEZ ROJAS (1964). in *The rôle of the gastrointestinal tract in protein metabolism,* p.269. Ed. H.N. Munro. Blackwell, Oxford
GLASS, G.B.J. and ISHIMORI, A. (1961). *Am. J. dig. Dis.,* **6,** 103
GOLDBERG, A. and GUGGENHEIM, K. (1962). *Biochem. J.,* **83,** 129
GRAY, G.M. and COOPER, H.L. (1971). *Gastroenterology,* **61,** 535
GREEN, G.M. and LYMAN, R.L. (1972). *Proc. Soc. exp. Biol. Med.,* **140,** 6
GROSSMAN, M.I., GREENGARD, H. and IVY, C.A. (1942). *Am. J. Physiol.,* **138,** 676
GRUHN, K. and HENNIG, A. (1972). *Arch. Tierernähr.,* **22,** 543
GUGGENHEIM, K., HALEVY, S. and FRIEDMANN, N. (1960). *Archs Biochem. Biophys.,* **91,** 6

GULLBERG, R. and OLHAGEN, B. (1959). *Nature,* **184,** 1848
HAGIHIRA, H., OGATA, M., TAKEDATSU, N. and SUDA, M. (1960). *J. Biochem. Tokyo,* **47,** 139
HARKINS, R.W., LONGENECKER, J.B. and SARETT, H.P. (1964). *Gastroenterology,* **47,** 65
HARTMANN, F., LENZ, H., LOPEZ CALLEJA, C. and MUNZENBERG, H. (1956). *Arch. exp. Path. Pharmak.,* **228,** 403
HEIZER, W.D., KERLEY, R.L. and ISSELBACHER, K.J. (1972). *BBA,* **264,** 450
HELLIER, M.D., HOLDSWORTH, C.D., McCOLL, I. and PERRETT, D. (1972). *Gut,* **13,** 965
HENDERSON, L.M., SCHURR, P.E. and ELVEHJEM, C.A. (1949). *J. biol. Chem.,* **177,** 815
HILL, K.J., NOAKES, D.E., and LOWE, R.A. (1969). in *Physiology of digestion and metabolism in the ruminant.* Ed. A.T. Phillipson. p.166. Proc. Third International Symposium, Cambridge Oriel Press Ltd., Newcastle upon Tyne
HIRSCHFIELD, J.S. and KERN, F. (1969). *J. clin. Invest.,* **48,** 1224
HOPKINS, A. (1966). *J. Physiol.,* **182,** 144
HORSZCZARUK, F. (1971a). *Biuletyn Instytut genetyki i Hodowli Zwierzat Polskiej. Akademii Nauk.,* **21,** 117
HORSZCZARUK, F. (1971b). *Biuletyn Instytut genetyki i Hodowli Zwierzat Polskiej. Akademii Nauk.,* **21,** 137
HOWARD, F. and YUDKIN, J. (1963). *Br. J. Nutr.,* **17,** 281
HOWELL, W.H., (1906-1907). *Am. J. Physiol.,* **17,** 273
HUANG, K.C. (1961). *Fedn. Proc. Fedn Am. Socs exp. Biol.,* **20,** 246
HUGO, E., GALLO TORRES, and LUDORF, J. (1973). *J. Nutr.,* **103,** 1745
HUNT, J.N. (1960). *J. Physiol.,* **154,** 170
HUNT, J.N. (1961). *Gastroenterology,* **41,** 49
HUNT, J.N. (1971). *Am. J. dig. Dis.,* **16,** 641
HUNT, J.N., and KNOX, M.T. (1968). *J. Physiol.,* **194,** 327
HUNT, J.N. and McDONALD, I. (1954). *J. Physiol.,* **126,** 459
HUNT, J.N. and PATHAK, J.D. (1960). *J. Physiol.,* **154,** 254
HUNT, J.N. and SPURRELL, W.R. (1951). *J. Physiol.,* **113,** 157
HUSBAND, J., HUSBAND, P. and MALLINSON, C.N. (1970). *Lancet,* **ii,** 290
JACOBS, F.A. (1965). *Fedn Proc. Fedn Am. Socs exp. Biol.,* **24,** 946
JACOBS, F.A. and LARGIS, E.E. (1969). *Proc. Soc. exp. Biol. Med.,* **130,** 692
JEEJEEBHOY, K.N. (1964). in *The rôle of gastrointestinal tract in protein metabolism,* pp.357. Ed. H.N. Munro. Blackwell, Oxford
JEFFRIES, G.H., and SLEISENGER, M.H. in *Handbook of physiology section 6: Alimentary Canal,* **Vol.5.** *Bile, digestion, ruminal physiology,* 2775. Ed. Ch.F. Code. American Physiological Society, Baltimore
JOHANSSON, C. (1973). *Scand. J. Gastroent.,* **8,** 533
JOHANSSON, C., LAGERLÖF, H.O., EKELUND, K., KULSDOM, N., LARSSON, I. and NYLIND, B. (1972). *Scand. J. Gastroent.,* **7,** 489

JU, J.S. and NASSET, E.S. (1959). *J. Nutr.*, **68**, 633
KALAFIAN, J.S. (1970). *Dissertation Absts. Internat.* (B), **30**, 3016B-3017B
KALMYKOFF, M.P. (1924). *Arch. ges. Physiol.*, **205**, 493
KIDDER, D.E. and MANNERS, M.J. (1968). *Proc. Nutr. Soc.*, **27**, 46A
KIDDER, D.E., MANNERS, M.J. and McCREA, M.R. (1961). *Res. Vet. Sci.*, **2**, 227
KIM, Y.S., BIRTWHISTLE, W., KIM, Y.W. and SLEISENGER, M.H. (1971). *Gastroenterology*, **60**, 685
KIRSCH, R.E., SAUNDERS, S.J. and BROCK, J.F. (1968). *Am. J. clin. Nutr.*, **21**, 1302
KLEMINA, E.A., SISOVA, O.A. and KASATOCKIN, V.I. (1965). *Vop. Pitan*, **24**, **6**, 31
KRATZER, F.H. (1944). *J. biol. Chem.*, **153**, 237
KRISTEN, H., POPPE, S. and MEIER, H. (1970). *Arch. Geflügelzucht Kleintierk*, **19**, 125
KRYSCIAK, J., OSTROWSKI, H. and RYS, R. (1966). *Acta Biochim. Pol.*, **13**, 229
KUIKEN, K.A. and LYMAN, C.M. (1948). *J. Nutr.*, **36**, 359
LAGERLÖF, H.O., JOHANSSON, C. and EKELUND, K. (1973). *Scand. J. Gastroent.*, **8**, 735
LAPLACE, J.P. (1971). *Xème Congr. Int. Zootech.*, Versailles
LAPLACE, J.P. (1972). *Recl. méd. Vét.*, **148**, 37
LAPLACE, J.P. (1974a). *Recl. méd. Vét.*, **150**, 121
LAPLACE, J.P. (1974b). *Ann. Zootech.*, **23**, 89
LAPLACE, J.P. and TOMASSONE, R. (1970). *Ann. Zootech.*, **19**, 303
LAZAROV, I. and IVANOV, N. (1973). *Zhivotnovudni Nauki*, **10**, 71
LEBEDEVA, L.K. and ALIEV, A.A. (1970). *Izvestiya Akademii Nauk Latviiskoi SSR*, **11**, 80
LEVENSON, S.M., ROSEN, H. and UPJOHN, H.L. (1959). *Proc. Soc. exp. Biol. Med.*, **101**, 178
LEVIN, R.J. (1970). *Life Sci.*, **9**, **2**, 61
LIS, M.T., CRAMPTON, R.F. and MATTHEWS, D.M. (1972a). *Br. J. Nutr.*, **27**, 159
LIS, M.T., MATTHEWS, D.M. and CRAMPTON, R.F. (1972b). *Br. J. Nutr.*, **28**, 443
LONGENECKER, J.B. and HAUSE, N.L. (1958). *Nature*, **182**, 1739
LUNDH, G. (1958). *Acta Chir. Scand. Suppl.*, **231**, 1
MATTHEWS, D.M. (1972). *Proc. Nutr. Soc.*, **31**, 171
MATTHEWS, D.M., CRAMPTON, R.E. and LIS, M.T. (1968). *Lancet*, **ii**, 639
MATTHEWS, D.M., LIS, M.T., CHENG, B. and CRAMPTON, R.F. (1969). *Clin. Sci.*, **37**, 751
MATTHEWS, D.M., and SMYTH, D.H. (1954). *J. Physiol.*, **126**, 96
MATTHEWS, D.M. and WISEMAN, G. (1953). *J. Physiol.*, **120**, 550
McLAUGHLAN, J.M., NOEL, F.J., MORRISON A.B. and CAMPBELL, J.A. (1963). *Can. J. Biochem. Physiol.*, **41**, 191
MEADE, R.J. (1972). *J. Anim. Sci.*, **35**, 713
MEHNERT, H. and FÖRSTER, H. (1968). *Diabetologia*, **4**, 26
MEIER, H., POPPE, S. and WIESEMULLER, W. (1970). *Arch. Tierernähr.*, **20**, (a) 557; (b) 567; (c) 575

METTRICK, D.F. (1970). *Comp. Biochem. Physiol.*, **37**, 517
MORRIS, B. (1956). *Q. Jl. exp. Physiol.*, **41**, 326
MORRIS, I.G. (1968). in *Handbook of Physiology. Alimentary canal*, section 6, vol.3, *Intestinal absorption*, 1491. Ed. C.F. Code and W. Heidel. American Physiological Society, Washington DC
MUNCK, B.G. (1965). *Biochim. biophys. Acta*, **94**, 136
MUNRO, H.N. (1966). in *Postgraduate Gastroenterology*, 58 Eds. T.J. Thompson and I.E. Gillipsie. Balliere, London
NAKAMURA, Y., YASUMOTO, K. and MITSUDA, H. (1972). *J. Nutr.*, **102**, 359
NASSET, E.S. (1962). *J. Nutr.*, **76**, 131
NASSET, E.S. (1965). *Fedn Proc. Fedn Am. Socs exp. Biol.*, **24**, 953
NASSET, E.S. (1972). *World Review of Nutrition and Dietetics*, **14**, 134
NASSET, E.S., GANAPATTY, S.N. and GOLDSWILL, D.P.J. (1963). *J. Nutr.*, **81**, 343
NASSET, E.S. and JU, J.S. (1961). *J. Nutr.*, **74**, 461
NASSET, E.S., SCHWARTZ, P. and WEISS, H.V. (1955). *J. Nutr.*, **56**, 83
NEALE, R.J. (1971). *Lancet*, **i**, 143
NEAME, K.D. and WISEMAN, G. (1957). *J. Physiol.*, **135**, 442
NEAME, K.D. and WISEMAN, G. (1958). *J. Physiol.*, **140**, 148
NEWEY, H., SANFORD, P.A. and SMYTH, D.H. (1970). *J. Physiol.*, **208**, 705
NEWEY, H. and SMYTH, D.H. (1959). *J. Physiol.*, **145**, 48
NEWEY, H. and SMYTH, D.H. (1962). *J. Physiol.*, **164**, 527
NIELSEN, J.J. (1966). Determination of the digestibility of amino acids. Licenciate Thesis, Royal Veterinary and Agricultural College, Copenhagen
NIXON, S.E. and MAWER, G.E. (1970). *Br. J. Nutr.*, **24**, (a) 227; (b) 241
NODA, K., MORITOKI, K., YAMAMOTO, S. and YOSHIDA, A. (1969). *J. Jap. Soc. Fd. Nutr.*, **22**, 102
NORDSTROM, J.W., WINDELS, H.F., TYPPO, J.T., MEADE, R.J. and STOCKLAND W.L. (1970). *J. Anim. Sci.*, **31**, 874
OCHOA-SOLANO, A. and GITLER, C. (1968). *J. Nutr.*, **94**, 249
OLMSTED, W.W., NASSET, E.S. and KELLEY, M.L. (1966). *J. Nutr.*, **90**, 291
ORTEN, A.V. (1959). *J. Mich. St. Med. Soc.*, **58**, 767
OSTROWSKI, H. (1969). *Anim. Prod.*, **11**, 521
PADALIKOVA, D. (1966). *Zentbl. Vet. Med.*, **13A**, 709
PARSHIN, A.N. and RUBEL, L.N. (1951). *Chem. Abstr.*, **45**, 7206
PEKAS, J.C., HAYS, V.W. and THOMPSON, A.M. (1964). *J. Nutr.*, **82**, 277
PERAINO, C. and HARPER, A.E. (1962). *Archs. Biochem. Biophys.*, **97**, 442
PERAINO, C. and HARPER, A.E. (1963). *J. Nutr.*, **80**, 270
PERAINO, C., ROGERS, Q.R., YOSHIDA, M., CHEN, M.L. and HARPER, A.E. (1959). *Can. J. Biochem. Physiol.*, **37**, 1475
PETERS, T.J. and McMAHON, M.T. (1970). *Clin. Sci.*, **39**, 811

PION, R. (1974). this symposium
PION, R., FAUCONNEAU, G. and RERAT, A. (1963). *Annls Biol. anim. Biochim. Biophys.*, **3**, HS, 31
PION, R., FAUCONNEAU, G. and RERAT, A. (1964a). *Annls Biol. anim . Biochim. Biophys.*, **4**, 383
PION, R., FAUCONNEAU, G. and RERAT, A. (1964b). in *The role of the gastrointestinal tract in protein metabolism*, 309. Ed. H.N. Munro. Alden Press, Oxford
PION, R., FAUCONNEAU, G. and RERAT, A. (1966). in *Amino acids, peptides, proteins*, **6**, 325. A.E.C. Commentary
PION, R. and RERAT, A. (1967). *C.R. Acad. Sci.*, **264D**, 632
PION, R. and RERAT, A. (1969a). *Z. Tierphysiol. Tierernähr. Futtermittelk. Band.*, **25**, H6
PION, R. and RERAT, A. (1969b). *J. Rech. Porcine en France, Paris*, 151. INRA, ITP
PION, R., PRUGNAUD, J. and RERAT, A. (1972). *Nutr. Repts Int.*, **6(6)**, 331
PIRK, F. and SKALA, I. (1970). *Digestion*, **3**, 73
POLIVODA, D.I. and GRITSENKO, N.M. (1973). *Sel' skokhozyaistvennaya Biologiya*, **8**, 108
POPPE, S. and MEIER, H. (1971a). *Arch. Tierernähr.*, **21**, 531
POPPE, S. and MEIER, H. (1971b). *Arch. Tierernähr.*, **21**, 619
POPPE, S., MEIER, H. and UHLEMANN, H. (1971). *Arch. Tierernähr.*, **21**, 161
PORTER, J.W.G. and ROLLS, B.A. (1971). *Proc. Nutr. Soc.*, **30**, 17
POULAKOS, L. and KENT, T.H. (1973). *Gastroenterology*, **64**, 962
PROCKOP, D.J., KEISER, H.R. and SJOERDSMA, A. (1962). *Lancet*, **2**, 527
PUCHAL, F., HAYS, V.W., SPEER, V.C., JONES, J.D. and CATRON, D.V. (1962). *J. Nutr.*, **76**, 11
RERAT, A. (1971a). *Annls Biol. anim. Biochim. Biophys.*, **11**, 277
RERAT, A. (1971b). *Ann. Zootech.*, **20**, 193
RERAT, A. (1973). *Proc. Nutr. Soc.*, **32**, 49A-51A
RERAT, A., and DUEE, P.H. (1973). *Annls Biol. anim. Biochim. Biophys.*, **13**, 788
RERAT, A. and LOUGNON, J. (1963). *Annls Biol. anim. Biochim. Biophys.*, **3**, HS, 21
REYNELL, P.C. and SPRAY, G.H. (1956). *J. Physiol. Lond.*, **131**, 452
RHODES, J., GOODALL, P. and APSIMON, H.T. (1966). *Gut*, **7**, 515
RICHARDSON, L.R., HALE, F. and RITCHEY, S.J. (1965). *J. Anim. Sci.*, **24**, 368
ROGERS, Q.R., CHEN, M.L., PERAINO, C. and HARPER, A.E. (1960). *J. Nutr.*, **72**, 331
ROGERS, Q.R. and HARPER, A.E. (1966). *World Rev. Nutr. Diet.*, **6**, 250
ROGERS, Q.R., SPOLTER, P.D. and HARPER, A.E. (1962). *Archs Biochem. Biophys.*, **97**, 497
ROLLS, B.A., PORTER, J.W.G. and WESTGARTH, D.R. (1972). *Br. J. Nutr.*, **28**, 283
ROLLS, B.A., WILLIAMS, A.P. and PORTER, J.W.G. (1969). *Proc. Nutr. Soc.*, **28**, 69A

ROSENTHAL, S. and NASSET, E.S. (1958). *J. Nutr.*, **66**, 91
RUBINO, A., FIELD, M. and SCHWACKMAN, H. (1971). *J. biol. Chem.*, **246**, 3542
RUNE, S.J. (1968). *Scand. J. Gastroent.*, **3**, 91
RUNE, S.J. (1972). *Gastroenterology*, **62**, 533
RÜSTOW, B. and HOCK, A. (1970). *Arch. Tierernähr.*, **20**, 211
RYLE, A.P. and PORTER, R.R. (1959). *Biochem. J.*, **73**, 75
SADIKALI, F. (1971). *Gut*, **12**, 276
SALTER, D.N. (1973). *Proc. Nutr. Soc.*, **32**, 65
SAUNDERS, S.J. and ISSELBACHER, K.J. (1965). *Biochim. biophys. Acta*, **102**, 397
SAUNDERS, S.J. and ISSELBACHER, K.J. (1966). *Gastroenterology*, **50**, 586
SCHARRER, E. (1971). *Archiv. Geflügelk.*, **35**, 21
SCHARRER, E. and BRUGGEMANN, J. (1971). *Z. Tierphysiol. Tierernähr. Futtermittelk.*, **27**, 327
SCHARRER, E., ERBERSDOBLER, H., ZUCKER, H. (1967). *Z. Tierphysiol. Tierernähr. Futtermittelk.*, **22**, 265
SCHARRER, E. and ZUCKER, H. (1967). *Z. Tierphysiol. Tierernähr. Futtermittelk.*, **23**, 169
SCHARRER, E. and ZUCKER, H. (1968). *Z. Tierphysiol. Tierernähr. Futtermittelk.*, **23**, 321
SCHLÜSSEL, H. (1953). *Klin. Wschr.*, **31**, 545
SCHLÜSSEL, H. (1956a). *Klin. Wschr.*, **34**, 1288
SCHLÜSSEL, H. (1956b). *Klin. Wschr.*, **34**, 1290
SCHLÜSSEL, H. (1959). *Clin. Chim. Acta*, **4**, 748
SCHURR, P.E., THOMPSON, H.T., HENDERSON, L.M., WILLIAMS, J.N. Jr., and ELVEHJEM, C.A. (1950). *J. biol. Chem.*, **182**, 39
SEIGFFERS, M.J., SEGAL, H.L. and MILLER, L.L. (1963). *Am. J. Physiol.*, **205**, 1106
SEVE, B. and LAPLACE, J.P. (1974). *Ann. Zootech.*, **23**
SHIMADA, M.A. and ZIMMERMAN, D.R. (1973). *J. Anim. Sci.*, **36**, 245
SHINGLETON, W.W., WELLS, M.H., BAYLIN, G.J., RUFFIN, J.M., SAUNDERS, A., and DURHAM, N.C. (1956). *Surgery*, **38**, 134
SIKOV, M.R., THOMAS, J.M. and MALHUM, D.D. (1969). *Growth*, **33**, 57
SIMMONDS, W.J. (1955). *Aust. J. exp. Biol. med. Sci.*, **33**, 305
SNOOK, J. (1973). *Wld. Rev. Nutr. Diet.*, **18**, 121
SNOOK, J.T., and MEYER, J.H. (1964). *J. Nutr.*, **82**, 409
STAROVOJTOV, A.M. and MIROSNICENKO, N.A. (1968). *Sel'skohoz Biol.*, **3**, 245
STEIN, W.H. and MOORE, S. (1954). *J. biol. Chem.*, **271**, 915
STEINER, M. and GRAY, S.J. (1969). *Am. J. Physiol.*, **217**, 747
STOCKLAND, W.L., MEADE, R.J. and MELLIERE, A.L. (1970). *J. Nutr.*, **100**, 925
STOCKLAND, W.L., MEADE, R.J., TUMBLESON, M.E. and PALM, B.W. (1971). *J. Anim. Sci.*, **32**, 1143
STORDY, S.N., GREIG, J.H. and BOGOCH, A. (1969). *Am. J. dig. Dis.*, **14**, 463
STOROZUK, P.G. (1968). *Vop. Pitan.*, **27**, 50
STOROZUK, P.G. (1970). *Vop. Pitan.*, **29**, 3

SWENSEID, M.E., YAMADA, C., VINYARD, E., FIGUEROA, W.G. and DRENICK, E.J. (1967). *Am. J. clin. Nutr.*, **20**, 52

TANG, J., WOLF, S., CAPUTTO, R. and TRUCCO, R.E. (1959). *J. biol. Chem.*, **234**, 1174

TAYLOR, W.H. (1959). *Biochem. J.*, **71**, 73

TAYLOR, W.H. (1968). in *Handbook of physiology. 6. Alimentary canal*, Vol.V, Amer. Physiol. Soc., Washington

TKACHEV, E.Z. and PAKHNO, V.S. (1970). *Sbornik Nauchnykh Rabot, Vsesoyuznyi Nauchno-Issledovatel'skii Institut Zhivotnovodstva*, **20**, 25

TWOMBLY, J. and MEYER, J.H. (1961). *J. Nutr.*, **74**, 453

TYPPO, J.T., MEADE, R.J., NORDSTROM, J.W. and STOCKLAND, W.C. (1970). *J. Anim. Sci.*, **31**, 885

UGOLEV, A.M. (1965). *Physiol. Rev.*, **45**, 555

UGOLEV, A.M. (1966). *Nahrung*, **10**, 483

UHLEMANN, H. and POPPE, S. (1970). *Arch. Tierernähr.*, **20**, 165

Van SLYKE, D.D. and MEYER, D.M. (1912). *J. biol. Chem.*, **12**, 399

WALDMAN, J.A., MORELL, A.G., WOEHNER, R.D., STROBER, W. and STERNLEIB, (1967). *J. clin. Invest.*, **46**, 10

WELBOURN, R.B. and DOGGART, J.R. (1956). *Br. J. Surg.*, **44**, 320

WELLER, L.A., MARGEN, S. and CALLOWAY, D.H. (1969). *Am. J. clin. Nutr.*, **22**, 1577

WELLS, C.F. (1968). *Dissertation Absts. (B)*, **29**, 825B

WEST, C.D., WILSON, J.L. and EYLES, R. (1946). *Am. J. dis. Child.*, **72**, 251

WHEELER, P. and MORGAN, A.F. (1958). *J. Nutr.*, **64**, 137

WHITAKER, T.R. and PATRICK, H. (1971). Bulletin, Agricultural Experiment Station, West Virginia University

WIEPKEMA, P.R., ALINGH-PRINS, A.J., and STEFFENS, A.B. (1972). *Physiol. Behav.*, **9**, 759

WIGGANS, D.S. and JOHNSTON, J.M. (1959). *Biochem. biophys. Acta*, **32**, 69

WILSON, T.H. and WISEMAN, G. (1954). *J. Physiol. Lond.*, **123**, 116

WISEMAN, G. (1951). *J. Physiol. Lond.*, **114**, 78P

WISEMAN, G. (1953). *J. Physiol. Lond.*, **120**, 63

WISEMAN, G. (1955). *J. Physiol. Lond.*, **127**, 414

WISEMAN, G. (1956). *J. Physiol. Lond.*, **133**, 626

WISEMAN, G. (1968). in *Handbook of Physiology, Alimentary canal*, section 6, vol.3, *Intestinal absorption*, 1277. Eds C.F. Code and W. Heidel. American Physiological Society, Washington, D.C.

WISS, O. (1948). *Helv. Chim. Acta*, **31**, 2148

ZEBROWSKA, T. (1968). *Br. J. Nutr.*, **22**, 483

ZEBROWSKA, T. (1971). *Roczniki Nauk. Rolniczych*, **93B**, 77

ZEBROWSKA, T. (1973). *Roczniki Nauk. Rolniczych*, **95B-1**, 15, 135

ZEBROWSKA, T. and BURACZEWSKA, L. (1972a). *Roczniki Nauk. Rolniczych*, **94B-1**, 81

ZEBROWSKA, T. and BURACZEWSKA, L. (1972b). *Roczniki Nauk. Rolniczych*, **94B-1**, 97

ZEBROWSKA, T. and BURACZEWSKA, L. (1972c). *Zeszyty Problemowe Postepow Nauk Rolniczych*, **126**, 115

ZIEMLANSKI, S., CIESLAKOWA, D. and PALASZEWSKA, M. (1970). *Bulletin de l'Académie Polonaise des Sciences. Série des Sciences Biologiques,* **18,** 649

ZIEMLANSKI, S., CIESLAKOWA, D. and PALASZEWSKA, M. (1972). *Pol. Med. J.,* **11(4),** 881

ZIEMLANSKI, S., CIESLAKOWA, D., PLISZKA, B. and SZCZYGIEL, A. (1967). *Nahrung,* **11,** 559

ZIMMERMANN, D.R., NIELSEN, C. and SCHOENHEYDER, F. (1962). *Biochim. biophys. Acta,* **63,** 201

7

AMINO ACID AVAILABILITY

H. ERBERSDOBLER,
Institute für Tierphysiologie der Universität München, West Germany

Introduction

An organism is usually highly capable of adapting its various mechanisms for protein utilisation to many dietary changes. But sometimes all these mechanisms are not able to overcome the difficulties arising from some special food proteins or those resulting from extreme processing of the food. The consequence is a gap between the calculated and the actual supply of utilisable amino acids. The total quantity of amino acids which are present in proteins can be determined by chemical analysis after acid hydrolysis and these are sometimes only partially utilisable by the living organism. Some of the amino acids are said to be 'unavailable'. This discrepancy between the total amount and the actual available amino acids is difficult to calculate and leads to various errors in predicting the protein quality of food and feedstuffs. Furthermore, the high economic losses arising during the processing of food and feedstuffs can be reduced if the mechanisms of heat damage are clearly understood and all the physiological consequences considered.

Definition of the Term 'Amino Acid Availability'

As we know, the majority of amino acids enter the portal blood as free amino acids. Only small amounts penetrate the gut wall as amino acid derivatives, peptides or proteins. The majority of amino acids are utilised for protein synthesis but a varying proportion are metabolised. All available amino acids must be of potential use for protein synthesis, even if, at the moment of the test, they are then used as a source of energy. Normally, only small amounts leave the organism without any utilisation, either unchanged or slightly altered by way of the kidneys or as a result of other excretory mechanisms. Therefore all the amino acids (potentially of use in protein synthesis) which are absorbed and utilised can be defined as being 'available'.

Factors Which Impair the Amino Acid Availability

The main factors which reduce the availability of the amino acids are summarised in the following synopsis:

INCOMPLETE DIGESTION AND ABSORPTION CAUSED BY:
(1) Inaccessibility of the total protein because of indigestible cell walls, bulky structure, or too many cross-linkages in the molecule.
(2) Inhibition of the enzyme binding-site in the protein.
(3) The presence of protease inhibitors.
(4) Inhibition of amino acid absorption by peptides or peptide-like compounds.

There are many reasons for the occurrence of incomplete digestion. Sometimes the proteases are not able to penetrate through a non-digestible cell wall or to infiltrate a bulky inaccessible protein molecule. Mainly native and non-denatured proteins (Kakade, Hoffa and Liener, 1973; Fukushima, 1968) or proteins whose structure is extremely distorted by technological manipulations, are resistant to the attack of the enzymes. Many peptidases show a high specificity depending on the structure of the binding sites in the substrate (Shotton and Hartley, 1970; Steitz, Henderson and Blow, 1969). If these sites of contact are somehow changed, the action of the proteases is inhibited. Thus it is possible to impair the action of trypsin by blocking the ϵ-amino group of lysine. The presence of protease inhibitors in some food proteins is another factor which reduces their digestibility.

Sometimes the digestion of the food proteins is possible, but the absorption of the amino acids is inhibited. Several peptides or peptide-like compounds seem to inhibit the absorption of amino acids, as was first assumed by Buraczewski (1966), Buraczewska and Ford (1967).

Many amino acids or amino acid-derivatives can be absorbed but they are not at all or are only partially utilisable, or they are incompletely transformed into the respective active and available amino acid. Thus they are not utilised and are deaminated, or are soon excreted by the kidneys. The problems arising in connection with protein synthesis should not be discussed at present since these factors are hard to define and calculate.

The principal reason for all these factors reducing the amino acid availability is an excessive heat treatment of the proteins during the processing of the food and feedstuffs. The mechanisms of the heat-damage of proteins are varied. A mild heat treatment improves the digestibility by a denaturation of the native proteins and an inactivation of certain protease inhibitors. Extensive damage arises if higher temperatures are used for a longer period of time. In this case interactions between functional groups within the protein or with other food components like reducing sugars may occur. Mecham and Olcott (1947) first studied the reactions in pure proteins and suggested that the amino group of lysine reacted with carboxylic groups in the protein, thus forming new linkages resistant to the digestive enzymes.

The most important interaction, however, making an amino acid 'unavailable' is the so-called Maillard reaction (Maillard, 1912). It has been well-known for several years that in the case of proteins mainly lysine is involved in these 'browning reactions' since lysine can react with its free and reactive ϵ-amino groups thus forming firstly an unavailable lysine-sugar complex and then leading in further stages to a total destruction of the lysine molecule (e.g. Hodge, 1953; Heyns and Paulsen, 1960; Reynolds, 1963; 1965).

Estimation of Available Amino Acids in Food and Feedstuffs

The gap between the chemically determined total amount of amino acids and the amount actually available to the organism, can only finally be determined by experiments with higher animals. Methods using higher animals are usually expensive, time-consuming and are complicated by many sources of error. Nevertheless all other methods ultimately have to be compared with the results of animal experiments. The most important procedures are growth assays, *in vivo* digestibility measurements and methods using the blood amino acid level as criterion.

The growth assays or feeding experiments are, in most cases, not highly specific. The technique depends on the amino acid content of the basic diet, which must also be determined by chemical methods. Furthermore it depends on the amino acid requirements of the animal used and on other various criteria such as the protein content, the amino acid balance, the sources of energy and other factors complicating the response of the animals in the test. Moreover this technique has the disadvantage of evaluating only one amino acid at a time. Results of earlier growth experiments compared with the chemical determination of available lysine made in our laboratory (Brüggemann, Erbersdobler and Zucker, 1969) can be seen in *Figure 7.1*. There was a good correlation between the fluor-dinitrobenzene (FDNB)-available lysine in fishmeals or meat and bone meals and the Protein Efficiency Ratio (PER) in rats, which were fed a lysine deficient cereal diet supplemented with the tested fishmeals or meat and bone meals of different quality. Thus it was possible to classify the protein quality of the tested protein sources. Similar results have been shown by many other laboratories (e.g. Boyne, Carpenter and Woodham, 1961; National Research Council, 1963; Erbersdobler and Zucker, 1965; Zucker, 1962). But in most cases only a good relative quality classification but not absolutely valid data were obtained from the growth assays.

Since it is possible to use mixtures of crystalline amino acids instead of proteins or to add an amino acid mixture to a certain

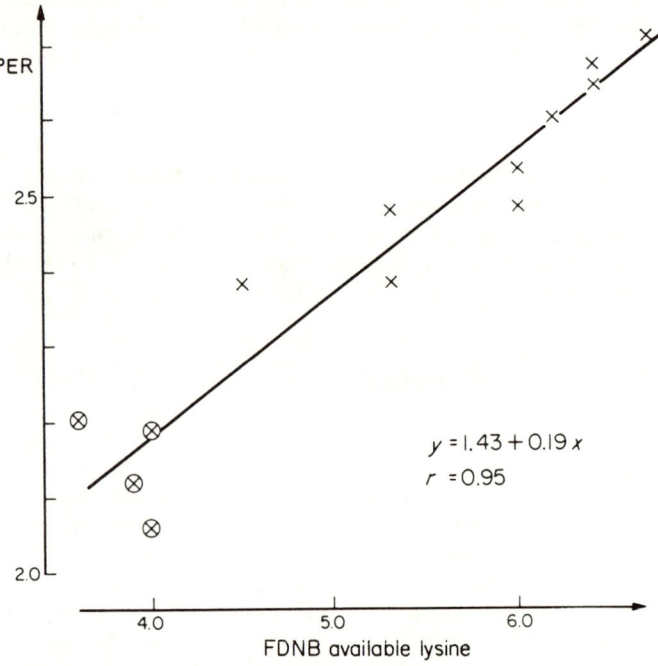

Figure 7.1 Relationship between FDNB-available lysine in nine fish meals (×) or four meat and bone meals (⊗) and the Protein Efficiency Ratio (PER) for rats given diets with 2.4 per cent protein from the fish meals or the meat and bone meals together with eight per cent protein from wheat and barley (Brüggemann et al., 1969)

protein some but not all of the previously mentioned problems could be solved (Dean and Scott, 1966; Dobson, Anderson and Warnick, 1964; König, Erbersdobler and Gropp, 1973; Velu, Scott and Baker, 1972).

For several years many authors have recommended the use of balance techniques in measuring the amino acid availability (e.g. Eggum, 1973; Kuiken and Lyman, 1948; Poppe and Meier, 1971). It must be realised, however, that the microflora of the large intestine may destroy or change some unavailable amino acids and they would not then be registered in this way (Erbersdobler, Gunsser and Weber, 1970; Salter and Coates, 1969; Zebrowska, 1973). Therefore the results of these so-called amino acid digestibility studies can be misleading if proteins with a reduced digestibility were tested (Erbersdobler, 1967; Erbersdobler, Weber and Gunsser, 1972). On the other hand, a fair estimate of the protein quality can be made if highly damaged proteins were tested or if special techniques measuring the digestibility of the amino acids have been used (Erbersdobler and Riedel, 1972; 1973; Nesheim and Carpenter, 1967; Payne et al., 1971; Soares et al., 1971; Varnish and Carpenter, 1971). Unavailable amino acids, which are absorbed but deaminated or excreted without utilisation, however, cannot or can only be partially detected in this way.

Another way of evaluating the amino acid requirements or the amino-acid availability is by blood amino acid measurements. By using amino acid levels in the peripheral blood, either a deficiency or an excess of an amino acid and sometimes also the requirements, can be determined. (Dean and Scott, 1966; Hewitt and Lewis, 1972; Longenecker and Hause, 1959; Smith and Scott, 1965). Measurements of the influx of amino acids into the portal blood after a test meal, were made mainly in order to estimate the amino acid availability in different protein sources (Dent and Schilling, 1949; Denton and Elvehjem, 1954; Erbersdobler, 1967; Erbersdobler, Dümmer and Zucker, 1968; Erbersdobler, 1969; Erbersdobler, Weber and Gunsser, 1972; Goldberg and Guggenheim, 1962).

Figure 7.2 shows, as an example, lysine concentrations in the portal plasma after a test meal of either unheated or heat-treated (90°C, 2 h) isolated soya protein. Although the example shows clear differences between both groups of rats, several troubles can arise during the test. The introduction of a-amino isobutyric acid as an internal standard in the diets has many advantages (Smith, 1966; Weber, 1972). High protein concentrations in the test meal and the adaptation of the animals to the conditions of the test plus a training for absolutely spontaneous feed consumption must also be recommended (Erbersdobler, 1967; Scharrer, Erbersdobler and Zucker, 1967). In most cases, therefore, only model proteins and not simple feedstuffs can be tested exactly. Further difficulties arise if exact values for the availability have to be calculated, since the fasting level in the portal blood is not the adequate basal reference value. Values obtained after feeding a protein-free meal (leading to decreased concentrations of the amino acids) would even be better for this purpose, as is shown by the example in *Figure 7.2*.

IN VITRO PROCEDURES

In order to simulate the biological digestion by enzymic hydrolysis *in vitro,* several proteases have been tested in the past (e.g. Akeson and Stahmann, 1964; Ford and Salter, 1966; Mauron *et al.,* 1955; Menden and Cremer, 1966). Above all, the complex digestion procedures with two or more successive enzymes and more recently, the utilisation of the very potent microbial enzymes, have been successful. In this way, qualitative estimations of the digestibility and the quality of the test material can be made. But absolutely valid data cannot be obtained. It seems to be impossible to imitate the concerted action of the digestive enzymes in the organism. Moreover it is impracticable, even with molecular gel filtration, to separate the available amino acids and peptides from the smaller unavailable peptides or peptide-like compounds. The final hydrolysis with hydrochloric acids before the amino acid analysis of the different fractions, also splits most of the unavailable linkages.

Table 7.1 shows a comparison of amino acid availability data obtained by different procedures of measurement with the same test material. The methods of *in vitro* 'digestion' with enzymes seem to underestimate the availability of the amino acids in most cases.

Table 7.1 Effect of heat treatment of casein on results of several protein quality measurements

Heat treatment: Method	Casein + 10% Glucose, 90°C, 24 h				Casein + 20% Glucose, 105°C, 24 h			
	Digestibility (rats)	Portal plasma test	Hydrolised pronase	Hydrolised pepsin papain	Digestibility (rats)	Portal plasma test	Hydrolised pronase	Hydrolised pepsin papain
Methionine	100	118	97	76	79	50	39	41
Lysine[1]	67	40	51	61	53	37	24	16
Threonine	90	110	85	84	75	45	34	47
Valine	93	111	72	88	90	84	33	47
Isoleucine	87	116	91	84	70	69	39	48
Leucine	96	126	93	84	74	85	37	46
Phenylalanine	92	104	83	80	74	87	32	48
Tyrosine	86	94	90	80	71	111	31	34
Arginine	87	106	75	70	78	43	22	26
Histidine	76	166	67	77	71	62	37	31
Aspartic acid	85	119	90	79	76	41	37	45
Glutamic acid	94		98	78	70		49	44
Serine	94	120	94	85	72	45	46	55
Glycine	81	90	95	85	95	74	30	45
Alanine	87	89	97	86	89	75	36	48
Proline	94	86	108	77	89	75	36	48

[1] The lysine content in the two proteins at 90°C and at 105°C was 78 per cent and 54 per cent respectively of the control; the FDNB available lysine was 55 per cent and 36 per cent of the control

All values are given as percentage of corresponding values in the unheated controls. Cystine and tryptophan were not determined

Some of these previously mentioned problems can be overcome by using certain microbial organisms, such as the bacterium *Streptococcus zymogenes*, which is used mainly for the determination of available methionine after a pre-treatment step of the test material with papain (Ford, 1962). With the protozoan *Tetrahymena pyriformis* (Fernell and Rosen, 1956; Baum, 1966) the availability of lysine may be tested (Shorrock and Ford, 1973). Many other methods, however, are not definitively established or they are either rather complicated or susceptible to various disturbances such as those caused by other food ingredients. Moreover, the specificity of the enzymic outfit of the microorganisms can be quite different from the conditions in higher animals (Haenel and Kharatyan, 1973).

Since lysine is the most important nutritionally damaged amino acid, many chemical procedures for the determination of available lysine have been developed. The fluorodinitrobenzene (FDNB) procedure for available lysine (Carpenter and Ellinger, 1955) has been found satisfactory when applied to fish meals, meat and bone meals (e.g. Carpenter, 1960; Erbersdobler and Zucker, 1964; Pritchard et al., 1964) and milk

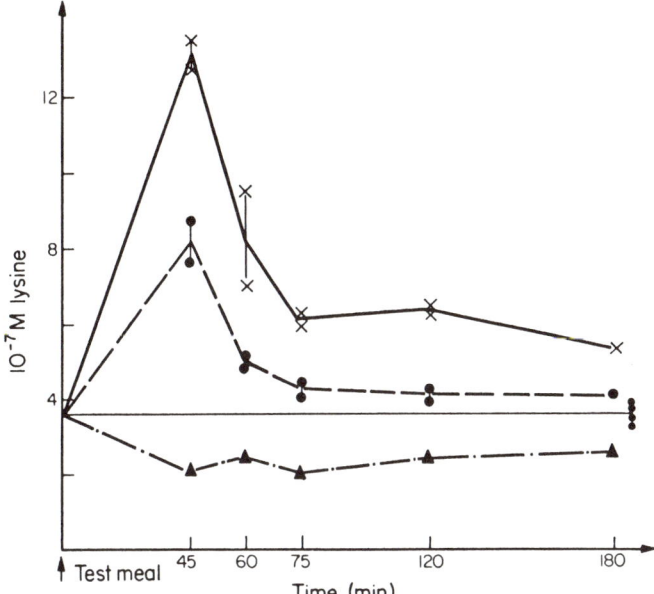

Figure 7.2 Concentrations of lysine in the portal plasma (each point = blood from three rats, which were bled within 10 minutes) after a single meal of 1.5 g protein from
× ———— × *'control' = unheated isolated soya protein*
● — — — ● *'heated' = soya protein heated together with glucose*
▲ — · — ▲ *'prot.fr.' = protein free diet*
Calculations using the mean values for the determination of available lysine

$$\frac{\text{sum of values 'heated'}}{\text{sum of values 'control'}} \cdot 100 = 65 \text{ per cent}$$

$$\frac{(\text{sum of values 'heated'} - \text{fasting levels}) \cdot 100}{(\text{sum of values 'control'} - \text{fasting levels})} = 36 \text{ per cent}$$

$$\frac{(\text{sum of values 'heated'} - \text{sum of 'prot.fr.' values}) \cdot 100}{(\text{sum of values 'control'} - \text{sum of 'prot.fr.' values})} = 50 \text{ per cent}$$

The chemically determined availability of lysine (FDNB) = 66 per cent

products (Erbersdobler and Zucker, 1966; Erbersdobler, 1970; Pion, 1961; Schober and Prinz, 1956). Useful values can also be obtained for vegetable proteins although some difficulties may arise, especially with carbohydrate rich material (Booth, 1971). The details of the FDNB-method and of other more or less varied procedures of labelling the free and therefore available amino group, have been reported most recently by Carpenter (1973).

A procedure directly measuring one of the most important reaction products of the Maillard condensation is the determination of furosine. This useful indicator was detected in acid hydrolysates of milk products some years ago and identified as a following product of fructoselysine and lactuloselysine (Erbersdobler and Zucker, 1966; Erbersdobler, 1967; Brüggemann and Erbersdobler, 1968). Heyns,

Heukeshoven and Brose, (1968) and Finot *et al.* (1968) determined the structure of the compound as N-(2-furosyl methyl)L-lysine and named it as furosine. A second derivative was also found in our chromatographs but was not definitely appointed to fructoselysine (Erbersdobler, 1967). Finot *et al.* (1969) identified the compound as another derivative of fructoselysine, determined the structure and named it as pyridosine.

The following scheme shows the initial steps of the Maillard reaction with the formation of furosine during the hydrolysis *(Figure 7.3).*

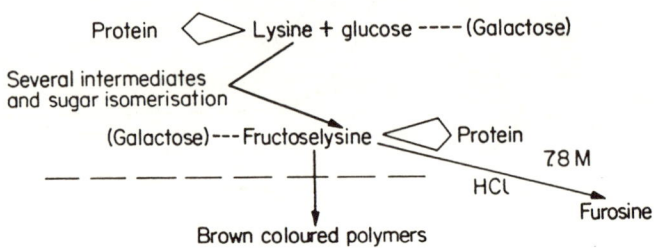

Figure 7.3 The Maillard reaction

Furosine proved to be a useful indicator for the determination of the unavailable lysine in milk products and other food proteins. It can be used as a qualitative indicator in order to find out if lysine-glucose (or -lactose) reactions have occurred in a certain food. Furthermore, quantitative estimations of the extent of the damage are possible too. The recovery of furosine out of fructose-lysine has been found to be about 30 per cent if the hydrolysis is done with 7.75 N hydrochloric acid under reflux (Finot, 1973). In our calculations (Brandt and Erbersdobler, 1973; Erbersdobler, 1970) using other ninhydrin equivalents, we have found somewhat higher values amounting to 40 per cent. In agreement with Finot (1973) we have found that during hydrolysis of milk products 50 per cent of the fructose-lysine is split to lysine, whereas the rest passes into pyridosine and some other intermediates. In other food proteins, the amount of pyridosine can be very high as Sulser and Buchi (1969) have shown in their work.

Regarding these relations, the following calculations can be made:

Lysine bound in furosine $\times$ 2.5 = inactivated lysine
Lysine content + 50 per cent of the inactivated lysine = total lysine
Lysine content − 50 per cent of the inactivated lysine = available lysine

These calculations were applied to a number of furosine values from different foods (Brandt and Erbersdobler, 1973; unpublished data). The results are given in *Table 7.2.* As can be seen from these values some foods contained considerable amounts of inactivated lysine connected with a decreased content of available lysine. In many cases the total

Table 7.2 The lysine and furosine contents and the calculated amounts of inactivated lysine, total lysine and available lysine (all values in g/16 gN) in some commercial milk products

	No. of samples	lysine	furosine	calculated mean values of		
				total lysine	inactivated lysine	available lysine
Spray dried skim milk	33	8.5±0.5	0.4±0.3	8.8	0.6	8.2
Roller dried skim milk	25	8.2±0.6	0.8±0.5	8.8	1.1	7.6
Spray dried whey	33	8.3±0.3	0.3±0.3	8.6	0.5	8.0
Roller dried whey	16	5.9±1.2	2.7±0.8	7.9	3.9	3.9
Baby food	80	8.0±0.6	0.5±0.4	8.3	0.7	7.7
Whey concentrates[1]	47	7.4±1.2	1.1±1.2	8.2	1.6	6.6
Diets for veal calves	35	8.2±0.5	0.5±0.3	8.6	0.7	7.8

[1] partially deprived with lactose

lysine content was low, indicating that not only an inactivation but also a destruction of lysine had happened. Similar findings and conclusions were reported from Finot (1973).

The Biological Significance of Unavailable Nitrogenous Compounds

As previously mentioned, the proteases do not attack a protein whose reactive sites are unavailable. It therefore seems to be of some importance to study the biological effects which arise if proteins are not digested or if unavailable amino acids or peptide-like compounds pass through the organism without utilisation. The toxicological questions, which are also very important and which are waiting to be solved in later experiments, cannot be discussed in this connection.

It can be demonstrated, as an example, that undigested proteins or unavailable peptides do not influence a special mechanism of stomach emptying (*Table 7.3*).

In experiments with adult rats previously adapted to the different diets, the stomach emptying occurred quite regularly up to a certain amount of protein (1.8 - 2.3 g dry matter with 40 per cent protein). Diets rich in protein (60-80 per cent) remained in the stomach in greater amounts than the test meals with the lower protein content. The decrease in the rate of stomach emptying was accompanied by highly increased amino acid concentrations in the portal and (not registered in *Table 7.3*) peripheral blood. This delayed stomach emptying could not be observed if protein-rich test meals with heat-damaged, poorly-digestible proteins were given. In diets with a high

Table 7.3 Rates of stomach emptying compared to the amino acid concentrations in the portal plasma of adapted rats fed test meals of 1.8 - 2.3 g dry matter with different protein content

Protein content of the diets	Values (Mean ± SD)				
	10%	20%	40%	60%	80%
mg dry matter released out of the stomach					
Soya protein unheated	1042±269	1210±150	1159±171	846±213	705±166[1]
Soya protein heat-treated	1005±112	1078±132	1142±306	1009±240	1091±187
Sum of nine amino acids[2] in the portal plasma (u Moles/100 ml)[3]					
Soya protein unheated	273-284	317-374	464±35	518-561	761±85
Soya protein heat-treated	244-272	243-259	318±52	317-392	491±32

[1] Significantly lower
[2] Threonine; Glycine; Alanine; Valine; Methionine; Isoleucine; Leucine; Phenylalanine; Tyrosine
[3] For the plasma amino acid measurements the blood of two rats in each group was pooled. Values with standard deviation = more than three single results

protein content of 80 per cent, the rate of stomach emptying perfectly corresponded to the amino acid levels in the portal blood and in this way also with the availability of the amino acids (Erbersdobler, 1973a; 1973b; Weber, 1972). This seems to prove that this special feedback mechanism (Crider and Walker, 1948; Peraino et al., 1959) depends on the available amino acids and is not related to the total supply of proteins, e.g. poorly digestible proteins.

FRUCTOSELYSINE AS A MODEL OF AN UNAVAILABLE COMPOUND
As shown previously and in Table 7.2, ε-fructoselysine is the most important lysine-sugar complex, representing the major part of blocked lysine after heat treatment of proteins with glucose or lactose. Some food proteins contained more lysine bound as fructoselysine or lactuloselysine (galactose-fructoselysine after the reaction with lactose) than available lysine.

The reaction of the ε-amino group with glucose seems to hinder the action of trypsin and possibly some other enzymes (Folk, 1956; Melnick and Oser, 1949). After feeding proteins rich in ε-fructoselysine to rats, the author could not detect a greater amount of the fructoselysine group in the portal blood (measured as furosine, after the acid hydrolysis - Brüggemann and Erbersdobler, 1968). The results of Ford and Shorrock (1971), Valle-Riestra and Barnes (1970) and Finot (1973) lead to the conclusion that less than 10 per cent of the lysine units in the protein substituted at the ε-amino group with fructose, are liberated out of the peptide chain.

Fructoselysine itself is not actively transported across the intestine, as can be seen in Table 7.4 (Alfke, 1974). An uptake by diffusion

Table 7.4 Transport of fructoselysine by everted sacs of the upper jejunum of rats (n = 9)

Dry weight of the everted sacs (mg)	75±11
Final volume of the serosal medium (ml)	0.89±0.05
s/m concentration gradients	0.68±0.04

Initial volume of the serosal medium = 1 ml;
Initial volume of the mucosal medium = 15 ml.

An incubation medium of 1 mMol fructoselysine with a specific activity of 56 µCi/Mol in Krebs Henseleit bicarbonate buffer (pH 7.3) was used. The fructoselysine proved to be stable during the time of incubation (1 h, 37°C)

however, is possible. *Figure 7.4* shows the amounts of ^{14}C labelled lysine and ε-fructoselysine remaining in jejunal segments of rats after an incubation for various periods of time (Alfke, 1974, Erbersdobler *et al.*, 1974). As can be seen in the figure, this lysine-sugar complex is absorbed at a relatively slow rate compared to lysine (Erbersdobler, 1971; 1973). The compound presumably has a considerable affinity for gut wall components, which possibly leads to the high concentration differences responsible for the delayed but significant diffusion rate. This is confirmed by the work of Finot, who (in balance studies) found a recovery of 60-70 per cent of orally administered ε-fructoselysine in urine. However, α-fructoselysine is only poorly absorbed, but has the same high affinity for the gut wall (Alfke, 1974, Erbersdobler *et al.*, 1974).

The distribution of absorbed fructoselysine also occurs by diffusion. In this way, fructoselysine also penetrates into the intracellular compartments and reaches the foetus via the placental barrier, as could be shown in experiments with pregnant guinea pigs. *Figure 7.5* shows the first results of our trials comparing the behaviour of fructoselysine with mannitol and α-amino isobutyric acid (Erbersdobler *et al.*, 1974).

Since an active re-absorption presumably does not take place in the kidneys, excretion will occur rapidly. In our experiments with guinea pigs, two hours after an intravenous injection of the labelled fructoselysine, about 20 per cent was excreted into the urinary bladder. In balance trials with rats (Erbersdobler and Schlecht, unpublished) about 90 per cent of the injected ^{14}C-fructoselysine was excreted after 12 hours while only small amounts were still retained in the organism (*Table 7.5*).

Fructoselysine, which is not liberated out of the peptide chain by digestion or not absorbed, reaches the hind gut, as we could show with germ-free chicks (unpublished data). In conventionally bred animals, the lysine-sugar complexes can be destroyed and presumably utilised by the micro-organisms in the hind gut (Erbersdobler, Gunsser and Weber, 1970). Thus misleading results of digestibility experiments can be obtained, since the unavailable compounds cannot be detected by faecal analysis.

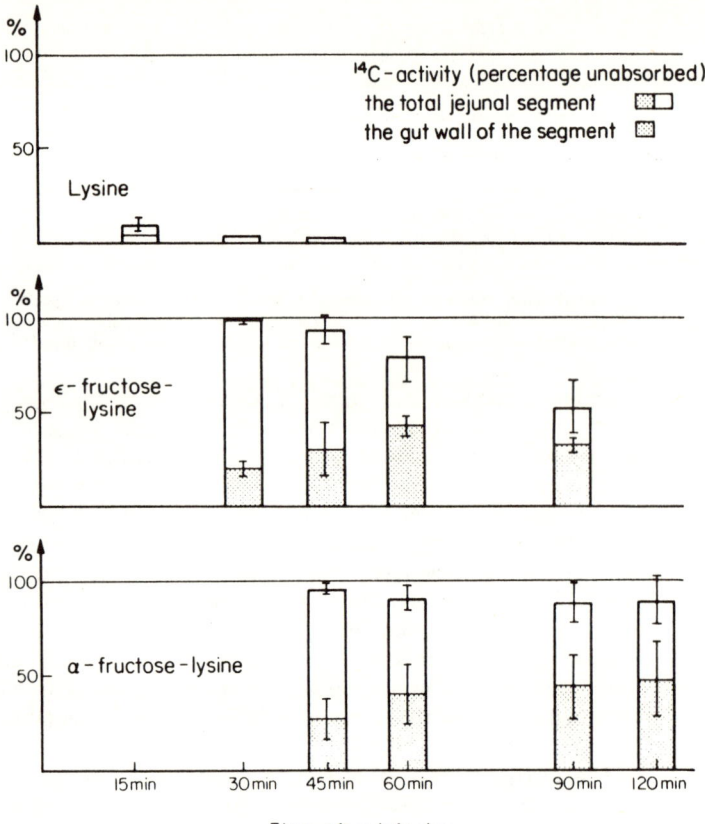

Figure 7.4 Intestinal absorption of lysine and fructose-lysines from jejunal segments of rats (each value = 5-7 animals).

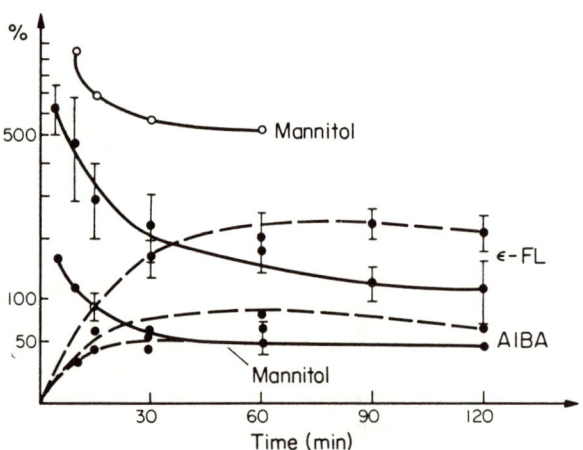

Figure 7.5 Distribution and transport across the placental wall of ε-fructoselysine in guinea pigs. Concentrations of ε-fructoselysine, mannitol and α-amino-isobutyric acid in 1 ml plasma of the maternal (———) and foetal (— — —) blood. The intravenous injected activity per g of the animal = 100 per cent

Table 7.5 Recovery of ^{14}C-activity[1] 12 hours after the intravenous injection of labelled ε-fructoselysine to three male rats of about 200 g live weight. Injected activity (4 μmoles ε-fructoselysine with a specific activity of 17 mCi/mol) = 100 per cent

Sample:	urine[1]	kidneys	faeces + gut content	liver	spleen
Activity in per cent	86-92	<0.5	1.4-2.0	1.0-1.6	<1.0

[1] Over 80 per cent of the specific activity (after ion-exchange chromatography) in the urine could be attributed to the intact fructoselysine molecule. Less than 5 per cent could be found as labelled lysine. Samples of blood and muscle showed only traces of activity

OTHER REACTIONS OF LYSINE

As has been shown, lysine having a non-protein-bound free ε-amino group can react with carbonyl groups of reducing sugars, thus forming unavailable complexes. Not all of the lysine derivatives substituted at the ε-amino group are unavailable, however. Smaller groups such as methyl- or acetyl-substitutes are at least partially utilisable (Bjarnason and Carpenter, 1969; Leclerc and Benoiton, 1968). Also, ε-propionyl-lysine is not utilised by the rat, but is partially hydrolysed to lysine in the kidneys of the chick (Varnish and Carpenter, 1970). In heat-damaged pure proteins, glutamyl-lysine linkages have been hypothesised (Bjarnason and Carpenter, 1970; Ford and Shorrock, 1971; Mecham and Olcott, 1947). Waibel and Carpenter (1972) however, have found that this compound is fully available as lysine.

On the other hand, these linkages seem to hinder the enzymic cleavage of protein during its digestion. This phenomenon does not limit the adverse effect to only glutamic acid and lysine, the two directly involved amino acids but presumably these cross-linkages also impair the access of the enzymes to greater sections of the protein molecule. In this way the availability of the other amino acids and the digestibility of the total protein may also be depressed. It can actually be shown by heating proteins alone, that the lysine damage is not so predominant as in proteins heated together with reducing sugars (Carpenter, 1973; Bjarnason and Carpenter, 1970; Erbersdobler, Weber and Gunsser, 1972).

Of the numerous compounds also capable of reacting with lysine, the complexes with gossypol in cotton-seed are well-known (Conkerton and Frampton, 1959) whereas the possibilities of reactions with fat components or with oxidised fat have only recently been fully understood (Andrews et al., 1965; Carpenter, 1973; Lea and Carpenter, 1960; Narayan, Sugai and Kummerow, 1964; Pokorny and Janicek, 1968; Roubal and Tappel, 1966; Weber, 1972).

DAMAGE OF OTHER AMINO ACIDS

Apart from the previously mentioned indirect damage to the digestibility of all amino acids caused by the cross-linkages of lysine with glutamic or aspartic acids or their amides, there is also direct damage to several amino acids. The hydroxyl group of threonine is said to react in a similar way to that of lysine with functional carboxylic groups (Ford, 1973; Mauron, 1970; Mecham and Olcott, 1947).

The use of amino acid supplementation of foods and feed mixtures also makes it necessary to consider the reactions of free amino acids with reducing sugars. Hagan et al. (1970) described fructoseglycine to be fully available as a source of nitrogen. Horn, Lichtenstein and Womack, (1968) have found that fructosemethionine is nearly (80 per cent) as effective as free methionine. Sgarbieri et al. (1973) worked with fructoseleucine and fructosetryptophan in rats, and found them to be highly unavailable. In our investigations α-fructoselysine proved to be only poorly absorbed compared to ϵ-fructoselysine, as can be seen in *Figure 7.4* (Alfke, 1974; Erbersdobler et al., 1974).

The sulphur-containing amino acids, especially cystine and cysteine are also highly sensitive to external influences. Cystine and cysteine are rapidly degraded if the protein in food and feedstuffs is heat-treated or oxidised (Bjarnason and Carpenter, 1970; Erbersdobler, Dümmer and Zucker, 1968; Erbersdobler, 1967; Miller, Carpenter and Milner, 1965; Schöberl, 1941). Heat treatment either results in a partial or total destruction of the structure or in a formation of new compounds. Alkali treatment of proteins, especially when combined with heat, leads to the formation of lysinoalanine, involving cystine and lysine in the reaction (Bohak, 1964; De Groot and Slump, 1969; Patchornik and Sokolovsky, 1964). Bjarnason and Carpenter (1970) found greater amounts of H_2S and other volatile sulphur compounds as break-down products of cystine, in heated proteins. Oxidative processes lead to the formation of cysteic acid which is nutritionally unavailable (Miller, Carpenter and Milner, 1968). Nearly all these reactions of cystine cause a total destruction of its former molecular structure. For this reason damage is easily detectable by analytical procedures.

Methionine, which is highly resistant to high temperatures (*Table 7.1*) is easily changed to methionine sulphoxides and further to methionine-sulphone by oxidation. Nutritional studies have shown that methionine sulphoxides, but not its sulphone, can replace methionine to a high degree (Miller, Carpenter and Milner, 1968; Njaa, 1962). In very young rats, the methionine sulphoxides also seem to be unavailable (Ellinger and Palmer, 1969). Methionine sulphoxides can be formed by the reaction of proteins with lipid peroxides (Tannenbaum, Barth and le Roux, 1969; Weber, 1972). A hydrogen peroxide treatment, commonly used for the sterilisation of milk or whey and for the bleaching of various foods, also leads to increasing quantities of methionine sulphoxides.

Cuq et al. (1973) conclude from their *in vitro* digestibility studies with pronase, that protein-bound methionine oxidised to sulphoxides is only poorly digested and therefore highly unavailable to the animal.

Table 7.6 *The concentrations of methionine and methionine sulphoxides in the portal plasma of rats (in µ Mol/100 ml) 1-4 h after a single meal of casein heated alone or with oil or oxidised oil (24 h. 110°C)*

	hours after giving the test meal				
	0	1	2	3	4
TRIAL 1					
Casein heated alone					
Methionine	5	15	15	13	-
Methionine sulphoxides	0[1]	0	0	0	-[2]
Casein heated with oil (Peroxide value = 208[3])					
Methionine	5	12	12	9	-
Methionine sulphoxides	0	7	6	4	-
TRIAL 2					
Casein heated alone					
Methionine	6	22	-	-	-
Methionine sulphoxides		0	-	-	-
Casein heated with oil (Peroxide value = 261[3])					
Methionine	6	18	19	15	12
Methionine sulphoxides	0	6	5	4	4
Casein heated with oxidised oil (Peroxide value = 481[3])					
Methionine	6	17	16	16	13
Methionine sulphoxides	0	6	7	8	6

[1] Not detectable or only small traces
[2] Not determined
[3] µ equivalents per g fat. The samples were heated in open air

Experiments in our laboratory (Erbersdobler, 1969; Erbersdobler, 1973; Weber, 1972) however, have shown that methionine sulphoxides appear in high amounts in the portal blood of rats after a test meal of casein that had previously been heated with oxidised oil in the open air. The results of the two experiments are given in *Table 7.6*. The rapid influx and the relatively high amounts of methionine sulphoxides indicate that the methionine derivative must have been liberated out of the peptide chain by digestion in the intestine, which is perhaps less specific than the pronase used by Cuq *et al.* It must be definitively proved by further experiments if the oxidation of the peptide-bound methionine to sulphoxides is such an important factor in the methionine availability as Ellinger and Palmer (1969), Cuq *et al.* (1973) and others (anonymous in Nutr. Rev. 1973) have suggested.

Summary

The amino acid availability is defined as that proportion of the amino acids present in a protein, which can be liberated by digestion, absorbed through the intestinal wall and which is not excreted without utilisation. The principal reason for the reduction of the amino acid availability is heat damage occurring during the processing or storage of food and feedstuffs.

Biological methods for determining a reduced availability are time-consuming, expensive and not always sufficiently specific. Sometimes they can be misleading, as is the case with balance techniques, which are subject to the effects of the activities of the microflora in the hind gut. The *in vitro* techniques and chemical procedures are either not specific enough or limited to only a few amino acids or special situations.

The reduction of the rate of stomach emptying that can be noticed after feeding high quantities of protein, cannot be observed if the amino acids of the protein are unavailable. The special feedback mechanism regulating the rate of stomach emptying seems therefore to be mainly related to the supply of available amino acids.

The typical behaviour of an unavailable compound is shown by fructoselysine. Fructoselysine is only poorly liberated by the digestive enzymes, it is absorbed by diffusion, not utilised by the organism, and soon excreted, mainly by the kidneys, without undergoing any changes. Therefore, fructoselysine can be regarded as the perfect model of an unavailable amino acid, demonstrating the biological significance of unavailable nitrogenous compounds. Lysine is involved in many other reactions influencing the digestion and, by this, the utilisation of proteins. In this way, the reaction involving lysine as well as certain other amino acids can reduce the availability of amino acids in general by disturbing the protein digestion as a whole.

The sulphur-containing amino acids are sensitive to an oxidative influence and in the case of cystine also to heat-damage. It has not been finally proved whether the oxidation of methionine to its sulphoxides is an important factor in methionine availability.

References

AKESON, W.R. and STAHMANN, M.A. (1964). *J. Nutr.*, **83**, 257
ALFKE, F. (1974). Dissertation; University of Munich, West Germany
ANDREWS, J.S., BJORKSTEN, F.J., TRENK, B.F., HENICK, A.S. and KOCH, R.B. (1965). *J. Am. Oil Chem. Soc.*, **42**, 779
BAUM, F. (1966). *Nahrung*, **10**, 453, 571
BJARNASON, J. and CARPENTER, K.J. (1969). *Br. J. Nutr.*, **23**, 859
BJARNASON, J. and CARPENTER, K.J. (1970). *Br. J. Nutr.*, **24**, 313
BOHAK, Z. (1964). *J.biol. Chem.*, **239**, 2878
BOOTH, V.H. (1971). *J. Sci. Fd. Agric.*, **22**, 658
BOYNE, A.W., CARPENTER, K.J. and WOODHAM, A.A. (1961). *J. Sci. Fd Agric.*, **12**, 832

BRANDT, A. and ERBERSDOBLER, H. (1973). *Landw. Forsch.*, **28/2**, *Sonderheft*, 115
BRUGGEMANN, J. and ERBERSDOBLER, H. (1968). *Z. Lebensmittelunters. u. Forsch.*, **137**, 137
BRUGGEMANN, J., ERBERSDOBLER, H. and ZUCKER, H. (1969). *Z. Tierphysiol. Tierernähr. Futtermittelk.* **25**, 128
BURACZEWSKI, S. (1966). Ph.D. Thesis; University of Reading, England
BURACZEWSKI, S., BURACZEWSKA, L. and FORD, J. (1967). *Acta biochim. pol.*, **14**, 121
CARPENTER, K.J. (1960). *Biochem. J.*, **77**, 604
CARPENTER, K.J., (1973). *Nutr. Abstr. Rev.*, **43**, 604
CARPENTER, K.J. and ELLINGER, G.M. (1955). *Biochem. J.*, **61**, 11
CONKERTON, E.J. and FRAMPTON, V.L. (1959). *Archs Biochem. Biophys.*, **81** 130
CRIDER, R.J. and WALKER, S.M. (1948). *Archs Surg.*, **57**, 10
CUQ, J.L., PROVANSAL, M., GUILLEUX, F. and CHEFTEL, C. (1973). *J. Fd Sci.*, **38**, 11
DEAN, W.F., and SCOTT, H.M. (1966). *J. Nutr.*, **88**, 75
DE GROT, A.P. and SLUMP, P. (1969). *J. Nutr.*, **98**, 45
DENT, C.E. and SCHILLING, J.A. (1949). *Biochem. J.*, **44**, 318
DENTON, A.E. and ELVEHJEM, C.A. (1954). *J. biol. Chem.*, **206**, 455 and 449
DOBSON, D.C., ANDERSON, J.O. and WARNICK, R.E. (1964). *J. Nutr.*, **82**, 67
EGGUM, B.O. (1973). Ph.D. Thesis; University of Copenhagen
ELLINGER, G.M. and PALMER, R. (1969). *Proc. Nutr. Soc.*, **28**, 42A
ERBERSDOBLER, H. (1967). *Habilitationsschrift; Munich*
ERBERSDOBLER, H. (1969). *Wiss. Z. Univ. Rostock*, **18**, M1/2, 211
ERBERSDOBLER, H. (1970). *Milchwissenschaft*, **25**, 280
ERBERSDOBLER, H. (1971). *Z. Tierphysiol. Tierernähr. Futtermittelk.*, **28**, 171
ERBERSDOBLER, H. (1973a). in *Proteins in Human Nutrition.* Eds J.W.G. Porter and B.A. Rolls p.453; Academic Press, London and New York
ERBERSDOBLER, H. (1973b). *Z. Tierphysiol. Tierernähr. Futtermittelk.*, **32**, 21
ERBERSDOBLER, H., DÜMMER, H. and ZUCKER, H. (1968). *Z. Tierphysiol. Tierernähr. Futtermittelk.*, **24**, 136
ERBERSDOBLER, H., GUNSSER, I. and WEBER, G. (1970). *Zentrl. Vet.Med. A.*, **17**, 573
ERBERSDOBLER, H., WEBER, G. and GUNSSER, I. (1972). *Z. Tierphysiol. Tierernähr. Futtermittelk.*, **29**, 325
ERBERSDOBLER, H., HUSSTEDT, W., ALFKE, F., BRANDT, A. and CHELIUS, H.H. (1974). *Z. Tierphysiol. Tierernähr. Futtermittelk.*, **33**, 202
ERBERSDOBLER, H. and RIEDEL, G. (1972). *Arch. Geflügelkde.*, **36**, 218
ERBERSDOBLER, H. and RIEDEL, G. (1973). *Proc. Int. Amino Acid. Symp. Brno, CSSR, Suppl. A*, **29**
ERBERSDOBLER, H. and ZUCKER, H. (1964). *Z. Tierphysiol. Tierernähr. Futtermittelk.*, **19**, 244

ERBERSDOBLER, H. and ZUCKER, H. (1965). *Kraftfutter,* **48,** 161
ERBERSDOBLER, H. and ZUCKER, H. (1966). *Milchwissenschaft,* **21,** 564
FERNELL, W.R. and ROSEN, G.D. (1956). *Br. J. Nutr.,* **10,** 143
FINOT, P.A. (1973). in *Proteins in Human Nutrition.* Eds. J.W.G. Porter and B.A. Rolls. p.501, Academic Press, London and New York
FINOT, P.A., BRICOUT, J., VIANI, R. and MAURON, J. (1968). *Experientia,* **24,** 1097
FINOT, P.A., BRICOUT, J., VIANI, R. and MAURON, J. (1969). *Experientia,* **25,** 134
FOLK, J.E. (1956). *Archs Biochem. Biophys.,* **64,** 6
FORD, J.E. (1962). *Br. J. Nutr.,* **16,** 409
FORD, J.E. (1973). in *Proteins in Human Nutrition.* Eds J.W.G. Porter and B.A. Rolls. p.515. Academic Press, London and New York
FORD, J.E. and SALTER, D.N. (1966). *Br. J. Nutr.,* **20,** 843
FORD, J.E. and SHORROCK, C. (1971). *Br. J. Nutr.,* **26,** 311
FUKUSHIMA, D. (1968). *Cereal Chem.,* **45,** 203
GOLDBERG, A. and GUGGENHEIM, K. (1962). *Biochem. J.,* **83,** 129
GRIMM, F. (1973). *Med. Vet. Dissertation,* University of Munich, West Germany
GRIMM, F. (1973). *Dissertation;* University of Munich, West Germany
HAENEL, H. and KHARATYAN, S.G. (1973). in *Proteins in Human Nutrition.* Eds. J.W.G. Porter and B.A. Rolls. p.195. Academic Press, London and New York
HAGAN, S.N., HORN, M.J., LIPTON, S.H. and WOMACK, M. (1970). *J. agric. Fd Chem.,* **18,** 273
HEYNS, K., HEUKESHOVEN, J. and BROSE, K.H. (1968). *Angew. Chemie,* **80,** 627
HEYNS, K. and PAULSEN, H. (1960). in *Veränderungen der Nahrung durch industrielle und haushaltsmässige Verarbeitung.* Ed. Deutsche Ges.f.Ernährung p.15; Dr. Dietrich Steinkopf, Verlag, Darmstadt
HEWITT, D. and LEWIS, D. (1972). *Br. Poult. Sci.,* **13,** 387 and 449
HODGE, J.E. (1953). *J. agric. Fd Chem.,* **1,** 928
HORN, M.J., LICHTENSTEIN, H. and WOMACK, M. (1968). *J. agric. Fd Chem.,* **16,** 741
KAKADE, M.L., HOFFA, D.E. and LIENER, I.E. (1973). *J. Nutr.,* **103,** 1772
KÖNIG, K., ERBERSDOBLER, H. and GROPP, J. (1973). *Z. Tierphysiol. Tierernähr. Futtermittelk.,* **32,** 7
KUIKEN, A. and LYMAN, C.M. (1948). *J. Nutr.,* **36,** 359
LEA, C.H. and CARPENTER, K.J. (1960). *Br. J. Nutr.,* **14,** 91
LECLERC, J. and BENOITON, L. (1968). *Can. J. Biochem.,* **46,** 471
LONGENECKER, J.B. and HAUSE, N.L. (1959). *Archs Biochem. Biophys.,* **84,** 46
MAILLARD, L.C. (1912). *C. r. hebd. Sèanc. Acad. Sci., Paris.* **154** 66,
MAURON, J. (1970). *J. Int. Vitaminol.,* **40,** 209
MAURON, J., MOTTU, F., BUJARD, E. and EGLI, R.H. (1955). *Archs Biochem. Biophys.,* **59,** 433
MECHAM, D.K. and OLCOTT, M.S. (1947). *Ind. Engng. Chem.,* **39,** 1023
MELNICK, D. and OSER, B.L. (1949). *Fd Technol.,* **3,** 57
MENDEN, E. and CREMER, H.D. (1966). *Nutr. Dieta,* **8,** 188

MILLER, E.L., CARPENTER, K.J. and MILNER, C.K. (1965). *Br. J. Nutr.*, **19**, 547
NARAYAN, K.A., SUGAI, M. and KUMMEROW, F.A. (1964). *J. Am. Oil Chem. Soc.*, **41**, 254
NATIONAL ACADEMY OF SCIENCES-NATIONAL RESEARCH COUNCIL (1963). *Evaluation of Protein Quality;* Publ. 1100, Washington, D.C.
NESHEIM, M.C. and CARPENTER, K.J. (1967). *Br. J. Nutr.*, **21**, 399
NJAA, L.R. (1962). *Br. J. Nutr.*, **16**, 571
NUTRITION REVIEWS (1973). Ed. D.M. Hegsted. **31**, 220
PATCHORNIK, A. and SOKOLOVSKY (1964). *J. Am. chem. Soc.*, **86**, 1860
PAYNE, W.L., KIFER, R.R., SNYDER, D.G. and COMBS, G.F. (1971). *Poult. Sci.*, **50**, 143
PERAINO, C., ROGERS, Q.R., YOSHIDA, M., CHEN, M.L. and HARPER, A.E. (1959). *Can. J. Biochem.*, **37**, 1475
PION, R. (1961). *Ann. Biol. Anim. Bioch. Biophys.*, **1**, 237
POKORNY, J. and JANICEK, G. (1968). *Nahrung*, **12**, 81
POPPE, S. and MEIER, H. (1971). *Arch. Tierernähr*, **21**, 447 and 515
PRITCHARD, H., McLARNON, J. and GILLIVRAY, R. (1964). *J. Sci. Fd. Agric.*, **8**, 544
REYNOLDS, T.M. (1963). *Adv. Fd Res.*, **12**, 1
REYNOLDS, T.M. (1965). *Adv. Fd Res.*, **14**, 167
ROUBAL, W.T. and TAPPEL, A.L. (1966). *Archs Biochem. Biophys.*, **113**, 5
SALTER, D.N. and COATES, M.E. (1969). in *Proc. 8th Int. Congr. Nutr. Prag.*, p.425
SCHARRER, E., ERBERSDOBLER, H. and ZUCKER, H. (1967). *Z. Tierphysiol. Tierernähr. Futtermittelk.*, **22**, 265
SCHOBER, R. and PRINZ, I. (1956). *Milchwissenschaft*, **12**, 466
SCHÖBERL, A. (1941). *Chem. Ber.*, **74B**, 1225
SGARBIERI, V.C., AMAYA, J., TANAKA, M. and CHICHESTER, C.O. (1973). *J. Nutr.*, **103**, 657
SHORROCK, C. and FORD, J.E. (1973). in *Proteins in Human Nutrition*, Eds: J.W.G. Porter and B.A. Rolls, p.207. Academic Press, London and New York
SHOTTON, D.M. and HARTLEY, B.S. (1970). *Nature*, **225**, 802
SMITH, R.E. (1966). *J. Nutr.*, **89**, 276
SMITH, R.E. and SCOTT, H.M. (1965). *J. Nutr.*, **86**, 37, 45
SOARES, J.H. and KIFER, R.R. (1971). *Poult. Sci.*, **50**, 41
SOARES, J.H., MILLER, D., FITZ, N. and SANDERS, M. (1971). *Poult. Sci.*, **50**, 1134
STEITZ, T.A., HENDERSON, R. and BLOW, D.M. (1969). *J. molec. Biol.*, **46**, 337
SULSER, H. and BÜCHI, W. (1969). *Fd Sci. Technol.*, **2**, 105
TANNENBAUM, S.R., BARTH, H. and LE ROUX, J.P. (1969). *J. agric. Fd Chem.*, **17**, 1353
VALLE-RIESTRA, J. and BARNES, R.H. (1970). *J. Nutr.*, **100**, 873
VARNISH, S.A. and CARPENTER, K.J. (1970). *Proc. Nutr. Soc.*, **29**, 45A
VARNISH, S.A. and CARPENTER, K.J. (1971). *Proc. Nutr. Soc.*, **30**, 70A

VELU, J.G., SCOTT, H.M. and BAKER, D.H. (1972). *J. Nutr.*, **102**, 741
WAIBEL, P.E. and CARPENTER, K.J. (1972). *Br. J. Nutr.*, **27**, 509
WEBER, G. (1972). *Med. Vet. Dissertation.* University of Munich, West Germany
ZEBROWSKA, T. (1973). *Proc. Int. Amino Acid Sympos,* Brno, CSSR, A34
ZUCKER, H. (1962). *Habilitationsschrift,* Munich

8

NITROGEN PASSAGE THROUGH THE WALL OF THE RUMINANT DIGESTIVE TRACT

M.I. CHALMERS, I. GRANT and F. WHITE
Rowett Research Institute, Bucksburn, Aberdeen

Introduction

As the first paper concerned with the ruminant animal, a diagrammatic representation of the differences and the similarities in the movement of nitrogen between ruminant and monogastric animals is given in *Figure 8.1*. The passage below the line is movement of nitrogen into and out of the digestive tract common to the monogastric and the ruminant animal (posterior to the forestomach). This paper is concerned with movements specific to the ruminant, illustrated above the line in the diagram. The passage of nitrogen is considered as movement into and out of sites in the digestive tract, with emphasis on the rumen. There is still controversy on the quantitative movement of urea, ammonia and amino acids through the rumen epithelium and in the qualitative changes imposed on the nitrogen components of the ingested food by rumen organisms.

Certain difficulties and experimental limitations must be borne in mind when evaluating the suitability of the various techniques used in measuring movements within biological systems. The anaesthesia in acute experiments and the surgical preparation of pouches and isolated organs interfere with the nervous control, and the flow and distribution of blood. The unusual posture has additional hazards in the ruminant, particularly since it may affect the gaseous exchange. Movements across membranes can be studied with isolated tissues but this is a completely artificial system. The use of perfusion fluids may be criticised since their composition, flow rate and pressure are not necessarily physiologically normal. In addition, any methods in which interference causes a change in the permeability of membranes are not likely to reflect movements that occur normally, e.g. permeability changes due to shock are well known. These comments apply throughout the following discussion on the movement of urea and ammonia.

Urea

It was almost a century ago that it was first suggested that nitrogen from sources other than protein could be used beneficially in rations

160 *Nitrogen passage*

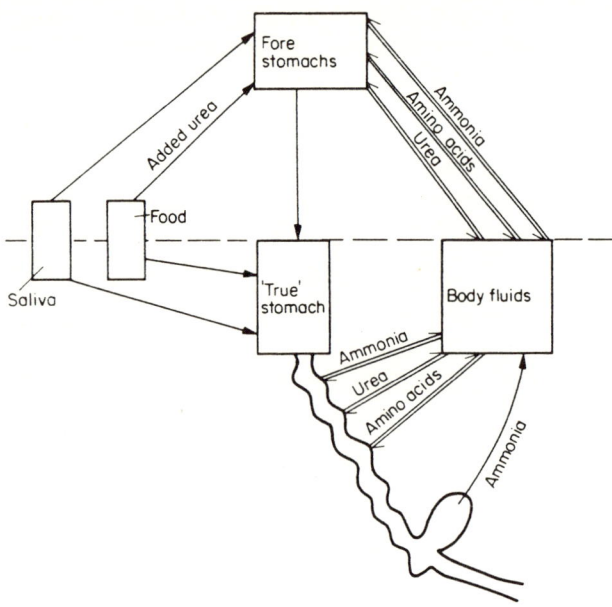

Figure 8.1 Diagrammatic representation of nitrogen movement between the digestive tract and body fluids. Below the line the passage is common to the monogastric and the ruminant animal posterior to the forestomachs

for ruminants, and as an anonymous writer in Lancet (1971) puts it 'hydrolysis of urea to ammonia and carbon dioxide by rumen bacteria is 'old hat' to physiologists and hard cash to stock raisers'. The expected benefit of releasing the urea slowly in the rumen to avoid accumulation of ammonia has produced a number of feeding-systems in which urea is either in chemical combination with a substance such as formaldehyde or encapsulated within a compound of low solubility. Under these conditions, feed urea can be present in the rumen as such, and the question arises - can urea be absorbed through the rumen wall?

The ruminant has the ability to survive longer than the monogastric under adverse nutritional conditions, an asset attributed to the recycling of nitrogen for re-utilisation and the synthesis of protein by the rumen organisms. The recycling of urea via the saliva is recognised as a contributing source of nitrogen for rumen bacteria and there are many estimates of the degree of magnitude of this nitrogen to the protein status of the animal. Since the early 1950's, a number of research groups have investigated the passage of urea through the epithelium from blood into rumen, and estimated the contribution of this passage to the total urea recycled. In 1970, Houpt reviewed this evidence, in particular that of transfer across rumen epithelium. The discussion centred on whether or not passage was by simple diffusion, a view favoured by Englehardt and Nickel (1965) and Juhasz (1963; 1965); by carrier system, as suggested by Decker (1961) and Gärtner (1962; 1963), or by a combination of both, (Várady *et al.*, 1969). A compilation

Table 8.1 A compilation of the literature on movement of urea across rumen epithelium

	Animal	Slices rumen mucosa	Whole animal experimental procedure						Urea movement	
			Anaesthesia	Ligated forestomach	Rumen pouch	Isolated rumen	AV difference across rumen	Perfusion of urea into blood	Blood to rumen	Rumen to blood
1957 Simmonet, Le Bars and Mollé	sheep		Yes	X					X	
1957 Tsuda	goat		No		X					X
1959 Houpt	sheep		Yes	X			X		X	
1962 Gärtner	cattle	X							X	
1963 Gärtner	cattle	X							X	
1963 Ash and Dobson	sheep		Yes	X					X	
1963 Juhász	sheep		No		X				X	
1964 Houpt and Houpt	goat		No		X			X	X	X
1965 Engelhardt and Nickel	sheep calves	X	No						X	X
1965 Juhász	sheep goats		No		X				X	
1965 Packett and Groves	sheep		Yes				X		X	
1966 Bod'a and Várady	sheep		No		X				X	
1967 Várady, Bod'a, Havassy Bajo and Tomás	starved sheep		No					X	X	
1968 Houpt and Houpt	goats sheep		No		X				X	X
1969 Várady, Bod'a, Havassy and Bajo	sheep		No			X			X	
1971 Thorlacius, Dobson and Sellers	cow		No			X			X	

of these experiments is given in *Table 8.1*. The majority of observations were done using Pavlov-type rumen pouches, and in a series of experiments in sheep and goats, Houpt and Houpt (1968) highlighted the difference in the rate of urea-nitrogen transfer into rumen pouches when the pouch was undisturbed compared with that after a vigorous washing treatment or treatment with anti-bacterial agents to eliminate urease activity. In the treated pouch system, there was a net transfer of urea *per se* across the epithelium in either direction, linearly related to the urea concentration difference between blood and pouch. Where the pouch was undisturbed, urea transfer was estimated by the increase in the amount of ammonia in the pouch, and indicated an increase in the net transport of nitrogen over that found in the well-rinsed pouches. It can be argued that the immediate hydrolysis of urea to ammonia in the untreated pouch would produce a urea concentration difference in favour of urea transfer to the pouch, giving an increased rate of urea transfer. Houpt goes a step further in nominating urease within the rumen epithelium as the controlling agent in the transfer of urea-nitrogen from blood to rumen, a hypothesis not accepted by Dobson (1970).

This far, the evidence for transfer across the rumen epithelium has been obtained from studies on isolated tissues and organs. It appears to favour the free diffusion of urea across the ruminal wall in both directions but there is inevitably some confusion caused by the action of urease associated with rumen bacteria. The validity of results from Pavlov-pouch type of experiments must be questioned as there is tissue continuity between the rumen and the pouch since only the mucosa is separated (Thomas, 1942) and although urea *per se* may not be present in the large rumen, there is a possibility of ammonia molecules diffusing into the emptied and washed-out pouch.

The next development is the use of normally fed whole animals in which urea entry rates are measured by the radio-isotope distribution technique. The method involves the conditioning of the animal to the test rations over a period of time, before introducing an intravenous infusion of (^{14}C) urea, and sampling blood and urine at regular intervals for determination of quantity and specific activity of urea. Cocimano and Leng (1967) demonstrated the net return of nitrogen as urea to the alimentary tract but could not specify the parts of the gastro-intestinal tract involved in degrading the urea. Various rations of different protein contents were used. In the next series of experiments (Nolan and Leng, 1972) sheep were fed 33 g of chaffed lucerne hourly to obtain constant concentrations of ammonia in ruminal fluid and of urea in plasma, these criteria were taken to indicate the system was in a steady-state. Both nitrogen and carbon isotopes were introduced to six sheep as indicated in *Table 8.2*. The single infusion experiments provided data on the proportion of urea-nitrogen contributed by ruminal ammonia, and the proportion of ruminal ammonia derived from plasma urea. Continuous infusion experiments extended these data to give the proportional rates of entry. From these data, together with analyses of rumen contents, urine and faeces, the authors proposed a model of nitrogen metabolism and estimated the flow of nitrogen where possible.

Table 8.2 *Plan of the nitrogen and carbon isotopes used in the experiments of Nolan and Leng (1972)*

Sheep	Infusion	Into rumen	Intravenous injection
A	S	$[^{15}N]NH_4Cl$	$[^{14}C]$ urea
B	S	$[^{15}N]NH_4Cl$	$[^{14}C]$ urea
C	S	nil	$[^{15}N]$ urea + $[^{14}C]$ urea
D	S	nil	$[^{15}N]$ urea + $[^{14}C]$ urea
E	C	$[^{15}N]$ NH_4Cl	$[^{14}C]$ urea
F	C	nil	$[^{15}N]$ urea + $[^{14}C]$ urea
H	C	$[^{15}N]$ $(NH_4)_2SO_4$	

S - single infusion
C - continuous infusion

Their conclusions within the context of this discussion are:-

(1) The major proportion of urea degradation occurs distal to the rumen.
(2) The quantity of urea appearing as ruminal ammonia could be accounted for by total salivary flow.
(3) It seemed unlikely that any appreciable quantity of urea crossed the rumen wall.

The same feeding system was used by Hecker and Nolan (1971) to investigate the passage of urea from blood to rumen by measuring arterio-venous differences in urea concentrations across the forestomachs of sheep. They found the mean urea-nitrogen concentration in venous blood was lower than in arterial blood, the difference being significant. The determinations of urea were on plasma only, which suggests that the authors have assumed there is no difference in urea determined on 100 ml whole blood and 100 ml blood plasma. We have found that this is not the case, urea per 100 ml plasma is always higher than it is per 100 ml blood (*Figure 8.2*). However, this does not invalidate the finding of Hecker and Nolan that when the urea-nitrogen concentrations are corrected for water movement obtained by measuring haemoglobin percentage in blood, the arterio-venous difference is negligible. Our data, extracted from 83 experiments on sheep with urea values ranging from 4 to 58 mg/100 ml blood, show that right ruminal blood has always a lower urea concentration than blood from the jugular vein or aorta. In the 43 experiments in which packed cell volume (PCV) was determined in addition to blood urea concentrations, the PCV in rumen vein blood was consistently lower than in all other blood vessels we have measured, indicating a movement of water. Correction of urea values confirmed the conclusion of Hecker and Nolan, that any movement of urea into the rumen by passage from blood through the rumen wall is insignificant when compared with the input of urea in saliva.

At the Rowett Institute we have followed a different approach in the use of the whole animal in studying nitrogen movement. Catheters have been implanted into various blood vessels of the vascular system and the

164 *Nitrogen passage*

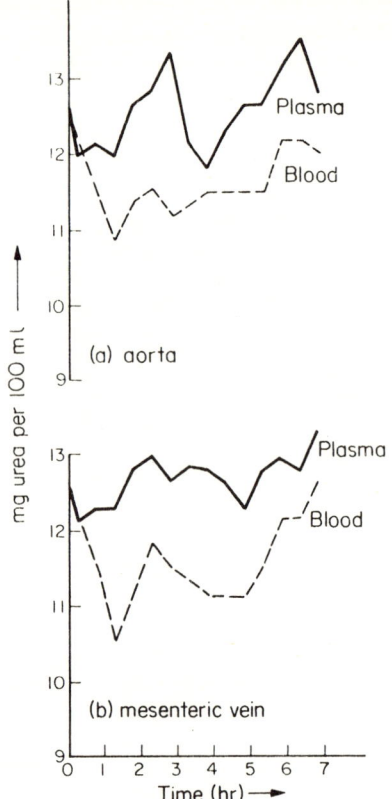

Figure 8.2 The concentration of urea per 100 ml of blood and plasma in (a) the aorta and (b) the mesenteric vein of a sheep

animal allowed to recover from the operation (Chalmers and White, 1969). In the course of experiments on ammonia toxicity arising from the introduction of urea into the digestive tract there are some interesting observations on the passage of urea. Urea introduced directly into the rumen passed into right ruminal vein blood within minutes (*Figure 8.3*) and on occasions reached the blood of the anterior mesenteric vein (Chalmers and White, 1969). We considered that urea introduced into the rumen diffused generally through the wall, not solely into the capillary bed of the veins draining the rumen. Any vein will show an increased urea concentration depending on its position in relation to the rumen wall at that point in time. In 33 experiments on sheep in which urea was introduced into the rumen, blood samples taken at minute intervals from the venous system of the viscera showed rapid rises in urea concentration which were not related to conditions in the rumen *viz* rumen pH, total ammonia concentration in rumen liquid, addition of acetic acid, hydrochloric acid or of protein (casein).

If the movement of urea is by simple diffusion only, then by definition, urea will cross the rumen wall in either direction in accordance

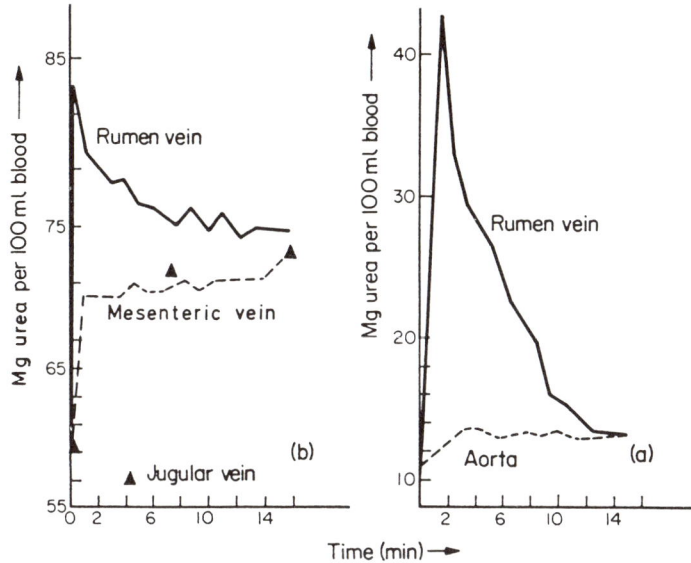

Figure 8.3 The movement of urea from the rumen into the vascular system of sheep. (a) 15 g urea into the rumen of sheep on basal diet at zero time. (b) 15 g urea into the rumen of sheep 3 hrs after an intake of basal diet + 200 g casein

with the concentration gradient. The evidence so far presented does not appear to be compatible with diffusion. However, the concentration gradient of urea between rumen and blood when urea is directly introduced into the rumen is vastly different from the gradient between blood and rumen of normal sheep, even when fed high-protein rations. Hecker and Nolan estimated the mean arterio-venous difference corrected for water movement as 0.015 mg urea-nitrogen per 100 ml. This, multiplied by the rate of flow of blood, estimates the rate of loss of urea from arterial blood.

The average blood flow in the portal vein of these sheep on a maintenance ration is 2300 ml per minute, and if the flow in the ruminal veins is only one-third of that, the passage of urea from blood to rumen over 24 hrs is 0.164 g urea-nitrogen. This is about 10 per cent of Nolan and Leng's finding that 1.2 g nitrogen derived from urea appeared in ruminal ammonia. Nolan and Leng considered that amount of ammonia could be accounted for by salivary flow, but in the absence of any direct evidence the contribution of a small quantity of urea to rumen by simple diffusion from blood is a possibility. However two sheep differed from the other 35 animals used in our work. One sheep did not recover her appetite after the operation and in an experiment under these conditions when 20 g urea was given into the rumen it remained at concentrations of 500 mg per 100 ml or over, during the six-hour period of the experiment. Urease activity was very low,

Table 8.3 *The variation within one sheep in the movement of nitrogen from urea administered into the rumen.*

	Date	Supplement given into rumen	Initial rumen pH	45 min after urea input			Jugular Blood
				Rumen liquor			
				Urea mg/100 ml	pH	Increment in ammonia mgN/100 ml	Increment in urea mg/100 ml
1	11.2.69	20 g urea + 20 ml acetic acid	6.1	2.0	6.2	86	4.6
2	12.2.69	20 g urea	6.2	8.0	7.1	54	4.8
3	13.2.69	30 g urea	6.4	3.3	7.3	73	7.9
4	20.2.69	40 g urea	6.0	18.1	6.7	50	9.3
5	07.2.69	40 g urea	5.9	61.8	6.7	45	8.1
6	10.6.69	20 g urea	6.7	27.1	7.7	104	5.8
7	08.7.69	30 g urea	6.8	7.3	7.0	22	10.3
8	25.9.69	30 g urea	6.8	91.3	8.1	100	5.1

which suggested a general debility of the rumen micro-organisms and explained the lack of breakdown of urea in the rumen. We cannot explain why in this case there was no measurable movement of urea into ruminal blood. The concentration of urea in jugular vein blood started to rise after three hours, i.e. at the time taken for digesta containing urea to reach the small intestine.

The history of the other sheep is given in *Table 8.3*. The sheep tolerated single inputs of 40 g urea into the rumen in experiments four and five, but was killed with 30 g urea some months later. Although this sheep was fed the same mixed ration throughout, the rumen flora in experiments one to five was similar to that of an animal fed barley pellets, i.e. absence of protozoa and low pH in the rumen. The rumen flora subsequently reverted to that typical of the ration, protozoa had re-appeared and rumen pH was higher in experiments six to eight. In all experiments there was little urea left 45 minutes after its administration into the rumen. The problem of the movement of urea into and out of the rumen is not solved but we consider the evidence favours passage by simple diffusion. We further suggest that this exchange is not specific between rumen and blood but includes all body fluids in contact with the rumen wall.

There is a body of evidence to suggest that urea is present in all body fluids at approximately equal concentrations. Values are recorded in a collection of biological data for, cerebral-spinal fluid (cattle, horse and man), milk (cow and man), pancreatic juice (man and dog), saliva (most species), seminal fluid (man), cervical lymph (dog), thoracic lymph (sheep), semen (cattle, horse, sheep and swine), tears (man), bile (man and swine), and humour of the eye both aqueous and vitreous (cat, horse, rabbit and dog). In humans, Rolls (1943) showed that urea concentration in blood plasma ranged from 13.46-14.81 mg/100 ml and that in corresponding blood cells was 10.99-11.96 mg/100 ml of cells. We have found a similar relationship in sheep. A blood plasma with a concentration of 12.64 mg/100 ml had a corresponding cell concentration of 10.98 mg/100 ml of cells.

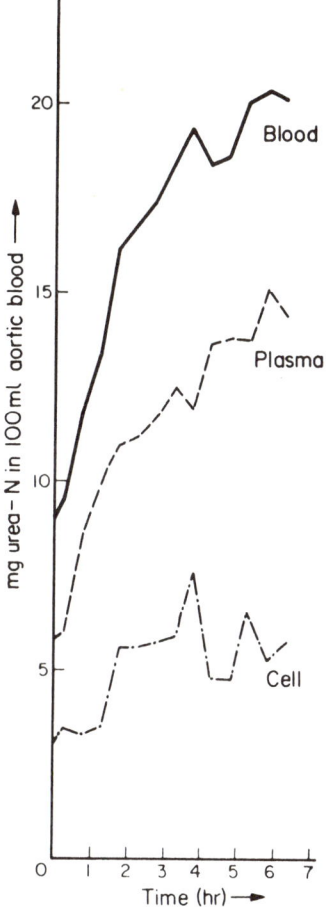

Figure 8.4 The distribution of blood urea between plasma and cells in the aorta of a sheep. 20 g urea into the abomasum at zero time

The concentration of urea in blood cells increased with the rise of concentration of plasma (*Figure 8.4*). Harmeyer *et al.* (1973) using a [^{14}C] urea isotope dilution technique, have shown that there is no difference in the concentration of urea between extracellular fluid (plasma) and intracellular fluid (muscle water) in sheep. We found that within 30 seconds of the injection of urea into the peritoneum of sheep, there was a significant increase in the concentration of urea in the vascular system and a sheep with 11.3 mg urea in 100 ml of jugular blood had a urea concentration of 10.6 mg in peritoneal fluid.

When heparinised Krebs-Ringer solution was run into the peritoneal cavity, the urea concentrations in the total fluid reached the level of that in jugular blood within 30-60 minutes. Urea was then introduced into the rumen and the rise in the concentration of urea in the fluid paralleled that in jugular blood (Chalmers, Jaffrey and White, 1971).

Nolan and Leng (1972) concluded from their model system that the major site of urea degradation was the digestive tract distal to the rumen and suggested that hind gut micro-organisms were involved in its breakdown. In the monogastric animal Combe et al. (1965) have studied the germ-free rat and compared it with the conventional rat. More soluble nitrogen was found in the caecum of germ-free rats than the caecum of conventional rats, and germ-free rats had an appreciable quantity of urea in the caecum which could only be of endogenous origin. No urea was found in the caecum of the conventional rat, but the greater quantity of ammonia, ten times that in germ-free rats, suggests that the urea had been hydrolysed. In a review article, Kornberg and Davies (1955) concluded that gastric urease is of bacterial origin and plays no essential role in gastric physiology. In experiments on sheep with isolated caecal pouches, saline was introduced into the washed out pouch and the urea concentration of the fluid measured. The concentration of urea slowly approached that of blood (Mason, Rattray and White, 1970), showing that urea can pass into the caecum when the concentration gradient is favourable.

Evidence of urea passing into and out of the lumen of the small intestine is shown in *Figure 8.5*. Urea introduced into the abomasum of a normally functioning sheep will move down the alimentary tract. The absorption of urea from the gut into the mesenteric vein is shown by the increased concentration of urea in the venous blood, compared with the aorta, from 45 mins to 4 hours 15 mins after the input of urea. From five hours onwards the situation is reversed, the arterio-venous difference illustrates the passage of urea into the digestive tract. Any urea remaining from the 20 g introduced must have passed on down the gut leaving a concentration gradient from blood to lumen in the section drained by the mesenteric vein.

The conclusion is that urea moves by free diffusion throughout the body water of the ruminant and can pass across membranes, the rate depending on the concentration gradient and the permeability of the membrane. In this respect the ruminant is similar to man (Walser and Bodenlos, 1959). It follows that urea nitrogen recycling takes place in all mammals but the usefulness of the protein regenerated by microorganisms from this nitrogen is limited to species in which the microbial synthesis occurs before the main digestive sites in the alimentary tract.

Ammonia

Ammonia is an end-product of protein degradation and has been studied in many biological systems. In mammals some small part of ammonia can be returned to tissue protein via the synthesis of non-essential amino

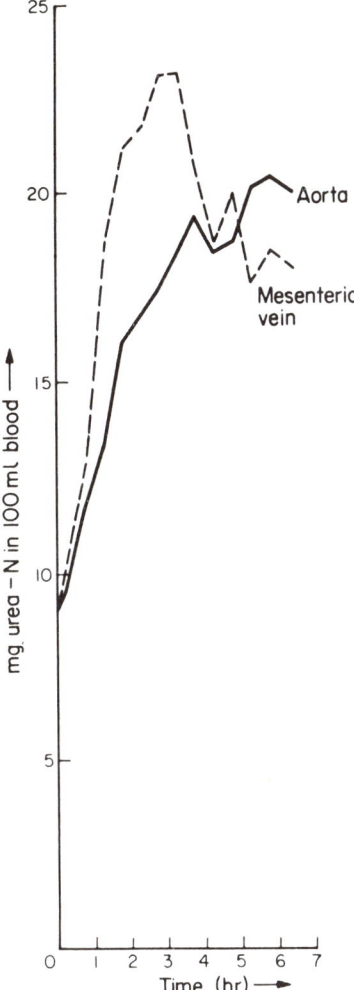

Figure 8.5 Concentration of urea in the blood of aorta and mesenteric vein in sheep given 20 g urea into the abomasum at zero time

acids (Jones *et al.*, 1969), but the great majority of it is a waste product. Any abnormality or interruption in the pathways of ammonia detoxication and excretion produces a toxicity which can be fatal. However the toxicity is reversible and no permanent damage remains after recovery.

The special case of the ruminant involves the bacterial generation of ammonia in the forestomach. McDonald (1948) was the first to demonstrate absorption of ammonia from the rumen by measuring the ammonia level of blood draining this organ, an experiment done under anaesthesia. In the intervening 25 years a mass of literature has been published on the relationships between protein fed ruminal ammonia levels and blood urea, but relatively little data on absorption of ammonia. The summation of evidence and opinion shows that the amount of ammonia absorbed from the rumen increases when either its concentration or the pH within the organ increases. This agrees with work on the absorption of ammonia from the intestine of the rat (Swales, Tange and Wrong, 1970), colonic absorption of ammonia in man (Down, *et al.*, 1972) and absorption of ammonia from human jejunum (Ewe and Summerskill, 1965). The debate is whether the transport of ammonia is active or by passive diffusion.

Ammonia, a weak electrolyte, exists in physiological media in a state of equilibrium as different molecular and ionic species. In a comprehensive review on ammonia toxicity in cells, Visek (1968) concluded that the movement of ammonia is by non-ionic diffusion of NH_3 across tissue barriers, and stressed the importance of pH gradients in the transfer of ammonia from extracellular fluid into cells.

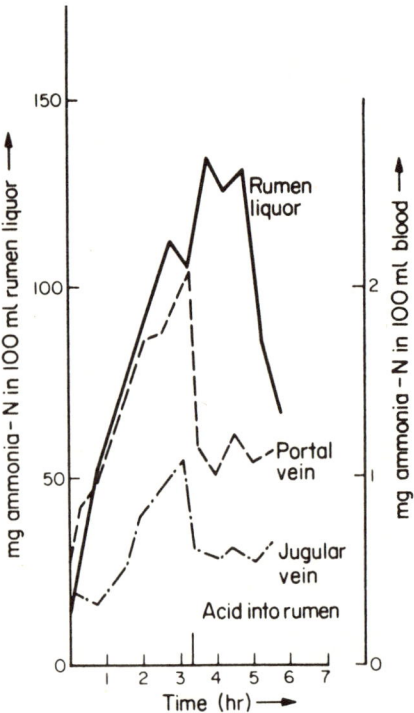

Figure 8.6 Comparison of the total ammonia concentration in rumen with the ammonia concentration in blood of portal and jugular veins. 20 g urea was given into the rumen at zero time and 20 ml acetic acid at 3¼ hours

In the first publication on ammonia toxicity in ruminants, Chalmers and White (1969) demonstrated the movement of ammonia from the rumen into the vascular system, and that it was pH dependent. An illustration of this is given in *Figure 8.6*, the movement of ammonia into the vascular system is decreased immediately acetic acid is introduced into the rumen, the pH of rumen fluid dropped from 7.6 to 5.8. The total ammonia concentration in rumen liquor was still rising after the addition of acid. Chalmers and White (1969) attributed the rapid movement of ammonia from the rumen when the fluid was at high pH to non-ionic diffusion and suggested that the NH_3 diffused across the peritoneum to produce the rapid rise in the ammonia level of mesenteric vein blood within 15 minutes of urea being administered into the rumen. Also the slower rise in blood urea concentrations when acid is administered with urea into rumen, indicates a slower rate of absorption of ammonia (*Figure 8.7*).

At this point our evidence did not conflict with the theory that if the amount of ammonia arriving in the portal blood exceeds the threshold value for ammonia in the liver, ammonia will pass into peripheral blood and cause a toxicity. After further work, we reported the relationship between rumen pH and the rise in jugular blood ammonia (Chalmers, Jaffray and White, 1971). Not only was there an immediate rise in jugular blood ammonia concentration when urea was administered into the rumen but as in *Figure 8.6* the concentration of ammonia in jugular blood always dropped with the addition of acid to the rumen, without any change in the level of ammonia in rumen fluid. We also showed that the movement of ammonia into peritoneal fluid was pH dependent and presented the hypothesis that non-ionic ammonia diffuses rapidly from the rumen into the peritoneum and arrives into jugular blood without traversing the liver. As peritoneum is drained by lymph which flows at the rate of 1.5 to 3 ml per minute, the contribution of ammonia by this route alone could not account for the observed rise in jugular blood ammonia.

Ammonia which leaves the rumen with the flow of digesta is in solution as the ionic form and, as shown in *Figure 8.6*, the lower the pH of rumen fluid, the more ammonia will pass with the digesta from a given input of an ammonia producing substrate. It was necessary to know how the ruminant dealt with the ammonia which reaches the abomasum and intestines. Abomasal pH is approximately 3.0 and the pH of the small intestine rises from 5.0 in the anterior small intestine to 7.0 in the terminal ileum. Ammonia solutions containing equi-molar amounts of ammonium as chloride, acetate and sulphate at pH 7.0 were introduced into the abomasum in a single dose and the movement of ammonia into the vascular system followed by taking serial samples of blood from portal, mesenteric and jugular veins. Sheep tolerated solutions containing 10 g ammonia-nitrogen without showing any signs of toxicity. When sheep 5854 was given 14.5 g ammonia-nitrogen, the ammonia concentration in mesenteric vein rose at 90 minutes after administration for a period of 150 minutes with a maximum concentration of 3.02 mg ammonia-nitrogen per 100 ml blood. The concentration of ammonia in jugular blood over this period did not exceed

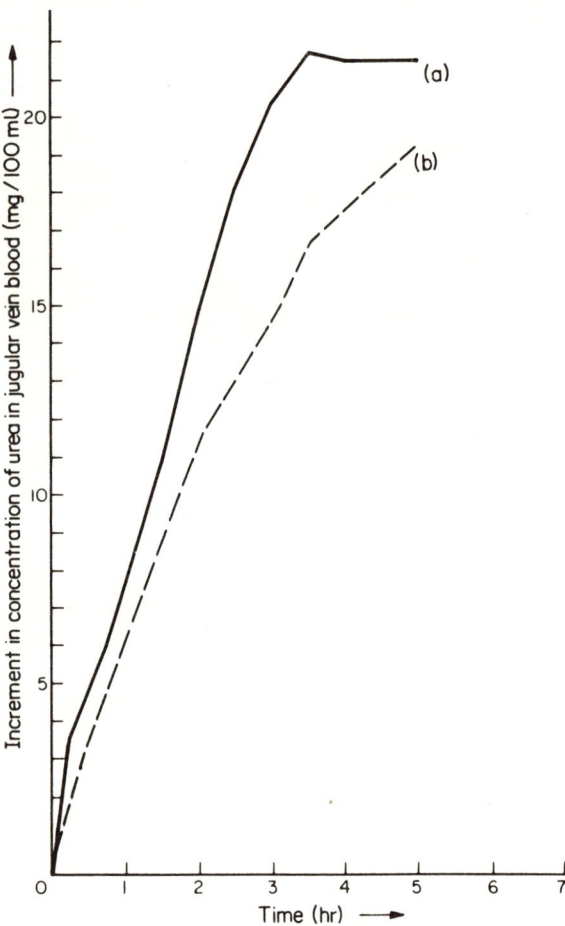

Figure 8.7 The effect of rumen pH on the rate of rise in urea concentration of jugular vein blood following administration of 20 g urea into rumen at zero time. (a) 20 g urea alone. (b) 20 g urea + 20 ml acetic acid

0.6 mg nitrogen per 100 ml. As in the monogastric animal, the normal liver of the ruminant effectively removed ammonia, delivered from the intestines. Ingle and Williams-Ashman (1962) showed that rats with 70 per cent of the liver removed could tolerate blood ammonia concentrations far in excess of those found under normal conditions of metabolism.

In the lower gut, the caecum is another site of bacterial generation of ammonia. The main source of this ammonia is endogenous urea discussed earlier in this paper, and to a lesser extent any protein present. The normal range of pH in caecal fluid is 7.4 to 7.6, if non-ionic diffusion of ammonia is the predominant means of transport, the introduction of ammonia directly into the caecum will produce a rapid movement of ammonia. A solution of ammonium salts which had caused no toxicity when introduced into the abomasum of sheep, produced acute ammonia toxicity when put directly into the caecum. The sheep collapsed in three minutes and died within 15 minutes. The administration of 20 g urea into the caecum caused severe ammonia toxicity in most sheep and death in some animals. Similar situations have been found in ponies (Hintz et al., 1970). It proved to be more difficult to control the conditions in the caecum sufficiently to allow the study of the movement of ammonia into the vascular system. However, a mixture of 10 g urea and 4.6 g nitrogen as ammonium salts introduced into the caecum produced an immediate toxicity in a sheep followed by a period of recovery. The sheep on the basal maintenance ration of cereals and hay had a ruminal ammonia concentration of 8-12 mg nitrogen per 100 ml rumen liquor throughout the experimental period. The movement of ammonia into the vascular system is shown in *Figure 8.8*. It is significant that the ammonia concentration in jugular vein blood rose rapidly and exceeded the concentration in the blood of the portal vein. This could only occur if considerable quantities of ammonia arrived in the vascular system by routes other than by collection into the portal vein and traversing the liver. There is also a significant increase of ammonia in the ruminal vein which we suggest is due to non-ionic diffusion of ammonia from the caecum, an organ normally situated close to the posterior wall of the rumen.

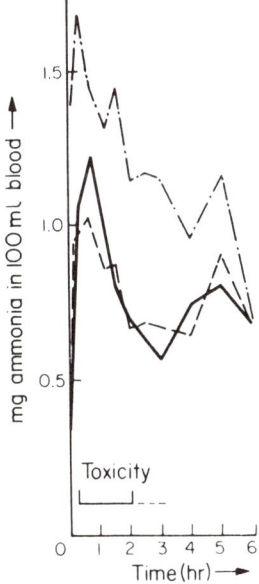

Figure 8.8 Ammonia concentration in blood of ruminal (—·—), portal (— —), and jugular (———) veins of a sheep given urea and ammonium salts into the caecum at zero time

It follows from the hypothesis of non-ionic diffusion that, if within the system there is a site in which the pH is suitable for the production of greater than normal amounts of non-ionic ammonia, then the level of ammonia in peripheral blood will not be solely dependent on the total quantity of ammonia escaping the detoxication by the liver. This is demonstrated in experiments shown in *Figure 8.9*. Ammonia

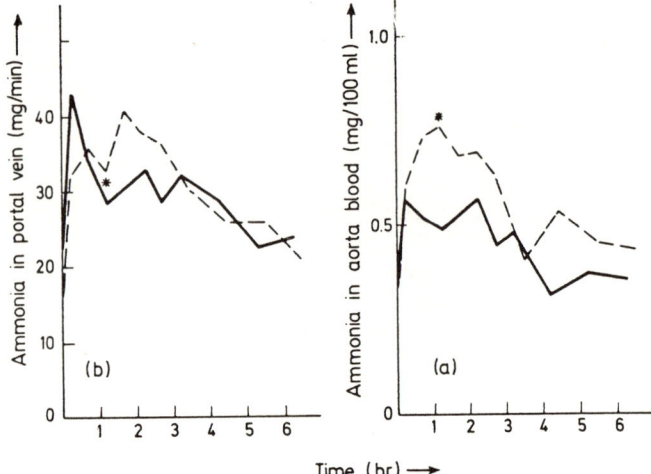

Figure 8.9 Comparison of the concentration of ammonia in aorta blood (a) with the total amount of ammonia in the blood of the portal vein (b), in one sheep given 15 g urea (-------) and 15 g urea with 15 ml acetic acid (————) into the rumen at zero time. The asterisk denotes toxicity occurred with urea alone

in aortic blood, and the rate at which blood flows in the portal vein measured by continuous thermal dilution (Webster and White, 1973) were additional estimations which increased the information available from one animal preparation. The total ammonia arriving into the portal vein calculated from ammonia concentration times blood flow (b) and the corresponding ammonia concentration in the aorta (a) are shown following the intra-ruminal administration of urea and urea with acetic acid; the data is from one sheep. Signs of ammonia toxicity started at 60 minutes after input of urea alone, the maximum effect was at 75 minutes, the animal apparently normal by 105 minutes. The toxicity coincided with the highest levels of ammonia in the aorta and not to the greatest quantity of ammonia present in the portal vein.

The net increase of ammonia in blood, after passage through portal drained viscera, is measured in mg per min by multiplying the difference in ammonia concentrations in portal vein blood and aorta blood (P-A) by the blood flow in the portal vein. Three experiments were done on the same sheep given different nitrogen sources by intraruminal administration. The net increase figures shown in *Table 8.4* are taken over a period of six hours following the input of 200 g casein daily, on a single input of 15 g urea or of 15 g urea with 15 ml glacial acetic acid at zero time. The corresponding values for concentrations

Table 8.4 Comparison of ammonia concentration in rumen with net uptake of ammonia by portal drained viscera and ammonia concentration in portal vein blood

Time	200 g Casein ≡ 26 g Nitrogen			15 g Urea ≡ 7.0 g Nitrogen			15 g Urea + 15 ml Acetic Acid ≡ 7.0 g Nitrogen		
	1	2	3	1	2	3	1	2	3
Before input	65.4	7.1	0.92	12.7	5.6	0.56	13.1	7.9	0.60
after - 15 min	45.4	17.7	0.98	54.5	17.8	1.37	67.2	27.4	1.58
45 min				74.8	19.4	1.63	87.4	21.3	1.46
1 h 15 min	53.9	16.6	1.19	90.1	17.4	1.64	69.8	16.3	1.19
45 min				78.4	23.3	1.63	-	-	-
2 h 15 min	81.5	21.5	1.21	70.5	19.4	1.44	52.6	17.3	1.21
45 min				71.9	20.6	1.49	50.8	16.6	1.10
3 h 15 min	85.9	23.9	1.08	64.6	19.8	1.20	51.7	18.7	1.18
4 h 15 min	68.4	24.7	1.09	58.4	12.8	1.07	47.0	19.3	1.02
5 h 15 min	103.4	26.1	0.96	48.0	15.3	1.12	46.6	12.6	0.85
6 h 15 min	111.5	26.3	1.12	46.6	10.3	0.88	46.5	13.1	0.80

1. Ammonia concentration in rumen liquor (mg Nitrogen per 100 ml)
2. Net uptake in portal vein blood (mg Nitrogen per minute)
3. Ammonia concentration in portal vein blood (mg Nitrogen per 100 ml)

Table 8.5 Net gain in the concentration of ammonia in the right ruminal vein (mg Nitrogen per 100 ml blood)

Time	150 g cereals 2 × 300 g lucerne	150 g cereals 2 × 300 g hay	150 g cereals 2 × 300 g hay + 200 g Casein	150 g cereals 2 × 300 g hay + 15 g urea	150 g cereals 2 × 300 g hay +15 g urea and 15 ml Acetic Acid
0	1.17	0.65	2.06	0.76	0.69
15 min	1.30	0.61	1.71	3.80	3.80
45 min				3.74	3.87
1 h 15 min	1.16	0.67	1.89	3.72	3.48
45 min				3.27	2.91
2 h 15 min	1.00	0.49	1.68	3.35	2.55
45 min				3.34	2.60
3 h 15 min	0.91	0.55	1.45	3.11	2.30
4 h 15 min	1.16	0.49	1.72	2.86	2.26
5 h 15 min	1.16		2.10	2.66	2.08
6 h 15 min	1.13	0.42	2.02	2.25	1.78

of total ammonia in rumen liquor and portal vein blood are given for comparison. It is of interest that the net increase of ammonia following an input of 26 g nitrogen as casein, was sustained at a higher level than at any point in the experiment when 15 g urea was the substrate, only the input of urea showed symptoms of stress from ammonia.

The relationship between the total amount of ammonia in blood of portal vein and concentration of ammonia in portal blood following the intra-ruminal administration of 200 g casein is given in *Figure 8.10*. There is little movement in the level of ammonia in the aorta, and these values are consistently below those in *Figure 8.9* from urea input. The finding that low levels of ammonia in aorta blood persist in conditions producing high ruminal ammonia concentrations provides an explanation for the observation that the symptoms of excess ammonia are not associated with the feeding of high levels of protein to animals.

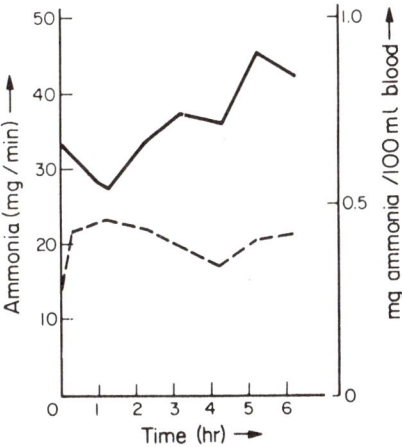

Figure 8.10 Comparison of the concentration of ammonia in aorta blood (-------) with the total ammonia in portal vein blood (———) in a sheep given a supplement of 200 g casein daily into the rumen at zero time

In a recent publication, Nolan, Norton and Leng (1973) using an isotope tracer technique in sheep fed 33 g lucerne at hourly intervals, concluded that negligible amounts of ammonia were absorbed through the rumen wall. We have always found measurable net gains of ammonia in the right ruminal vein using our technique which measures arterio-venous differences across the rumen. The data from five different dietary regimes is shown in *Table 8.5*. The ration with a comparable intake of lucerne to that of Nolan *et al.*, averaged a net gain of 1 mg ammonia per 100 ml blood.

References

ANON. (1971). *Lancet,* **1407**
ASH, R.W. and DOBSON, A. (1963). *J. Physiol. Lond.,* **169,** 39
BOD'A, K. and VÁRADY, J. (1966). *Vet. Med., Praha,* **11,** 597
CHALMERS, M.I., JAFFRAY, A.E. and WHITE, F. (1971). *Proc. Nutr. Soc.,* **30,** 7

CHALMERS, M.I. and WHITE, F. (1969). in *Urea and Other Substitutes for Natural Protein Sources*. F. Hoffmann-La Roche & Co. Ltd., Basle
COCIMANO, M.R. and LENG, R.A. (1967). *Br. J. Nutr.*, **21**, 353
COMBE, É., PENOT, Ê, CHARLIER, H. and SACQUET, E. (1965). *Annls Biol. anim. Biochim. Biophys.*, **5**, *189*
DECKER, P. (1961). *Zentbl. Vet. Med.*, **8**, 39
DOBSON, A. (1970). in *Physiology of Digestion and Metabolism in the Ruminant*, p.147. Ed. A.T. Phillipson, Oriel Press Limited, Newcastle upon Tyne
DOWN, P.F., AGOSTINI, L., MURISON, J. and WRONG, O.M. (1972). *Clin. Sci.*, **43**, 101
ENGELHARDT, W.V. and NICKEL, W. (1965). *Pflügers Arch. ges. Physiol.* **286**, 57
EWE, K. and SUMMERSKILL, W.H.J. (1965). *J. Lab. clin. Med.*, **65**, 839
GÄRTNER, K. (1962). *Pflügers Arch. ges. Physiol.*, **276**, 292
GÄRTNER, K. (1963). *Dr. tierärztl. Wschr.*, **70**, 16
HARMEYER, J., VÁRADY, J., BIRCK, R. and MARTENS, H. (1973). *Arch. Tierernähr.*, **23**, 537
HECKER, J.F. and NOLAN, J.V. (1971). *Aust. J. biol. Sci.*, **24**, 403
HINTZ, H.F., LOWE, J.E., CLIFFORD, A.J. and VISEK, W.J. (1970). *J. Am. vet. med. Ass.*, **157**, 963
HOUPT, T.R. (1959). *Am. J. Physiol.*, **197**, 115
HOUPT, T.R. (1970). in *Physiology of Digestion and Metabolism in the Ruminant*, p.119. Ed. A.T. Phillipson, Oriel Press Limited, Newcastle upon Tyne
HOUPT, T.R. and HOUPT, K.A. (1964). *Fedn Proc. Fedn Am. Socs exp. Biol.*, **23**, 262
HOUPT, T.R. and HOUPT, K.A. (1968). *Am. J. Physiol.*, **214**, 1296
INGLE, D.J. and WILLIAMS-ASHMAN, H.G. (1962). *Archs. Path.*, **73**, 343
JONES, E.A., SMALLWOOD, R.A., CRAIGIE, A. and ROSENOER, V.M. (1969). *Clin. Sci.*, **37**, 825
JUHÁSZ, B. (1963). *Magy, Allatov, Lap.*, **18**, 30
JUHÁSZ, B. (1965). *Acta vet. hung.*, **15**, 25
KORNBERG, H.L. and DAVIES, R.E. (1955). *Physiol. Rev.*, **35**, 169
MASON, V.C., RATTRAY, E. and WHITE, F. personal communication
MCDONALD, I.W. (1948). *Biochem. J.*, **42**, 584
NOLAN, J.V. and LENG, R.A. (1972). *Br. J. Nutr.*, **27**, 177
NOLAN, J.V., NORTON, B.W. and LENG, R.A. (1973). *Proc. Nutr. Soc.*, **32**, 93
PACKETT, L.V. and GROVES, T.D.D. (1965). *J. Anim. Sci.*, **24**, 341
RALLS, J.O. (1943). *J. biol. Chem.*, **151**, 529
SIMONNET, H., LE BARS, H. and MOLLÉ, J. (1957). *C.r. hebd. Séanc. Acad. Sci. Paris*, **244**, 943
SWALES, J.D., TANGE, J.D. and WRONG, O.M. (1970). *Clin. Sci.*, **39**, 769
THOMAS, J.E. (1942). *Proc. Soc. exp. Biol. Med.*, **50**, 58
THORLACIUS, S.O., DOBSON, A. and SELLERS, A.F. (1971). *Am. J. Physiol.*, **220**, 162
TSUDA, T. (1957). *Tohoku J. agric. Res.*, **7**, 241
VÁRADY, J., BOD'A, K., HAVASSY, I., BAJO, M. and TOMÁS, J. (1967). *Physiologia bohemoslov*, **16**, 571

VÁRADY, J., BOD'A, K., HAVASSY, I. and BAJO, M. (1969). *Physiologia bohemoslov*, **18**, 23
VISEK, W.J. (1968). *J. Dairy Sci.*, **51**, 286
WALSER, M. and BODENLOS, L.J. (1959). *J. clin. Invest.*, **38**, 1617
WEBSTER, A.J.F. and WHITE, F. (1973). *Br. J. Nutr.*, **29**, 279

III
NITROGEN-ENERGY RELATIONSHIPS

9

AMINO ACIDS AS SOURCES OF ENERGY

D.B. LINDSAY
Institute of Animal Physiology, Babraham, Cambridge

Introduction

A distinction is often made between the protein and energy components of a diet. Yet in general the greater part of protein intake serves as an energy source. According to Munro (1964) in adult man as little as 5 per cent of energy intake need be protein (40-50 g/d) of average quality in order to reach nitrogen equilibrium, and on a typical Western diet 100 g/d protein is a moderate intake. It is true, as Munro emphasised, that not all the additional protein is catabolised - more body protein is synthesised as protein intake is increased.

Nevertheless, in adults the greater part of absorbed amino acid nitrogen is fairly promptly excreted as urea. In dogs on a high protein diet at least 75 per cent of daily urea-nitrogen production is synthesised within 12 hr of feeding (Elwyn, Parikh and Shoemaker, 1968). Even in growing animals, more nitrogen is lost than is retained. Black and Tribe (1973) showed about 24 per cent of dietary nitrogen of growing lambs was retained, i.e. 76 per cent was either not absorbed or excreted. In a recent study of the foetal lamb (Battaglia and Meschia, 1973) it is claimed that only 60 per cent of the nitrogen crossing the placenta from the mother near term, is retained by the foetus for growth. On the basis of the estimated rate of urea production, 25 per cent of the foetal oxygen consumption could be accounted for by the catabolism of amino acids.

Excretion of nitrogen, while showing that protein catabolism is initiated, need not imply that all protein-carbon is oxidised. Some could be retained as fat or carbohydrate. While there appears to be no evidence of direct conversion to fat, it has long been supposed that conversion to carbohydrate is substantial. Janney (1915) suggested that 'dextrose must thus be considered one of the chief intermediary products of protein metabolism, and of certain proteins the most important. Formation of large amounts of glucose in protein metabolism may be considered a normal process'.

The original basis for this supposition was the work of Minkowski (1893). Minkowski claimed that in pancreatectomised dogs there was a constant ratio between glucose and nitrogen excreted in urine, although later work (Soskin, 1930) failed to support this. Janney's proposal, however, was based on his studies with dogs treated with phlorizin, in

Table 9.1 Glucose (g) recovered per 100 g protein fed to phlorizin treated dogs (Janney, 1915). Values ± s.e.m., with number of experiments in parenthesis. Calculated values are on the basis of the content of alanine, serine, cystine, aspartate, glutamate, proline, hydroxyproline and glycine (Block and Bolling, 1951; Tristram and Smith, 1963)

Protein	Experimental	Calculated
Zein	53 ± 1.6 (5)	47
Gliadin	80 ± 1.5 (5)	45
Edestin	65 ± 1.0 (5)	37
Fibrin	53 ± 1.0 (4)	40
Serum albumin	55 ± 1.5 (5)	26
Gelatin	65 ± 1.8 (5)	61
Ovalbumin	54 ± 1.2 (4)	38
Casein	48 ± 2.0 (8)	44

which he related extra glucose excreted in urine to the amount of protein fed. His results, shown in *Table 9.1* suggest that the yield of glucose for a given protein is constant. Janney (1915) suggested that this yield agreed well with the content of glycogenic amino acids as he knew them (glycine, alanine, serine, cystine, aspartic and glutamic acids and proline), although when more recent analytical data are used, agreement is not so impressive (*Table 9.1*); it becomes also necessary to suppose conversion of several of the essential amino acids.

It could be argued that such findings relate only to diabetic organisms, although in phlorizin diabetes the pancreas is not affected and retains the capacity to secrete insulin. Such animals are however hypoglycaemic; Haft, Tennen and Mehtalia (1972) suggest that net glucose production from amino acids by the isolated perfused rat liver is appreciable only when the perfusate glucose concentration is low.

The effect of an increased supply of protein or protein hydrolysate on glucose production has been studied in normal sheep by Lindsay and Williams (1971) and Judson and Leng (1973). The rate of glucose production was determined in sheep fed hourly, from the rate of dilution of continuously infused ^{3}H-glucose. Casein or a casein hydrolysate was infused into the abomasum at a constant rate and the increase in glucose production determined. Lindsay and Williams obtained an increase in glucose production of about 30 g glucose following infusion of 100 g casein. Judson and Leng (1973) obtained an increase in glucose production that was linearly related to the amount of casein hydrolysate infused into the abomasum, for 100 g casein the increase was about 12 g glucose. Lindsay and Dyke (1974) have shown that this difference can be explained by the length of the period of infusion of casein. Over a period of 5-10 hours the increase was about 10 g glucose, while over 24-48 hours of infusion the response was 28-30 g. Even this response is only about half that obtained by Janney.

Another approach was adopted by Hoogenraad *et al.* (1970). They exploited the fact that in ruminants on many diets, a large fraction of the amino acids absorbed from the gastrointestinal tract is derived from microbial protein. They therefore prepared ^{14}C-labelled bacteria (or fractions from these organisms), injected the material into the abomasum of sheep and estimated the mean recovery of ^{14}C in plasma glucose over the following 24 hours. On average 5.8 per cent of bacterial carbon passed through the glucose pool in 24 hours. They assumed that rumen micro-organisms were utilised in a similar way, and from their estimate of the average amount of such organisms synthesised, they calculated that about 17 g glucose/24 hours would be derived from bacterial carbon (mostly protein), or about 18-19 per cent of the total glucose synthesised per 24 hours. Clearly the reliability of this estimate depends on the validity of the estimate of the rate of bacterial protein synthesis. In terms of an estimate of the amount of glucose derived from amino acids, it also depends on the proportion of the relevant amino acid pool which is derived from bacterial protein. Both of these estimates carry a degree of uncertainty.

Hunter and Millson (1964) injected a mixture of ^{14}C-labelled amino acids intravenously into a lactating cow and compared the mean specific activity of milk protein and lactose over the period 0-16 hours after injection of isotope. Since the mean specific activity of lactose was about 12 per cent of that of milk protein, it was argued that amino acids supplied about 12 per cent of the glucose used for lactose synthesis. There are at least two reservations about this technique, it assumes that the amino acids used for protein synthesis in the mammary gland are derived from the same pool as the amino acids used for glucose synthesis in the liver, and that all amino acids may be used for both lactose formation and milk protein synthesis. The first point is clearly not valid, the second is untrue, at least for leucine or lysine which between them account for probably 15 per cent by weight of milk protein.

Ford and Reilly (1969; 1970) infused a mixture of ^{14}C-amino acids intravenously into sheep at a constant rate until the mean specific activity of mixed plasma amino acids and of glucose was constant. The ratio of the specific activity of plasma glucose to that of plasma amino acids gives the fraction of glucose derived from amino acids. This fraction ranged from 0.11-0.17 in four non-pregnant and 0.13-0.27 in four pregnant sheep. There are a number of difficulties in assessing this technique. Perhaps the most obvious is to decide the specific activity of the amino acid pool that is used for glucose synthesis. Reilly and Ford (1971) showed that they were well aware of this problem, the fraction of glucose derived from amino acids varied depending on the site chosen for estimating the specific activity of the amino acids. This specific activity was only slightly different in arterial and jugular blood, but was appreciably less in portal blood and least within the liver. It is clear that the plasma amino acids do not constitute a single well-mixed pool, moreover even within the liver it is probable that amino acids do not constitute a single pool (Mortimore, Woodside and Henry, 1972). Thus the percentage of glucose derived from amino acids varied from 28-73, depending on the assumed specific activity of amino acid carbon.

Table 9.2 presents some estimates of the amount of protein carbon converted to glucose on the basis of the studies described. Rough as these estimates are, they do not seem consistent with the notion that the greater part of protein carbon is catabolised through conversion to glucose. It seems therefore worth examining the evidence that individual amino acids are glucogenic.

Table 9.2 *Fraction of absorbed protein converted to glucose. In calculating the last two values it is assumed 500-1000 g protein for cows and 100-150 g for sheep are absorbed per 24 h; and that total glucose fluxes were 1 kg/d and 100 g/d respectively*

Animal		
Phlorizin-treated dogs	0.44	(Janney, 1915)
Normal sheep (response to extra protein)	0.10	(Judson and Leng, 1973)
-	0.24	(Lindsay and Williams, 1971)
-	0.22	(Lindsay and Dyke, 1974)
- (from 'bacterial protein')	0.02-0.08	(Hoogenraad et al., 1970)
Lactating cow	0.01-0.2	(Hunter and Millson, 1964)
Non-pregnant and pregnant sheep	0.07-0.22	(Ford and Reilly, 1969, 1970)

Evidence for Glucogenic Nature of Amino Acids

Reactions have been demonstrated which permit nearly all amino acids (at least in part) to be converted to intermediates such as pyruvate, oxalo-acetate, α-oxo-glutarate or propionyl-CoA, all of which are known to be convertible to glucose. The exceptions are leucine and lysine for which only reactions leading to acetoacetate and acetyl CoA have been demonstrated. Phenylalanine and tyrosine can yield fumarate (also potentially a glucose precursor) as well as acetoacetate; isoleucine can yield propionyl CoA as well as acetyl CoA, while the alanyl side chain of tryptophan is also potentially a glucose precursor. It is plausible that the fate of a major fraction of catabolised protein is a conversion to glucose. The likelihood of such a conversion is however very different for different amino acids. For some, such as alanine, glutamate and aspartate the probability is high, whereas for others such as histidine, phenylalanine or tyrosine the complexity of the route renders the extent of such conversion much less certain.

The early work relied on estimation of the extra glucose excreted in urine of dogs treated with phlorhizin, after feeding them single amino acids. On the basis of this work (e.g. Dakin, 1913) what were subsequently recognised as non-essential amino acids (glycine, alanine, serine, cystine, aspartic and glutamic acids, ornithine and proline) were classed as glycogenic, whereas the essential amino acids (valine, leucine, isoleucine, histidine, lysine, phenylalanine, tyrosine and tryptophan) were not. Although presented in this form the results look quite decisive, in fact there is much uncertainty about them. The amounts of amino acid given were small (7-20 g) so that the extra glucose was not large, measured against a substantial production of glucose from endogenous sources. It was occasionally quite variable (for arginine, for example,

35-70 g glucose/100 g amino acid) and even for some 'non-glucogenic' amino acids, not negligible (both histidine and leucine yield an average of 26 g/100 g amino acid). Moreover, different authors have obtained different results, whereas Dakin found no conversion of valine to glucose, Rose, Johnson and Haines (1942) found a substantial conversion. The results of Gurin, Delluva and Wilson (1947) also make it difficult to give much weight to results obtained in this way. They fed phlorizin-treated rats ^{13}C-alanine. Although extra glucose appeared in urine equivalent to 60-70 per cent of that expected, only 1-5 per cent of isotope appeared in urinary glucose. In contrast, 25-30 per cent of ^{14}C-lactate administered, appeared in urine within six hours. Thus, when a relatively large dose of a single amino acid is fed, most of the extra glucose is not derived from the amino acid injected.

In another technique that was used to assess the glycogenic ability of amino acids, the increase in liver glycogen in starved rats was measured after feeding individual amino acids. Apart from the fundamental difficulty that it is not possible to make any quantitative assessment, the findings of Olsen, Hemingway and Nier (1943) that after injection of ^{13}C-glycine into mice the (increased) liver glycogen contained less than one per cent of fed isotope, makes it clear that the increased liver glycogen is not necessarily derived from the fed amino acid.

Studies have also been made of the effect of individual amino acids in stimulating glucose production by isolated liver or kidney tissue.

LIVER

Studies of glucose production from amino acids by slices of liver are of dubious value since it is known that the gluconeogenic capacity is substantially below that of the perfused liver (Ross, Hems and Krebs, 1967). In studies with the perfused rat liver, Exton and Park (1967) obtained a moderate stimulation of glucose production when L-alanine or L-serine were present in the medium at 10 mM concentration. Other amino acids, *viz.* L-glutamate, L-aspartate, L-cysteine, L-threonine, L-glycine, L-isoleucine, L-arginine, L-proline, or L-ornithine were ineffective when added singly at 10 mM concentration. Ross, Hems and Krebs, (1967) who also used perfused rat livers, obtained comparable increases in glucose production from alanine or serine. Significant increases in glucose production were also obtained by perfusion with L-glutamine, L-asparagine and L-glutamate individually at 10 mM concentration. For no amino acid was the stimulus to glucose production as high as that from lactate or pyruvate, although the concentration of amino acids was much higher than is likely to occur physiologically. This point is important since Mallette, Exton and Park (1969) have shown that incorporation of ^{14}C in glucose from ^{14}C-alanine by the perfused rat liver is increased in proportion to circulating concentration up to about 10 mM. However, these workers also showed that when a mixture of amino acids, simulating the amino acid composition of rat plasma, was infused, glucose production was increased from 0.12 to 0.4 μmole/min/g; on doubling the concentrations, there was a further increase to 0.7 μmole/min/g. This increase in glucose production following 'physiological' amino acid infusion was demonstrable only with a 'non-recirculating' perfusion system. Mallette, Exton and Park, (1969) consider this is because only a few amino acids (especially alanine and serine) stimulate glucose production. In a recirculating system these amino acids are present in insufficient amounts to produce a significant increase in glucose production. Another difference, not considered by the authors, is that non-recirculating perfusion medium reaching the liver is free of glucose. Haft, Tennen and Mehtalia (1972) who also used perfused rat livers, observed that net glucose production resulting from infusion of an amino acid mixture occurred only when the medium was hypoglycaemic.

KIDNEY

Glucose production by the kidney has been related to its role in acidosis. Hems (1972) using an isolated perfused kidney preparation from acidotic rats obtained a rate of glucose production of 100-113 μmole/g dry weight/h when glutamate or glutamine (5 mM) was precursor. Glutamate was equally effective when kidneys taken from normal rats starved 48 h were used, but in these conditions glutamine was only 40 per cent as effective. Kidney slices are rather more effective than liver slices for the study of gluconeogenesis. Krebs, Notton and Hems (1966) reported rates of glucose production from glutamate in mouse kidney slices, of 160 μmole/g dry weight/h. Apart from proline, other amino acids were poor substrates for glucose production.

It might be thought that the use of isotopically labelled amino acids, and examination of the extent of incorporation of label into glucose or glycogen would give a clear indication of the relative glucogenic contribution of individual amino acids; in practice such experiments are fraught with difficulties of interpretation. Aikawa, et al. (1972) for example, injected a number of ^{14}C-labelled amino acids into the portal (and femoral) vein of fed or fasted rats and determined radioactivity in blood glucose 0.5-10 min after isotope injection. Taking incorporation of DL-2-^{14}C-lactate as 100, the incorporation into glucose from amino acids (corrected for variation in amounts of isotope injected) were: aspartate 66, alanine 62, glutamate 30, serine 7, glycine 3, and histidine 2. These values were for fasted animals, for fed animals values were too small to make comparison possible. While the radioactivity in glucose (c.p.m./ml blood) was reasonably constant from 2-10 min following injection of aspartate, alanine and glutamate, it was continuing to rise after injection of serine, glycine or histidine. Thus the relatively low values from these amino acids might simply reflect a different pool size within the liver or increased numbers of steps between precursor and product. The specific activity within the portal vein, of each amino acid would clearly vary with its concentration. Aikawa et al. (1972) showed that adding unlabelled precursor substantially affected incorporation of glutamate and aspartate, but had little effect, within moderate limits, on the incorporation of alanine and serine. They suggested that glutamate and aspartate might therefore be used less effectively for hepatic gluconeogenesis at high plasma concentrations. However, although the fraction used for glucose synthesis might be reduced, it could still be that the absolute amount converted was not greatly changed. Even for alanine or aspartate, the quantity of injected isotope that was in the circulating glucose was less than 5 per cent. Wolff and Bergman (1972) have used a technique similar to that described earlier by Reilly and Ford (1971) but using constant intravenous infusion of single ^{14}C-labelled amino acids. They determined the (constant) specific activity of the appropriate amino acid as the (weighted) average of that in hepatic arterial and portal plasma. On this basis, the percentage of glucose derived from the amino acids studied was: alanine 5.5 per cent, aspartate 0.6 per cent, glycine 0.9 per cent, glutamate 3.4 per cent and serine 0.7 per cent, a total of 11 per cent. Heitman, Hoover and Sniffly (1973) used a similar technique but used the specific activity of jugular plasma amino acids in their estimation. They attempted serial isotope experiments in close succession, but from their results it seems clear that it is only the initial experiment of each series that is reliable. On this basis, the estimate of the percentage of glucose derived from serine is 0.86 per cent and from glutamate, 3.7 per cent, values which correspond closely with those obtained by Wolff and Bergman.

Black and workers (Egan and Black, 1968; Egan, Moller and Black, 1970; Black, *et al.*, 1968) have used lactating cows and goats to study gluconeogenesis from amino acids. They have injected a number of ^{14}C-labelled amino acids individually, and determined the proportion of injected isotope recovered in milk casein and lactose, over the 48 hours after injection. Their results are shown in *Table 9.3*. Recovery of label in lactose is greater from non-essential amino acids (aspartate, alanine, serine and glutamate) than from essential amino acids, whereas for casein the reverse is true. It is reasonable to regard lactose in milk as a kind of 'metabolic trap' for labelling of glucose, analogous to urine glucose in studies of phlorhizin diabetes, since milk lactose is at least 85 per cent derived from blood glucose in cows (Bickerstaffe, Annison and Linzell, 1974). The proportion of ^{14}C from amino acid that is recovered in lactose however, does not represent the proportion of glucose derived from the amino acid but the proportion of the amino acid flux through the body which is converted to glucose in the time under consideration (more precisely, it is the proportion of that part of the flux which has come into isotopic equilibrium with the injected isotope). Thus the proportions of isotope recovered in lactose, as shown in *Table 9.3*, represent a 'ranking' of glucogenicity only to the extent that the flux rates of different amino acids are comparable.

Black *et al.* (1970) have suggested a way of estimating the flux rate of essential amino acids. They assume that after injection, for example of [^{14}C]-phenylalanine, the fraction of injected ^{14}C that was recovered over seven days in the phenylalanine of casein represented the fraction of the flux that was used for the synthesis of phenylalanine of casein. The output of phenylalanine in

190 *Amino acids as sources of energy*

Table 9.3   Recovery of ^{14}C-labelled amino acids in milk products, and estimated conversion of essential amino acids to glucose (see text for details). Data from Egan and Black (1968); Black (1968); Egan, Moller and Black (1970); and Black, et al., (1970). Values are for cows, except those for valine and arginine, which are for goats

Amino acid	Average per cent ^{14}C recovered over 48 h or longer, in lactose	Average per cent ^{14}C recovered over 48 h or longer, in casein	Estimated flux (g/24 h)	Estimated g amino acid converted to lactose per day
Glutamate	6.7 (4)	4.2	–	–
Aspartate	6.6 (2)	4.8	–	–
Alanine	7.7 (2)	9.9	–	–
Serine	7.0 (2)	14.8	–	–
Arginine	1.0 (2)	16.7	–	–
Valine	2.0 (2)	17.0	214	4.3
Threonine	2.6 (1)	18.8	113	2.9
Leucine	2.1 (1)	40.9	115	2.4
Isoleucine	2.2 (1)	21.2	167	3.7
Phenylalanine	2.7 (2)	25.0	116	3.2
Tyrosine	5.8 (1)	27.1	118	6.8

casein divided by the fractional contribution gives an estimate of the phenylalanine flux. If we assume, for the essential amino acids of *Table 9.3*, that about 85 per cent of ^{14}C in casein is in the specific administered amino acid, as Black and his colleagues found for phenylalanine, flux rates for the essential amino acids listed are about 100-200 g/24 h. Thus the contribution for the six essential amino acids listed, to lactose synthesis, is only 2-4 g/day each (7 g for tyrosine), out of a total lactose production of about 700 g/day. While this calculation is only approximate, it does suggest that the contribution of essential amino acids to gluconeogenesis is small in the lactating cow. Indeed it may even over-estimate this contribution, since we know that ^{14}C is transferred from CO_2, through fixation reactions, and also from acetyl CoA through exchange reactions in the tricarboxylic acid cycle. This technique cannot be applied to the non-essential amino acids. It is known that these amino acids in casein are derived not only from the plasma amino acid, but also in part within the mammary gland from other sources. Uptake, across the mammary gland, as a fraction of output in milk, is variable (Bickerstaffe *et al.*, 1974). The fraction of isotope recovered in a non-essential amino acid of casein does not represent the fraction of the flux of that acid that is used for the synthesis of the amino acid in protein.

Studies of individual amino acids then, are consistent with the notion that alanine and probably serine are substantially converted to glucose. For the other non-essential amino acids some conversion certainly occurs although to what extent is uncertain. For the essential amino acids, the little evidence available suggests conversion to glucose is not substantial. Similar conclusions may be drawn from studies of the splanchnic (predominantly hepatic) uptake of amino acids in post-absorptive man (Felig and Wahren, 1971). Alanine accounts for 50 per cent and serine at least 20 per cent of the total amino acids extracted by the liver. In diabetics (Wahren *et al.,* 1972) and obese patients (Felig, *et al.,* 1974) where total splanchnic amino acid uptake is almost doubled, alanine plus serine still accounted for 70-80 per cent of total amino acid uptake. The large uptake of alanine reflects in part the substantial release of alanine by muscle, an amount greater than can be due to muscle proteolysis. However, even in dogs on a high protein diet (Elwyn, Parikh and Shoemaker, 1968) alanine accounted for 24 per cent and serine 8 per cent of the large hepatic uptake of amino acids in the period following a meal, while in fed sheep they accounted for 19 and 7 per cent respectively of total amino acid uptake (Wolff, Bergman and Williams, 1972).

If all potentially glucogenic amino acids were converted to glucose in post-absorptive man, they could account for only a small part of the glucose-produced 7 per cent in normal, 12 per cent in obese and 11 per cent in diabetic subjects. Most of the glucose released is derived from glycogenolysis (Nilsson, Furst and Hultman, 1973). Amino acids removed by the liver in the absorptive phase could be converted to glycogen, thus part of the glycogenolytic glucose might ultimately be derived from amino acids. From the study by Elwyn, Parikh and Shoemaker (1968) in dogs, it may be calculated that glycogenic amino acids removed by the liver in the absorptive phase could account for about half the glucose secreted over a period of 24 hours. In their studies with fed sheep, Wolff and Bergman (1972) calculate that a maximum of 29 per cent of glucose could be derived from the total amino acid taken up by the liver. Their studies with ^{14}C-amino acids (alanine, glycine, serine, glutamine/glutamate and asparagine/aspartate) suggest that on average, only 60 per cent of the hepatic uptake of these amino acids was used for glucose synthesis.

Role of Hormones

GLUCAGON

There is evidence that, *in vivo,* injected glucagon stimulates hepatic uptake of amino acids (Shoemaker and Itallie, 1960) and urinary nitrogen excretion (Izzo and Glasser, 1961). In addition to a well documented stimulation of gluconeogenesis from lactate (e.g. Schimmasek and Mitzkat, 1963; Ross, Hems and Krebs, 1967; Exton and Park, 1968) there is evidence that in the perfused rat liver, glucagon stimulates gluconeogenesis from alanine (Garcia, Williamson and Cahill, 1966; Mallette, Exton and Park, 1969a) and from endogenous hepatic glutamate (Uri, Exton and Park, 1973). Mallette, Exton and Park (1969b) have also shown that glucagon stimulates uptake of some of the amino acids added to the perfusion medium, the effect being much the greatest for alanine and glycine. In the absence of added amino acids, glucagon altered the medium and intracellular concentrations of several amino acids in a manner consistent with increased hepatic proteolysis and increased utilisation of alanine, glycine and glutamate. Sokal (1966) showed earlier that glucagon stimulates gluconeogenesis from endogenous liver protein, although the effect was not large (about

Table 9.4 *Amounts of injected ^{14}C amino acids appearing as $^{14}CO_2$ in rats, sheep and lactating cows. The period studied varied from 1-6 hours after injection, although most were about three hours*

Animal	Amino acid	Per cent appearing as $^{14}CO_2$	Estimated per cent converted to glucose	Reference
Sheep	mixed ^{14}C-amino acids	62	5-7	Ford and Reilly, 1969
Pregnant sheep	mixed ^{14}C-amino acids	31	5-10	Ford and Reilly, 1970
Rat	mixed ^{14}C-amino acids	9-12	-	Neale, 1971
Sheep	^{14}C-bacteria	12-20	2-8	Hoogenraad et al., 1970
Rat	DL-2-^{14}C-glutamate	40	-	McFarlane and Von Holt (1969)
Sheep	L-U-^{14}C-glutamate	54-60	9-10	Heitman, Hoover and Sniffley (1973)
Cow	L-U-^{14}C-glutamate	34-40	3	Egan and Black (1968)
Cow	L-U-^{14}C-aspartate	33-34	1	Black (1968)
Rat	DL-U-^{14}C-alanine	80	-	McFarlane and Von Holt (1969)
Cow	L-U-^{14}C-alanine	30-31	1-2	Black (1968)
Sheep	L-U-^{14}C-serine	20	5	Heitman, Hoover and Sniffley (1973)
Cow	L-U-^{14}C-phenylalanine	8	1	Black, Thompson, Anand and Chapman (1970)
Rat	L-U-^{14}C-phenylalanine	10	-	McFarlane and Von Holt (1969)
Cow	L-U-^{14}C-tyrosine	12	1	Black (1968)
Cow	L-U-^{14}C-threonine	6	1	Black (1968)
Cow	L-U-^{14}C-arginine	7-10	1	Black (1968)
Sheep	L-U-^{14}C-lysine	4-8	-	Brookes, Owens, Brown and Garrigus (1973)
Rat	L-U-^{14}C-lysine	2-10	-	Brookes, Owens and Garrigus (1972)
Cow	L-U-^{14}C-valine	5	1	Black (1968)
Rat	L-U-^{14}C-valine	4-8		Neale (1971)
Rat	L-U-^{14}C-valine	10-11		Reed (1974)
Cow	L-U-^{14}C-isoleucine	9	1	Black (1968)
Rat	L-[1]^{14}C-leucine	16	-	Sketcher, Fern and James (1974)
Cow	L-U-^{14}C-leucine	5	-	Black (1968)

60 μmole glucose/100 g rat/hour). Mallette, Exton and Park, (1969a) have shown that when glucagon stimulates gluconeogenesis from exogenous amino acids, the maximum increase is about 50 μmole/100 g rat/hour. Thus, the effect of glucagon on gluconeogenesis is quantitatively limited; and there is at present little evidence for an effect on other than two or three amino acids.

INSULIN

Although insulin markedly inhibits glucose output by the liver (Steele, 1966) the effect on splanchnic amino acid uptake in post-absorptive man is quite small. Felig and Wahren (1971) observed no change with a doubling of plasma insulin (due to infused glucose) and only a moderate effect (mainly on alanine and glycine) with a five-fold insulin rise. Mondon and Mortimore (1967) showed that insulin did not affect uptake of, or urea formation from, added amino acid by the perfused rat liver; but endogenous urea formation was suppressed by about 30 µmole/100 g rat/hour, i.e. about half the stimulating effect of glucagon.

The 'Direct' Oxidation of Amino Acid

It is clear that it cannot be supposed that the major catabolic fate of most amino acids involves conversion to glucose. The evidence presented in *Table 9.4* in contrast, suggests there is prompt oxidation of many amino acids and from the limited results available much of this does not occur through conversion to glucose. This may account for the findings of Annison *et al.* (1967) that oxidation of glucose and long and short-chain fatty acids accounted for only 76 per cent of the carbon dioxide produced in fed sheep. In general there is oxidation of a greater fraction of non-essential than of essential amino acids. This may be because oxidation of non-essential amino acids occurs in both muscle and liver, while oxidation of essential amino acids occurs predominantly in either liver or (the branch-chain acids) in muscle (Miller, 1962).

The intake of an amino acid could well affect the amount oxidised. This is confirmed for ^{14}C-lysine in rats (Brookes, Owens and Garrigus, 1972) and sheep (Brooks *et al.*, 1973) when the amino acid intake, apart from lysine, is constant. More varied results however have been obtained in rats receiving a low-protein diet. While the oxidation of phenylalanine (McFarlane and Von Holt, 1969), [1-^{14}C]-leucine (Sketcher, Fern and James, 1974) and [1-^{14}C]-valine (Reed, 1974) is reduced there is no decreased oxidation of glutamate and alanine (McFarlane and Von Holt, 1969); while Neale (1971) observed in fed protein-deficient rats an increased oxidation of uniformly labelled lysine, leucine and valine. Irrespective of the origin of these differences, they do suggest there is at best a limited capacity to suppress the oxidation of amino acids.

For this reason the distinction between 'glycogenic' and other amino acids (leucine, lysine, tryptophan) may be important even for amino acids which are not appreciably 'gluconeogenic', since during oxidation such amino acids contribute citric acid cycle intermediates and thus replace at least one of the functions of glucose.

References

AIKAWA, T., MATSUTAKA, H., TAKEZAWA, K. and ISHIKAWA, E. (1972). *Biochim. biophys. Acta,* **279,** 234

ANNISON, E.F., BROWN, R.E., LENG, R.A., LINDSAY, D.B. and WEST, C.E. (1967). *Biochem. J.,* **104,** 135

BATTAGLIA, F.C. and MESCHIA, G. (1973). in *Foetal and Neonatal Physiology*, Barcroft Centenary Symposium, p.382; Cambridge University Press, Cambridge

BICKERSTAFFE, R., ANNISON, E.F. and LINZELL, J.L. (1974). *J. agric. Sci.*, **82**, 71
BLACK, A.L. (1968). in *Isotope studies on the Nitrogen Chain*, Symposium sponsored by International Atomic Energy Agency, Vienna. p.287
BLACK, A.L., EGAN, A.R., ANAND, R.S. and CHAPMAN, T.E. (1968). in *Isotope studies on the Nitrogen Chain*, symposium sponsored by International Atomic Energy Agency, Vienna, p.247
BLACK, A.L., THOMPSON, J.R., ANAND, R.S. and CHAPMAN, T.E. (1970). in *Energy Metabolism of Farm Animals*. European Association Animal Production Publication No.13
BLACK, J.L. and TRIBE, D.E. (1973). *Aus. J. agric. Res.*, **24**, 763
BLOCK, R.J. and BOLLING, D. (1951). *The Amino Acid Composition of Proteins and Food* (2nd Ed.), C.C. Thomas, Illinois
BROOKES, I.M., OWENS, F.N. and GARRIGUS, U.S. (1972). *J. Nutr.*, **102**, 27
BROOKES, I.M., OWENS, F.N., BROWN, R.E. and GARRIGUS, U.S. (1973). *J. Anim. Sci.*, **36**, 965
DAKIN, H.D. (1913). *J. biol. Chem.*, **14**, 321
EGAN, A.R. and BLACK, A.L. (1968). *J. Nutr.*, **96**, 450
EGAN, A.R., MOLLER, F. and BLACK, A.L. (1970). *J. Nutr.*, **100**, 419
ELWYN, D.H., PARIKH, H.C. and SHOEMAKER, W.C. (1968). *Am. J. Physiol.*, **215**, 1260
EXTON, J.H. and PARK, C.R. (1967). *J. biol. Chem.*, **242**, 2622
EXTON, J.H. and PARK, C.R. (1968). *J. biol. Chem.*, **243**, 4189
FELIG, P. and WAHREN, J. (1971). *J. clin. Invest.*, **50**, 1702
FELIG, P., WAHREN, J., HENDLER, H. and BRUNDIN, T. (1974). *J. clin. Invest.*, **51**, 1870
FORD, E.J.H. and REILLY, P.E.B. (1969). *Res. vet. Sci.*, **10**, 409
FORD, E.J.H. and REILLY, P.E.B. (1970). *Res. vet. Sci.*, **11**, 575
GARCIA, A., WILLIAMSON, J.R. and CAHILL, G.F. (1966). *Diabetes*, **15**, 188
GURIN, S., DELLUVA, A.M. and WILSON, D.W. (1947). *J. biol. Chem.*, **171**, 101
HAFT, D.E., TENNEN, E. and MEHTALIA, S. (1972). *Am. J. Physiol.*, **222**, 365
HEITMAN, R.N., HOOVER, W.H. and SNIFFLY, C.J. (1973). *J. Nutr.*, **103**, 1587
HEMS, D.A. (1972). *Biochem. J.*, **130**, 671
HOOGENRAAD, N.J., HIRD, F.J.R., WHITE, R.G. and LENG, R.A. (1970). *Br. J. Nutr.*, **24**, 129
HUNTER, G.D. and MILLSON, G.C. (1964). *Res. vet. Sci.*, **5**, 1
IZZO, J.L. and GLASSER, J.R. (1961). *Endocrinol.*, **68**, 189
JANNEY, N.W. (1915). *J. biol. Chem.*, **20**, 321
JUDSON, G.J. and LENG, R.A. (1973). *Br. J. Nutr.*, **29**, 175
KREBS, H.A., NOTTON, B.M. and HEMS, R. (1966). *Biochem. J.*, **101**, 607
LINDSAY, D.B. and WILLIAMS, R.L. (1971). *Proc. Nutr. Soc.*, **30**, 35A
LINDSAY, D.B. and DYKE, C.S. (1974). *Proc. Nutr. Soc.*, **33**, 39A
MALLETTE, L.E., EXTON, J.H. and PARK, C.R. (1969a). *J. biol. Chem.*, **244**, 5713

MALLETTE, L.E., EXTON, J.H. and PARK, C.R. (1969b). *J. biol. Chem.*, **244**, 5724

McFARLANE, I.G. and VON HOLT, C. (1969). *Biochem. J.*, **111**, 557

MILLER, L.L. (1962). in *Amino Acid Pools*, Ed. J.T. Holden, p.708, Elsevier, Amsterdam

MINKOWSKI, O. (1893). *Arch. exp. Path. Pharmak.* **31**, 85

MONDON, C.E. and MORTIMORE, G.E. (1967). *Am. J. Physiol.*, **212**, 173

MORTIMORE, G.E., WOODSIDE, K.H. and HENRY, J.E. (1972). *J. biol. Chem.*, **247**, 2776

MUNRO, H.N. (1964). in *Mammalian Protein Metabolism*, Vol.1, p.381, Eds. H.N. Munro and J.B. Allison, Academic Press, New York and London

NEALE, R.J. (1971). *Nature, New Biology*, **231**, 117

NILSSON, L.H., FURST, P. and HULTMAN, E. (1973). *Scand. J. clin. Lab. Invest.*, **32**, 331

OLSEN, N.S., HEMINGWAY, A. and NIER, A.O. (1943). *J. biol. Chem.*, **148**, 611

REED, P.J. (1974). *Br. J. Nutr.*, **31**, 259

REILLY, P.E.B. and FORD, E.J.H. (1971). *Br. J. Nutr.*, **26**, 249

ROSE, W.C., JOHNSON, J.E. and HAINES, W.J. (1942). *J. biol. Chem.*, **145**, 679

ROSS, B.D., HEMS, R. and KREBS, H.A. (1967). *Biochem. J.*, **102**, 942

SCHIMMASEK, H. and MITZKAT, H.J. (1963). *Biochem. Z.*, **332**, 510

SHOEMAKER, W.C. and VAN ITALLIE, T.B. (1960). *Endocrinol.*, **66**, 260

SKETCHER, R.D., FERN, E.B. and JAMES, W.P.T. (1974). *Br. J. Nutr.*, **31**, 333

SOKAL, J.E. (1966). *Endocrinol.*, **538**

SOSKIN, S. (1930). *J. Nutr.*, **3**, 99

STEELE, R. (1966). *Ergebn. Physiol.*, **57**, 91

TRISTRAM, G.R. and SMITH, R.H. (1963). in *The Proteins*, 2nd Edn., Ed. Neurath, p.49, Academic Press, London

URI, M., EXTON, J.H. and PARK, C.R. (1973). *J. biol. Chem.*, **248**, 5350

WAHREN, J., FELIG, P., CERASI, E. and LUFT, R. (1972). *J. clin. Invest.*, **51**, 1870

WOLFF, J.E. and BERGMAN, E.N. (1972). *Am. J. Physiol.*, **223**, 455

WOLFF, J.E., BERGMAN, E.N. and WILLIAMS, H.H. (1972). *Am. J. Physiol.*, **223**, 438

10

THE ENERGETIC EFFICIENCY OF AMINO ACID METABOLISM

P.J. BUTTERY and K.N. BOORMAN
Department of Applied Biochemistry and Nutrition, University of Nottingham

Introduction

Amino acids serve a variety of functions in the animal body in addition to their major role (the provision of monomers from which proteins are synthesised). Examples of the secondary role of amino acids include the provision of energy particularly by gluconeogenesis, the synthesis of hormones (e.g. adrenalin) and the synthesis of certain constituents of bile. The complexity of the integration of amino acid metabolism makes theoretical calculations of efficiency of nitrogen metabolism difficult, indeed perhaps meaningless. In this review an attempt has been made to estimate from theoretical considerations the energy consumption associated with synthesis of protein and excretion of nitrogen. The results are compared with those obtained from whole animal studies.

The Excretion of Excess Amino Nitrogen

The energy cost of urea synthesis would appear to be two high energy phosphate bonds for each atom of nitrogen converted to urea, see *Figure 10.1;* however an additional allowance must be made for the transport of ornithine through the mitochondrial membrane, although the transport of citrulline through the mitochondrial membrane does not appear to be energy linked (Gamble and Lehninger, 1973). The energy associated with the regeneration of aspartate and the release of ammonia from the excess amino acids also requires consideration.

Studies with perfused livers measuring the oxygen uptake associated with urea synthesis have yielded some interesting results. Hems *et al.* (1966) calculated that the energy consumption associated with the synthesis of urea from ammonia in rat liver is nearly three times the theoretical value of 2ATP/g atom nitrogen converted to urea. A similar observation was reported by Lundsgaard (1942) using perfused cat liver.

Using intact sheep, Martin and Blaxter (1965) estimated the energy cost of urea synthesis from ammonium chloride to be 4.8ATP ± 0.71 mole of urea synthesised. These authors suggested that the difference between the then accepted theoretical value of 4ATP/mole and their

198 *Energetic efficiency*

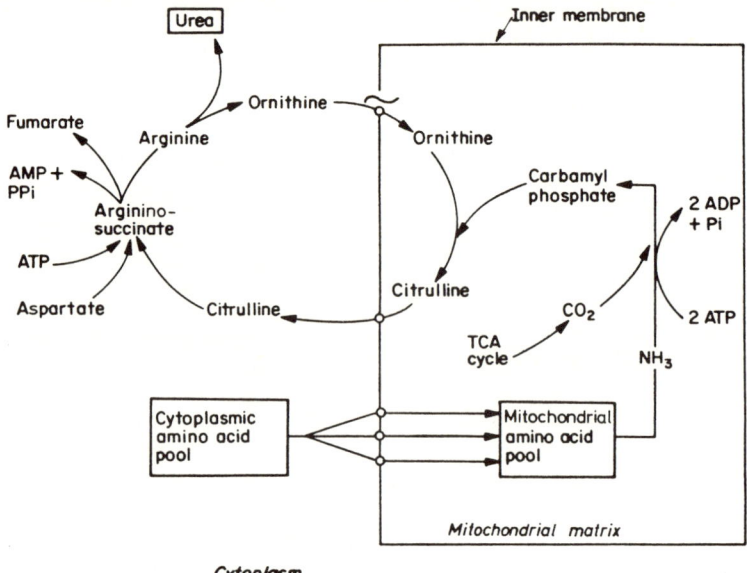

Figure 10.1   The urea cycle

observed value was due to experimental error. However, in the light of the observations mentioned above, it may be that even 4.8 ATP/mole is an underestimate of the energy costs of urea synthesis.

The total energy drain associated with urea synthesis, assuming glucose to be the energy source, would therefore be:-

$$\frac{2.87}{38} \times 4.8 = 0.36 \text{ MJ/mole urea synthesised}$$

assuming that 38 ATP are synthesised during the oxidation of glucose:-

$$C_6H_{12}O_6 + 6O_2 \longrightarrow 6CO_2 + 6H_2O \qquad \Delta G = -2.87 \text{ MJ}$$

Since Martin and Blaxter (1965) suggested that a further 0.9 ATP are consumed during the excretion of urea, the total energy cost may be as high as 0.43 MJ/mole. These calculations neglect consideration of urea re-cycling following hydrolysis to ammonia in the digestive tract; a process taking place both in the ruminant (Chalupa, 1974) and in the non-ruminant (Visek, 1974). The re-cycling of urea would of course present an extra energy demand.

The synthesis of uric acid is a much more complicated process than the synthesis of urea and assessment, on theoretical grounds, of the energy consumption during the detoxification of nitrogen by this route is difficult. Studies with the isolated perfused chick liver (Buttery, Boorman and Barratt, 1973; Barratt, Buttery and Boorman, 1974) have indicated that the production of uric acid from glycine was accompanied by an increased oxygen consumption equivalent to 9 ATP/mole while in the presence of glutamine this figure was decreased to 7 ATP/mole. This difference is of interest since two amide nitrogens are used in the synthesis of uric acid and the synthesis of each amide group consumes 1ATP. This difference between glycine and glutamine does therefore give some confidence in the results obtained, although it is possible that the addition of glutamine and glycine may be stimulating pathways other than uric acid synthesis. The direct incorporation of glycine into the uric acid molecule may put an additional energy demand on the animal. If it is assumed that this glycine in the chick is normally metabolised *via* serine to yield pyruvate which in turn is oxidised by the tricarboxylic acid cycle (Richert, Amberg and Wilson, 1962) then the catabolism of one mole of glycine would yield 9ATP. The effective total energy consumption associated with uric acid synthesis might then be as high as 18ATP/mole or equivalent to:-

$$\frac{18 \times 2.87}{38} = 1.4 \text{ MJ/mole}$$

if glucose provides the energy. Great care must be taken in comparing this result with those obtained by Martin and Blaxter (1965) and Hems *et al.,* (1966) for urea synthesis; however it would appear that uric acid synthesis is a more expensive method of excretion of excess nitrogen. Perhaps this is the penalty the bird has to pay for the convenience of excreting an insoluble nitrogenous product.

THE ENERGETICS OF PEPTIDE BOND SYNTHESIS

The mechanism of protein synthesis in animals is gradually being elucidated and current knowledge has been discussed elsewhere in this symposium. In the present context only those aspects associated with consumption of energy will be discussed.

The activation of each amino acid requires two high energy phosphate bonds:-

(1) Amino acid + activating enzyme + ATP $\rightleftarrows$ [enzyme-amino acyl-AMP] + P~P

(2) [enzyme-amino acyl-AMP] + tRNA $\rightleftarrows$ enzyme + AMP + amino acyl-tRNA

(3) AMP + ATP $\rightleftarrows$ 2ADP

P~P $\rightarrow$ 2Pi
___
(4) Amino acid + 2ATP $\rightarrow$ amino acyl-trRNA + 2ADP + 2Pi

It is interesting to note that the free energy of hydrolysis of the amino acyl tRNA (-29 kJ/mole^{-1}) is more than twice that of the peptide bond (-13 kJ/mole^{-1}) (Lengyel, 1969). The polymerisation of the activated amino acids requires the hydrolysis of GTP. The exact number of high energy phosphate bonds hydrolysed has been the subject of much study and speculation, although current views would suggest that two GTP are hydrolysed for each peptide bond synthesised (Haselkorn and Rothman-Denes, 1973). The total energy consumption would therefore be 4ATP for each peptide bond synthesised. This value does not allow for any energy consumption associated with the synthesis of the protein synthetic apparatus. Calculation of such requirements are difficult owing to the lack of information on the half lives of the many components of the protein synthetic apparatus. In bacteria it has been estimated that less than 5 per cent and more than likely less than 1 per cent of the total energy associated with peptide bond synthesis can be attributed to mRNA synthesis (Salser et al., 1969). Calculation of the cost of synthesis in animal systems has also suggested that the cost of synthesis of the biosynthetic apparatus is insignificant (for example, Buttery and Annison, 1973) although the energy cost associated with the rapid turnover of the various precursor RNA species is difficult to assess (Clark, 1974). No allowance has been made for the transport of the amino acids required for protein synthesis. Most amino acids are transported against concentration gradients and this must require the expenditure of energy. The exact amount of energy associated with this process is difficult to assess. It has however been suggested that the transport of metabolites may account for a considerable proportion of total body energy expenditure (Milligan, 1971) for example in brain slices 52 per cent of total aerobic energy expenditure can be accounted for by the operation of the Na^+/K^+ pump (Whittam, 1964).

The energy consumption associated with peptide bond synthesis would therefore appear to be approximately equivalent to the hydrolysis of four high energy phosphate bonds, i.e:-

$$\frac{4}{38} \times 2.87 = 0.30 \text{ MJ}$$

assuming the energy to be generated from glucose catabolism. Thus the theoretical energy cost of protein deposition would be 3 kJ/g protein plus the energy of the deposited protein, 23.3 kJ/g.

THE SIGNIFICANCE OF PROTEIN TURNOVER

The energy cost of protein deposition is increased by the operation of protein turnover. The largest proportion of protein synthesis is for the replacement of degraded protein rather than for increasing body protein mass. In a 50 kg sheep, mixed muscle protein of the *l. dorsi* muscle was found to have a fractional synthetic rate of 0.018 day^{-1} (Buttery et al., 1975) which would indicate that in a sheep of 50 kg, 86 g (dry weight) of skeletal muscle protein are synthesised per day, while the net deposition of protein would only be expected to be a few grams. In the rapidly growing broiler cockerel (900 g) it has been similarly calculated that 6.8 g of skeletal muscle protein are synthesised per day while skeletal muscle protein deposition was found to be 3.9 g/day (Buttery, Boorman and Barratt, 1973). In young rats (100 g) 713 mg muscle protein is synthesised/day while the net deposition is only 149 mg/day (Millward, 1970).

The possibility of the breakdown of protein requiring energy must not be excluded (for example, Brostrom and Jeffay, 1970; Schimke, 1974). The breakdown of protein should theoretically render all the released amino acids available for the resynthesis of proteins. If this were entirely true one might expect the excretion of urea nitrogen to fall to zero on a nitrogen-free diet but this does not occur (Peret and Jacquot, 1972). What proportion of this endogenous nitrogen excretion

Table 10.1 Energy cost of nitrogen metabolism in sheep (50 kg)

(a)	[1]Turnover muscle protein (85 g)	=	> 0.27 MJ/day
(b)	[1]Turnover liver protein (23 g)	=	> 0.07 MJ/day
(c)	[2]Urea synthesis 13.5 g nitrogen [8.0 excreted, 5.5 recycled]	=	0.17 MJ/day
(d)	Urea excretion	=	0.02 MJ/day
	[Total daily ME intake ≃ 10 MJ]		

[1] Buttery et al. (1975)
[2] Nolan, Norton and Leng, (1973)

Table 10.2 Energy cost of nitrogen metabolism in chick (900 g)

Deposition of skeletal muscle (3.9 g)	=	94.2 kJ/day
Protein turnover skeletal muscle (6.8 g)	=	> 18.8 kJ/day
Uric acid synthesis (1 g nitrogen)	=	26.8 kJ/day
[Total daily ME intake 1.26 MJ]		

From Buttery et al., 1973

can be attributed to the 'inadvertent' breakdown of amino acids as they pass through the intracellular amino acid pools during the process of protein turnover is difficult to assess. The inability to re-utilise hydroxyproline, methyl histidine and methyl lysine (Munro, 1970; Young et al., 1972) must also be considered. The energy consumption associated with protein turnover in sheep and chicks is shown in Tables 10.1 and 10.2.

THE BALANCE OF AMINO ACID SUPPLY

For efficient protein synthesis, amino acids must be available at the site of synthesis in proportions similar to those found in the protein. The effects of feeding amino acids in grossly unbalanced proportions are discussed elsewhere in these proceedings but perhaps more relevant in terms of practical agriculture are the effects of feeding diets only marginally deficient in one amino acid. It has been observed that chicks fed on a diet marginally deficient in methionine have an impaired food conversion compared with animals fed on an adequate diet, although little difference is observed in weight gain (Carew and Hill, 1961; Anonymous, 1962; Solberg, Buttery and Boorman, 1971). The fate of this extra energy intake must either be an increased energy content of the carcass or an increased heat output. Carew and Hill (1961) obtained evidence for an increased energy content of the carcass and calculated that there was little extra heat production. In contrast, using a different experimental design, Shoji, Totsuka and Tajima (1966) found a decreased deposition of energy in the carcass and an increased oxygen uptake. Studies in this laboratory (Solberg, Buttery and Boorman, 1971) have shown that a diet marginally deficient in methionine results in an increased uric acid production (Table 10.3). Using a figure of 1.4 MJ ME per mole of uric acid synthesised,

Table 10.3 Uric acid excretion of chicks fed a diet marginally deficient in methionine

	Basal	Basal + 0.15 per cent methionine
Excreted uric acid (m mol/day)	7.74 ± 0.38	5.36 ± 0.15[1]
Uric acid excreted as per cent nitrogen in diet	33.5 ± 1.6	24.9 ± 0.8[1]
∴ Extra uric acid synthesised = 2.38 m mole/day ≡ 3.4 KJ/day ≡ 67 per cent of increased ME intake		
Total ME intake	590 KJ/day	

[1] Results statistically different, i.e. ($P < 0.001$)

See text for details and Solberg *et al.* (1971)

67 per cent of the increased energy intake (5.1 kJ per chick per day) of the deficient birds can be accounted for by increased uric acid synthesis. As each bird was eating 590 kJ/day, this increased uric acid production accounts for a little less than one per cent of the total energy intake. Presumably a similar situation arises with other amino acids (for example, Thomas, Davidson and Boyne, 1969).

Energy Cost of Protein Deposition - Whole Animal Studies

Estimates of the energy expended in depositing protein may be obtained from energy balance information obtained with the whole animal. Usually experiments are arranged so that different animals receive different intakes of protein and energy over as wide a range as possible within a positive energy balance. After a period of time or when the animals reach a fixed weight, the animals are killed and the energy stored as protein and fat in the carcass is measured. The energy required for protein deposition can then be calculated from an equation partitioning energy intake into maintenance, protein deposition and fat deposition, derived by multiple regression analysis,

$$E = a.M^n + b.P + c.F + i$$

Where E = ME intake (kJ/d); M^n is a maintenance term expressed either in terms of body weight or a function thereof or in terms of days to reach a fixed weight; P = protein deposited either in g or kJ; F = fat deposited (g or kJ); a, b, c and i are constants: b representing the cost of depositing a unit of protein (kJ/g or kJ/kJ), c, the similar coefficient for fat and i representing an intercept term.

Several modifications of this method have been used. Energy balance data may be collected by respirometry (Thorbek, 1970; Bøndsdorff-Petersen, 1970) and modifications of the partition equation may be used (Kielanowski and Kotarbinska, 1970).

In an early attempt to apply such methods, Kielanowski (1965) found values of about 30 kJ energy expended per g protein deposited in pre-ruminant lambs and young piglets. As the author pointed out, this value is little in excess of the theoretical value of about 26 kJ/g calculated by Blaxter (1962) and Schiemann (1963). However Kielanowski and Kotarbinska (1970) reported that re-calculation of the data for piglets by a modified procedure which assumed a 100 per cent conversion of dietary fat to body fat, yielded a value of 48 kJ/g for protein deposition. In the same report, data from Kotarbinska and Kielanowski (1969) for the growing pig, which had previously yielded a value of 46 kJ/g, were re-calculated and yielded a value of 67 kJ/g. The authors appeared to place more reliance on these higher values. Other values for the growing pig (Oslage, Eadeken and Fliegel, 1970; Thorbeck, 1970), growing chick (Bønsdorff Petersen, 1970) and growing rat (Schiemann, 1970; Schiemann, Chudy and Hereg, 1969) are in broad agreement with the re-calculated value (48 kJ/g) quoted by Kielanowski and Kotarbinska (1970). Values outside this range, however, have also been obtained by Ørskov and McDonald (1970) for the growing lamb (68 kJ/d) and by McCracken (1973) for the growing rat (27 kJ/d). A summary of some of the available values is shown in *Table 10.4*, and Kielanowski presents further values elsewhere in these proceedings.

Table 10.4 Metabolisable energy expended in protein and fat deposition found from energy balance experiments

Animal	Energy cost (kJ/g) Protein	Fat	Source
Pre-ruminant lamb	29.6^a	62.6	Kielanowski, (1965)
Piglet	31.4^b	48.7	Ibid
Growing pig	46.2^b	56.3^c	Kotarbinska and Kielanowski, (1969)
Growing rat	50.8^1	55.6^2	Schiemann et al. (1969)
Piglet (re-calculated)	48.1^a	54.2^c	Kielanowski and Kotarbinska, (1970)
Growing pig (re-calculated)	66.8^b	-	Ibid
Growing pig	54.8	51.9	Thorbek, (1970)
Growing pig	44.5^1	55.6^2	Oslage et al. (1970)
Growing lamb	68.0	47.9	Ørskov and McDonald, (1970)
Growing chick	46.9	51.0	Bønsdorff Petersen, (1970)
Growing rat	48.0	55.6	Schiemann (1970)
Growing rat	26.9	50.9	McCracken (1973)

^{a,b,c}values with the same superscript were calculated from the same original data
^{1} assuming energy content of deposited protein to be 23.3 kJ/g
^{2} assuming energy content of deposited fat to be 38.9 kJ/g

Evidently there is variability amongst the values obtained for the energetic efficiency of protein deposition. That there has been dissatisfaction with the method is indicated by the several modifications of the partition equation that have been used. The necessity for the intercept term (McCracken, 1973), the meaning of the maintenance term and the interdependence of the coefficients (Ørskov and McDonald, 1970) have been commented upon. However it is notable that the method does produce a degree of consistency when used to estimate the energy expenditure of fat deposition. In *Table 10.4* the values for protein deposition, omitting original estimates which were re-calculated, range from 26.9 to 66.8 with a standard deviation of 13.35 kJ/g; those for fat deposition range from 47.9 to 62.6 with a standard deviation of 4.3 kJ/g. This raises the question, is the variation in the values for protein deposition indicative of an intrinsically more variable process?

The question of whether the energetic efficiency of protein deposition decreases with age has been raised (Kotarbinska and Kielanowski, 1969). However, there has been no systematic examination of this question and the data available do not allow such an examination. An alternative source of variation might be found in the amino acid pattern of the diet. Theoretical calculations pre-suppose an ideal dietary supply for tissue syntheses. Departure from such an ideal supply will necessitate catabolism of amino acids in excess of the needs for the protein deposition allowed by the supply of the most deficient amino acid. This will result in decreased energetic efficiency of protein deposition. Ideal dietary balance of amino acids is rarely achieved, except possibly when milk proteins are fed to very young animals, and requirement patterns change with age. Even quite small imbalances have quite marked effects (Solberg, Buttery and Boorman, 1971). It is possible, therefore, that all estimates of energy expenditure in protein deposition include an additional inefficiency, due to the factors described above. The extent by which these factors reduce the efficiency of protein deposition is difficult to assess.

Conclusions

From whole animal studies it would appear that the energy cost of protein deposition is very variable although most values are in the range 42-54 kJ/g protein deposited. The theoretical cost of peptide bond synthesis plus the energy content of the protein deposited is approximately 26 kJ/g. The discrepancy between the theoretical value may be accounted for partially by the process of protein turnover, the excretion of excess nitrogen, the unlikelihood of achieving a completely balanced amino acid supply and the transport of amino acids. Although protein synthesis is relatively inefficient in energetic terms it shows a remarkable efficiency in other directions; the production of many thousands of proteins from twenty different amino acids is a process where an error could lead to disastrous results.

Acknowledgement

We wish to acknowledge helpful discussions with Dr. K.J. McCracken.

References

Anonymous, (1962). *Nutr. Rev.,* **20,** 122
BARRATT, E., BUTTERY, P.J. and BOORMAN, K.N. (1974). *Biochem. J.,* **144,** 189
BLAXTER, K.L. (1962). *The Energy Metabolism of Ruminants,* p.277. Hutchinson, London
BØNSDORFF-PETERSEN, Ch.R. (1970). in *Energy Metabolism of Farm Animals.* Eds. A. Schurch and C.Wenk. p.205. Juris-Verlag, Zurich
BUTTERY, P.J. and ANNISON, E.F. (1973). in *The Biological Efficiency of Protein Production.* Ed. J.W.C. Jones, p.141. University Press, Cambridge
BUTTERY, P.J., BECKERTON, A., MITCHELL, R.M., DAVIES, K. and ANNISON, E.F. (1975). *Proc. Nutr. Soc.* **34,** 91A
BUTTERY, P.J., BOORMAN, K.N. and BARRATT, E. (1973). *Proc. Nutr. Soc.,* **32,** 80A
BROSTROM, C.O. and JEFFAY, H. (1970). *J. biol. Chem.,* **245,** 4001
CAREW, L.E. and HILL, F.W. (1961). *J. Nutr.,* **74,** 185
CHALUPA, W. (1974). *Fedn Proc. Fedn Am. Socs exp. Biol.,* **31,** 1152
CLARK, B.F.C. (1974). in *Companion to Biochemistry.* Eds. A.T. Bull et al., p.1. Longman Group, London
GAMBLE, J.G. and LEHNINGER, A.L. (1973). *J. biol. Chem.,* **248,** 610
HASELKORN, R. and ROTHMAN-DENES, L.B. (1973). *Ann. Rev. Biochem.,* **42,** 397
HEMS, R., ROSS, B.D., BERRY, M.N. and KREBS, H.A. (1966). *Biochem. J.* **101,** 284
KIELANOWSKI, J. (1965). in *Energy Metabolism.* Ed. K.L. Blaxter. Academic Press, London
KIELANOWSKI, J. and KOTARBINSKA, M. (1970). in *Energy Metabolism of Farm Animals.* Eds A. Schurch and C. Wenk. p.145. Juris Verlag, Zurich
KOTARBINSKA, M. and KIELANOWSKI, J. (1969). in *Energy Metabolism of Farm Animals.* Eds K.L. Blaxter, G. Thorbek and J. Kielanowski. p.299. Oriel Press, Newcastle-upon-Tyne
LENGYEL, P. (1969). *Cold Spring Harbor Symposium on Quantitative Biology,* **34,** 828
LUNDSGAARD, E. (1942). *Acta Physiol. Scand.,* **4,** 330
McCRACKEN, K.J. (1973). *Proc. Nutr. Soc.,* **32,** 66A
MARTIN, A.K. and BLAXTER, K.L. (1965). in *Energy Metabolism.* Ed. K.L. Blaxter. p.83. Academic Press, New York and London
MILLIGAN, L.P. (1971). *Fedn Proc. Fedn Am. Socs exp. Biol.,* **30,** 1454
MILLWARD, D.J. (1970). unpublished observation
MUNRO, H.N. (1970). in *Mammalian Protein Metabolism,* **Vol.4.** Ed. H.N. Munro. p.90. Academic Press, New York and London
NOLAN, J.V., NORTON, B.W. and LENG, R.A. (1973). *Proc. Nutr. Soc.,* **32,** 93
ØRSKOV, E.R. and McDONALD, I. (1970). in *Energy Metabolism of Farm Animals.* Eds. A. Schurch and C. Wenk. p.121. Juris Verlag, Zurich
OSLAGE, H.J., EADEKEN, D. and FLIEGEL, H. (1970). *ibid,* p.133
PERET, J. and JACQUOT, R. (1972). in *Protein and Amino Acid Functions.* Ed. E.J. Bigwood. Pergamon Press, Oxford
RICHERT, D.A., AMBERG, R. and WILSON, M. (1962). *J. biol. Chem.,* **237,** 99

SALSER, W., GESTELAND, R.F. and RICARD, B. (1969). *Cold Spring Harbor Symposium on Quantitative Biology*, **34**, 771

SCHIEMANN, R. (1963). *Deutsche Akademic der Landw. Sitzungsberichte*, **12**, 39

SCHIEMANN, R. (1970). *Wiss. Ztschr. Humbddt-Univ. Berl., Math-Natariviss. Reihe*, **19**, 35

SCHIEMANN, R., CHUDY, A. and HEREG, O. (1969). *Arch. Tierernähr.*, **19**, 395

SCHIMKE, R.T. (1974). in *Mammalian Protein Metabolism*, Vol.4, p.177. Ed. H.N. Munro. Academic Press, New York and London

SHOJI, K., TOTSUKA, H. and TAJIMA, M. (1966). *Jap. J. Zootech. Sci.*, **37**, 246

SOLBERG, J., BUTTERY, P.J. and BOORMAN, K.N. (1971). *Br. Poult. Sci.*, **12**, 297

THOMAS, O.A., DAVIDSON, J. and BOYNE, A.W. (1969). *Br. Poult. Sci.*, **10**, 67

THORBEK, G. (1970). in *Energy Metabolism of Farm Animals*. Eds. A. Schurch and C. Wenk. p.129. Juris Verlag, Zurich

VISEK, W.J. (1974). *Fedn Proc. Fedn Am. Socs exp. Biol.*, **31**, 1178

WHITTAM, R. (1964). *Nature, Lond.*, **191**, 603

YOUNG, V.R. (1970). in *Mammalian Protein Metabolism*, Vol.4, p.584. Ed. H.N. Munro, Academic Press, New York and London

YOUNG, V.R., ALEXIS, S.D., BALIGA, B.S. and MUNRO, H.N. (1972). *J. biol. Chem.*, **247**, 3592

11

ENERGY COST OF PROTEIN DEPOSITION

J. KIELANOWSKI,
Institute of Animal Physiology and Nutrition, Jablonna, Poland

The feed energy cost of fat deposition in the animal body has been measured accurately in respiration trials since the beginning of this century and reliable estimates of the maintenance requirements had been obtained long ago; whereas until recently very little was known about the energy cost of protein deposition in the tissues. This gap was due mainly to difficulties connected with direct or indirect determinations of heat production related to the rather small amounts of protein deposited, but also to insufficient knowledge of changes of the chemical body composition during the growth of animals. Only a decade ago was the first attempt made to estimate the energy cost of protein deposition statistically (Kielanowski, 1965). It was assumed that the total metabolisable energy of the food ingested by growing animals could be partitioned by means of a regression equation into three portions: one proportional to the live weight of animals (representing maintenance requirement), another proportional to the mass of fat deposited in the body, and a third one, proportional to nitrogen retained. The amount of energy estimated in this way, corresponding to 1 g of nitrogen × 6.25 retained, has been called the unitary energy cost of protein deposition (ECPD). A similar approach was adopted later in numerous experiments with several species of animals. Some investigators determined the accretion of protein and fat in the body by comparative slaughter tests; others based their computations on data collected in respiration trials or applied these two methods simultaneously, and in a few experiments heat production was measured directly. In all cases, however, the principle was the same, i.e. regressing the live weight (often raised to a fractional power) and the deposition of protein and fat on the energy of ingested feed, with the stipulation that the independent variables are not highly correlated. The results of a number of these experiments are shown in *Table 11.1*.

Regression equations represent simply a line fitting a set of data most closely; ascribing to the regression coefficients any actual meaning, in isolation of experimental conditions, can be misleading. Hence, estimates of ECPD computed in the described way, together with factors expressing costs of fat deposition and maintenance, might be understood only as multipliers helping to compute the total energy requirement of animals treated like those in the experiment from which the

Table 11.1 Estimates of the energy cost of crude protein deposition in growing animals

Description and live weight of animals	N[2]	Method applied	Energy cost of protein deposition		Reference
			Metabolisable energy[1] (kJ/g)	(Moles ATP/mole amino acid)	
Milk-fed lambs, 5.5 to 10.5 kg	6	comparative slaughter	45.1 ± 6.74	28	Kielanowski (1965) (recomputed)
Lambs, 15 to 40 kg		comparative slaughter	68.0	53	Ørskov and McDonald (1970)
Lambs, 32 to 58 kg	6	respiration trials	68.7	52	Bickel and Durrer (1973)
Bull-calves, 32 to 145 kg	32	comparative slaughter	56.7 ± 20.9	41	Osinska (1974)
Milk-fed piglets, 2.5 to 8.5 kg	8	comparative slaughter	48.1 ± 1.97	32	Kielanowski and Kotarbinska (1970)
Milk-fed piglets, 4 to 12 kg	60	comparative slaughter	50.8	36	Müller and Kirchgessner (1973)
Castrated male pigs, 20 to 40 kg	28	nutritional balance and direct calorimetry	41.2	23	Close, Verstegen and Mount (1973)
Castrated male pigs, 20 to 90 kg	48	respiration trials	54.9 ± 4.77	41	Thorbek (1970)
Castrated male pigs, 30 to 90 kg	54	comparative slaughter	66.8 ± 8.03	57	Kotarbinska (1969)
Castrated male pigs, 25 to 110 kg	4	respiration trials	45.6	29	Oslage, Gädeken and Fliegel (1970)
Castrated male pigs, 30 to 110 kg	16	respiration trials	44.1	27	Gädeken, Oslage and Fliegel (1973)
Male rats, 80 to 200 g	9	respiration trials and comparative slaughter	52.0	38	Schiemann, Chudy and Herceg (1969)
Cockerels, 0.1 to 0.7 kg	72	comparative slaughter	48.1 ± 16.32	32	Bønsdorff-Petersen et al. (1970)
Cockerels, 0.4 to 1.5 kg	16	respiration trials	46.7 ± 0.63	31	Bønsdorff-Petersen (1970)

estimates were obtained. However, in *Table 11.1*, only those estimates of ECPD have been collected, which were accompanied by simultaneously obtained estimates of the energy cost of fat deposition not discrepant from values accepted generally, as well as by reliable estimates of maintenance requirement (i.e. close to 418 kJ of metabolisable energy per $kg^{0.75}$ of the metabolic live weight). It seems, therefore, that the values presented can be regarded as fair empirical indicators of energy expenditure connected with protein deposition, when commonly assumed factors of energy utilisation for maintenance and fat deposition are adopted.

In the original papers, estimates of ECPD were given in units of energy per mass unit of protein deposited or as multiples of the heat of combustion of protein. Assuming this heat as 23.85 kJ (unless otherwise indicated), all values of ECPD in *Table 11.1* have been expressed uniformly in kJ per g of crude protein deposited. In order to facilitate comparisons between ruminant and non-ruminant animals, estimates of ECPD have also been expressed in moles of high-energy phosphate bonds per mole amino acid incorporated into the deposited protein. For this purpose it has been assumed (Armstrong, 1969) that in ruminating animals, 86 kJ of a mixture of volatile fatty acids corresponds on average to one mole high-energy bond, and that the heat of rumen fermentation is 10 per cent of the combustion heat of ingested carbohydrates (Blaxter, 1962). With these assumptions, in a ruminant animal 96 kJ of metabolisable energy would correspond to one mole of high-energy bond. For non-ruminants and milk-fed ruminants it has been assumed that 84 kJ of metabolisable energy correspond to one mole of high-energy bond. It has been assumed further that the average molecular weight of amino acids incorporated into protein is 112.

In most experiments quoted in *Table 11.1*, nitrogen retention has been determined by comparative slaughter tests. This method seems to be more trustworthy than nutritional balance trials in which nitrogen retention is often over-estimated, and this would cause an under-estimation of ECPD. It is probably for this reason, that the estimates of Oslage, Gädeken and Fliegel (1970), Gädeken, Oslage and Fliegel (1973) and Close, Verstegen and Mount (1973) are the lowest in *Table 11.1*. If these estimates are omitted it can be seen that the values of ECPD obtained with very young animals are lower than those obtained with older ones. This could possibly be explained by the finding of Buraczewski and Pastuszewska (1974) that the proportion of amino nitrogen in total nitrogen in the body increases with age in rats (86 per cent at 21 and 91 per cent at 88 days of age) and pigs (78 per cent at 10 and 89 per cent at 100 kg live weight). It also seems probable that the fraction of free amino acids in the body decreases with advancing age proportionally to the relative diminution of body fluids. The same amount of nitrogen retained, therefore, would correspond to smaller amounts of true protein in younger than in older animals, and this might be reflected by the values of ECPD.

Data presented in *Table 11.1* demonstrate that there is a definite relationship between the deposition of crude protein in the tissues of growing animals and their energy expenditure. Values of ECPD are useful for the design and interpretation of feeding trials and can be applied for the compilation of feeding standards. For such purposes ECPD values of about 45 to 50 kJ would be suitable for milk-fed animals and chickens, 55 to 65 for growing-fattening pigs, and 60 to 70 for growing ruminants.

Although estimates of ECPD can be regarded as practical empirical indices, little can be said about their true physiological meaning. The energy cost of fat deposition in animal tissues, measured in respiration trials or estimated statistically, is consistent with the energy cost of fat synthesis, deduced from biochemical considerations (Armstrong, 1969). However, it has been deduced similarly that 8 to 10 moles of high-energy phosphate bonds are needed for the incorporation of one mole of amino acid into protein (Armstrong, 1969), and this is considerably less than has been computed from the quoted values of ECPD. In *Table 11.2* estimates of the energy cost of protein secretion in milk and eggs are presented. The energy cost of secretion of egg protein is very similar to ECPD for growing chickens, but the energy cost of secretion of milk protein, expressed in moles of high-energy bond, is lower than any value of ECPD, although still about twice as high as the deduced cost of protein synthesis. This deduced cost of about 10 high-energy bonds seems to be over-estimated rather than under-estimated, but it could be argued that it is a purely theoretical value and that the actual cost of protein synthesis has never been determined directly. Nevertheless, even the discrepancy between the energy cost of secretion of milk protein and the estimates of ECPD justifies the conclusion that ECPD cannot be identified with the cost of protein synthesis. At the present state of knowledge, therefore, the physiological meaning of ECPD can only be speculated upon.

First of all it should be remembered that the processes of synthesis and breakdown of protein are going on continuously in the organism and that the net gain of protein in a growing animal should be regarded as a balance of these two processes (Millward and Garlick, 1972). At nitrogen equilibrium, the energy expenditure for re-synthesis of body proteins must be included in the energy cost of maintenance. However, the rate of turnover of protein in some tissues could be correlated with protein deposition, and this would be reflected in the statistically computed value of ECPD. It is also quite probable that the cost of renewal of the mammary gland tissue in lactating animals is included in the overall cost of secretion of milk protein.

Greater protein deposition often results from a higher content of protein in the ration. It could be deduced, therefore, that the costs of assimilation of feed protein enter into the account of ECPD, though the finding (Zebrowska and Buraczewska, 1972) that the amount of endogenous nitrogen secreted into the pig's gut is practically independent of the protein concentration in feed, does not support this supposition.

Interesting and difficult to answer, is the question of the relationship between the quality of feed protein and ECPD in simple-stomached animals. Since protein synthesis would obviously not go on at any site if all amino acids needed were not present in the right proportion,

Table 11.2 Estimates of the energy cost of protein secretion in milk and eggs

Animals	N[1]	Method applied	Energy cost of protein secretion		Reference
			Metabolisable energy (kJ/g)	(Moles ATP/mole amino acid)	
Dairy cows	134	respiration trials	43.0	22	Hoffmann et al. (1972)
Laying hens	155	respiration trials	52.0	38	Hoffmann and Schiemann (1973)

[1] number of individual respiration trials

and since the energy loss caused by deamination of superfluous amino acids is accounted for by the determination of metabolisable energy, it seems that ECPD should not depend on the quality of feed protein. However, premises for this conclusion may not be adequate and the possibility cannot be excluded that the utilisation of poor quality protein causes an additional energy loss, resulting in a higher value of ECPD. Investigations by Thorbek (1970), who obtained ECPD values ranging from about 42 to 67 kJ for animals fed on rations including protein from different sources, seem to suggest that the variance due to this factor might have been quite important.

In all values of ECPD presented in *Table 11.1*, crude protein has been understood as the product of nitrogen balance multiplied by 6.25. However, the content of nitrogen in the amino acids found in the whole body of pigs (Buraczewski, 1973) amounts to about 15 per cent. The multiplier, therefore, should be 6.67 rather than 6.25, and this would diminish the values of ECPD by about 6 per cent. This correction, together with the corrections connected with the aforementioned changes of proportion between the amino nitrogen and total nitrogen, should be taken into account when sufficient information has been accumulated, but they would only slightly shift the values of ECPD and would not contribute significantly to the disappearance of the discrepancy discussed. This probably also concerns corrections due to the efficiency of formation of high-energy bonds, which, at least in some cases, could be less than had been assumed.

Moe, Tyrrell and Flatt (1970) found that the maintenance requirements of lactating cows were about 18 to 28 per cent higher than in non-lactating ones, and it could be assumed that this increase of the rate of metabolism was related to the synthesis of milk protein. By analogy, it could also be assumed that in growing animals, synthesis of protein is associated with an acceleration of the rate of metabolism, and if it were so, this would be reflected by the regression coefficient for protein deposition. The supposition that there are endocrine factors involved in the process of intensified protein synthesis simultaneously increasing the heat production is plausible. For example, it has been shown (Sokoloff and Kaufmann, 1961) that thyroxine, known as a thermogenic agent, stimulated the incorporation of leucine into a cell-free liver homogenate, and it has been found recently (Witkowska, 1974) that the level of tri-iodothyromine in the blood serum of growing pigs was positively correlated with the nitrogen balance.

Symbolising the maintenance requirement at zero nitrogen balance by E_M, the energy cost of synthesis plus heat of combustion of a weight unit of protein by E_P, the amount of protein synthesised by P, and a function of P by which E_M is multiplied in effect of protein synthesis by f_P; the metabolisable energy (ME) corresponding to the sum of the increased maintenance requirement and the cost of synthesis of protein would then be:

$$ME = E_M f_P + E_P P \qquad (1)$$

The unitary energy cost of protein deposition, i.e. ECPD, would obviously equal the derivative of ME with respect to P.

If for non-ruminating animals 55 kJ are assumed as ECPD, 31 kJ as E_P and 418 kJ as E_M per $kg^{0.75}$ metabolic weight, the following hypothetical equation will be obtained from equation (1):

$$ME \text{ (kJ)} = 418\ e^{0.05P} + 31\ P \qquad (2)$$

and the derivative of *equation (1a)*, corresponding to ECPD, will be:

$$\frac{d\ ME}{d\ P} \text{ (kJ/g)} = 20.9\ e^{0.05P} + 31 \qquad (3)$$

It can be computed further from *equations (1a)* and *(2)* that an increase of the maintenance requirement by 0.22 to 28 per cent, corresponding to an increment of 4 to 5 g protein per $kg^{0.75}$ metabolic weight, would account for the difference between the value of 55 kJ found empirically as ECPD and 31 kJ deduced as the energy cost of protein synthesis. The increase of the maintenance requirement assumed corresponds to the increase found actually by Moe, Tyrrell and Flatt, (1970) for lactating cows. If, on the other hand, in the energy balance of a cow weighing 600 kg and producing 20 kg milk daily, about 14,200 kJ (i.e. 28 per cent of the maintenance requirement of dry cows) were added to the energy costs of secretion of milk protein computed from *Table 11.2*, it would result in a joint unitary cost amounting to about 66 kJ/g of milk protein secreted, consistent with ECPD for growing ruminants.

All factors discussed can be responsible partly for the discrepancy between the estimated ECPD and the apparent cost of protein synthesis, but the assumption that it is due mainly to an increase of the rate of metabolism, as well as an increase of protein turnover correlated with the rate of protein deposition, seems to be most reasonable. The verification of this hypothesis needs further investigation, in which nutritional and endocrinological studies should be combined.

Energy Cost of Milk Protein Secretion

The calculation has been based mainly on the results of extensive investigations carried out at the Oskar-Kellner Institute of Animal Nutrition in Rostock, summarised by Hoffmann *et al.* (1972). These authors found that lactating cows converted metabolisable energy into milk energy with an average efficiency of 61.9 ± 5.1 per cent, and this agrees very closely with the findings of Moe, Tyrrell and Flatt, (1970). The authors expressed the opinion that the conversion of metabolisable energy into milk-fat energy is approximately as efficient as that into the energy of fat deposited in the tissues, and could be assumed to be 55 per cent. The efficiency of the formation of lactose must be very high, and according to Nehring and Schiemann (1966) could be assumed to be 90 per cent. These findings and assumptions permit the calculation of the unitary energy cost of the secretion of milk protein.

Energy content of 1 kg milk (average composition as reported by Hoffmann et al., 1972):

$$
\begin{array}{rcl}
37.3 \text{ g fat} \times 38.5 \text{ kJ} & = & 1436.0 \text{ kJ} \\
49.0 \text{ g lactose} \times 16.5 \text{ kJ} & = & 808.5 \text{ kJ} \\
32.1 \text{ g protein} \times 24.5 \text{ kJ} & = & \underline{786.5 \text{ kJ}} \\
& & 3031.0 \text{ kJ}
\end{array}
$$

Requirement of metabolisable energy for the synthesis of fat and lactose:

$$
\begin{array}{rcl}
1436.0 \text{ kJ of fat} : 0.55 & = & 2610.9 \text{ kJ} \\
808.5 \text{ kJ of lactose} : 0.90 & = & \underline{898.3 \text{ kJ}} \\
& & 3509.2 \text{ kJ}
\end{array}
$$

Metabolisable energy used for the secretion of 3031.0 kJ in milk is 3031.0 kJ : 0.62 = 4888.7 kJ. It means that for the secretion of 32.1 g protein 1359.5 kJ were needed, and that the cost of secretion of 1 g protein was 43.0 kJ.

References

ARMSTRONG, D.G. (1969). in *Handbuch der Tierernährung*. Eds. W. Lenkeit, K. Breirem and E. Crasemann, vol.1, p.385. Paul Parey Verlag, Hamburg and Berlin

BICKEL, H. and DURRER, A. (1974). in *Energy Metabolism of Farm Animals*. Eds K.H. Menke, H.J. Lantzsch and J.R. Reichl, p.119, Universität Hohenheim

BLAXTER, K.L. (1962). *The Energy Metabolism of Ruminants*, Hutchinson, London

BØNSDORFF-PETERSEN, Ch.R. (1970). in *Energy Metabolism of Farm Animals*. Eds A. Schürch and C. Wenk, p.205. Juris Druck, Zurich

BØNSDORFF-PETERSEN, Ch.R., ZNANIECKA, G., POCZOPKO, P. and FRYDRYCHEWICZ, A. (1970). in *Landøkonomisk Forsøgslaboratoriums efterarsmøde*, p.238. Frederiksberg Bogtrykkeri, Copenhagen

BURACZEWSKI, S. (1973). in *Proceedings of the Symposium on Amino Acids, Brno*, Czechoslovakia, C-D, 1 (mimeographed)

BURACZEWSKI, S. and PASTUSZEWSKA, B. (1974)., unpublished data, Inst. of Animal Physiology and Nutrition, Jablonna, Poland

CLOSE, W.H., VERSTEGEN, M.W.A. and MOUNT, L.E. (1973). *Proc. Nutr. Soc.*, **32**, 72A

GÄDEKEN, D., OSLAGE, H.J. and FLIEGEL, H. (1973). in *Energy Metabolism of Farm Animals*. Eds K.H. Menke, H.J. Lantzsch and J.R. Reichl, p.169. Universität Hohenheim

HOFFMANN, L., JENTSCH, W., WITTENBURG, H. and SCHIEMANN, R. (1972). *Arch. Tierernähr*, **22**, 721

HOFFMANN, L. and SCHIEMANN, R. (1973). *Arch. Tierernähr.*, **23**, 105

KIELANOWSKI, J. (1965). in *Energy Metabolism*. Ed. K.L. Blaxter, p.13, Academic Press, London

KIELANOWSKI, J. and KOTARBINSKA, M. (1970). in *Energy Metabolism of Farm Animals*. Eds A. Schürch and C. Wenk, p.145, Juris Druck, Zurich

KOTARBINSKA, M. (1969). *Badania nad Przemiana Energii u Rosnacych Swiń,* Instytut Zootechniki, Wroclaw

MILLWARD, D.J. and GARLICK, P.J. (1972). *Proc. Nutr. Soc.,* **31,** 257

MOE, P.W., TYRRELL, H.F. and FLATT, W.P. (1970). in *Energy Metabolism of Farm Animals.* Eds A. Schürch and C. Wenk, p.65. Juris Druck, Zurich

MÜLLER, H.L. and KIRCHGESSNER, M. (1973). in *Energy Metabolism of Farm Animals.* Eds K.H. Menke, H.J. Lantzsch and J.R. Reichl, p.185. Universität Hohenheim

NEHRING, K. and SCHIEMANN, R. (1966). in *Vergleichende Ernährungslehre des Menschen und seiner Tiere.* Ed. A. Hock, p.581, VEB Gustav Fischer Verlag, Jena

OSINSKA, Z. (1974). unpublished data, Institute of Animal Physiology and Nutrition, Jablonna, Poland

OSLAGE, H.J., GÄDEKEN, D. and FLIEGEL, H. (1970). in *Energy Metabolism of Farm Animals.* Eds A. Schürch and C. Wenk, p.133, Juris Druck, Zurich

ØRSKOV, E.R. and McDONALD, I. (1970). in *Energy Metabolism of Farm Animals.* Eds A. Schürch and C. Wenk, p.121, Juris Druck, Zurich

SCHIEMANN, R., CHUDY, A. and HERCEG, O. (1969). *Arch. Tierernähr.* **19,** 395

SOKOLOFF, L. and KAUFMAN, S. (1961). *J. biol. Chem.,* **236,** 795

THORBEK, G. (1970). in *Energy Metabolism of Farm Animals.* Eds A. Schürch and C. Wenk, p.129, Juris Druck, Zurich

WITKOWSKA, A. (1974). unpublished data, Institute of Animal Physiology and Nutrition, Jablonna, Poland

ZEBROWSKA, T. and BURACZEWSKA, L. (1972). *Rocz. Nauk rol.,* **B-94-1,** 81

12

NITROGEN-ENERGY INTERACTIONS IN RUMEN FERMENTATION

N.P. McMENIMAN, D. BEN-GHEDALIA and D.G. ARMSTRONG
Department of Agricultural Biochemistry, University of Newcastle upon Tyne

Introduction

The central role of microbial fermentation occurring within the forestomach of the ruminant animal is universally accepted. The major part of the host animal's supply of amino acids arises by subsequent digestion within the small intestine of the microbial biomass synthesised within the rumen; furthermore, the volatile fatty acids arising as end products of the anaerobic fermentation provide the major part of the non-nitrogen energy yielding substrates available to the animal. Since essentially all the energy for microbial growth is derived from the fermentation of dietary carbohydrate and the nitrogenous constituents provide the nitrogen requirements of the micro-organisms, nitrogen-carbohydrate interrelationships occurring within the rumen are of considerable importance to overall rumen metabolism. Certain aspects of this subject have been reviewed recently (Hobson, 1972; Thomson, 1972).

The Composition of the Microbial Population

While relatively little is known concerning the influence of different types and amounts of dietary protein on the composition and concentration of the individual species of bacteria and protozoa comprising the rumen microbial biomass, this is not so for the principal energy-yielding dietary constituents, i.e. carbohydrates. Thus Latham, Sharpe and Sutton (1971) showed that the predominant species of bacteria present in the rumen of cows fed a concentrate diet (high in starch) were *Selenomonas* spp, *Streptococcus bovis* and *Peptostreptococcus Elsdenii* which are lactic acid producers and/or fermenters. *Butyrivibrio fibrisolvens*, which under the conditions of this experiment was the main cellulolytic bacterium present, comprised only 10 per cent of the total bacterial numbers. When roughage diets (high in cellulose and hemicellulose) were fed, *B.fibrisolvens* comprised 38 per cent of total bacterial numbers. On the concentrate diet, characterised by a high propionate fermentation, the number of ciliate protozoa in the rumen liquor was greatly reduced.

The high susceptibility of protozoa to the low rumen pH found when large amounts of concentrates are fed, is well known (Purser and Moir, 1959) and low protozoal numbers have been shown to be associated with high bacterial concentrations (Eadie and Hobson, 1962). The feeding of concentrate diets, however, is not always associated with low protozoal numbers. Eadie *et al.* (1970) fed heifers on a restricted all-barley diet and found exceptionally high rumen ciliate populations and a bacterial flora typical of roughage fed animals; the microbial population was associated with a higher proportion of butyric acid than of propionic acid in the rumen fluid. When the barley diet was fed *ad libitum*, however, there was a decrease in rumen pH and a complete loss of rumen ciliates. It is interesting to note that reduction of the particle size of roughages by grinding prior to feeding leads to the establishment of a microbial population characteristic of that observed when a high concentrate diet is fed (Thorley, Sharpe and Bryant, 1968).

Requirements for Microbial Growth

The nutritional requirements of the microbial flora and fauna comprise a considerable number of micronutrients such as minerals and vitamins, in addition to energy, carbon and nitrogen sources. However, in this paper reference will only be made to certain aspects of their requirement for nitrogen and energy.

NITROGEN

Bryant and Robinson (1962) have shown that 82 per cent of the strains of bacteria isolated from the rumen grew with ammonia nitrogen as their principal source of nitrogen. Indeed, for some bacteria, e.g. *Bacteroides succinogenes, Ruminococcus flavefaciens, Ruminococcus albus, Bacteroides amylophilus, Methanobacterium ruminantium* and *Eubacterium ruminantium,* ammonia is an essential nutrient (Bryant, 1963; Hungate, 1966). Growth of the cellulose digesters *Ruminococcus* spp, and *Selenomonas ruminantium,* which ferment both starch and lactic acid, is stimulated by amino acids. Bacteria such as *Peptostreptococcus Elsdenii* and *Butyrivibrio fibrisolvens* require some amino acid nitrogen for growth (Hungate, 1966); other bacteria use peptides (Pitman and Bryant, 1964).

From studies on nitrogen cycling in the rumen of sheep fed lucerne chaff, Nolan and Leng (1972) concluded that some 29 per cent of the nitrogen digested in the rumen was incorporated directly as amino acid nitrogen into the bacteria. Similar levels of incorporation of amino acid nitrogen into bacterial protein could be inferred from the reports of Pilgrim *et al.* (1970) and Mathison and Milligan (1971). From studies with sheep involving ^{35}S, it appears that from 20 to 38 per cent of the sulphur incorporated into bacteria is not derived from the H_2S pool, thus indicating a direct incorporation of sulphur amino acids of food origin into bacterial protein (Ben-Ghedalia and McMeniman, unpublished).

The addition of natural proteins to a purified diet in which urea was the only source of nitrogen, increased microbial production within the rumen (Hume, 1970). Dietary proteins also serve as sources of branched-chain and higher volatile fatty acids, which are growth stimulants, particularly for cellulolytic bacteria and essentially for a large proportion of the rumen bacteria (Annison, 1954; Hungate, 1966). *Peptostreptococcus elsdenii* can obtain branched-chain and higher volatile fatty acids from carbohydrates as well as proteins (Hobson, Mann and Oxford, 1958). The addition of urea and branched-chain volatile fatty acids to a teff hay diet, stimulated the growth of the cellulolytic ruminococci and increased cellulolysis (van Gylswych and Roche, 1970).

The rumen gases contain an appreciable quantity of nitrogen gas; a value of 7 per cent is given by McArthur and Meltimore (1961). Mathison and Milligan (1971) suggested that some nitrogen gas fixation may occur in order to account for certain of their findings relating to nitrogen cycling in the rumen. Support for this view can be found in reports that nitrogen assimilating bacteria inhabit the intestines of humans, pigs and guinea pigs (Bergensen and Hipsley, 1970). More recent *in vitro* work that has also been confirmed *in vivo,* however, (Hobson *et al.*, 1973) showed that a negligible amount (about 1 mg/day) of nitrogen gas is fixed in the rumen of sheep.

The nitrogen requirements of the protozoa have been less well documented but present evidence indicates that their major source is bacterial protein; they can also use dietary proteins (Hungate, 1966).

Peptides, amino acids and ammonia arise as a result of the action of proteases, peptidases and deaminases on dietary protein. Rumen ammonia also arises as a result of bacterial urease acting on dietary and endogenous urea, the latter recycled into the rumen via the saliva (Houpt, 1970). A further source of ammonia nitrogen is from the degradation within the rumen of microbial cell material (Nolan and Leng, 1972). Protozoa produce ammonia as an end product of their intermediary metabolism and Eadie and Gill (1971) have shown that rumen ammonia levels of defaunated lambs are lower than those of lambs with a normal microbial flora and fauna.

ENERGY AND CARBON SKELETONS

It is appropriate to group these together since the catabolic reactions of fermentation provide not only the ATP used in meeting the microbial energy requirements for maintenance and synthesis, but also the carbon skeletons for the synthesised biomass; these skeletons arise as intermediary (e.g. pyruvate, succinate) or final (e.g. carbon dioxide and acetate) products of the catabolic reactions. While it is generally accepted that dietary carbohydrates are the principal fermentation substrates, proteins are also used as sources of energy; the extent of their contribution being governed by their solubility as well as their concentration in dietary dry matter. Rates of proteolysis and deamination are independent of rates of microbial protein synthesis. It is interesting to note that *Acidaminococcus fermentas* is unable to ferment carbohydrate and obtains energy from amino acid fermentation (Elias, 1971, cited by Hobson, 1972). The pathways of carbohydrate fermentation within the rumen have been well documented (Leng, 1970; 1973).

With reference to energy requirements, direct measurements of those for maintenance of growing bacterial cells are not available but determinations that have been made on non-growing cells of *Escherichia coli* and *Streptococcus faecalis* (Forrest and Walker, 1963; McGrew and Mallette, 1965) suggest that such requirements are very limited compared with those for growth. Forrest and Walker (1971) suggest that some 75 per cent of the energy required for growth is required for the synthesis of protein.

Quantitative Aspects of Microbial Growth

The theoretical energy input required for the synthesis of 100 g of bacterial cells has been calculated to be 3.62 moles of ATP (Forrest and Walker, 1971). This corresponds to a value of 28 for the Y_{ATP} yield (g dry bacterial cells/mole ATP made available from the energy released in fermentation). Many *in vitro* growth experiments have been performed with anaerobic bacteria growing on a wide range of substrates; growth has been directly measured and ATP yield calculated from the energy-yielding pathways followed in fermentation. Such experiments have given Y_{ATP} yields less than the theoretical value of 28. Thus, growth of anaerobic bacteria under energy-limiting conditions gave a value of approximately 10.5 (Bauchop and Elsden, 1960). Y_{ATP} yields summarised by Forrest and Walker (1971) for a number of anaerobic organisms gave a mean value of 10.6 ± 0.1 (range 8.5-13.1). However, the continuous culture technique (which more closely resembles fermentation in the rumen) gave a Y_{ATP} of 20 for *Selenomonas ruminantium* and *Bacteroides amylophilus* (Hobson, 1965; Hobson and Summers 1967). Hobson and Summers (1972) and de Vries, Kapteijn and Oosterhuis (1974) have suggested that the Y_{ATP} value for *S.ruminantium* may not be correct; additional ATP may have been produced by proposed new pathways. However, high apparent ATP yields have also been found with other bacteria (Payne, 1970; de Vries *et al.*, 1970).

Hobson (1972) has discussed factors that can influence cell yield from the fermentation of carbohydrate. Two, of particular importance to fermentation processes, are dilution rate and the phenomenon of uncoupled fermentation. In studies with pure cultures, increasing the dilution rate results in more microbial growth (Hobson and Summers, 1967; 1972).

Whilst under ideal conditions fermentation is coupled to cell growth, there are circumstances in which catabolism (i.e. fermentation) of substrate proceeds without microbial growth occurring. Forrest and Walker (1965) suggested that uncoupled fermentation could occur when certain nutrients essential for microbial growth were lacking.

Pantothenate starvation of *Zymomonas mobilis* resulted in a decrease in the microbial growth without a decrease in the amount of glucose fermented per weight of organisms present (Belaich, Belaich and

Simonpietri, 1972). Ladzunski and Belaich (1972) further showed that under these conditions the pool of ATP increased, suggesting that ATP was produced despite the decrease in microbial growth. Assuming that ATP storage in bacteria is limited and that maintenance requirements of bacteria are low, the fate of the excess ATP produced in uncoupled fermentations has to be explained. Ladzunski and Belaich (1972) found an ATPase specific for nucleotide triphosphates associated with the cell membrane of *Zymomonas mobilis* and suggested that its action assisted in the dissipation of the excess energy. Another possible means of disposal of the excess ATP would be to use it as an energy source for the synthesis of reserve materials such as polysaccharides and lipids in the microbiota.

Nitrogen deficiency would also affect the microbial growth in the rumen. Hume, Moir and Somers (1970) fed sheep protein-free, purified diets containing 0.54, 0.95, 1.82 and 3.29 per cent nitrogen as urea and observed that microbial production was increased as the nitrogen level rose, with no change in the amount of organic matter fermented or presumably in ATP produced. However, Henderson, Hobson and Summers (1969) in studies with *Bacteroides amylophylus* have shown that when nitrogen was deficient, the synthesis of hydrolytic enzymes was reduced. It is also known that the addition of urea to low quality roughage diets increases the amount of organic matter digested (Campling, Freer and Balch, 1962). This would seem to indicate that there is a critical level of nitrogen supply below which the catabolic activity is affected and that there is another higher level above which normal catabolism and growth is supported. Uncoupled fermentation may occur when the nitrogen supply is between these two levels.

There are a number of difficulties associated with obtaining reliable values for Y_{ATP} in the rumen. Firstly, while the general biochemical pathways of rumen fermentation are known, particular reactions and electron transfer systems associated with the generation of ATP are not so well understood. This aspect has been discussed by Leng (1970; 1973) who noted, for example, that uncertainty attaches to amounts of ATP produced when pyruvate is metabolised to propionate and butyrate. The existence of cytochromes in anaerobic bacteria (de Vries, Kapteijn and Oosterhuis, 1974) raises further questions regarding the production of ATP.

The possible uptake of oxygen gas by the rumen microbes should not be neglected. Although it is known that the amounts of oxygen gas entering with the food and possibly by exchange from the blood across the rumen epithelium are small and that such oxygen gas is rapidly removed by the surface layers of the biomass, ensuring the maintenance of an oxidation reduction potential in the neighbourhood of −0.35 volts throughout the large mass of rumen contents (Hungate, 1966), some oxidative metabolism could result. Barry (1967) has shown that in sheep the amount of oxygen in rumen gases can, for short periods of time, reach quite high levels viz. 12.6 per cent and 16.0 per cent for forage and concentrate diets respectively, with mean 24 h concentrations of 3.6 ± 0.4 per cent and 3.3 ± 0.3 per cent.

Notwithstanding the foregoing, Walker (1965), Baldwin, Lucas and Cabrera (1970) and Leng (1970) have discussed the stoichiometry of rumen fermentations and it appears from their calculations that the net yield of ATP per mole of volatile fatty acid produced is of the order 2.0-2.8 moles.

Another difficulty in obtaining reliable Y_{ATP} yields, lies in the determination of the amounts of volatile fatty acids produced and has led to the use of organic matter fermented within the rumen as a parameter against which to judge microbial cell yields. The determination of cell yields from *in vivo* rumen fermentations also presents major problems. The substance a-diaminopimelic acid (DAPA), which is peculiar to gram negative bacterial cells, has been used to estimate bacterial flow from the rumen (Hutton, Bailey and Annison, 1971). Recently, however, Nickolic and Jovanovic (1973) have shown that the DAPA content of whole rumen contents was as high or higher than the content of bacterial cells. Bacterial mutants which build up DAPA in their cytosol and under certain circumstances can release it into the surrounding medium, have been isolated (Hagino, Hiroski and Nakayama, 1970; Kase, Hagino and Nakayama, 1970) although few of these species have been reported in the rumen (Hungate, 1966). However, DAPA does not of course give an indication of protozoal flow from the rumen. The substance 2-amino-ethylphosphoric acid is peculiar to rumen protozoa (Abou Akkada *et al.*, 1968) but does not yet appear to have been used to measure the flow of protozoa from the rumen.

The use of nucleic acids, particularly RNA, as microbial markers, has been advocated by Smith and McAllan (1970). The nucleic acid nitrogen (RNA or DNA) to total nitrogen ratios in bacteria and protozoa can vary with diet, species of host, time of sampling and environment in which the host animal is kept (Smith, 1969; Smith and McAllan, 1974; Papasolomontos and Givens, unpublished data). Determining the nucleic acid content of bacteria and protozoa for each animal on each experiment would eliminate some of the errors, but until the amount of protozoa present in fluid passing from the rumen can be determined, reliable results for microbial production rates will not be available. The incorporation of ^{35}S into microbial protein (Roberts and Miller, 1969; Harrison, Beever and Thomson, 1972) appears to provide realistic estimates of microbial nitrogen flow from the rumen (Hume, 1974) and does not suffer from the disadvantage of not measuring protozoal nitrogen. As yet very few results are available using this method.

Leng (1973), using the data of Walker and Nader (1968) and Hogan and Weston (1970), estimated the Y_{ATP} yield in the rumen to be 10.1, which is close to that proposed by Forrest and Walker (1971). Beever, Thomson and Harrison (1974) have obtained Y_{ATP} values for rye grass silage and for two similar silages treated with formaldehyde of 19.6, 5.8 and 6.3 respectively. Variability in the efficiency of microbial cell yield can also be seen in data in which yields are expressed as g nitrogen per 100 g organic matter digested in the reticulo rumen (organic matter digested = organic matter apparently disappearing in the rumen plus organic matter in microbial cells

Table 12.1 *The influence of diet on the amount of microbial nitrogen[1] synthesised in the rumen*

Diet		[2] g microbial nitrogen per 100 g organic matter actually digested in the rumen	[3] g food nitrogen digested in rumen/100 g organic matter actually digested in rumen
Fresh grass	(i)	3.0	3.1
Dried grass	(i)	3.1	2.2
Chopped lucerne	(ii)	3.4	1.7
Wafered lucerne	(ii)	3.5	2.5
Pelleted lucerne	(ii)	3.5	1.4
Early cut chopped grass, low intake	(iii)	3.3	3.1
Early cut pelleted grass, low intake	(iii)	3.4	2.6
Medium cut chopped grass, low intake	(iii)	3.7	3.3
Medium cut pelleted grass, low intake	(iii)	3.6	3.1
Early cut chopped grass, high intake	(iii)	3.5	2.9
Early cut pelleted grass, high intake	(iii)	2.8	2.1
Mean for forages		3.3 ± 0.1	2.5 ± 0.2
80% ground maize	(iv)	2.9	2.3
80% flaked maize	(iv)	2.9	1.9
70% rolled barley	(v)	1.8	1.5
70% flaked barley	(v)	2.0	1.0
70% micronised barley	(v)	2.2	1.0
70% rolled maize	(v)	1.8	0.9
70% flaked maize	(v)	1.7	0.9
70% micronised maize	(v)	2.2	1.4
Mean for concentrates		2.2 ± 0.2	1.4 ± 0.2

(i) Proud (1972)
(ii) Coelho da Silva, Seeley, Thomson, Beever and Armstrong (1972)
(iii) Coelho da Silva, Seeley, Beever, Prescott and Armstrong (1972)
(iv) Coelho da Silva (1971)
(v) Papasolomontos and Givens (unpublished data)

[1] Microbial nitrogen was calculated using the assumption that RNA-nitrogen to bacterial nitrogen ratio was 0.127 (Papasolomontos and Givens, unpublished data)

[2] Bacterial organic matter was calculated using the assumption that bacteria contained 7.8 per cent nitrogen (Papasolomontos and Givens, unpublished data)

[3] Food nitrogen digested = Food intake − (total nitrogen flow through duodenum − microbial nitrogen). This value would be an overestimate because it does not allow for NH_3-nitrogen and endogenous nitrogen in the duodenal fluid

produced). Hume (1970) found that the flow of protein through the omasum when sheep were fed a protein-free, urea-containing diet corresponded to a yield of 2.7 g nitrogen. Similar values have been obtained by Hume and Bird (1970), Ørskov, Fraser and McDonald (1972) and Miller (1973). *Table 12.1* lists some values which give a mean value for g microbial nitrogen/100 g organic matter digested in the rumen of 3.3 ± 0.1 for forage diets and 2.2 ± 0.2 for diets high in cereals.

The foregoing indicates that considerable variability appears to exist in the efficiency of microbial protein production in the rumen and hence implies different ratios of volatile fatty acids energy to protein energy subsequently available to the host animal. Some comments concerning the possible causes of such differences will now be referred to.

The relationship between the bacteria and protozoa within the rumen may play an important part in contributing to nitrogen cycling within the rumen, a process that would lower Y_{ATP} yield. It is known that the protozoa obtain a major part of their requirements for growth from engulfed bacteria. In the detailed studies of nitrogen cycling in sheep fed chopped lucerne (Nolan and Leng, 1972), 3.1 g of microbial nitrogen was recycled within the rumen per day, indicating extensive engulfment of bacteria by protozoa (Coleman, 1967) or lysis of viable bacterial cells owing to bacteriophage activity (Adams *et al.*, 1966; Hoogenraad *et al.*, 1967). Since in this study some 12.0 g microbial nitrogen left the rumen per day, the recycling of microbial nitrogen within the reticulo rumen is a process of some considerable magnitude. That the protozoa may play a considerable part in the recycling process is indirectly indicated by the results of Eadie and Gill (1971), who showed that rumen ammonia levels of defaunated lambs were lower than those with a normal protozoal population.

Another factor contributing to variations in efficiency of microbial synthesis within the reticulo rumen is undoubtedly the concentration of dietary nitrogen, and in particular that part of it that is capable of being made available to the rumen micro-organisms. The level must be adequate to ensure that the nitrogen requirements of the micro-organisms are fully met, otherwise uncoupled fermentation may occur. Mention has already been made of the studies of Hume *et al.* (1970) with sheep fed urea-containing purified diets, in which microbial protein synthesis increased and ammonia concentration remained low up to a dietary crude protein (nitrogen × 6.25) level of approximately 11 per cent. Above this level, ammonia concentration increased while microbial synthesis remained constant. In this study and that of Satter and Slyter (1972) using continuous *in vitro* fermenters, there was no relationship between microbial protein synthesis and fermentation activity when ammonia nitrogen was limiting.

The importance of an adequate level of nitrogen supply to the rumen micro-organisms is also indicated in the studies of Beever, Thomson and Harrison (1974) already referred to. When the untreated silage, which gave a high Y_{ATP} yield was fed, some 78 per cent of the crude protein in the silage was degraded within the rumen, whereas with the formaldehyde-treated silages in which low Y_{ATP} yields were observed, only some 10 per cent of the ingested nitrogen was released within the rumen. Similar comparisons can be made with the data in Table 12.1. With forage diets, where the microbial nitrogen yield was 3.3 g/100 g organic matter digested within the rumen, the amount of dietary nitrogen digested per 100 g of organic matter digested in the rumen was 2.5 g, whereas with the concentrate diets where microbial yield was 2.2 g, the amount of nitrogen digested was also low viz. 1.4 g. It has often been assumed that urea-nitrogen recycled into the rumen via the saliva could make good such deficiencies. However, the studies of Nolan and Leng (1972) suggest that such nitrogen cycling is directed more towards the caecum and colon than to the rumen, at least in sheep fed lucerne diets.

NUTRITIONAL IMPLICATIONS

In the final section reference will be made to some of the nutritional implications for the host animal that arise as a result of nitrogen/carbohydrate interactions occurring within the rumen.

The first of these relates to the importance of ensuring an adequate supply of nitrogen to meet the growth requirement of the rumen microorganisms. The key role of ammonia-nitrogen and the evidence obtained by Hume, Moir and Somers (1970) suggesting that a dietary crude protein equivalent of 11 per cent is necessary, has already been referred to. That this level cannot be taken as being applicable to all rations, seems probable for several reasons. Natural proteins used in ruminant feeds show a wide range of solubility in rumen liquor and it is known that solubility is directly correlated with ammonia release (Armstrong and Hutton, 1972). Furthermore, the solubility of a protein can be markedly affected by processes such as grass drying (Harrison et al., 1973) or treatment with formaldehyde (Ferguson, Hemsley and Reis, 1967; Reis and Tunks, 1969). Some values for the percentage of protein escaping digestion in the rumen of sheep are: sunflower seed meal 25; Peruvian fish meal 70 (Miller, 1973); peanut meal 36; and soya bean meal 61 (Hume, 1974).

In addition, the rate of release of ammonia-nitrogen from urea is so rapid that the level observed by Hume et al. (1970) to maximise microbial protein synthesis may still have been associated with inefficient use of some of the ammonia nitrogen released. Furthermore, the high digestibility of the purified diet is not typical of many ruminant diets; with those of lower digestibility a lower level of protein may well be adequate. Indeed, on the basis of evidence then available, the Agricultural Research Council (1965) concluded that a level of 9 per cent crude protein in the dry matter was adequate to ensure normal digestion in ruminants fed forage diets. Harrison et al. (1973) infused urea at the rate of 12.6 g/day into the rumen of sheep consuming a pelleted hay diet with a crude protein content of 7.5 per cent and found no increase in the amount of crude protein or bacterial nitrogen leaving the rumen. It would seem reasonable to postulate that the best guide to optimising microbial cell yield is to ensure an intake of dietary nitrogen that maintains the ammonia nitrogen levels in rumen liquor at levels which are optimal for microbial growth. However, there appears to be some confusion as to what this level is; Buttery and Annison (1972) suggest that it may be as low as $4\text{-}5 \times 10^{-3}$ M ($\equiv$ 5.6-7.0 mg nitrogen/100 ml) while Allen and Miller (1972) found that the greatest flow of microbial nitrogen from the rumen of lambs fed a high-energy low-protein diet supplemented with increments of urea, occurred when the rumen ammonia concentration was 17×10^{-3} M ($\equiv$ 23.8 mg nitrogen/100 ml).

A second nutritional implication relates to the fact that although high microbial yields can be obtained in the rumens of sheep fed protein-free urea-containing diets, there is evidence that supplements of natural proteins give significant responses (Hume, 1970). In this connection, it is noteworthy that in the feeding of cattle on molasses/urea diets (Preston and Willis, 1970) or on sugar cane pith and tops with urea (Donefer, James and Laurie, 1973) the importance of including some natural protein in the diet has been clearly demonstrated. The value of such inclusions is most probably associated with increased uptake of amino acid nitrogen from the small intestine but such increased uptake may well result, at least in part, from increased microbial biosynthesis within the rumen and not solely from intact protein escaping degradation within the rumen.

The processing of materials before they are fed also has implications with respect to the quantitative aspects of rumen fermentation. The effect of altering the solubility of protein on its digestion in the rumen has already been referred to. The processing of roughages by grinding and pelleting usually leads to a reduction in the amount of organic matter digested in the rumen (Thomson, 1972), while processing of cereals, usually by some form of heat treatment, increases the proportion of organic matter digested in the rumen and this can result in a greater flow of non-ammonia crude protein from the rumen (Armstrong, 1972). Reichl and Baldwin (1971) have shown that microbial synthesis per unit of carbohydrate was raised when the frequency of feeding was increased. With continuous feeding, the microbia would have a more constant supply of nutrients which may avoid the establishment of conditions favourable for uncoupled fermentation.

Finally, experiments in this laboratory (I. Johnson, unpublished data) suggest that to optimise microbial biosynthesis with a cereal ration it is necessary to match up the rate of release of ammonia nitrogen with the rate of microbial attack on the cereal starch.

References

ABOU AKKADA, A.R., MESSMER, D.A., FINA, L.R. and BARTLEY, E.E. (1968). *J. Dairy Sci.*, **51**, 78

ADAMS, J.C., GAZAWAY, J.A., BRACKSFORD, M.D., HARTMAN, P.A. and JACOBSON, N.L. (1966). *Experimenta*, **22**, 717

AGRICULTURAL RESEARCH COUNCIL (1965). *The Nutrient Requirements of Farm Livestock: No.2, Ruminants;* Agricultural Research Council, London

ALLEN, S.A. and MILLER, E.L. (1972). *Proc. Nutr. Soc.*, **31**, 26A

ANNISON, E.F. (1954). *Biochem. J.*, **57**, 400

ARMSTRONG, D.G. (1972). in *Cereal Processing and Digestion*, p.9; Technical publication of U.S. Feed Grains Council, London

ARMSTRONG, D.G. and HUTTON, K. (1972). *Proc. 2nd Wld. Congr. Anim. Feeding*, **4**, 219

BALDWIN, R.L., LUCAS, H.L. and CABRERA, R. (1970). in *Physiology of Digestion and Metabolism in the Ruminant*, Ed. A.T. Phillipson, p.319; Oriel Press, Newcastle upon Tyne

BARRY, T.N. (1967). *The validity of observations made in an in vitro rumen system by comparison with those obtaining in vivo;* Ph.D. Thesis, University of Newcastle upon Tyne

BAUCHOP, T. and ELSDEN, S.R. (1960). *J. gen. Microbiol.*, **23**, 457

BEEVER, D.E., THOMSON, D.J. and HARRISON, D.G. (1974). *12th Int. Grasslds. Congr., Moscow*. Section 5, p.69

BELAICH, J.P., BELAICH, A. and SIMONPIETRI, P. (1972). *J. gen. Microbiol.*, **70**, 179

BERGENSEN, F.J. and HIPSLEY, E.H. (1970). *J. gen. Microbiol.*, **60**, 61

BRYANT, M.P. (1963). *J. Anim. Sci.*, **22**, 801

BRYANT, M.P. and ROBINSON, I.M. (1962). *J. Bact.*, **84**, 605

BUTTERY, P.J. and ANNISON, E.F. (1972). in *Biological efficiency of protein production*. Ed. J.W.G. Jones; Academic Press, London

CAMPLING, R.C., FREER, M. and BALCH, C.C. (1962). *Br. J. Nutr.*, **16**, 115

COELHO DA SILVA, J.F. (1971). *The digestion of nitrogenous constituents in forage and forage-cereal diets by adult sheep;* Ph.D. Thesis, University of Newcastle upon Tyne

COELHO DA SILVA, J.F., SEELEY, R.C., BEEVER, D.E., PRESCOTT, J.H.D. and ARMSTRONG, D.G. (1972). *Br. J. Nutr.*, **28**, 357

COELHO DA SILVA, J.F., SEELEY, R.C., THOMSON, D.J., BEEVER, D.E. and ARMSTRONG, D.G. (1972). *Br. J. Nutr.*, **28**, 43

COLEMAN, C.S. (1967). *J. gen. Microbiol.*, **47**, 449

DONEFER, E., JAMES, L.A. and LAURIE, C.K. (1973). *3rd World Conf. Anim. Prod.;* Melbourne, Australia (in press)

EADIE, J. Margaret and GILL, J.C. (1971). *Br. J. Nutr.*, **26**, 155

EADIE, J. Margaret and HOBSON, P.N. (1962). *Nature, London*, **193**, 503

EADIE, J. Margaret, HYLDEGAARD-JENSEN, J., MANN, S.O., REID, R.S. and WHITELAW, F.G. (1970). *Br. J. Nutr.*, **24**, 157

ELIAS, A. (1971). *The rumen bacteria of animals fed on a high molasses urea diet*, Ph.D. Thesis, University of Aberdeen

FERGUSON, K.A., HEMSLEY, J.A. and REIS, P.J. (1967). *Aust. J. Sci.*, **30**, 215

FORREST, W.W. and WALKER, D.J. (1963). *Biochem. Biophys. Res. Commun.*, **13**, 217

FORREST, W.W. and WALKER, D.J. (1965). *J. Bact.*, **89**, 1448

FORREST, W.W. and WALKER, D.J. (1971). *Adv. in Microbiol. Physiol.*, **5**, 213

HAGINO, H., HIROSKI and NAKAYAMA, K. (1970). *Nippon Nogei Kogaka Kaishi*, **44**, 422

HARRISON, D.G., BEEVER, D.E. and THOMSON, D.J. (1972). *Proc. Nutr. Soc.*, **31**, 60A

HARRISON, D.G., BEEVER, D.E. and THOMPSON, D.J. (1973). *J. agric. Sci., Camb.*, **81**, 391
HENDERSON, C., HOBSON, P.N. and SUMMERS, R. (1969). *Continuous Culture of Microorganisms*, p.189. Ed. I. Malek; Academic Press, London and New York
HOBSON, P.N. (1965). *J. gen. Microbiol.*, **38**, 167
HOBSON, P.N. (1972). *Proc. Nutr. Soc.*, **31**, 135
HOBSON, P.N., MANN, S.O. and OXFORD A.E. (1958). *J. gen. Microbiol.*, **19**, 462
HOBSON, P.N. and SUMMERS, R. (1967). *J. gen. Microbiol.*, **47**, 53
HOBSON, P.N. and SUMMERS, R. (1972). *J. gen. Microbiol.*, **70**, 351
HOBSON, P.N., SUMMERS, R., PESTGATE, J.R. and WARE, D.A. (1973). *J. gen. Microbiol.*, **77**, 255
HOGAN, J.P. and WESTON, R.H. (1970). *Physiology of Digestion and Metabolism in the Ruminant*, p.474. Ed. A.T. Phillipson, Oriel Press, Newcastle upon Tyne
HOOGENRAAD, N.J., HIRD, F.J.R., HOLMES, I. and MILLIS, N.F. (1967). *J. gen. Virol.*, **1**, 575
HOUPT, T.R. (1970). in *Physiology of Digestion and Metabolism in the Ruminant*, p.119. Ed. A.T. Phillipson, Oriel Press, Newcastle upon Tyne
HUME, I.D. (1970). *Aust. J. agric. Res.*, **21**, 305
HUME, I.D. (1974). *Aust. J. agric. Res.*, **25**, 155
HUME, I.D. and BIRD, P.R. (1970). *Aust. J. agric. Res.*, **21**, 315
HUME, I.D., MOIR, R.J. and SOMERS, M. (1970). *Aust. J. agric. Res.*, **21**, 283
HUNGATE, R.E. (1966). *The Rumen and its Microbes;* Academic Press Inc., New York
HUTTON, K., BAILEY, F.J. and ANNISON, E.F. (1971). *Br. J. Nutr.*, **25**, 165
KASE, K., HAGINO, H. and NAKAYAMA, K. (1970). *Nippon Nogei Kogaka Koishi*, **44**, 457
LADZUNSKI, A. and BELAICH, J.P. (1972). *J. gen. Microbiol.*, **70**, 187
LATHAM, M.J., SHARPE, M.E. and SUTTON, J.D. (1971). *J. appl. Bact.*, **34**, 425
LENG, R.A. (1970). in *Physiology of Digestion and Metabolism in the Ruminant*, p.406. Ed. A.T. Phillipson, Oriel Press, Newcastle upon Tyne
LENG, R.A. (1973). in *Chemistry and Biochemistry of Herbage, Vol.3*, p.81. Eds G.W. Butler and R.W. Bailey, Academic Press, London and New York
MATHISON, G.W. and MILLIGAN, L.P. (1971). *Br. J. Nutr.*, **25**, 351
MILLER, E.L. (1973). *Proc. Nutr. Soc.*, **32**, 79
McARTHUR, J.M. and MELTIMORE, J.E. (1961). *Can. J. anim. Sci.*, **41**, 187
McGREW, S.B. and MALLETTE, M.F. (1965). *Nature, London*, **208**, 1096
NICKOLIC, J. Anna and JOVANOVIC, M. (1973). *J. agric. Sci., Camb.*, **81**, 1
NOLAN, J.V. and LENG, R.A. (1972). *Br. J. Nutr.*, **27**, 177
ØRSKOV, E.R., FRASER, C. and McDONALD, I. (1972). *Br. J. Nutr.*, **27**, 491

PAYNE, W.J. (1970). *Ann. Rev. Microbiol.*, **24**, 17
PILGRIM, A.F., GRAY, F.V., WELLER, R.A. and BELLING, C.B. (1970). *Br. J. Nutr.*, **24**, 589
PITMAN, K.A. and BRYANT, M.P. (1964). *J. Bacteriol.*, **88**, 401
PRESTON, T.R. and WILLIS, M.B. (1970). *Intensive Beef Production;* Pergamon Press
PROUD, C.J. (1972). *A study of the digestion of nitrogen in the adult sheep;* Ph.D. Thesis, University of Newcastle upon Tyne
PURSER, D.B. and MOIR, R.J. (1959). *Aust. J. agric. Res.*, **10**, 555
REIS, P.J. and TUNKS, D.A. (1969). *Aust. J. agric. Res.*, **20**, 775
REICHL, J.R. and BALDWIN, R.L. (1971). *J. Dairy Sci.*, **54**, 770
ROBERTS, S.A. and MILLER, E.L. (1969). *Proc. Nutr. Soc.*, **28**, 32A
SATTER, L.D. and SLYTER, L.L. (1972). *J. Anim. Sci.*, **35**, 273
SMITH, R.H. (1969). *J. Dairy Res.*, **36**, 313
SMITH, R.H. and McALLAN, A.B. (1970). *Br. J. Nutr.*, **24**, 545
SMITH, R.H. and McALLAN, A.B. (1974). *Br. J. Nutr.*, **31**, 27
THOMSON, D.J. (1972). *Proc. Nutr. Soc.*, **31**, 127
THORLEY, C.M., SHARPE, M.E. and BRYANT, M.P. (1968). *J. Dairy Sci.*, **51**, 184
VAN GYLSWYCK, N.O. and ROCHE, C.E.G. (1970). *J. gen. Microbiol.*, **64**, 11
VRIES, W. DE, KAPTEIJN, W.M.C. and OOSTERHUIS, S.K.H. (1974). *J. gen. Microbiol.*, **81**, 69
VRIES, W. DE, KAPTEIJN, W.M.C., VAN DER BECK, E.G. and STOUTHAMER, A.H. (1970). *J. gen. Microbiol.*, **63**, 333
WALKER, D.J. (1965). in *Physiology of Digestion in the Ruminant*, p.296. Eds R.W. Dougherty, R.S. Allen, W. Burroughs, N.L. Jacobson and A.D. McGilliard; Butterworths, London
WALKER, D.J. and NADER, C.J. (1968). *Appl. Microbiol.*, **16**, 1124

IV

MEASUREMENT OF PROTEIN ADEQUACY

13

BIOLOGICAL CRITERIA OF PROTEIN EVALUATION

J. DELORT-LAVAL
Centre National de Recherches Zootechniques, Jouy-en-Josas, France

Introduction

The ability of a protein to comply with a given physiological requirement is mainly a function of the availability and balance of its amino acids in relation to the renewal and synthesis of tissue proteins. However, the biological efficiency of dietary protein utilisation depends not only on the balance of available amino acids but also on the nitrogen and energy intake and upon the species and physiological state of the animal.

Methods of protein quality evaluation have been frequently reviewed previously (Frost, 1959; MacLaughlan and Campbell, 1970; Rerat, 1971; Den Hartog and Pol, 1972; Meade, 1972; Carpenter, 1973). Among the older methods are those based on weight gain (Osborne and Mendel, 1917); the nitrogen balance method of Mitchell (1924) or body nitrogen retention (Miller and Bender, 1955). Our knowledge of nitrogen metabolism and of amino acid requirements, the notion of limiting factor (Mitchell and Block, 1946; Oser, 1959) are based on these techniques. Recently, more careful attention has been paid to amino acid availability by means of chemical (Carpenter and Ellinger, 1955), enzymatic (Scheffner, Eckfeld and Spector, 1956) or microbiological (Ford, 1962) methods. These methods have been compared with *in vivo* studies of amino acid digestion, by determination of the disappearance from the gut (Bayley, Cho and Holmes, 1974), faecal amino acid determination (Kuiken and Lyman, 1948), studies of plasma amino acids (Longenecker, 1963), muscle free amino acids (Pawlak and Pion, 1967; Larbier and Guillaume, 1972), plasma protein or urea in the blood (Eggum, 1970), liver enzyme activities and nitrogenous urinary components (Kiriyama, 1970).

Some of these techniques are dealt with elsewhere in this volume. This paper will mainly be concerned with biological methods of protein evaluation.

Weight Gain Methods

The various criteria of protein efficiency deduced from weight gain in relation to protein intake were clearly illustrated by Allison (1964) (*Figure 13.1*) who showed that the relationship between weight gain and protein intake is linear only for low protein intakes. The slope of this curve defines the nitrogen growth index, an index similar to net protein ratio (NPR), obtained from the difference in gain of two groups of animals, fed on either the experimental or a protein-free diet.

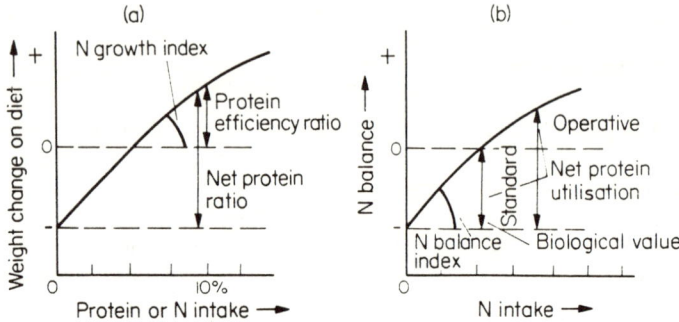

Figure 13.1 Procedures for measuring the nutritive values of proteins (Allison, 1964)

The protein efficiency ratio (PER) relates weight gain to ingested protein and the value obtained depends mainly on the initial slope of the linear portion of the curve and of its curvature at higher protein intakes. The effect of protein intake on PER is well known the maximum gain allowed by a given protein decreases with its quality (*Table 13.1*) and is only reached at a higher protein intake with an unbalanced protein. In the same way, supplementing a protein of low value, such as groundnut, with free amino acids, increases the maximum

Table 13.1 Maximum growth supported by various proteins and inclusion rate of protein at which maximum growth occurred (Rerat, 1971)

	Maximum gain (g/day)	Proportion of protein for maximum growth (%)
Fish meal	5.0	16
Sunflower meal	4.7	20
Peanut meal	4.4	24
Wheat gluten	4.0	44

PER which is obtained at a lower protein intake (*Figure 13.2*) (Sundaravalli and Rao, 1969). On this basis, Barnes and Bosshardt (1946) defined the 'optimum protein ratio' as the relation of maximum weight gain to the digestible protein intake. Despite the limitations of the PER index, its value generally correlates well with NPR, when

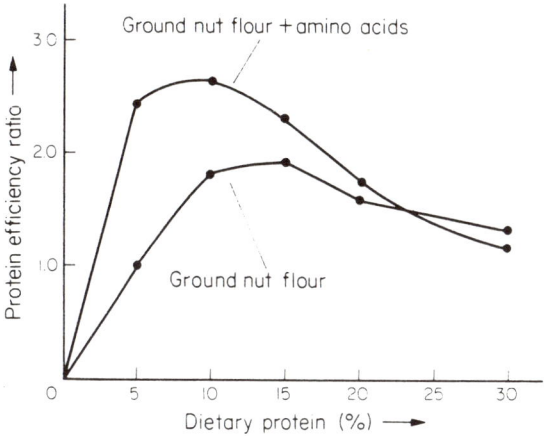

Figure 13.2 Effect of amino acid supplementation of a poor protein on PER at different dietary protein contents (Sundaravalli and Rao, 1969)

growth rates are high. It would, however, show a larger discrepancy when compared with the 'nitrogen growth index' for a poor protein (Hegsted and Chang, 1965). This defect is further illustrated by the data of Ferro-Luzzi, Mariani and Migliaccio (1970), demonstrating that the PER of gluten is twice as high at 14 than at 10 per cent protein in the diet, whereas NPU on the same diets, decreases from 42 to 33 *(Table 13.2)*.

Table 13.2 PER and NPU values for wheat gluten and wheat gluten plus casein determined at 10 or 14 per cent dietary protein (Ferro-Luzzi et al., 1970)

	PER		NPU	
	10%	14%	10%	14%
Wheat gluten	0.37 ± 0.10	0.83 ± 0.14	42	33
Wheat gluten plus casein	2.83 ± 0.31	2.19 ± 0.36	61	57

In a clear discussion of the influence of dietary protein content on the criteria of protein evaluation, MacLaughlan (1972) pointed out that the values of a protein for maintenance and growth are quite different in the rat because of the very low maintenance requirements for lysine and leucine. This effect is illustrated by the high NPU value of proteins devoid of lysine (Bender, 1961) or by the calculation made by MacLaughlan and Campbell (1970) indicating that it would be possible for a protein having an amino acid composition ideal for maintenance to have a chemical score of only 17 for growth. Thus, if two proteins which differ in lysine content gave similar responses near maintenance level, a large difference would occur at higher protein intakes, due to the high requirement of lysine for growth.

Considering the influence of dietary energy on protein deposition all the factors which tend to modify either the energy concentration of the diet or the food intake can influence the utilisation of dietary nitrogen (Rerat, 1971). The demonstration of the role of the protein/energy ratio on the biological efficiency of a protein is made possible by separate feeding of the nitrogen and energy-providing components of the diet. The use of the principle of separate feeding in the growing animal arises from an observation of Calet, Jouandet and Baratou (1961), who showed that the nutritional value of a protein is linked to the method of feeding. When the animals were fed free-choice, the growth achieved with fish meal and a separate protein-free diet offered *ad libitum* was better than when the same quantity of protein was provided in a mixed diet. With peanut meal, the two methods gave similar results. It seems therefore that the animal adjusts its energy intake to the biological value of the protein offered. With regard to the rat and chick, however, free-choice feeding of the protein and of the protein-free diet simultaneously often results in excess protein intake and it is therefore difficult to determine the proportion of this excess which is metabolised for energy production or for protein synthesis. The method of separate feeding of the protein and non-protein fractions successfully avoids this drawback; it limits the amount of protein (0.18 g N in one or two meals) offered to the animal and allows a free consumption of the protein-free portion of the diet (Peretianu and Abraham, 1963).

It has been demonstrated that energy intake is directly related to protein intake in the rat, chicken and pig and to the biological value of that protein; there is a direct relationship, on separate feeding, between weight gain and energy intake. This phenomenon has been described as 'caloric-nitrogen adjustment' by Jacquot and Peret (1972) and provides a value which is in good agreement with PER measured under similar conditions (*Table 13.3*). The use of separate feeding also improves the agreement between classical measures of protein adequacy and observed performance in the ranking of different proteins (*Table 13.4*). The separate feeding method has been described and discussed by Jacquot and Peret (1972). It has been used in studies on amino acid (lysine, methionine) supplementation in the growing rat (Peret and Jacquot, 1972). From a practical viewpoint, this method is easy to use, the only measured response being weight gain, provided that the animals consume the protein food in a short time. In the pig this technique is of more limited interest. In contrast to the rat and chicken, the pig does not spontaneously consume large amounts of protein. At lower nitrogen intakes, the gilt can adjust energy consumption better than the castrated male; however, the pig in general tends to overconsume energy, because of an apparently uncontrolled ability to deposit fat (Henry and Rerat, 1972).

Table 13.3 *Variations of the PER and of the ratio, energy:nitrogen when protein is fed separately (Abraham, Morin-Jomain and Peretianu, 1964)*

Experimental animals	Rats		Chickens	
Diet	PER	[1] energy consumed (kJ) / nitrogen (g)	PER	[1] energy consumed (kJ) / nitrogen (g)
Fishmeal	2.3	627	3.0	874
Casein + cystine	2.1	552	-	-
Soya bean	2.0	510	2.0	752
Peanut	1.8	456	1.5	702
Wheat gluten	0.8	410	-	-

[1] Gross energy of food consumed per g nitrogen consumed

Table 13.4 *Gain in body weight, PER and NPU with mixed feeding of a 15 per cent protein diet, or with protein fed separately (Peretianu and Abraham, 1963)*

Diet	Gain in body weight		PER		NPU	
	g/day per rat	relative value	absolute value	relative value	absolute value	relative value
	Mixed feeding					
Casein + 0.15 % cystine	3.8	100	2.16	100	37.8	100
Fish meal	4.4	116	1.89	88	33.1	88
Soyabean meal	3.8	100	1.70	79	28.9	76
	Separate feeding					
Casein + 0.15 % cystine	2.9	100	2.70	100	48.5	100
Fish meal	2.3	79	2.12	79	37.6	78
Soyabean meal	2.1	72	1.94	72	34.6	71

Repletion Method

The nitrogen repletion method (Cannon *et al.*, 1944) is based on the response to a protein intake of an animal previously fed on a protein-free diet for a period of time sufficient to deplete part of its labile protein reserves. This state of protein depletion increases the efficiency of protein utilisation (Allison, 1959) and shortens the time of response. Under strictly defined conditions, Frazier *et al.* (1949) demonstrated that the quantity and balance of amino acids required for growth were the same as those necessary for repletion in the protein-deprived adult rat.

This technique has been used with growing animals (chickens, piglets). Frost and Sandy (1949) underlined the excellent reproducibility of the method when applied to the evaluation of lactalbumin in growing rats, during ten consecutive periods of depletion and repletion.

In his critical study of the proposed method, Frost (1959) emphasised the effect of protein intake and advocated the separate feeding of the protein and the other components of the diet. In the depleted animal, receiving a fixed quantity of nitrogen (0.2 g/day), the nutritive value of the protein fed correlates well with the spontaneous intake of the protein-free diet offered *ad libitum*. This 'index of the degree of appetite' has probably the same physiological significance as in the growing animal, but growth and nitrogen retention are higher after a period of depletion than in the normal animal and they are independent of the status of the protein reserves (*Figure 13.3*).

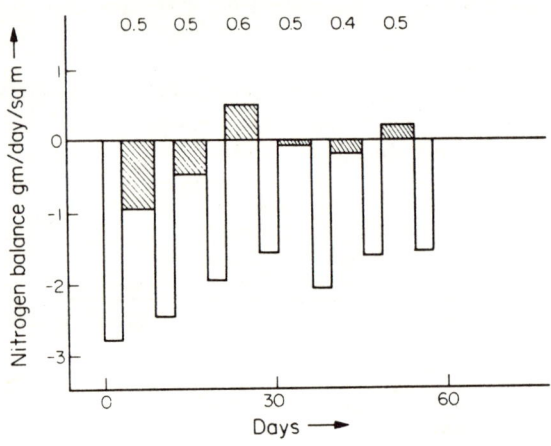

Figure 13.3 Effect on nitrogen balance of feeding a protein-free diet (open bars) alternating with wheat gluten nitrogen (shaded bars). Figures above bars are indices of nitrogen balance in relation to nitrogen intake. Allison (1959)

In the growing pig simply the determination of urinary nitrogen during eight days of repletion after two weeks on a protein-free diet, shows larger differences than the usual balance method on the same animals: this difference arises from a smaller nitrogen excretion by the pigs fed on the best protein (Delort-Laval, Charlet-Lery and Zelter, 1966), even when protein intake is high (*Figure 13.4*) thus reflecting the greater sensitivity of pigs in a depleted state.

The repletion value of proteins can be determined indirectly by regeneration of plasma or liver proteins, or by increased activity of enzymes related to protein metabolism. It is, however, not proved that such data can be correlated with the overall regenerative value of a test protein; in fact, the needs for regeneration of protein reserves are determined by the specific requirements of a tissue and by the order of priority in which various tissues are synthesised. This could explain the discrepancies observed between the growth promoting effect and, for example, regeneration of plasma proteins, Allison (1959).

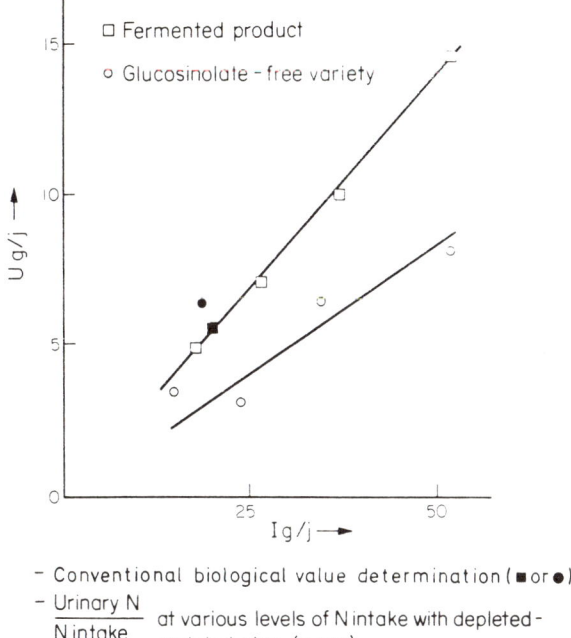

Figure 13.4 Comparison of protein qualities of two rapeseed meals using protein depletion-repletion method

According to Den Hartog and Pol (1972) in depletion-repletion experiments it is the reversible breakdown and synthesis of proteins per unit of tissue DNA that is involved. Protein synthesis is very active in some tissues and much greater than the amount retained in the tissues of a growing animal (Arnal, Fauconneau and Pech, 1971). Even skeletal muscle protein can be used as a source of proteins for more essential syntheses in the depleted state (Arroyave, 1972). In this respect, it is interesting to note the agreement between the rate of protein synthesis in rat muscle after six days of repletion and the biological value of the protein (Omstedt and Von der Decken, 1972). If the processes of synthesis and breakdown are truly reversible, as pointed out by Den Hartog and Pol (1972), it should be equally as justifiable to measure the generation of new tissue proteins as to measure the regeneration of protein reserves and there should be no need, in selecting the response to be measured, to limit oneself to the synthesis of certain tissue proteins or to the increased activity of particular enzymes.

The method of depletion-repletion, based on weight gain (or on nitrogen balance in the pig) offers the following advantages: homogeneity and sensitivity; linear response, even at high protein intakes; easy execution and rapid response. With separate feeding, it allows an adaptation of the energy intake to the quantity and quality of the protein.

Balance Methods

The efficiency of utilisation of dietary protein may be estimated from the difference between nitrogen intake and excretion, or from the increase of body proteins. The latter, mainly carried out with small animals (rat, chicken) can be simplified by taking into account the constant nitrogen/water ratio in the body (Miller and Bender, 1955), which may be accepted within given limits (Rerat, 1971; MacLaughlan, 1972).

Nitrogen balance can be expressed as a percentage of ingested or digested nitrogen. The slope of the response line to different protein intakes defines, on the linear portion of the curve (*Figure 13.1b*) the nitrogen balance index (Allison, 1964). The nitrogen excretion of dietary origin is estimated by allowing for the nitrogen of endogenous origin. This is represented by the body nitrogen loss (Bender and Doell, 1957) or by the nitrogen in the urine and faeces of animals fed on a nitrogen-free diet. This technique may be criticised (Peret and Jacquot, 1972b), but data obtained using a protein-free diet fed for a reasonable length of time: 12 to 25 days to the growing rat (Frost, 1959) or 16 days to the growing pig (Zelter and Charlet-Lery, 1961) are related to the basal metabolism in many species (Terroine and Sorg-Matter, 1927) and can be expressed as a function of the metabolic weight of the animal. The importance of the energy intake on urinary nitrogen excretion during this period cannot be neglected; when fed on a protein-free diet, the growing rat reduces its food intake and eliminates more nitrogen in urine than when fed on a diet supplemented with a small quantity of egg protein. There is no direct sparing effect of egg protein on the body nitrogen reserves, but only an indirect influence on the spontaneous intake of the supplemented diet (Hugot and Causeret, 1966). The metabolic nitrogen in the faeces varies with the composition and digestibility of the diet and has often been related to the dry matter intake. Zelter and Charlet-Lery (1961) showed that the faecal nitrogen of pigs fed on a protein-free diet is directly proportional to the faecal dry matter. The value of these estimations of endogenous losses is indicated by the good agreement between the direct determination and those deduced from the change in nitrogen balance at different nitrogen intakes, as recently confirmed by Hoffmann and Gebhardt (1973) in the chicken.

Like growth responses, nitrogen balance methods (net protein utilisation: NPU; biological value: BV) depend on quality and quantity of protein, as pointed out by MacLaughlan (1972). Mitchell (1924) had already discussed the reasons for this fall in BV at higher protein contents and stated that 'this decrease is probably due to a lower utilisation (of protein) for growth than for maintenance and on the other hand, to an increased use of the protein fed at a high level for covering the energy demand of the body'.

Systematic differences have been observed by several authors (Rerat, 1971) between results based on carcass analyses and those based on nitrogen balance data. They can be minimised by careful consideration of the numerous causes of errors involved in both types of method.

Among these factors, not enough attention has been paid to the fact that a change in protein content or quality results in an adaptation of the animal. This was observed long ago by Voit (1867) in adult animals. In the adult dog, at maintenance, adaptation to a new protein intake is obtained only after 30 days (Allison, 1959); in the growing pig, 12 days are necessary after a protein-free diet for animals subsequently fed on a constant diet (Armstrong and Mitchell, 1955), and up to 20 days if the food intake is adjusted to gain and live-weight during the repletion period (Delort-Laval *et al.*, 1966), on the basis of change in urinary nitrogen and urea. In the pig, blood urea concentration cannot be considered as constant before the 12th day of feeding a given diet (Bergner, Münchow, Reischuck, 1971). Recently, Said, Hegsted and Hayes (1974) showed that BV or NPU, as determined by the usual methods over a short period, tended to overestimate the value of the protein, especially for proteins of low quality, poor in lysine. In fact, the rat cannot maintain its protein stores, when fed on a diet limiting in various essential amino acids. Protein quality may thus be considered as an important factor in the adaptation of an animal to a new diet (see also Allison, 1959).

Other Criteria of Protein Quality

The balance technique allows for a separation and quantitative determination of nitrogen in urine and faeces. These two routes of nitrogen loss have quite different physiological significances and the determination of their nitrogenous components may give a more complete picture of the digestion and metabolic efficiency of the protein.

DIGESTION AND FAECAL LOSS OF AMINO ACIDS

Faecal nitrogen includes endogenous nitrogen and undigested residues of ingested protein, partially taken up and catabolised by the microorganisms of the alimentary tract. Faecal amino acids have been determined in rats (Kuiken and Lyman, 1948), pigs (Dammers, 1964) and chickens (Poppe and Meier, 1971a). In the chicken, larger differences occur in amino acid absorption than in rat or pig. The influence of the gut microflora on this process cannot be underestimated, as shown by comparison of the digestion of intact or heat-damaged protein by germ-free and conventional chickens (Salter and Coates, 1971).

In the pig, endogenous nitrogen is determined either directly from the amino acid excretion resulting from a protein-free diet (Dammers, 1964; Carlson and Bayley, 1970), or from the relationship between nitrogen loss and nitrogen intake, extrapolated to zero intake, proposed by Mitchell and Bert (1954). Both techniques give comparable results (Poppe and Meier, 1971b): the direct determination reduces the risk of error, arising from the large variations which occur at low protein intakes.

The interference of microbial fermentation, acting on the undigested part of the protein, can be partly eliminated by measurement of the undigested material in the ileum (Bayley, Cho and Holmes, 1974). The values thus obtained are lower than those found by faecal analysis, but allow for a clear distinction of heat-damaged protein (Varnish and Carpenter, 1971). This technique is also of evident interest for the ruminant, whose dietary nitrogen is greatly modified in the rumen; it has been broadly used, in studies on amino acid absorption from urea or tanin-protected protein diets in growing lambs (Faichney, 1970) or dairy cows (Hagemeister and Pfeffer, 1973).

URINARY NITROGEN EXCRETION

As pointed out by Kiriyama (1970), urinary nitrogen includes various metabolites, deriving directly or indirectly from protein metabolism, and whose summation, irrespective of their origin, cannot be used to express one parameter such as protein quality. Urinary concentration of urea, of degradation products of nucleic acids (allantoin, pseudouridine), of creatinine and amino acids, would allow the detection of biochemical changes, which would not be apparent from nitrogen balance data alone. For example, by studying the value of urea nitrogen per cent urinary nitrogen in the rat fed on a diet containing variable amounts of an amino acid but balanced in other respects, Kiriyama (1970) clearly showed that at the minimal value of this ratio the supplementation by the limiting amino acid was optimal (*Figure 13.5*). Also, the value:

$$\frac{\text{allantoin}}{\text{urine nitrogen}} \times \text{nitrogen intake}$$

has a much sharper maximum than the biological value, determined according to Njaa (1959).

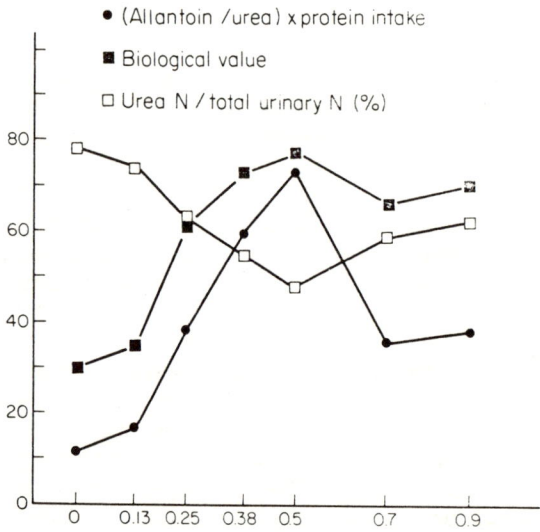

Figure 13.5 Changes in values of (Allantoin/urea) × protein intake, urea nitrogen per cent total urinary nitrogen and in biological value when weanling rats were fed on diets varying in threonine level (Kiriyama, 1970)

The importance of urea excretion can be related to its formation and concentration in the blood, which correlates with the biological value of the protein (Münchow and Bergner, 1968 ; Eggum, 1970). Though the relative variation in amino acid excretions can be important, their quantities generally remain low under normal conditions when compared with amino acid intakes, even for heat-damaged proteins (Ford and Shorrock, 1971).

Choice of Parameters in Bio-assay Procedures

Various techniques, some of them strictly standardised, are in use to determine the growth promoting value of a dietary protein or its effect on the nitrogen balance. As pointed out by MacLaughlan (1972), the agreement between various techniques is generally good for high quality proteins and progressively worsens as the quality of the protein deteriorates (Henry, 1965). The repeatability of results within a given laboratory and reproducibility from one place to another are important criteria for selecting a method. Data collected in collaborative assays of PER (Derse, 1965) or NPU (Niess and Müller, 1972) show fairly large variation, whereas the results obtained in one laboratory allow a more uniform ranking of products (Carpenter, 1973). Jacquot and Peret (1972) stress in this respect the high repeatability of growth tests using separate feeding

If classical growth or balance methods can be routinely used for screening of food ingredients, they are of little value in assessing the protein quality of a component in mixed diets, especially in compound feeds for intensive animal production. As pointed out by Lewis and Boorman (1969), protein evaluation has to be performed near or at the level of maximum response. Data derived from experiments on small animals cannot be applied without caution to other species and direct studies, though far more difficult to perform, have to be made for each species and each physiological state where protein synthesis plays an important part (growth, pregnancy, milk production, etc.).

The dose/response curve is not linear in the zone of maximum performance and the response can depend in this zone on the balance of amino acids and other components of the diet. An imbalance of amino acids is frequently linked with a decrease in food intake (Rerat, 1971), which does not appear in practice if restricted feeding is used (Uhlemann and Poppe, 1970).

The proposition of Lewis and Boorman (1969) has been criticised by Miller (1970), who prefers experimental conditions where the relation between dose and response is linear. However, the result of the bioassay can be distorted by the large relative increase in amino acid requirements, especially lysine, for rapid growth (MacLaughlan, 1972). On the other hand, Miller (1970) and Carpenter (1973) rightly insist on the effect of adjustment of protein level in bioassay on the availability of amino acids in protein concentrates.

Another problem arising in biological tests for protein evaluation results from dilution of the protein in a given ingredient and from the possible presence of anti-nutritional factors which may act directly or indirectly on protein digestion and efficiency. Soya bean meal is a good example of the influence of anti-nutritional factors on growth and protein efficiency; untoasted meal, still containing a high anti-tryptic activity, lowers protein digestion and biological value, thus reducing the growth promoting value of this meal. On the other hand, haemagglutinin, in the raw product, also has a negative effect on growth, not directly related to the biological value of the protein. Protein dilution in a concentrate rich in starch (cereals, tubers) also limits the possibility of a correct biological evaluation, extracted potato protein contains large amounts of essential amino acids and is of good biological value. When included in a diet, raw potato decreases nitrogen digestion and retention; it has been demonstrated that raw potato starch is only partly digested in the intestinal tract and that its consumption greatly increases the loss of endogenous nitrogen in faeces and urine. Efficiency of protein utilisation of a diet, containing a high percentage of raw potato starch is reduced accordingly (Zelter, Charlet-Lery and Delort-Lavel, 1966). Thus, a given food cannot be substituted without care for dietary components of the same type in the reference diet. It is also known (Buraczewski *et al.*, 1971) that the type of dietary sugar can modify the rate of passage and the absorption of the protein.

Interpretation of biological tests must take into account all these limitations, but the bioassay is the only absolute source, from which our knowledge of protein quality can be derived.

Conclusion

Estimation of the quality of dietary protein relies mainly on growth and nitrogen retention in the young animal. It cannot be extended to other species and other physiological states, whose specific amino acid requirements are not precisely known. These requirements must be expressed in terms of available amino acids. Assessment of availability by appropriate biochemical or enzymatic tests must be checked by *in vivo* assay using various animal species.

If the growth criterion can, in this respect, be considered satisfactory, under experimental conditions where body composition remains unchanged, the direct or indirect determination of nitrogen balance is of more general value especially in the range of optimum performance, or when protein synthesis is not related to changes in body weight.

In bioassays, variability among animals can be partly reduced and their sensitivity increased by the use of a depletion-repletion technique. Lastly, on account of the fundamental influence of the energy intake on the efficiency of utilisation of dietary protein, the principle of spontaneous energy/protein adjustment, allowed by separate feeding of the protein fraction of the diet, could be more extensively used in protein evaluation studies.

References

ABRAHAM, J., MORIN-JOMAIN, M. and PERETIANU, J. (1964). *Bull. Soc. Chim. biol.*, **46**, 755

ALLISON, J.B. (1959). in *Protein and Amino Acid Nutrition.* Ed. A.A. Albanese, Academic Press, London

ALLISON, J.B. (1964). in *Mammalian protein metabolism.* Eds. H.N. Munro and J.B. Allison. Vol.II, p.41. Academic Press, London

ARMSTRONG, D.G. and MITCHELL, H.H. (1955). *J. Anim. Sci.*, **14**, 49

ARNAL, M., FAUCONNEAU, G. and PECH, R. (1971). *Annls. Biol. anim. Biochim. Biophys.*, **11**, 245

ARROYAVE, G. (1962). *Am. J. clin. Nutr.*, **11**, 447

BARNES, R.H. and BOSSHARDT, D.K. (1946). *Ann. N.Y. Acad. Sci.*, **47**, 273

BAYLEY, H.S., CHO, C.Y. and HOLMES, J.H.G. (1974). *Fedn Proc. Fedn Am. Socs exp. Biol.*, **33**, 94

BENDER, A.E. (1961). *Natn Acad. Sci. Nat. Res. Council Publ.*, **843**, 407

BENDER, A.E. and DOELL, B.H. (1957). *Br. J. Nutr.*, **11**, 140

BERGNER, H., MUNCHOW, H. and BEISCHUCK, M. (1971). *Arch. Tierernähr.*, **21**, 133

BURACZEWSKI, S., PORTER, J.W.G., ROLLS, B.A. and ZEBROWSKA, T. (1971). *Br. J. Nutr.*, **25**, 299

CALET, C., JOUANDET, C. and BARATOU, J. (1961). *Annls. Biol. anim. Biochim. Biophys.*, **1**, 1

CANNON, P.R., HUMPHREYS, E.M., WISSLER, R.W. and FRAZIER, L.E. (1944). *J. clin. Invest.*, **23**, 601

CARLSON, K.H. and BAYLEY, H.S. (1970). *J. Nutr.*, **100**, 1353

CARPENTER, K.J. (1973). *Nutr. Abstr. Rev.*, **43**, **6**, 424

CARPENTER, K.J. and ELLINGER, G.M. (1955). *Biochem. J.*, **61**, 11

DAMMERS, J. (1964). in *Verteringsstudies Bij Het Varken. Faktoren van Invloed de Vertering der Voeder-componenten en de Verteerbaar-Reid der Aminozuren. Inst. Veevoedingsonderzoek*, Hoorn, p.152

DELORT-LAVAL, J., CHARLET-LERY, G. and ZELTER, S.Z. (1966). *Proc. 7th Intern. Congr. Nutr.*, Hamburg. **4**, p.350. Pergamon Press, Oxford

DEN HARTOG, C. and POL, G. (1972). in *Protein and amino acid functions*, p.311. Ed. E.J. Bigwood. Pergamon Press, Oxford

DERSE, P.H. (1965). *J. Ass. off. agric. Chem.*, **48**, 847

EGGUM, B.O. (1970). *J. Nutr.*, **24**, 983

FAICHNEY, G.J. (1970). in *Feeding Protected Protein to Sheep and Cattle.* Ed. D.W. Horwood, Austr. Soc. Anim. Prod., Sydney,

FERRO-LUZZI, A., MARIANI, A. and MIGLIACCIO, P.A. (1970). *Nutr. Metab.*, **12**, 306

FORD, J.E. (1962). *Br. J. Nutr.*, **16**, 409

FORD, J.E. and SHORROCK, C. (1971). *Br. J. Nutr.*, **26**, 311

FRAZIER, L.E., WOOLRIDGE, R.L., STEFFE, C.H. and BENDITT, E.P. (1949). *Fedn Proc. Fedn Am. Socs exp. Biol.*, **8**, 355

FROST, D.V. (1959). in *Protein and Amino acid nutrition.* Ed. A.A. Albanese, Academic Press, London

FROST, D.V. and SANDY, H.R. (1949). *J. Nutr.*, **39**, 247

HAGEMEISTER, H. and PFEFFER, E. (1973). *Z. Tierphysiol., Tierernähr., Futtermittelk.*, **31**, 275
HEGSTED, D.M. and CHANG, Y. (1965). *J. Nutr.*, **85**, 159
HENRY, K.M. (1965). *Br. J. Nutr.*, **19**, 125
HENRY, Y. and RERAT, A. (1972). *Ninth Intern. Congr. of Nutr.*, Mexico
HOFFMANN, M. and GEBHARDT, G. (1973). *Arch. Tierernähr.*, **23** 273
HUGOT, D. and CAUSERET, J. (1966). *Annls. Biol. anim. Biochim. Biophys.*, **6 (2)**, 179
JACQUOT, R. and PERET, J. (1972). in *Protein and Amino acid functions*. Ed. E.J. Bigwood. Pergamon Press, Oxford
KIRIYAMA, S. (1970). in *Newer methods of Nutritional Biochemistry*. Ed. A.A. Albanese, **Vol.4**, p.37. Academic Press, London
KUIKEN, K.A. and LYMAN, C.M. (1948). *J. Nutr.*, **36**, 359
LARBIER, M. and GUILLAUME, J. (1972). *Annls Biol. anim. Biochim. Biophys.*, **12**, 637
LEWIS, D. and BOORMAN, K.N. (1969). in *Proteins as human foods*. Ed. R.A. Lawrie. Butterworths, London
LONGENECKER, J.B. (1963). in *Newer methods of Nutritional Biochemistry*. Ed. A.A. Albanese, **Vol.1**, p.113. Academic Press, London
MacLAUGHLAN, J.M. (1972). in *Newer methods of Nutritional Biochemistry*. Ed. A.A. Albanese, **Vol.5**, 33. Academic Press, London
MacLAUGHLAN, J.M. and CAMPBELL, J.A. (1970). in *Mammalian Protein Metabolism*. Ed. H.N. Munro, **3**, p.391. Academic Press, New York
MEADE, R.J. (1972). *J. Anim. Sci.*, **35**, 713
MILLER, D.S. and BENDER, A.E. (1955). *Br. J. Nutr.*, **8**, 382
MILLER, E.L. (1970). *F.A.O. Fish. Rep.*, **92**, 66
MITCHELL, H.H. (1924). *Physiol. Rev.*, **4**, 424
MITCHELL, H.H. and BERT, M.H. (1954). *J. Nutr.*, **52**, 483
MITCHELL, H.H. and BLOCK, R.J. (1946). *J. biol. Chem.*, **163**, 599
MÜNCHOW, H. and BERGNER, H. (1968). *Arch. Tierernähr.*, **18**, 222
NIESS, von E. AND MÜLLER, R. (1972). *Z. Tierphysiol., Tierernähr. Futtermittelk.*, **30**, 177
NJAA, L.R. (1959). *Br. J. Nutr.*, **13**, 137
OMSTEDT, P.T. and VON DER DECKEN, (1972). *Br. J. Nutr.*, **27**, 467
OSBORNE, T.B. and MENDEL, L.B. (1917). *J. biol. Chem.*, **26**, 1
OSER, B.L. (1959). in *Protein and Amino Acid Nutrition*. Ed. A.A. Albanese. Academic Press, London
PAWLAK, M. and PION, R. (1967). *C.R. Acad. Sci.*, **264 D**, 380
PERET, J. and JACQUOT, R. (1972a). in *Newer methods of nutritional biochemistry*. Ed. A.A. Albanese, **Vol.5**, p.197. Academic Press, London
PERET, J. and JACQUOT, R. (1972b). in *Protein and Amino Acid Functions*. Ed. E.J. Bigwood. Pergamon Press, Oxford
PERETIANU, J. and ABRAHAM, J. (1963). *Ann. Nutr. Alim.*, **17**, 81
POPPE, S. and MEIER, H. (1971a). *Arch. Tierernähr.*, **21**, 447
POPPE, S. and MEIER, H. (1971b). *Arch. Tierernähr.*, **21**, 619
RERAT, A. (1971). *Ann. Zootech.*, **20**, 193

SAID, A.K., HEGSTED, D.M. and HAYES, K.C. (1974). *Br. J. Nutr.*, **31**, 47
SALTER, D.N. and COATES, M.E. (1971). *Br. J. Nutr.*, **26**, 55
SCHEFFNER, A.L., ECKFELD, G.A. and SPECTOR, H. (1956). *J. Nutr.*, **60**, 105
SUNDARAVALLI, O.E. and RAO, M.N. (1969). *Nutr. Dieta,* **11**, 101
TERROINE, E.F. and SORG-MATTER, H. (1927). *Archs int. Physiol.*, **29**, 121
UHLEMANN, H. and POPPE, S. (1970). *Arch. Tierernähr.*, **20**, 517
VARNISH, S.A. and CARPENTER, K.J. (1971). *Proc. Nutr. Soc.*, **30**, 70A
VOIT, C. (1867). in *Le besoinazoté.* Eds R. Jacquot and H. Vigneron, (1958). A.E.C. Commentary, 03
ZELTER, S.Z. and CHARLET-LERY, G. (1961). *Annls Biol. anim. Biochim. Biophys.*, **1**, 29
ZELTER, S.Z., CHARLET-LERY, G. and DELORT-LAVAL, J. (1966). *C.R. Acad. agric. France,* 567

14

INDIRECT MEASURES OF PROTEIN ADEQUACY

B.O. EGGUM
Department of Animal Physiology and Chemistry, Rolighedsvej 25, Copenhagen, Denmark

Introduction

Protein metabolism in whole animals and man can be studied by a variety of indirect criteria. It is impossible, within the scope of this paper, to provide a comprehensive survey of all indirect measures of protein adequacy. This discussion will therefore be limited to certain aspects of the study of protein adequacy in the individual. Measures based on total nitrogen balance will not be considered directly, however they will be used for comparison with indirect measures.

Blood Proteins and Amino Acids

It is established that in severe protein malnutrition in man, the total protein concentration in the plasma is greatly decreased often to as little as 4 g per 100 ml (Waterlow, Cravioto and Stephen, 1960). The decrease is mainly a decrease in the albumin fraction and Brock (1961) stated that 'in the present state of our knowledge, the earliest and most sensitive biochemical index of mild or impending protein deficiency is a drop in serum albumin into the marginal range'. The total globulins show little change and the γ-globulins may even be increased. According to Waterlow (1969), a fall in serum albumin concentration would be a relatively late event in protein depletion, since there seem to be mechanisms which tend to protect the total circulating albumin mass when protein supplies are short. Schendel, Hansen and Brock (1962) found that, even with a poor quality food, such as maize, serum albumin concentration was maintained for several weeks before it began to decrease. In children from families of four different incomes (Wittmann *et al.*, 1967) a small, but significant, difference between the albumin concentrations of the lowest and the highest income groups was found, but there was a much greater difference in body weights. The results suggest that the serum albumin concentration is probably not a very sensitive index of protein adequacy.

The quantitative relationship between plasma amino acid concentration and dietary protein has been reviewed by McLaughlan and Morrison (1968) and Eggum (1973a) and it appears that some investigators have failed to establish a relationship between dietary concentrations of amino acids and increases in plasma concentrations during the absorption period. There may be several reasons for this discrepancy, one of them being the differences in the availabilities of the amino acids (Eggum, 1973b). Another reason may be the differences in the rate of absorption. It was demonstrated by Delhumeau, Velez Pratt and Gitler (1962) that the rate of absorption of lysine was only 63 per cent of that of cystine. It has also been reported by some workers (Lebedeva and Nuriakhmetova, 1964; Nielsen, 1965) that excitement of experimental animals markedly affects plasma concentrations of amino acids.

Zimmerman and Scott (1965) found a definite relationship exists between plasma amino acid concentrations in chicks and the amino acid adequacy of the diet. The first limiting amino acid remained at a very low concentration in the blood, irrespective of the severity of the amino acid deficiency. They suggested that this helps to explain why, in some instances, supplementing diets with the first limiting amino acid has failed to increase the concentration of this amino acid in the plasma (Richardson, Blaylock and Lyman, 1953; Owings and Balloun, 1961). According to Zimmerman and Scott, no increase would be anticipated until the dietary concentration exceeded that needed for maximum growth. It is of interest that severe deficiencies of either lysine or arginine markedly increased plasma threonine. According to the authors this did not appear to be unique for threonine since many of the other amino acids behaved in this manner, but to a lesser degree.

Direct relationships between dietary and blood amino acids are also obscured by differences in the rates of removal of amino acids from the plasma to meet the requirements of the tissues. Moreover, amino acids may be metabolised within the epithelium of the small intestine. Finch and Hird (1960) observed that substantial amounts of arginine were converted to ornithine and that transamination of glutamic and aspartic acids with pyruvate also resulted in low recoveries of these amino acids from the mucosal cells, significant amounts of alanine being formed. A recent study by Knipfel *et al.* (1972) showed that feeding swine on single meals resulted in fluctuations in serum amino acid concentrations that could not be accounted for by the diet alone. These workers proposed a short interval feeding which resulted in relative stability of serum amino acid concentrations.

Holt *et al.* (1963) described the pattern of free amino acids in the plasma of 64 patients with kwashiorkor from nine different countries. The pattern was surprisingly uniform and the changes corresponded closely with those found earlier by Arroyave *et al.* (1962) in Guatemala, by Vis (1963) in the Congo, and by Edozien, Phillips and Collis (1960) in Nigeria. The salient deviations from the normal aminogram reported by all these authors were summarised by Waterlow (1969) as follows:

(1) There was usually a fall in total α-amino nitrogen.
(2) There was a fall in most of the essential amino acids, particularly the branched-chain amino acids. Phenylalanine and lysine were much less affected.

(3) There was a fall in some of the non-essential amino acids, particularly tyrosine.
(4) The concentrations of most of the non-essentials, notably alanine, proline, histidine, serine and aspartic acid were maintained or even increased.

Holt et al. (1963) were unable to detect a close correlation between alterations in plasma amino acid concentrations and the clinical severity of the disease in the patients studied.

As a result of his investigations, Eggum (1973b) concluded that the plasma amino acid method may provide valuable information regarding the nutritional status, limiting amino acid and amino acid requirements. McLaughlan and Campbell (1969) considered that convincing evidence has not as yet been presented to indicate unique advantages for the use of plasma amino acid concentrations for rating the relative qualities of various protein foods. These workers further concluded that future research on plasma amino acid concentrations may prove this criterion to be a rapid and reliable means of assessing the adequacy of dietary amino acid contents. However, no-one (Gitler, 1964) has yet presented convincing evidence that there are unique advantages in using plasma amino acid concentrations as indicators of protein adequacy.

Blood Urea Concentration

In experiments with rats, Kumta and Harper (1961) demonstrated that an amino acid imbalance caused an increase in the blood urea concentration and that the concentration could be decreased by restoring the dietary balance. The urea concentration reached a maximum three hours after feeding. A similar dependence of blood urea on time after feeding was found by Anderson and Edney (1969) in experiments with dogs and by Eggum (1970) in an experiment with a pig.

In experiments with growing rats and pigs, Münchow and Bergner (1968) and Bergner, Münchow and Reischuck, (1971) found a high negative correlation between the biological value of the dietary protein and blood urea concentration. The same relationship was obtained in experiments with rats by Eggum (1970) and Prabhu (1971). Nielsen (1973) also found a significant relationship between blood urea concentration and the quality of bacon pigs after slaughter, but the correlation was not as high as in the other cases. Münchow and Bergner (1968) and Eggum (1970) found that blood urea concentration also increased with the protein content of the diet. This correlation is consistent with the demonstration that an increase in dietary protein content caused a decline in biological value (Eggum, 1973b).

Thus at least three factors influence blood urea concentration, the quality and quantity of protein in the diet and the time of sampling after feeding. By standardising technique it is possible to eliminate the effects of both protein content and time after feeding (Eggum, 1970). Thus, protein quality should become the decisive factor influencing blood urea concentration and thus become a good indirect measure of the protein adequacy of the diet of an individual.

Enzymes in Blood or Liver

In the 1950's there were several studies of the concentrations of various enzymes in the plasma or serum of malnourished patients (Waterlow, 1969). The activities of several enzymes decrease in patients suffering from protein inadequacy. Serum transaminase activities are usually normal in infants with malnutrition. According to McLean (1962) high values probably reflect liver damage, and have no direct bearing on the state of protein nutrition. Barrows (1958) found that the effect of dietary protein on plasma cholinesterase activity was only approximately consistent with the effect on rate of growth in female rats. In experiments with bacon pigs Just-Nielsen (1973) found a positive correlation ($r = 0.38$) between creatine kinase activity and per cent lean in the side of bacon.

Wirthgen, Bergner and Münchow (1967) investigated rat liver for the activities of glutamate-pyruvate transaminase (GPT), glutamate-oxalacetate transaminase (GOT), ornithine-carbamyl transferase (OCT), and arginase. The activities of GPT, OCT, and arginase decreased linearly with the increasing biological value of the protein fed, whereas the enzyme GOT showed the highest activity with high quality protein, and a decrease with proteins of poorer quality. However, the decrease in GOT activity was not maintained with proteins of 50 or less biological value. In this work significant correlations were found between the activities of several enzymes and biological values found using the Thomas-Mitchell method with rats. However, the correlations were not as good as the correlation between biological value and the reciprocal of blood urea concentrations discussed above.

So far, none of the enzyme activities which have been measured seem to give a more sensitive index of the state of protein nutrition than methods already discussed.

Urinary Nitrogen

Many studies have been made regarding the amount and partition of urinary nitrogen in malnourished subjects and particularly of the relation of urea nitrogen to total nitrogen or to creatinine. Waterlow (1969) said that the same general conclusions have been reached in his work. The total urinary nitrogen is low, and the amount of urea nitrogen, both absolute and relative, is less in malnourished subjects; urea may account for only 50 per cent of the total nitrogen, instead of 80-90 per cent, as in a normal subject on a normal diet. It has been suggested that a low proportion of urea nitrogen might be regarded as evidence of protein depletion (Platt and Heard, 1958).

Scrimshaw et al. (1966) used urinary nitrogen as a measure of protein utilisation. Young men were given an egg diet (providing the approximate minimum protein requirement) until a constant rate of urinary nitrogen excretion was attained. Iso-nitrogenous test diets were fed and negative nitrogen balance was indicated by an increase in the excretion of urinary nitrogen. This study indicated that a simple measure of urinary nitrogen might provide a satisfactory method for evaluating the adequacy of proteins for humans.

In attempts to evaluate the dietary protein quality with rats, Kiriyama and Ashida (1964); Kiriyama and Iwao (1964) and Kiriyama et al. (1967) reported that the ratio of urinary allantoin to urea multiplied by protein intake (A/K × Ip) changes in proportion to the dietary amino acid balance. This index was found to be more sensitive than those based on growth rate or nitrogen balance. In these studies, the excretory pattern of urinary nitrogen compounds and other criteria for assessing the nutritive value of proteins were examined simultaneously. When rats were fed on a 'threonine-imbalanced' or 'corrected' diet, the differences in urea or allantoin excretions were significant between the two groups while the differences in body weight gain or nitrogen retention (per cent) were almost the same. In recent work by Kiriyama, Suzuki and Iwao (1971) it was shown that urea excretion was the main variable in total urinary nitrogen output, causing the significant differences in the proportion of nitrogen retained. Liver arginase activity changed inversely with urea excretion but proportionately with the qualitative improvement in dietary protein. Changes in liver glutamate dehydrogenase activity were also consistent with the improvement of dietary protein quality.

Experiments with pigs have been conducted by Brown and Cline (1972a,b) to determine the feasibility of using urinary or plasma urea as a measure of amino acid adequacy. In the first experiment, 25 kg pigs were fed on a diet consisting of maize supplemented with tryptophan and with or without lysine. Food intake was equalised and daily urea excretion was measured for ten days and plasma urea was determined on day ten. Addition of lysine decreased daily urea excretion (3.97 vs 2.89 g/day). Maximal difference between diets was obtained by day three. Lysine supplementation decreased plasma urea from 13.4 to 10.0 mg/100 ml. In a second experiment similar results were obtained when four concentrations of lysine were added to a tryptophan-supplemented maize diet. The diets were fed both on an *ad libitum* and equalised intake basis and plasma and urinary urea were measured on day three. In experiments with 20 kg pigs Brown and Cline (1972b) compared two methods of determining amino acid requirements. The technique of measuring total urinary urea production was compared with the standard growth assay. In both assays increasing amounts of lysine were added to lysine-deficient diets. In both types of assay, optimal response was obtained when 0.36 per cent lysine was added to the basal diet.

On the evidence of these experiments, measurements on urinary urea seem to give valuable information on the protein status of the animal. Therefore similar measurements have been used with rats at our Institute, where urinary urea has been compared with biological value.

Alternative Methods

MUSCLE FIBRE DIAMETER
Hammond and Appleton (1932) and Jubert (1954; 1956) measured the average diameter of fibres in different muscles in lambs and sheep. They found that the postnatal growth of the muscle was largely due to an increase in the length and diameter of the muscle fibres. A high correlation was also found between diameter and the quantity of meat in each of the animals. McMeekan (1940; 1941) found that nutrition greatly influenced the diameter of the muscle fibres in pigs and further that the number of muscle fibres did not change between birth and maturity.

In Denmark this technique was studied extensively by Staun (1963) in experiments with bacon pigs. One of his experiments comprised 15 groups of eight pigs each (four gilts and four castrates). Three different protein contents were combined with five different feeding rates, from a low, to a very high plane of nutrition. The plan of the experiment and some results obtained are shown in *Table 14.1*. The results showed that fibre diameter increased with increasing amounts of protein in the diet. However, increasing food intake caused a decrease in the diameter of the muscle fibres. Moderate feeding produced the greatest fibre diameter, while a low rate of feeding produced muscle fibres of a slightly smaller diameter. The results seem to prove that increasing amounts of protein in the diet, up to a certain limit, will increase the diameter of muscle fibres and at the same time increase meat production. In experiments with rats (Eggum, 1969a) positive correlation between muscle fibre diameter and NPU was also established.

This method deserves special attention as it also can be used on living animals. Biopsies of muscle samples can be taken relatively easily. For this reason the method might be of special interest in the control of protein quality and quantity for the breeding animal.

PARTIAL CARCASS MEASURES
In a study by Mortensen and Madsen (1970) the correlation coefficients between the total meat content in the bacon side and different indirect measures of the protein content in pigs were calculated. The experiment included 60 female pigs slaughtered at 90 kg body weight. The results are shown in *Table 14.2*. It appears that significant relationships between meat in the bacon side and several indirect measures of meat content exist. The high negative correlation (−0.98) between fat and lean in the bacon side is to be expected. However, to measure fat in the bacon side is just as difficult as to measure the lean content. The side fat thickness measured 8 cm down on the side behind the last rib, seems to give a good indication of the lean content. This criterion is also one of the very best indirect measures of slaughter quality together with the criteria of the areas of fat and lean of the longissimus dorsi.

Table 14.1 Number and diameter of muscle fibres in m. longissimus dorsi in pigs fed on different intakes of diets of different protein content

Protein regime	Low	Normal	High
Skimmed milk (kg/pig/day)	0.75	1.5	2.1
Protein supplement[1] (g/pig/day)	55	110	155

Rate of feeding	Protein regime				Protein regime			
	Low	Normal	High	Mean	Low	Normal	High	Mean
	S.F.U.[2] pig/day				Digestible true protein g/day			
1	1.65	1.71	1.69	1.68	134	174	198	169
2	1.89	1.91	1.88	1.89	149	185	211	182
3	2.16	2.17	2.22	2.18	165	201	231	199
4	2.45	2.45	2.55	2.48	182	217	251	217
5	2.50	2.60	2.71	2.60	185	226	259	223
mean	2.13	2.17	2.21		163	201	230	
	Muscle fibre diameter (μ)				No. of fibres per sq. mm			
1	69.8	71.9	75.8	72.5	359	314	288	320
2	72.4	76.3	75.4	74.7	313	314	317	315
3	66.4	72.7	73.4	70.8	390	336	329	352
4	61.5	68.1	65.0	64.9	419	373	386	393
5	63.7	62.4	65.5	63.9	418	392	375	395
mean	66.8	70.3	71.0		380	346	339	

[1] Protein supplement: 2/3 soyabean meal + 1/3 meat and bone meal
[2] Scandinavian feed unit

Table 14.2 The correlation coefficient between lean meat (per cent) in the bacon side and different indirect measures of meat content

Back fat thickness × meat in bacon side	-0.55
Side fat thickness × meat in bacon side	-0.78
Area of fat in musc. dorsi × meat in bacon side	-0.72
Fat in bacon side × meat in bacon side	-0.98
Wt. of inner musculus iliopsoas × meat in bacon side	0.56
Area of musc. dorsi × meat in bacon side	0.70
Wt. of musc. dorsi × meat in bacon side	0.72

$p < 0.05 \sim r = 0.26$ $p < 0.01 \sim r = 0.34$

In extensive work by Nielsen (1973) the effect of different amounts of protein given to early-weaned pigs (18 days) on the development of different organs to 20 kg body weight was studied. Significant increases in the weights of lungs, kidneys, spleen and intestine were obtained with increasing protein concentration in the diet. Highly significant increases in the weights of several muscles (M.semitendinosus, M.semimembranosus, M.biceps femoris, M.gracilis, M.vastus, M.iliopsoas, M.longissimus dorsi) with increasing protein concentration in the diet were also obtained

Lachance and Miller (1973) studied the relationship of nitrogen concentration in the leg of rats to that in whole carcass. It was found that hind limb nitrogen content could be used as an indicator of whole carcass nitrogen. According to the authors this simplified procedure permits easier, quicker and more economical Net Protein Utilisation measurements to which valid statistical analyses can be applied. The authors found this partial carcass method to be comparable with the classical method of total carcass nitrogen analysis. The method is being further investigated in our laboratory.

Eggum *et al.* (1971) found in experiments with rats, chickens and bacon pigs that NER (nitrogen efficiency ratio) as a criterion of protein adequacy gave the best results in experiments with chickens. A correlation of 0.99 was obtained between NER and total lean in the body. In rat experiments, the correlation coefficient between NER and NPU was only 0.72. NER could not be used as a criterion of protein quality for pigs.

Several methods based on advanced techniques have not been discussed but attention should be paid to work with labelled amino acids in the search for indirect measurements of protein adequacy.

Conclusions

By their nature the tests or indices which have been discussed can give only indirect information about the state of protein nutrition. Results with blood proteins and amino acids suggest that the serum albumin concentration is probably not a very sensitive index of protein adequacy. Plasma amino acid concentrations may provide valuable indications regarding the protein status but there is no evidence for unique advantages in using blood amino acids as an indicator of protein adequacy. Blood urea concentration measured under standardised conditions can give good indirect measurements of protein adequacy of the individual animal. Various enzyme activities do not seem to give convincing indices of the state of protein nutrition. A simple measure of urinary nitrogen might provide a satisfactory method for evaluating adequacy of protein for a subject. Measurements on urinary urea, however, seem to be a more valuable indicator as to the protein status. The diameter of muscle fibres is dependent on the protein adequacy of the diet. Similarly, the weight of several muscles and different organs is responsive to the protein supply. The relationship of the nitrogen content in one leg of rats to the whole carcass might also be developed as an eventual method of indirect measurement of protein adequacy.

References

ANDERSON, R.S. and EDNEY, A.T.B. (1969). *Vet. Rec.*, **84**, 348
ARROYAVE, G., WILSON, D., DE FUNES, C. and BÉHAR, M. (1962). *Am. J. clin. Nutr.*, **11**, 517
BARROWS, C. (1958). *J. Nutr.*, **66**, 515
BERGNER, H., MÜNCHOW, H. and REISCHUCK, M. (1971). *Arch. Tierernähr.*, **21**, 133
BROCK, J.F. (1961). *Recent Advances in Human Nutrition.* Little, Brown, Boston, Massachusetts
BROWN, J.A. and CLINE, T.R. (1972a). *J. Anim. Sci.*, **35**, 211
BROWN, J.A. and CLINE, T.R. (1972b). *J. Anim. Sci.*, **35**, 1102
DELHUMEAU, G., VELEZ PRATT, G. and GITLER, C. (1962). *J. Nutr.*, **88**, 75
EDOZIEN, J.C., PHILLIPS, E.J. and COLLIS, W.R.F. (1960). *Lancet*, **I**, 615
EGGUM, B.O. (1969). unpublished data
EGGUM, B.O. (1970). *Br. J. Nutr.*, **24**, 983
EGGUM, B.O. (1973a). *Proteins in Human Nutrition.* Eds. J.W.G. Porter and B.A. Rolls, Academic Press, London and New York
EGGUM, B.O. (1973b). *Beretn.*, **406**, Forsøgslaboratoriet, Copenhagen, Denmark
EGGUM, B.O., PETERSEN, V.E., MADSEN, A. and MORTENSEN, H.P. (1971). *Yearbook. Royal Veterinary and Agricultural University*, Copenhagen, Denmark
FINCH, L.R. and HIRD, F.J.R. (1960). *Biochim. biophys. Acta.*, **43**, 278
GITLER, C. (1964). *Mammalian Protein Metabolism.* Eds. H.N. Munro and J.B. Allison, **Vol.I.** Academic Press, New York
HAMMOND, J. and APPLETON, A.P. (1932). *Growth and Development of Mutton Qualities in the Sheep.* Oliver and Boyd, London
HOLT, L.E. Jr., SNYDERMAN, S.E., NORTON, P.M., ROITMAN, E. and FINCH, J. (1963). *Lancet*, **II**, 1343
JUBERT, D.M. (1954). *Proc. Br. Soc. Anim. Prod.*, **49**
JUBERT, D.M. (1956). *J. agric. Sci.*, **47**, 59
JUST NIELSEN, A. (1973). *Bilag til Landøkonomisk Forsøgslaboratoriums efterårsmøde, Årbog*
KIRIYAMA, S. and ASHIDA, K. (1964). *J. Nutr.*, **82**, 127
KIRIYAMA, S. and IWAO, H. (1964). *Agric. Biol. Chem.*, **28**, 307
KIRIYAMA, S., SUZUKI, T. and IWAO, H. (1971). *Agric. Biol. Chem.*, **35**, 1844
KIRIYAMA, S., YAGISHITA, Y., SUZUKI, T. and IWAO, H. (1967). *Agric. Biol. Chem.*, **31**, 743
KNIPFEL, J.E., KEITH, M.O., CHRISTENSEN, D.A. and OWEN, B.D. (1972). *Can. J. Anim. Sci.*, **52**, 143
KUMTA, U.S. and HARPER, A.E. (1961). *J. Nutr.*, **74**, 139
LACHANCE, P.A. and MILLER, G.A. (1973). *Nutrition reports international*, **7**, 25
LEBEDEVA, Z.N. and NURIAKHMETOVA, Z. Yu. (1964). *Vop. Pitan.*, **23**, 23
McLAUGHLAN, J.M. and CAMPBELL, J.A. (1969). *Mammalian Protein Metabolism.* Ed. H.N. Munro. **Vol.III.** Academic Press, New York and London

McLAUGHLAN, J.M. and MORRISON, A.B. (1968). *Protein Nutrition and Free Amino Acid Patterns.* Ed. J.H. Leathem. Rutgers University Press, New Brunswick, New Jersey
McLEAN, A.E.M. (1962). *Lancet,* **II**, 1294
McMEEKAN, C.P. (1940). *J. agric. Sci.,* **30**, 276
McMEEKAN, C.P. (1941). *J. agric. Sci.,* **31**, 1(I-V)
MORTENSEN, H.P. and MADSEN, A. (1970). *NJF-symposium,* Nyborg, Denmark
MÜNCHOW, H. and BERGNER, H. (1968). *Arch. Tierernähr.,* **18**, 222
NIELSEN, H.E. (1973). *Beretn.,* **405**, Forsøgslaboratoriet. Copenhagen, Denmark
NIELSEN, J.J. (1965). *Licentiat Thesis.* The Royal Veterinary and Agricultural College, Copenhagen, Denmark
OWINGS, W.J. and BALLOUN, S.L. (1961). *Poult. Sci.,* **40**, 1718
PLATT, B.S. and HEARD, C.R.C. (1958). *Proc. Nutr. Soc.,* **17**, ii
PRABHU, G.A. (1971). personal communication
RICHARDSON, L.R., BLAYLOCK, L.G. and LYMAN, C.M. (1953). *J. Nutr.,* **51**, 515
SCHENDEL, H.E., HANSEN, J.D.L. and BROCK, J.F. (1962). *S. Afr. J. Lab. clin. Med.,* **8**, 23
SCRIMSHAW, N.S., YOUNG, V.R., SCHWARTZ, R., PICHE M.L. and DAS, J.B. (1966). *J. Nutr.,* **89**, 9
STAUN, H. (1963). *Acta Agric. scand.,* **XIII**, 293
VIS, H.L. (1963). *Aspects of mécanismes des hyperaminoaciduries de l'enfance.* Editions Arscia, Brussels
WATERLOW, J.C. (1969). in *Mammalian Protein Metabolism.* Ed. H.N. Munro, **Vol.III**. Academic Press, New York and London
WATERLOW, J.C., CRAVIOTO, J. and STEPHEN, J.M.L. (1960). *Adv. Protein Chem.,* **15**, 131
WIRTHGEN, B., BERGNER, H. and MÜNCHOW, H. (1967). *Arch. Tierernähr.* **17**, 281
WITTMANN, W., MOODIE, A.D., FELLINGHAM, S.A. and HANSEN, J.D.L. (1967). *S. Afr. Med. J.,* **41**, 664
ZIMMERMAN, R.A. and SCOTT, H.M. (1965). *J. Nutr.,* **87**, 13

15

DIETARY EFFECTS AND AMINO ACIDS IN TISSUES

R. PION
Centre de Recherches Zootechniques, Beaumont, France

Introduction

The products of digestion of dietary proteins are mainly carried by the blood as free amino acids. Their concentration in each tissue is the result of a balance between the input from the blood supply and the catabolism of protein and the loss due to protein synthesis and various catabolic pathways. Other reports in this symposium deal with some of the factors which affect amino acid uptake, such as protein synthesis and its regulation, protein turnover, hormonal control, and amino acid supply by the digestive tract. Our own purpose is to discuss the relationship between the free amino acid content of tissues and the ability of ingested nitrogenous matter to meet the nitrogen and the essential amino acid requirements.

Free Amino Acids in Tissues

Only a very small part of total body amino acid content is present as free amino acid and the concentration of free essential amino acids tends to be lower than that of many of the non essential amino acids. Therefore, it may be expected that the free amino acid contents of the tissues may be modified by the dietary amino acid supply or by the rate of uptake for protein synthesis. Indeed, many authors have studied the relationship between free amino acids in tissues and the ability of dietary proteins and amino acids to meet the requirements of an animal.

Since free amino acid pools in whole tissues and in sub-cellular structures have been extensively studied by Munro (1970), we shall merely discuss here the nutritional significance of the total free amino acid content in each tissue. Furthermore we shall only discuss the effects of diets without very large deficiencies or imbalances.

BLOOD
Sampling of blood in the living animal is very easy, and many authors have studied the influence of dietary factors on its free amino acid profile, following the work of Longenecker and Hause (1959) with

dogs; Morrisson, Middleton and McLaughlan (1961) with rats; Hill and Olsen (1963) and Zimmerman and Scott (1965) with chicks. This subject has been reviewed by several authors in the book of Leathem (1968), and by Young and Scrimshaw (1972); by Munro (1970), and by Featherston (1972), with regard to the chick.

Since the variation in free amino acid concentrations in the portal blood during digestion are discussed by Rerat (this symposium), we shall only discuss the relationship between the amino acid composition of diets used during long term feeding trials and blood amino acid concentrations.

It may be inferred from the work of Dawson and Porter (1962) with rats, using radioactive *Chlorella,* that the free amino acids coming from the digestion of proteins may be carried by the blood cells as well as by the plasma. Elwyn (1966) and Elwyn et al. (1972) studied the transport of amino acids by the plasma and by the blood cells in the dog and concluded that erythrocytes and plasma play distinct roles in the transport of amino acids, and that there may be a direct transfer of amino acids between erythrocytes and tissue cells. Squibb (1971). Stephens and Evans (1971), also pointed out the role of the blood cells in the transport of amino acids. Therefore, it may be useful in nutritional studies to estimate the whole blood amino acid level instead of the plasma amino acid level and we have been using this technique for many years (Pion, Fauconneau and Rerat, 1964).

Blood amino acid levels may be used to estimate the supplies of amino acid from dietary proteins if the requirements of the experimental animals are accurately known, or to estimate the needs of the animal if the availability of dietary amino acid is accurately known. They may also be used in order to estimate the results of the digestion of nitrogenous matter in the rumen. As pointed out by Munro (1970); Munro and Portugal (1972), the blood free amino acid profiles are not only related to the amino acid content of the diet, but they may be altered by its energy, vitamin and mineral status, since many of these nutrients may affect the amino acid utilisation by the tissues.

Many authors working with different species have been able to plot the concentration of one amino acid in the blood or in the plasma against the amount ingested for animals receiving graded levels of the amino acid. There is generally a sharp increase in the plasma amino acid when the ingested amount is high enough to support maximum growth (Young and Scrimshaw, 1972). Similar results were observed in the whole blood for lysine (Pawlak and Pion, 1968a) threonine (Pawlak and Pion, 1968b) and methionine (Pion, 1973), but not for isoleucine. There are often inverse relationships between two amino acids; a deficiency in one amino acid may result in an increase in some other blood amino acid, especially lysine or threonine. Increasing or decreasing the protein level may modify the blood concentration of the amino acid under study, particularly as far as lysine or threonine are concerned (Harker, Allen and Clark, 1968; Pawlak and Pion, 1968c; 1970; Richardson, Cannon and Webb, 1965).

The feeding schedule (Grizard, Prugnaud and Pion, 1974) the length of time of fasting before feeding (Harker, Allen and Clark, 1968; Stockland et al., 1971) and the time between ingestion of the meal and sampling of the blood (Patureau-Mirand, Prugnaud and Pion, 1971) may also affect the results.

Large differences between systemic and portal values may be expected due to the regulatory role of the liver. Many amino acids may be taken from the blood by the liver (Bloxam, 1972; Elwyn, 1968; 1972; Wolff, Bergmann and Williams, 1972) for protein synthesis or for catabolic purposes, one notable exception being the branched chain amino acids. However, the uptake of amino acids by the liver may be relatively small when compared with the supplies from the digestive tract during digestion, as may be seen from the results of an experiment with young growing pigs (Pion and Rerat, *Table 15.1*). The animals, fitted with permanent portal and jugular catheters, received a balanced diet containing sunflower meal supplemented with lysine, tryptophan and methionine. Blood samples were taken from

Table 15.1 Mean portal to jugular blood amino acid concentration ratios in the growing pig

Time after feeding (h)		0.5	1	1.5	2	2.5	3	4	5
Threonine	A	1.25	1.55	1.20	1.35	1.17	1.02	1.09	1.06
	B	1.08	1.03	1.11	1.32	1.11	0.91	0.75	0.77
Valine	A	1.05	1.44	1.04	1.08	1.13	0.98	1.10	1.02
	B	0.91	0.92	1.01	1.01	1.08	0.87	0.58	0.69
Isoleucine	A	1.12	1.50	1.08	1.20	1.18	1.06	1.16	1.47
	B	1.06	0.97	1.14	1.21	1.12	0.9	0.61	0.68
Leucine	A	1.17	1.63	1.10	1.23	1.13	1.05	1.24	1.27
	B	1.15	1.05	1.21	1.25	1.13	0.92	0.62	0.79
Tyrosine	A	1.37	1.43	1.06	1.36	1.24	0.94	1.25	1.18
	B	0.97	1.20	1.19	1.19	1.07	0.89	0.43	0.60
Phenylalanine	A	1.17	1.50	1.08	1.35	1.18	1.10	1.25	1.16
	B	1.06	1.51	1.13	1.11	1.14	0.88	0.47	0.65
Lysine	A	1.15	1.41	1.05	1.10	1.21	0.96	1.12	1.20
	B	0.98	**1.08**	1.08	0.93	0.74	0.84	0.76	0.88

Pion and Rerat, unpublished results
Diet A: sunflower meal + L lysine + DL methionine + L tryptophan
Diet B: sunflower meal + DL methionine + L tryptophan

the catheters during the five hours following the ingestion of test meals of equal amounts of either the basal diet, or the same diet without added lysine (Pion and Rerat, 1969). The ratios of portal to jugular amino acids in the blood samples taken after the ingestion of the balanced diet were not very high, except at one hour after feeding when they rose to about 1.5. These ratios were always lower in the blood taken after the test meal of the lysine deficient diet; indeed the amino acid concentrations were higher in the jugular than in the portal blood three to five hours after the meal. This was due to small amounts of lysine in the portal blood and to large amounts of other essential amino acids in the jugular blood. Thus the amounts of amino acids which are not used for protein synthesis, due to the lack of lysine, are not as quickly catabolised as they are when coming from the intestine.

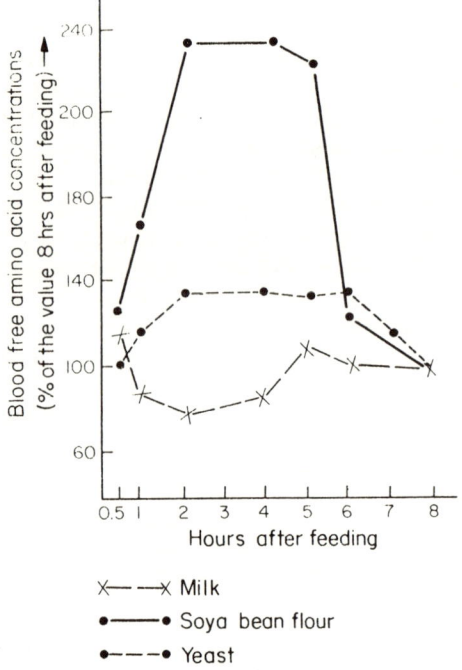

X———X Milk
●———● Soya bean flour
●– – –● Yeast

Figure 15.1 Changes in blood amino acid after feeding in the pre-ruminant calf (Patureau-Mirand, Prugnaud and Pion, 1971)

The high levels of amino acids in the jugular blood of the pre-ruminant calf during digestion, observed by Patureau-Mirand, Prugnaud and Pion, 1971 *(Figure 15.1)* leads to a similar conclusion.

LIVER

There are only a few data demonstrating the relationship between ingested amino acids and free amino acids in the liver. Richardson, Cannon and Webb (1965) found a lower amount of free lysine in the liver of chicks receiving a lysine deficient diet than in the liver of the animals given a lysine supplemented diet, free liver threonine was higher with the lysine deficient diet. Jacobs and Crandall (1972) also found a relationship between the lysine content of the diet and the free lysine content of the liver; however, in these two experiments the differences between the free lysine content of the liver of the animals given the high and low lysine diets were smaller than the differences between their muscle free lysine content. Leung, Rogers and Harper (1968) found similar results with threonine imbalanced diets.

In spite of the difference between the catabolism of the amino acids in birds and mammals, the relationship between liver free amino acids and ingested amino acids seems to be similar in these two kinds of animal, except for arginine. The concentrations of arginine found in the liver by Richardson, Cannon and Webb (1965) were higher than the corresponding concentration in the blood plasma, whereas most authors have found only trace amounts of this amino acid in the liver of mammals.

INTESTINAL MUCOSA

There is also a relationship between the lysine content of the diet and the free lysine and free threonine concentrations in the intestinal mucosa (Jacobs and Crandall, 1972), but the differences between the animals given the lysine rich and the lysine deficient diets were lower than those observed in the plasma or in the liver.

MUSCLE

Muscle is the tissue that has the largest need for essential amino acids, and that contains the largest free amino acid pool (Munro, 1970). Richardson, Cannon and Webb (1965) concluded from a chick study that there was no clear relationship between tissue (including muscle) amino acids and amino acid supplies. When the dietary amino acid intake was modified, it was by changing the protein level of the diet. In contrast, the lysine content of the tissues, especially muscle, was strongly increased when lysine was added to a lysine deficient diet. It was also observed (Pawlak and Pion, 1967) with rats eating either a low lysine diet or a balanced one, that the differences between the lysine content of muscle of the rats given the two diets were much larger than the differences between the blood lysine concentrations. Therefore we studied the effect on blood and muscle free amino acids in the *ad libitum* fed rat of graded levels of lysine (Pawlak and Pion, 1968a) threonine (Pawlak and Pion, 1968b) isoleucine (Pawlak and Pion, 1971) and sulphur amino acids. Most of our results were summarised by Pion (1972; 1973). Most of the non-essential amino acids were found in higher concentrations in the muscle than in the blood, and that was also true for lysine, threonine, arginine, and methionine when the dietary supply of sulphur amino acids was high. Free lysine, free threonine and to a lesser extent free methionine concentrations showed greater changes in the muscle than in the blood. Free isoleucine concentrations in the blood and muscle were not very different, but dietary isoleucine deficiencies led to high muscle levels of lysine and threonine. Muscle free amino acids seem to be a more sensitive test of the adequacy of dietary amounts than blood amino acids, at least for amino acids like lysine and threonine. Larbier, Guillaume and Blum (1971); Larbier and Guillaume (1972) found relationships between dietary lysine and methionine and muscle free lysine and methionine with chicks, only if the animals received a single daily meal instead of being fed *ad libitum*. Jacobs and Crandall (1972) also found that there was a very large increase in the free lysine content of muscle with rats when wheat protein was supplemented by lysine.

It may be concluded that blood and free amino acid studies may be of value in studying the ability of dietary proteins and amino acids to meet the requirements of the animals, and to estimate the magnitude of some of these requirements. Muscle is a more sensitive indicator for some amino acids than is blood; but blood is easiest to sample from living animals. One must always keep in mind that there are many interactions between amino acids and that a single amino acid concentration may be of poor value, because it may result not only from an excess or lack of the amino acid under study but also from an excess or lack of other amino acids.

Estimation of Amino Acid Availability

Many workers have studied the relationship between amino acid composition of ingested protein and free blood amino acids. Their results have been summarised by McLaughlan and Morrisson (1968); Featherston (1972) and Young and Scrimshaw (1972). As the amino acid content of feedstuffs may be accurately estimated, it is very useful to apply the results of these studies to the estimation of availability of amino acids from natural or processed feeds.

Using plasma amino acid studies, Stockland and Meade (1970) found that some amino acids like isoleucine, threonine and phenylalanine were not fully available for the rat from meat and bone meals. De Vuyst *et al.* (1971; 1972) found that the heating of a commercial rat diet impaired the availability of its amino acids. Hill and Olsen (1967) found a decline in plasma lysine of chicks when the soya bean meal they ate had been autoclaved. Smith and Scott (1965) observed that autoclaving had deleterious effects on the availability of some amino acids, mainly of lysine. The results of Puchal *et al.* (1962) with pigs, suggest that some amino acids may not be fully available in the feeds under study. Combs *et al.* (1967), used plasma amino acids as a criterion of availability of amino acids for pigs to study the effect of heat treatment on soya bean meal.

Larbier, Guillaume and Calet (1971), studied the effect of heating and of storage of maize grains, and found that free lysine in chick muscle was a sensitive test for the estimation of the availability of this amino acid. We have studied the nutritive value of some feeds by estimating the blood and muscle free amino acids of rats receiving diets containing the feed under study supplemented with amino acids (Pawlak and Pion, 1970a,b; Pion, Mendes-Pereira and Prugnaud, 1974). We found an effect of drying on the availability of sulphur amino acids in wheat grains (Pion and Pawlak, 1973).

Estimation of the Amino Acid Requirements

Many authors have used blood or muscle response to estimate the requirements of various animal species:

RAT
It is possible to estimate the needs of growing rats for some amino acids from the muscle and blood values observed in the previously described experiments. McLaughlan and Illman (1967) estimated the need for lysine, tryptophan, threonine, leucine, isoleucine and histidine from plasma amino acid studies, while Stockland, Meade and Melliere (1970) studied the lysine requirement. Longenecker (1973) used his PAA ratios method to estimate these amino acid requirements.

CHICK
Most of the workers do not emphasise the use of the plasma amino acid results to estimate the needs of the animal. However, D'Mello (1974) found a good agreement between growth rate and efficiency of food conversion response and plasma amino acids, in a study on the requirements of the chick for leucine, isoleucine and valine.

PIG
Mitchell (1965) used plasma amino acids to study the requirements of the young pig for leucine, isoleucine, lysine and histidine. Further estimations of the need for isoleucine were made by Bravo *et al.* (1970) and Oestemer, Hanson and Meade (1973); while a study on high lysine maize by Pick and Meade (1970) led to an estimation of lysine and isoleucine requirements. Pion *et al.* (1971) were able to estimate the threonine requirement of growing pigs from the threonine levels in the whole blood.

PRE-RUMINANT CALF
As very little information is available on amino acid requirements of the pre-ruminant calf, the study of the influence of dietary amino acids on their blood amino acid levels may be of value. It may be seen from *Figure 15.1* that the shape of the free amino acid concentration curve in the jugular blood is related to the kind of dietary protein; there is a large increase from two to five hours with soya bean flour, a small increase with yeast and a decrease from one to four hours with milk. This decrease was also observed by Williams and Smith (1973).

The addition of graded levels of either lysine (Patureau-Mirand, Prugnaud and Pion, 1973a) or methionine (Patureau-Mirand, Prugnaud and Pion, 1973b) was used to estimate the lysine and methionine requirements. It led to the accumulation in the blood of the corresponding amino acid when the dietary levels were respectively

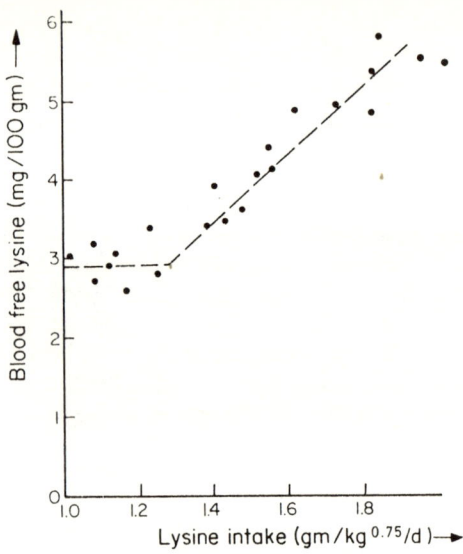

Figure 15.2 Relationship between blood free lysine in the pre-ruminant calf and daily lysine intake (Patureau-Mirand, Prugnaud and Pion, 1973a)

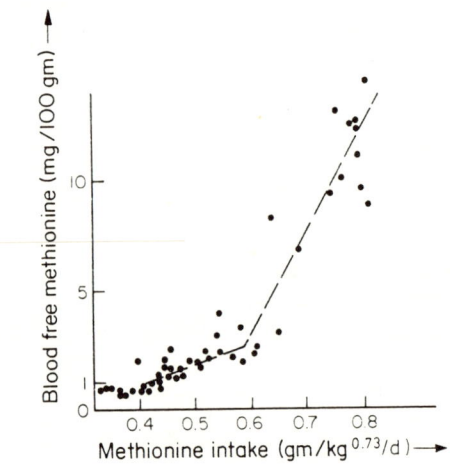

Figure 15.3 Relationship between blood free methionine in the pre-ruminant calf and daily methionine intake (Patureau-Mirand, Prugnaud and Pion, 1973b)

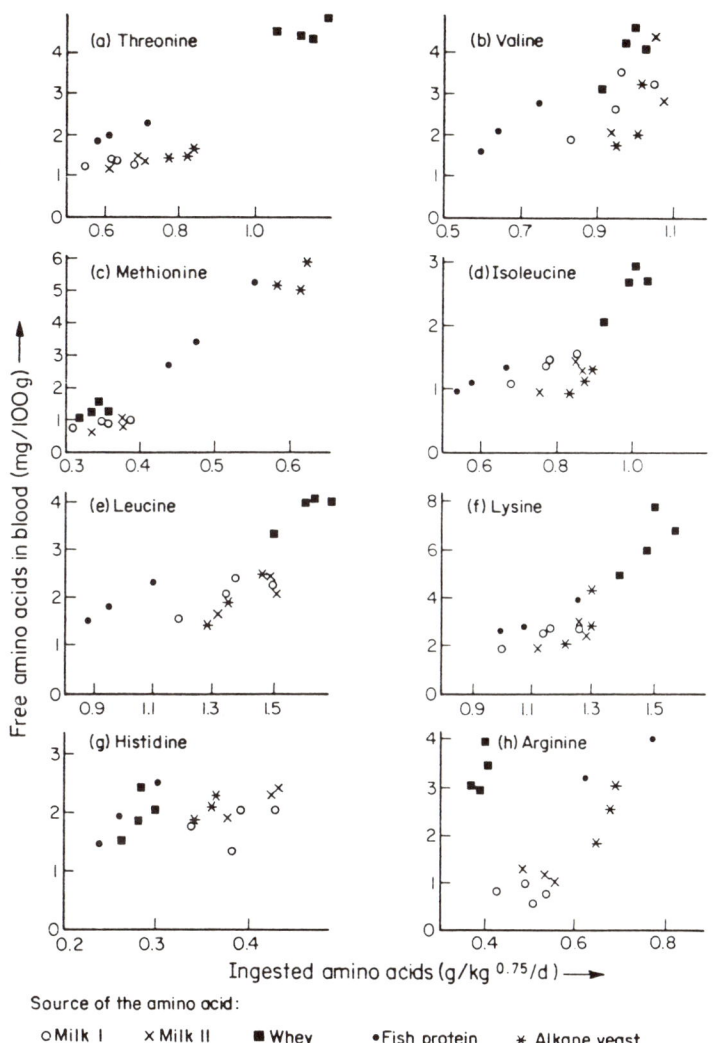

Figure 15.4 Relationship between free amino acids (mg/100 g) in blood and ingested amino acids (g/kg$^{0.75}$/day). (Patureau-Mirand et al., 1974)

Table 15.2 Amino acid requirements of the pre-ruminant calf, growing pig and growing rat (Patureau-Mirand et al. 1974)

Amino acids	Calf (1)	Calf (2)	Pig (2) (3)	Rat (2) (4)
Threonine	0.8	10.2	10.55	9.15
Valine	0.9	11.5	11.75	10.1
Methionine } Cystine }	0.65	8.3	11.75	11.0
Isoleucine	0.85	10.9	11.75	10.1
Leucine	1.3	16.6	14.1	13.4
Tyrosine } Phenylalanine }	0.9	11.5	11.75	13.95
Lysine	1.3	16.6	16.4	17.4
Histidine	0.4	5.1	4.25	4.6
Arginine	0.55	7.0	4.7	7.5
Tryptophan	-	2.2	3.05	2.75

(1) $(g/w^{0.75}/day)$
(2) Per cent of the sum of tyrosine, cystine and essential amino acids
(3) Fauconneau and Pion (1972)
(4) Ranhotra after Fauconneau (1967)

Table 15.3 The comparison between essential amino acid content of the diet and essential amino acid concentrations in the blood (Patureau-Mirand, Theriez and Prugnaud, 1975)

		Ewe's milk	Cow's milk	Whey protein concentrate	Fish protein concentrate	Alkane yeast
Crude protein (N × 6.25)		27.0	26.2	23.6	25.7	24.5
Threonine	(1)	1.310	1.205	1.097	1.162	1.252
	(2)	6.12	5.39	5.07	3.76	3.31
Valine	(1)	2.093	1.873	-	1.424	1.496
	(2)	3.80	3.74	2.56	1.97	2.03
Cystine	(1)	0.284	0.235	0.472	0.245	0.276
Methionine	(1)	0.810	0.681	0.602	1.040	0.849
	(2)	0.85	1.12	1.08	3.93	2.78
Isoleucine	(1)	1.539	1.506	1.298	1.212	1.321
	(2)	1.26	1.45	1.31	0.93	1.29
Leucine	(1)	2.592	2.620	2.549	2.053	2.034
	(2)	2.61	2.70	2.56	1.6	1.77
Tyrosine + phenylalanine	(1)	2.754	2.738	1.970	1.882	2.174
Tyrosine/ phenylalanine	(3)	1.46	1.37	1.08	1.00	1.21
Lysine	(1)	2.417	2.227	1.994	2.087	1.938
	(2)	3.02	3.94	3.89	3.97	3.06
Histidine	(1)	0.837	0.760	0.602	0.580	0.536
	(2)	1.98	2.52	2.11	1.45	1.23
Arginine	(1)	1.053	0.930	0.755	1.209	0.958
	(2)	2.26	2.78	2.55	3.17	2.34
Tryptophan	(1)	-	0.302	-	0.237	0.265

(1) Per cent amino acid in the diet (dry matter)
(2) Free amino acid concentration in the blood (mg/100 g)
(3) Ratio of blood tyrosine to blood phenylalanine

1.3 $g/kg^{0.75}$/day for lysine and 0.5 $g/kg^{0.73}$/day for methionine (*Figures 15.2* and *15.3*). Some less accurate information may be inferred from blood amino acid values of animals given dietary proteins of various amino acid compositions (Patureau-Mirand *et al.*, 1974) (*Figure 15.4*). For most essential amino acids, it is possible to have a rough estimation of the minimum dietary amount needed to bring about an accumulation in the blood. It may be seen from the *Table 15.2* that the tentative requirements deduced from these estimations are similar to the needs of growing pigs and rats.

PRE-RUMINANT LAMB

A similar study was made with young lambs by Patureau-Mirand, Theriez and Prugnaud, (1975) (*Table 15.3*). Blood free amino acids are generally related to the dietary amounts and it may be expected that further studies may lead to a tentative estimation of the amino acid requirements.

RUMINANT NUTRITION

Most dietary amino acids are degraded by bacteria in the rumen, and the products are used as well as non-protein nitrogen for microbial protein synthesis. Therefore, there is no direct relationship between the amino acid composition of the ingested proteins and the amino acid supply to the tissues of the ruminant animal, but plasma or blood free amino acid patterns may be of value in assessing the ability of the ingested diets to meet the amino acid requirements of the animal.

The influence of the time after feeding was studied by several authors. Leibholz (1965) using sheep receiving a single daily meal, found large increases in plasma amino acids one hour after the meal. On the contrary, Freitag, Smith and Beesson (1968) observed decreases in plasma free amino acids after meals, while Mangan and Wright (1973) found very small increases for some amino acids but decreases for most of them. Barej and Szczygiel (1973) found either decreases in plasma amino acids or no change, while Halfpenny, Rook and Smith (1969) and Champredon, Pion and Fauconneau (1969) observed only small variations during the day.

The changes in blood amino acid levels after parturition were studied by Champredon and Pion (1972) with goats and with cows by Verbeke, Roets and Peeters (1972); Champredon, Pion and Remond (1974) (*Table 15.4*). Decreases in some blood essential amino acids after parturition reflect the temporary inability of the digestive tract to meet the increased requirements of the mammary gland. The increase in blood glycine is probably related to tissue catabolism.

Many experiments have been carried out in order to estimate the effect of the composition of the diet on blood free amino acids. Bergen, Henneman and Magee (1973), studied the effect of diets containing several crude protein levels and several protein supplements or non-protein nitrogen upon plasma, liver and muscle amino acids of

Table 15.4 Free amino acids (mg/100 g of total blood) in blood of dairy cows before and after calving (Champredon, Pion and Remond, 1974)

After calving (weeks)	-2	+1	+3	+9
Threonine	1.07	0.94	1.13	1.05
Glycine	2.74	3.31	4.52	3.35
Alanine	1.87	1.21	1.96	2.23
Valine	3.15	1.81	2.32	3.02
Methionine	0.61	0.60	0.73	0.85
Isoleucine	1.96	1.39	1.51	1.73
Leucine	2.66	1.68	2.08	2.58
Tyrosine	1.37	0.82	1.11	1.36
Phenylalanine	0.92	0.74	0.79	0.88
Lysine	1.66	1.29	1.47	1.66
Histidine	1.56	1.17	1.39	1.52
Arginine	0.52	0.45	0.44	0.57
Sum of total amino acids	28.7	21.0	25.9	28.4
Sum of essential amino acids and tyrosine	15.5	10.9	13.0	15.2

Table 15.5 Free amino acids and urea (mg/100 g of total blood) in blood of lambs receiving diets containing different amounts of nitrogen (Champredon, Pion and Martin-Rosset, 1974)

Crude protein (per cent DM)	12.0	14.4	16.6	19.8
Threonine	2.15	2.56	2.91	2.52
Valine	1.98	2.41	2.48	3.51
Methionine	0.76	0.82	0.72	0.79
Isoleucine	1.14	1.32	1.25	1.56
Leucine	2.63	3.00	3.01	3.64
Tyrosine	1.70	2.24	1.99	2.01
Phenylalanine	1.17	1.25	1.41	1.21
Lysine	1.97	1.97	2.19	2.11
Histidine	1.58	1.33	1.48	1.31
Arginine	1.19	1.50	1.04	1.01
Urea	22.0	28.0	36.0	56.0
Sum of total amino acids	36.9	36.6	39.1	40.4
Sum of essential amino acids and tyrosine	16.3	18.4	18.5	19.7

sheep. They found a direct relationship between many plasma and muscle amino acids, but not between plasma and liver amino acids. Plasma amino acids seem to be related more closely to protein level than to the kind of protein ingested, due to the importance of microbial proteins.

The effect of the dietary protein level was studied by Schelling, Hinds and Hatfield (1967) with lambs. Increasing the nitrogen level of the diet increased most plasma free amino acids, including branched chain amino acids. Plasma lysine and histidine did not increase, and plasma methionine decreased. Similar results were found by Champredon

Pion and Martin-Rosset (1974) (*Table 15.5*); while Van Horn, Jacobson and Graden (1969) found only small differences between the plasma amino acid levels of dairy cows given diets differing in their protein level.

Since the work of Virtanen (1966) with dairy cows, many authors have studied the effect of substituting non-protein nitrogen for dietary protein nitrogen, for example: Boling, Young and Bradley (1972); Champredon and Pion (1972); Champredon, Pion and Journet (1970); Clifford and Tillman (1968); Freitag, Smith and Beeson (1968); Oltjen and Putnam (1966); Schelling, Hinds and Hatfield (1967). Many of them observed decreases in some blood essential amino acid levels, but the results varied with the actual experimental conditions.

Table 15.6 Free amino acids in blood of lactating cows (mg/100 g total blood) (Remond et al., 1971).

Diet	Control	With added methionine	
Number of weeks after calving	5	5	7
Threonine	1.24	1.06	1.15
Valine	3.53	2.61	3.32
Methionine	0.50	0.46	0.44
Isoleucine	1.72	1.37	1.73
Leucine	2.40	1.90	2.39
Tyrosine	1.32	1.11	1.33
Phenylalanine	1.08	0.80	1.03
Lysine	1.56	1.60	1.86
Histidine	1.05	1.38	1.53
Arginine	0.71	0.65	0.84
Sum of total amino acid	30.62	27.41	31.01
Sum of essential amino acids + tyrosine	15.11	12.96	15.62
Urea-cycle amino acids	3.14	2.59	3.24

The supplementation of ruminant diets with methionine and lysine (Schelling, Hinds and Hatfield, 1967), or with methionine (Whiting *et al.*, 1972; Remond *et al.*, 1971) (*Table 15.6*) did not increase the corresponding blood amino acids, but it may be seen from *Table 15.6* that most of the essential amino acids decreased and histidine and (to a lesser extent) lysine increased.

It may be concluded from all these results that there is no direct relationship between the amount and the kind of protein or non-protein nitrogen ingested and blood free amino acids, but that differences between their amounts in the blood of an animal given different diets may reflect differences in the ability of these diets to meet their protein requirements.

The need to have direct relationships between amino acid supplies and blood free amino acids has led many authors to infuse protein directly into the duodenum or amino acids into the duodenum, the abomasum or the blood.

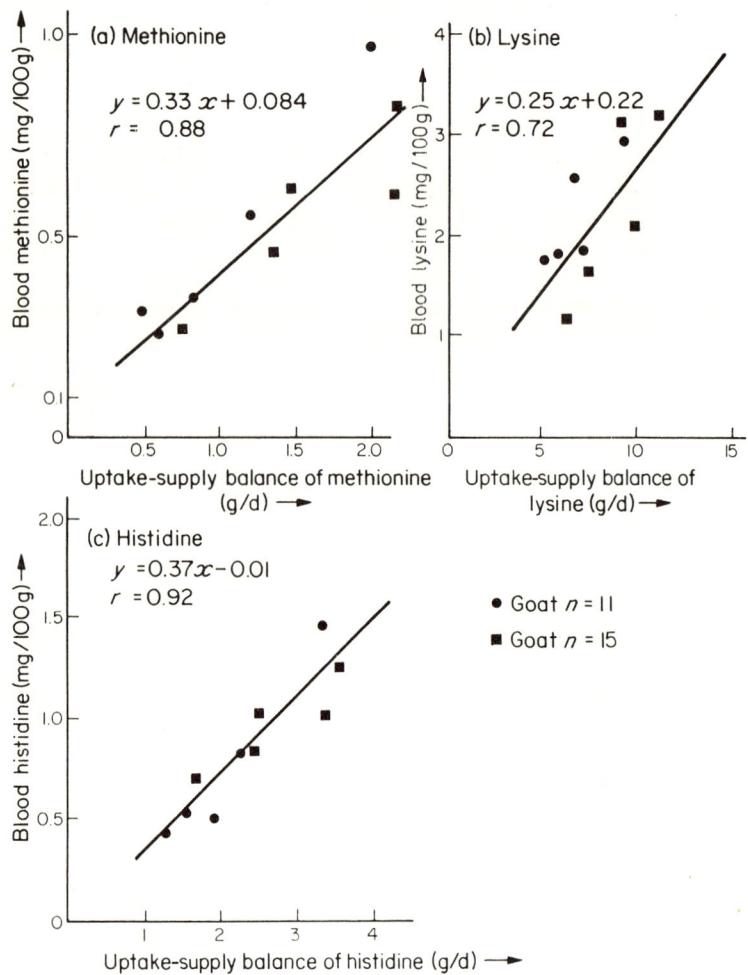

Figure 15.5 Relationship between free amino acids in blood of lactating goats and uptake-supply balance of amino acids.
'Uptake-supply balance = infused amino acid + absorbed amino acid − output in milk protein.
(Champredon, 1972; see also text)

Hogan, Weston and Lindsay (1968) infused into the abomasum of sheep graded levels of casein and observed that most plasma essential amino acids increased. Similar results were obtained by Potter and Bergen (1973) with casein and egg proteins. Increases in plasma amino acids after abomasal infusion of mixtures of amino acids were observed by Chalupa, Chandler and Brown (1973).

More specific responses were obtained by infusing single amino acids; Fisher (1972) infused methionine, histidine or lysine intravenously to lactating cows. The plasma concentration of the infused amino acid increased except when low levels of histidine and high levels of methionine were infused. Infusion of methionine or lysine decreased the plasma levels of other amino acids. Ely, Little and Mitchell (1969) injected lysine into the systemic blood of lambs and found an increase in plasma lysine. Schelling, Chandler and Scott (1973) found an increase in the plasma methionine levels of sheep after abomasal infusion of methionine. Champredon (1972) found increases in whole blood lysine, methionine and histidine after intravenous infusion of a mixture of these amino acids. He was able to calculate a linear relationship (*Figure 15.5*) between the blood levels of these amino acids and their 'uptake-supply balances' calculated from the outputs into milk proteins and the inputs from the digestive tract and from the infusion.

Wakeling, Lewis and Annison (1970) calculated the requirement for methionine of sheep from plasma methionine concentrations after duodenal infusion of graded levels of methionine. They were not able to calculate the need for lysine since the duodenal digesta meet the requirements of the test animals. Chalupa and Chandler (1972) also calculated the methionine requirement of growing sheep from the response of plasma methionine to the abomasal infusion of this amino acid. Other results are summarised and discussed by Armstrong and Annison (1973).

It may be concluded from all these results that blood free amino acids may be of value in determining the essential amino acid requirements of the tissues of the ruminant, and that there is the same relationship between the blood amino acid concentrations and amino acid requirements of the ruminant as exists for monogastric animals.

References

ARMSTRONG, D.G. and ANNISON, E.I. (1973). *Proc. Nutr. Soc.*, **32**, 113
BAREJ, W. and SZCZYGIEL, M. (1973). *Z. Tierphysiol. Tierernähr. Futtermittelk.*, **32**, 164
BERGEN, W.G., HENNEMAN, H.A. and MAGEE, W.T. (1973). *J. Nutr.*, **103**, 575
BLOXAM, D.L. (1972). *Br. J. Nutr.*, **27**, 233
BOLING, J.A., YOUNG, A.W. and BRADLEY, N.W. (1972). *J. Nutr.*, **102**, 1247
BRAVO, F.O., MEADE, R.J., STOCKLAND, W.L. and NORDSTROM, J.W. (1970). *J. Anim. Sci.*, **31**, 1137

BROOKES, I.M., OWENS, F.N., BROWN, R.E. and GARRIGUS, U.S. (1973). *J. Anim. Sci.*, **36**, 965

CHALUPA, W. and CHANDLER, J.E. (1972). in *Tracer Studies on non-protein nitrogen for ruminants*. International Atomic Energy Agency, p.107, Vienna

CHALUPA, W., CHANDLER, J.E. and BROWN, R.E. (1973). *J. Anim. Sci.*, **37**, 339

CHAMPREDON, C. (1972). *Thèse Doctorat de spécialité (Physiologie)*, Clermont-Ferrand, France

CHAMPREDON, C. and PION, R. (1972). *C.R. Soc. Biol.*, **166**, 378

CHAMPREDON, C., PION, R. and FAUCONNEAU, G. (1969). *C.R. Acad. Sci. Paris.*, **269**, 2029

CHAMPREDON, C., PION, R. and JOURNET, M. (1970). *Annls Biol. anim. Biochim. Biophys.*, **10**, 517

CHAMPREDON, C., PION, R. and MARTIN-ROSSET, W. (1974). unpublished results

CHAMPREDON, C., PION, R. and REMOND, B. (1974). unpublished results

CLIFFORD, A.J. and TILLMAN, A.D. (1968). *J. Anim. Sci.*, **27**, 484

COMBS, G.E., CONNESS, R.G., BERRY, T.H. and WALLACE, H.D. (1967). *J. Anim. Sci.*, **26**, 1067

DAWSON, R. and PORTER, J.W.G. (1962). *Br. J. Nutr.*, **16**, 27

D'MELLO, J.P.F. (1974). *J. Sci. Fd Agric.*, **25**, 187

ELWYN, D.H. (1966). *Fed. Proc.*, **25**, 854

ELWYN, D.H. (1968). in *Protein Nutrition and Free Amino Acid Patterns*. Ed. J.H. Leathem. Rutgers University Press, New Brunswick

ELWYN, D.H. (1972). in *Mammalian Protein Metabolism*. **Vol.IV**. Ed. H.N. Munro. p.523. Academic Press, New York

ELWYN, D.H., LAUNDER, W.J., PARIKH, H.C. and WISE, E.M. (1972). *Am. J. Physiol.*, **222**, 1333

ELY, D.G., LITTLE, C.O. and MITCHELL, G.E. (1969). *Can. J. Physiol. Pharmac.*, **47**, 929

FAUCONNEAU, G. (1967). *9th Internat. Congr. Anim. Prod.* p.62, Oliver and Boyd, Edinburgh

FAUCONNEAU, G. and PION, R. (1972). *Ann. Zootech.*, **21**, 275

FEATHERSTON, W.R. (1972). *Poultry Sci.*, **51**, 17

FISHER, L.J. (1972). *Can. J. Anim. Sci.*, **52**, 377

FREITAG, R.R., SMITH, W.H. and BEESON, W.M. (1968). *J. Anim. Sci.*, **27**, 478

GRIZARD, J., PRUGNAUD, J. and PION, R. (1974). *C.R. Soc. Biol.* (In press)

HALFPENNY, A.F., ROOK, J.A.F. and SMITH, G.H. (1969). *Br. J. Nutr.*, **23**, 547

HARKER, C.S., ALLEN, P.E. and CLARK, H.E. (1968). *J. Nutr.*, **94**, 495

HILL, D.C. and OLSEN, E.M. (1963). *J. Nutr.*, **79**, 303

HILL, D.C. and OLSEN, E.M. (1967). *Poultry Sci.*, **46**, 93

HOGAN, J.P., WESTON, R.H. and LINDSAY, J.R. (1968). *Aust. J. Biol. Sci.*, **21**, 1263

JACOBS, F.A. and CRANDALL, J.C. (1972). *Nutr. Rep. Intern.*, **5**, 27

LARBIER, M. and GUILLAUME, J. (1972). *Annls Biol. anim. Biochim. Biophys.*, **12**, 637

LARBIER, M., GUILLAUME, J. and BLUM, J.C. (1971). *Nutr. Rep. Intern.*, **3**, 273

LARBIER, M., GUILLAUME, J. and CALET, C. (1971). *Ann. Zootech.*, **20**, 653

LEATHEM, J.H. (1968). *Protein Nutrition and Free Amino Acid Patterns.* Rutgers University Press, New Brunswick

LEIBHOLZ, J. (1965). *Austr. J. Agric. Res.*, **16**, 973

LEUNG, P.M.B., ROGERS, Q.R. and HARPER, A.E. (1968). *J. Nutr.*, **96**, 303

LONGENECKER, J.B. (1973). in *Proteins in Human Nutrition.* Eds. J.W.G. Porter and B.A. Rolls. p.155. Academic Press, New York

LONGENECKER, J.B. and HAUSE, N.L. (1959). *Archs Biochem. Biophys.*, **84**, 46

McLAUGHLAN, J.M. and ILLMAN, W.I. (1967). *J. Nutr.*, **93**, 21

McLAUGHLAN, J.M. and MORRISSON, A.B. (1968). in *Protein Nutrition and Free Amino Acid Patterns.* Ed. J.H. Leathem. p.3. Rutgers University Press, New Brunswick

MANGAN, J.L. and WRIGHT, R.C. (1973). *Proc. Nutr. Soc.*, **32**, 53A

MITCHELL, J.R. (1965). Thesis, University of Illinois

MORRISSON, A.B,, MIDDLETON, E.J. and McLAUGHLAN, J.M. (1961). *Can. J. Biochem. Biophys.*, **39**, 1675

MUNRO, H.N. (1970). in *Mammalian Protein Metabolism.* **Vol.IV.** Ed. H.N. Munro. p.299. Academic Press, New York

MUNRO, H.N. and PORTUGAL, F.H. (1972). in *Protein and Amino Acid functions.* Ed. E.J. Bigwood. p.197. Pergamon Press, New York

OESTEMER, G.A., HANSON, L.E. and MEADE, R.J. (1973). *J. Anim. Sci.*, **36**, 679

OLTJEN, R.R. and PUTNAM, P.A. (1966). *J. Nutr.*, **89**, 385

PATUREAU-MIRAND, P., PRUGNAUD, J. and PION, R. (1971). *Communication.* Congrès de Zootechnie, Paris

PATUREAU-MIRAND, P., PRUGNAUD, J. and PION, R. (1973a). *Annls Biol. anim. Biochim. Biophys.*, **13**, 93

PATUREAU-MIRAND, P., PRUGNAUD, J. and PION, R. (1973b). *Annls Biol. anim. Biochim. Biophys.*, **13**, 683

PATUREAU-MIRAND, P., THERIEZ, M. and PRUGNAUD, J. (1975). *Annls Biol. anim. Biochim. Biophys.*, **15**, 95

PATUREAU-MIRAND, P., TOULLEC, R., PARUELLE, J.L., PRUGNAUD, J. and PION, R. (1974). *Ann. Zootech.*, **23**, 343

PAWLAK, M. and PION, R. (1967). *C.R. Acad. Sci.*, **264**, 380

PAWLAK, M. and PION, R. (1968a). *Annls Biol. anim. Biochim. Biophys.*, **8**, 517

PAWLAK, M. and PION, R. (1968b). *C.R. Acad. Sci. Paris*, **266**, 1993

PAWLAK, M. and PION, R. (1968c). *Annls Biol. anim. Biochim. Biophys.*, **8**, 457

PAWLAK, M. and PION, R. (1970a). *Annls Biol. anim. Biochim. Biophys.*, **10**, 171

PAWLAK, M. and PION, R. (1970b). *Annls Biol. anim. Biochim. Biophys.*, **10**, 317

PAWLAK, M. and PION, R. (1971). *Annls Biol. anim. Biochim. Biophys.*, **11**, 505

PICK, R.I. and MEADE, R.J. (1970). *J. Anim. Sci.*, **31**, 509

PION, R. (1972). in *Protéines et acides aminés en Nutrition Humaine et Animale.* Eds. C.L. de Cuenca and M. Vanbelle. p.529. Garsi, Madrid

PION, R. (1973). in *Proteins in Human Nutrition.* Eds. J.W.G. Porter and B.A. Rolls. p.329. Academic Press, New York

PION, R., FAUCONNEAU, G. and RERAT, A. (1964). *Annls Biol. anim. Biochim. Biophys.*, **383**

PION, R., MENDES-PEREIRA, E. and PRUGNAUD, J. (1974). 2nd Symp. 'Alimentation et Travail'. Vittel

PION, R. and PAWLAK, M. (1973). *Annls. Technol. Agric.*, **22**, 823

PION, R., PRUGNAUD, J., HENRY, Y. and RERAT, A. (1971). Xeme Congrès International de Zootechnie, Paris-Versailles

PION, R. and RERAT, A. (1969). Journées de la Recherche Porcine, INRA, Paris

POTTER, E.L. and BERGEN, W.G. (1973). *J. Anim. Sci.*, **37**, 353

PUCHAL, F., HAYS, V.W., SPEER, V.C., JONES, J.D. and CATRON, D.V. (1962). *J. Nutr.*, **76**, 11

REMOND, B., CHAMPREDON, C., DECAEN, C., PION, R. and JOURNET, M. (1971). *Annls Biol. anim. Biochim. Biophys.*, **11**, 455

RICHARDSON, L.R., CANNON, M.L. and WEBB, B.D. (1965). *Poultry Sci.*, **44**, 248

SCHELLING, G.T., CHANDLER, J.E. and SCOTT, G.C. (1973). *J. Anim. Sci.*, **37**, 1034

SCHELLING, G.T., HINDS, F.C. and HATFIELD, E.E. (1967). **92**, 339

SMITH, R.E. and SCOTT, H.M. (1965). *J. Nutr.*, **86**, 37

SQUIBB, R.L. (1971). *Poult. Sci.*, **50**, 491

STEPHENS, A.G. and EVANS, R.A. (1971). *Proc. Nutr. Soc.*, **30**, 49A

STOCKLAND, W.L. and MEADE, R.J. (1970). *J. Anim. Sci.*, **31**, 1156

STOCKLAND, W.L., MEADE, R.S. and MELLIERE, A.L. (1970). *J. Nutr.*, **100**, 925

STOCKLAND, W.L., MEADE, R.J., TUMBLESON, M.E. and PALM, B.W. (1971). *J. Anim. Sci.*, **32**, 1143

VAN HORN, H.H., JACOBSON, D.R. and GRADEN, A.P. (1969). *J. Dairy Sci.*, **52**, 1395

VERBECKE, R., ROETS, E. and PEETERS, G. (1972). *J. Dairy Res.*, **39**, 355

VIRTANEN, A.I. (1966). *Science*, **153**, 3744

DE VUYST, A., CHARLIER, H., VERVACK, W. and JADIN, V. (1971). *Rev. Ferment. Indus. Alim.*, **26**, 259

DE VUYST, A., VERVACK, W., CHARLIER, H. and JADIN, V. (1972). *Rev. Ferment. Indus. Alim.*, **27**, 5

WAKELING, A.E., LEWIS, D. and ANNISON, E.F. (1970). *Proc. Nutr. Soc.*, **29**, 60A

WHITING, F.M., STULL, J.W., BROWN, W.H. and REID, B.L. (1972). *J. Dairy Sci.*, **55**, 983

WILLIAMS, A.R. and SMITH, R.H. (1973). *Proc. Nutr. Soc.*, **32**, 52A

WOLFF, J.E., BERGMAN, E.N. and WILLIAMS, H.H. (1972). *Am. J. Physiol.*, **223**, 438

YOUNG, V.R. and SCRIMSHAW, N.S. (1972). in *Protein and Amino Acid Functions*. **Vol.II.** Ed. E.J. Bigwood. p.541. Pergamon Press New York

ZIMMERMAN, R.A. and SCOTT, H.M. (1965). *J. Nutr.*, **87,** 13

16

THE NUTRITIONAL AND METABOLIC EFFECTS OF AMINO ACID IMBALANCES

Q.R. ROGERS
University of California, Davis, California, U.S.A.

Nutritional Response to an Amino Acid Imbalance

Although various workers had shown that certain amino acids when added in excess to animal diets inhibited growth, it was not until after the last essential amino acid had been discovered and the amino acid requirements had been determined that the effect of balance of amino acids could be effectively examined. The concept of amino acid imbalance evolved during the late 40's and early 50's (Hier, Graham and Klein, 1944; Elvehjem and Krehl, 1947; 1955), and was based primarily on observations of the growth-depressing effect of a mixture of one or more amino acids lacking tryptophan to a niacin-free diet. An example of the tryptophan imbalance (niacin-free diet) is shown in *Figure 16.1* (Pant, Rogers and Harper, 1972a). Note that there is no adaptation to the imbalanced diet as measured by growth and after a month the animals begin to show signs of a niacin deficiency. The results in *Figure 16.2* (Pant, Rogers and Harper, 1972a) show that when a diet adequate in niacin but tryptophan-imbalanced is fed, the growth depressing effect occurs much more rapidly but after a few days the animals adapt and begin to grow. The extensive work of Harper and co-workers (Harper, Benevenga Wohlhueter, 1970) has clearly shown that amino acid imbalances are a general phenomena and can be produced whether the limiting amino acid is tryptophan or some other essential amino acid. Indeed, in a general way amino acid imbalances have been shown to result from the addition to a low-protein diet of one or more amino acids, other than the growth-limiting one, in amounts that are not individually toxic and cause depressions in food intake and growth that are readily prevented by a supplement of the growth-limiting amino acid (Harper, Benevenga and Wohlhueter, 1970). It should be emphasised that this is a nutritional description (i.e., based on the response and not a particular degree of disproportionality among the amino acids). Since the tryptophan imbalance in a niacin-free diet results metabolically in niacin deficiency symptoms (Pant, Rogers and Harper, 1972a; 1972b) rather than a 'typical' amino acid imbalance response, it might appear that the apparent adverse effect has a different metabolic origin. However, it now appears that there is one common initial metabolic response to an amino acid imbalance; that is, a positive anabolic

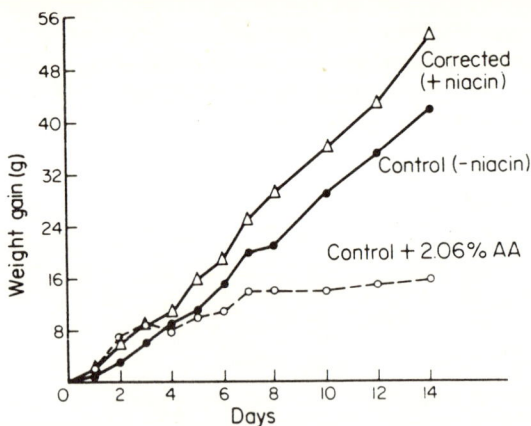

Figure 16.1 Weight gain of weanling rats (six rats/group) fed a niacin-free, tryptophan-imbalanced diet. Control: 8 per cent casein + 0.3 per cent L-methionine (niacin-free). Imbalanced: control + 2.06 per cent amino acid mixture minus tryptophan. Corrected: imbalanced + 2.5 mg niacinamide/100 g diet. Used with permission from Pant, Rogers and Harper (1972a)

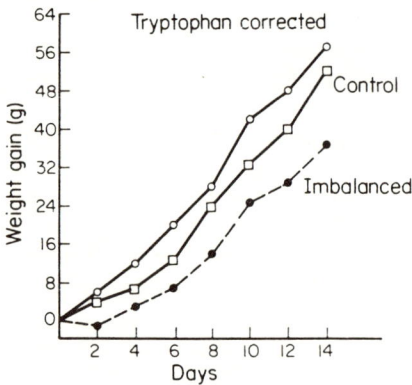

Figure 16.2 Weight gain of male weanling rats (five rats/group, initial weight 58 g) fed a tryptophan-imbalanced diet adequate in niacin. Control: 8 per cent casein + 0.3 per cent L-methionine (+ niacin). Imbalanced: control + 7.8 per cent amino acid mixture minus tryptophan. Tryptophan corrected: imbalanced + 0.1 per cent L-tryptophan. Used with permission from Pant, Rogers and Harper (1972a)

response resulting in a stimulation of protein synthesis (primarily hepatic). In general this results in a drop in the body pool of the limiting amino acid; whereas, in the animal fed the niacin-free tryptophan-imbalanced diet, the net effect also results in a niacin deficiency.

Before passing on to discussions of behavioural or metabolic effects of amino acid imbalances it is worth asking the question: are amino acid imbalances seen routinely or are they simply an experimental artifact? Since the concept and definition are nutritionally derived, based on the response of the animal to the imbalanced diet, compared with the response to a more balanced low protein diet, there is no way to be sure whether a given protein is imbalanced or not. However, if almost any series of multifactorial experiments is examined in which a given protein is being supplemented with various amino acids to improve the quality of the protein, amino acid imbalances often become apparent. Taking an example from a recent publication (*Table 16.1*, Mehansho *et al.*, 1973) it can be seen that when noog protein was supplemented with the second-limiting amino acid (lysine) rather than the first-limiting (threonine), rats fed the supplemented diet gained less weight than those receiving no supplement. Weight gains were restored, and indeed exceeded the control when threonine (the first-limiting amino acid) was added to the diet. This illustrates the nature

Table 16.1 Effect of lysine and threonine supplements on the growth of rats fed noog protein[1,2]

		Weight gain (g/10 day)
Control	noog-meal (10 per cent protein)	52
	noog-meal + 0.3 per cent threonine	56
Imbalance	noog-meal + 0.5 per cent lysine	39
Corrected	noog-meal + 0.3 per cent threonine + 0.5 per cent lysine	68

[1] noog (*Guizottia abyssinca*)
[2] Calculated from Mehansho *et al.* (1973)

and widespread existence of the phenomenon in supplementation studies. Theoretically an amino acid mixture could be designed to mimic a protein (e.g. wheat gluten, zein, etc.) and then the growth response of this mixture could be compared with that of the same mixture with all the excess amino acids removed. Although it might be predicted that growth would be improved, there does not appear to be any practical solution to this problem with intact proteins, except by the supplementation of deficient (or possibly imbalanced) proteins with the needed amino acids or with complementary proteins.

Although it has been helpful to think about unifying concepts in studying amino acid imbalances, it should not be forgotten that in amino acid nutrition there are 8-10 indispensable amino acids, each of which is involved not only to a different quantitative degree in protein synthesis but each of which has its unique catabolic pathway.

In addition, several of the indispensable amino acids also provide for other unique biosynthetic needs for the animal body. It should not be surprising therefore that even if there is one primary metabolic response to an imbalanced mixture of amino acids, there should be differences in degree of severity, and in extent and rapidity of adaptation, to various amino acid imbalances. If the general properties of the enzymes involved in the first step of protein synthesis are examined and compared with those of the enzymes of the catabolic pathways (*Table 16.2*) it is important to notice that the Michaelis constants (Km) of the catabolic enzymes for the essential amino acids are generally about 20-100 times greater than those of the respective amino acid activating enzyme. These general properties allow for highly efficient use of the limiting amino acid for tissue formation while providing the metabolic machinery for catabolising the excesses of non-limiting essential amino acids. This occurs because the tissue amino acid concentrations normally fall in between the Km values for each respective catabolic enzyme and amino acyl synthetase. The idea of control of amino acid metabolism by enzyme characteristics (Km) has been discussed by Krebs (1972) who emphasises that storage of amino acids does not occur, in contrast to carbohydrate and fat. The increased amino acid concentrations after a meal, therefore, automatically cause an increase in the rate of amino acid degradation. The stimulation of protein synthesis by an increase in the amino acid supply has 'first priority' (because of the low Km of amino acyl synthetases) and appears to be an essential part of the control of amino acid metabolism in order to replace proteins which are known to turn over at varying rates.

The adaptive increases for most of the catabolic enzymes are not caused by increases of their respective substrates in the diet but by the total protein in the diet or by glucocorticoids (Harper, 1965; Anderson, Benevenga and Harper, 1969; Nakano *et al.*, 1970; Kaplan and Pitot, 1970). The lack of capability to respond to excesses of individual amino acids results in severe metabolic aberrations, mental illnesses and even death, as evidenced by the multitude of genetic defects now known to be the result of deficiencies in single enzymes involved in amino acid catabolism (Nyhan, 1974). Less is known concerning the effect of increases of individual amino acids on the rate of charging of individual tRNAs for protein synthesis. This will be discussed later.

Often the response of an animal to an excess of one or more amino acids or protein, or to an amino acid imbalance or deficiency is referred to as an adverse effect. At this point I would like to re-emphasise a point made by Harper and co-workers (Harper, Benevenger and Wohlhueter, 1970) that the metabolic and behavioural response is not an adverse response; it is a protective response and undoubtedly is to the benefit of the animal. Perhaps the diet, or more specifically the amino acid pattern of the diet, should be referred to as an adverse pattern of dietary amino acids for meeting the animal's amino acid requirements. This follows even into the behavioural response of the animal. If the animal is in a cage and has

Table 16.2 Comparison of Michaelis constants for specific amino acid catabolic enzymes and amino acyl synthetases[1]

Amino acid	Catabolic enzyme	Km catabolic μM	Km acyl synthetase μM
Isoleucine	BCT[2]	840[3]	5[4]
	αKβMIVDH[5]	200[6]	
Methionine	Met-Ad. Tf[7]	900-2200[8]	500[9]
Homocysteine	Cysta Syn[10]	11000[11]	
Leucine	BCT[2], LAT[12]	750-25000[3]	0.01-100[13]
	αKICDH[14]	200[6]	
Lysine	Lys DH[15]	18000-200000[16]	5-25[17]
Phenylalanine	Phe Hydrox[18]	150-1100[19]	0.8-25[20]
Threonine	Thr DH$_2$O[21]	84000-93000[22]	4.3[23]
Tryptophan	Trp Diox[24]	50-300[25]	1.0*[26]
Valine	BCT[2]	4300[3]	100-190[4]
	αKIVDH[27]	200[6]	
Alanine	GPT[28]	9000-42000[29]	1000-4000*[30]
Arginine	Arginase	3000[31]	1.2-3.0[32]
Ornithine	OδAT[33]	600-7200[34]	
Asparagine	Asparaginase	4700[35]	(-)[36]
Aspartic acid	GOT[37]	360-3700[38]	(-)[39]
Cysteine	Cys DH$_2$S[40]	5500[41]	(-)[42]
	Cys Oxygenase[43]	740[44]	
Glutamic acid	GDH[45]	4350[46]	130[47]
Glutamine	Glutaminase	30000-100000[48]	18.5[49]
	Glm AT[50]	2000-2200[51]	
Glycine	Gly-methylene[52]	2800-10000[53]	600[54]
Histidine	Histidase	1800-3000[55]	(-)[56]
Proline	Pro DH[57]	2000[58]	430[59]
Tyrosine	TAT[60]	1500[61]	40*[62]
Serine	Ser DH$_2$O[63]	52000-91000[22]	500[64]
	Ser OH-me-tf[65]	540[66]	

[1] All values in this table are from rat liver except those listed with an asterisk in which case the source is indicated in the appropriate footnotes. The references for the source of each Km or 'apparent Km' are also listed in the footnotes. The enzymes listed for the catabolic enzyme were selected because they are thought to be the first step in the major catabolic pathway. When the first step is biologically reversible, the second enzyme is also given. [2] Branched Chain Transaminase. [3] Aki, Ogawa and Ichihara, 1968. [4] Hele and Barth, 1966; Yoshida, 1968. [5] Alpha keto beta methyl isovaleric dehydrogenase. [6] Wohlhueter and Harper, 1970. [7] Methionine adenosyl transferase. [8] Pan and Tarver, 1967. [9] Glenn, 1961. [10] Cystathionine synthetase. [11] Suda, Nakagawa and Kimura, 1971. [12] Leucine-glutamic amino transferase. [13] Vescia, 1967; Hele and Barth, 1966. [14] Alpha keto isocaproic dehydrogenase. [15] Lysine dehydrogenase. [16] Weiss and Pitot, 1969; Eggleston and Krebs, cited in Krebs, 1972. [17] Korzhov, 1965; Hele and Barth, 1966. [18] Phenylalanine hydroxylase. [19] Kaufman, 1970. [20] Shenoy, 1973; Lanks et al. (1971). [21] Threonine dehydratase. [22] Hoshino and Kröger, 1969 and Goldstein, Knox and Behrnean, 1962; Sayre and Greenberg (1966). [23] Allende et al., 1966. [24] Tryptophan dioxygenase. [25] Knox, Piras and Tokuyama, 1966; Schimke, 1970. [26] Davie, Konigsberger and Lipmann, 1956; from beef pancreas. [27] Alpha keto isovaleric dehydrogenase. [28] Alanine-glutamic amino transferase. [29] Swick et al., 1965. [30] Estimated from results of Webster, 1961; from pig liver. [31] Shimke, 1970. [32] Allende and Allende, 1964; Ikegami and Griffin, 1969. [33] Ornithine delta aminotransferase. [34] Peraino and Pitot (1963); Strecker (1965); Katunuma et al. (1964). [35] McGee, Greengard and Knox, 1971. [36] No mammalian Km for asparagine could be found. [37] Aspartic-glutamic amino transferase. [38] Boyd, 1961. [39] No mammalian Km for aspartic could be found. [40] Cysteine desulphydratase. [41] Estimated from results of Smyth, 1942. [42] No mammalian Km for cysteine could be found. [43] Cysteine oxygenase. [44] Lombardini et al., 1969. [45] Glutamic dehydrogenase. [46] diPrisco, Banay-Schwartz and Strecker, 1968. [47] Deutscher, 1967. [48] Katunuma, Huzino and Tomino, 1967. [49] Alford et al., 1963. [50] Glutamine amino transferase. [51] Yoshida, 1967. [52] Glycine-methylene THFA dehydrogenase. [53] Yoshida and Kikuchi, 1972. [54] Boyko and Fraser, 1964. [55] Spolter and Baldridge, 1963; Noda, 1970. [56] No mammalian Km for histidine could be found. [57] Proline dehydrogenase. [58] Kramar, 1971. [59] Fraser and Klass, 1963. [60] Tyrosine-glutamic amino transferase. [61] Litwack, 1966. [62] Clark and Eyzaguirre, 1962; Schweet, 1958; from hog pancreas. [63] Serine dehydratase. [64] Rouge, 1969. [65] Serine hydroxymethylase. [66] Nakano, Fujicka and Wada, 1968

284 Amino acid imbalances

no choice of diets, the protective response of the animal is to reduce the quantity of food ingested (*Figure 16.3*) (Leung, Rogers and Harper, 1968a, 1969). If the animal has a choice it will select a diet containing a balanced amino acid pattern in preference to a diet containing an imbalanced amino acid mixture (*Figure 16.4*). The rat will even select a protein-free diet rather than an amino acid-imbalanced diet (*Figure 16.4*, Leung, Rogers and Harper, 1968b; Harper, 1967). The animal shows a greater sensitivity to an amino acid imbalance in making these dietary

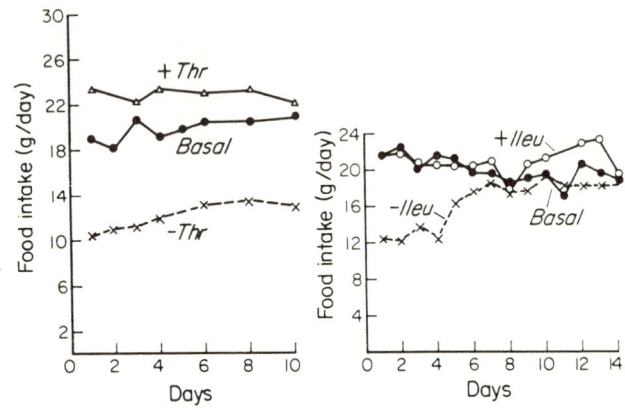

Figure 16.3 Food intake of rats fed ad libitum *a basal diet, threonine-imbalanced (-thr) or corrected (+thr) diets, isoleucine-imbalanced (-ileu) or corrected (+ileu) diets. Used with permission from Leung and Rogers (1969)*

choices than it does in responding through a depression in food intake; i.e., a change in dietary choice occurs with a lower proportion of a given imbalanced mixture of amino acids (Harper and Rogers, 1966). This would appear to have survival value if an animal in nature chooses foods containing a more adequate balance of amino acids and avoids the foods which result in a marginal deficiency of an essential amino acid. More detailed reviews are available concerning food intake depressions and dietary choices (Harper, 1967; Harper, 1970; Harper, Benevenga and Wohlhueter, 1970; Rogers and Leung, 1973).

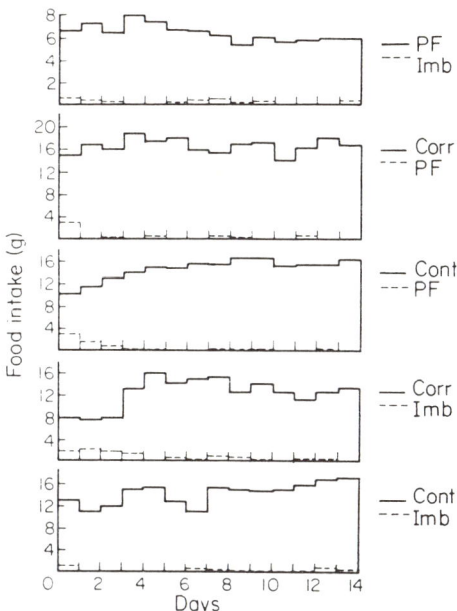

Figure 16.4   1. Food selection by rats fed imbalanced (Imb.) diet low in threonine and protein-free (PF) diet. 2. Food selection by rats fed the corrected (Corr.) diet and protein-free (PF) diet. 3. Food selection by rats fed the control (Cont.) diet and protein-free (PF) diet. 4. Food selection by rats fed imbalanced (Imb.) diet low in threonine and corrected (Corr.) diet. 5. Food selection by rats fed control (Cont.) diet and the threonine imbalanced (Imb.) diet. Control diet: 6 per cent casein plus 0.3 per cent DL-methionine; imbalanced diet: control plus 5.4 per cent amino acid mixture lacking threonine; corrected diet: imbalanced plus 0.45 per cent DL-threonine; and protein-free diet: same composition as the control diet except that protein and methionine were replaced by carbohydrate. Used with permission from Leung et al. (1968b)

Biochemical Response to an Amino Acid Imbalance

The key biochemical response of an animal to an imbalanced mixture of amino acids has not been easy to elucidate. One major problem complicating attempts to define clearly the major biochemical response is a common problem in biochemical nutrition studies in the whole animal, i.e. the rapid effect of the experimental condition on food intake. Often, the food intake depression can have a more profound effect on a given biochemical parameter than the nutritional change under study. This is illustrated in a nitrogen balance experiment (Kumta, Harper and Elvehjem, 1958) shown in *Table 16.3* in which it might first appear that there is a decreased efficiency of utilisation of the amino acid components of the diet (*ad libitum* imbalanced group). However, if the pair-fed controls are compared to the imbalanced group it appears as though there is also a lower efficiency of utilisation of

Table 16.3 Effect of amino acid imbalance on nitrogen balance of rats[1]

		Nitrogen retained (%)		
		Ad lib	Protein depleted	Meal-fed
6 per cent Fibrin	Control	66.4	59.9	65.9
6 per cent Fibrin + 0.4 per cent DL met + 0.6 per cent DL phe	Imbalance	44.6	43.7	53.3
Control, pair-fed	Pair-fed control	41.5	32.9	45.3

[1] Mean of six male (120-150 g) rats fed on a basal diet 4-5 d *ad libitum*, followed by the imbalanced diet. Used with permission from Kumta, Harper and Elvehjem (1958).

nitrogen. These types of experiments have not been very conclusive or satisfying. It soon became apparent that to avoid this problem it would be necessary to look for some biochemical change that would occur prior to the food intake response. In retrospect this is a logical approach since the food intake depression must be a result of the metabolic response of the animal to the amino acid imbalanced diet. One effective way to study the acute effect of a change of diet on a variety of metabolic parameters is to train animals to become meal-eaters and then substitute an experimental diet for a normal meal. Although some metabolic parameters change with a meal-feeding regime, (e.g., lipogenesis; Tepperman and Tepperman, 1958) the meal-feeding regime does allow pair-feeding the experimental group a normal size meal. This procedure has been used successfully by several research groups (Yoshida *et al.*, 1966; Benevenga, Harper and Rogers, 1968; Soliman and King, 1969; and Pronczuk, Rogers and Munro, 1970).

The most consistent early change in a biochemical parameter has been the decrease in the concentration of the limiting amino acid in blood plasma and in muscle. A marked drop in the concentration of the limiting amino acid in blood plasma has been found in both *ad libitum*-fed rats (Hill and Olsen, 1963; Sanahuja and Harper, 1963; Anderson, Benevenga and Harper, 1969; Peng, Tews and Harper, 1972) and meal-fed rats (Kumta and Harper, 1962; Leung, Rogers and Harper, 1968c). The decrease in plasma threonine concentration in the meal-fed rats fed a threonine-imbalanced diet (Leung, Rogers and Harper, 1968c) is not as great as the decrease in plasma histidine concentration in rats fed a histidine-imbalanced diet (Kumta and Harper, 1962), however, the decrease occurred early and in the same experiment the threonine concentration dropped even more in muscle (*Figure 16.5*). From these results it is not possible to tell whether the plasma threonine concentration is low as a result of the low muscle threonine or vice-versa. There is not as much consistency in the decrease of the limiting amino acid in visceral organs but results are accumulating which indicate

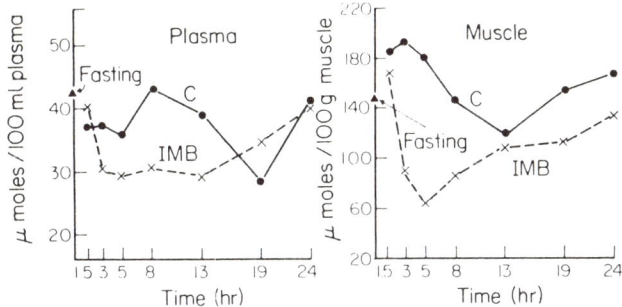

Figure 16.5 Effect of amino acid imbalance on threonine concentrations in plasma and muscle of rats fed the control diet (C) and the Imbalanced diet (IMB).   Control diet: 6 per cent casein plus 0.3 per cent DL-methionine; and imbalanced diet: 6 per cent casein plus 0.3 per cent DL-methionine and 5.4 per cent amino acid mixture lacking threonine.   Used with permission from Leung, Rogers and Harper (1968c)

a decrease in liver and in brain also (Ellison and King, 1968; Ip and Harper, 1974b; Shenoy, 1973; Peng, Tews and Harper, 1972).

In attempts to determine what the fate of the limiting amino acid is in rats fed amino acid-imbalanced diets, several investigations have followed the fate of carbon-14 or nitrogen-15 from a labelled limiting amino acid (Wilson et al., 1962; Florentino and Pearson, 1962; Yoshida et al., 1966; Hartman and King, 1967; Benevenga, Harper and Rogers, 1968; Soliman and King, 1969). In the first two studies referred to above, there was no measurement of the incorporation of the limiting amino acid into protein. The results are therefore difficult to reconcile since one group found a decrease in the catabolism of the limiting amino acid (tryptophan) to CO_2 (Wilson et al., 1962) while the other found an increase (Florentino and Pearson, 1962). Both of these imbalances were niacin-free tryptophan imbalances. Results from experiments uncomplicated by niacin deficiency have been more consistent concerning the short-term effect of an imbalanced diet on the catabolism of the limiting amino acid (Yoshida et al., 1966; Benevenga, Harper and Rogers, 1968; Soliman and King, 1969) i.e., there was a small decrease in the catabolism of the most limiting amino acid in rats fed the imbalanced diet. The distribution of the limiting amino acid six hours after feeding the control, imbalanced and corrected diets is shown in *Table 16.4*. There was somewhat less radioactivity from the limiting amino acid in the muscle and carcass of the rat fed the imbalanced diet at six hours whereas there was more radioactivity in liver, kidney, intestine and blood plasma of rats fed the imbalanced diet compared with that found in rats fed the control or corrected diets. As can be seen in *Figure 16.6,* between two and four hours after a meal, there is a cross-over of radioactivity from the limiting amino acid between the soluble free amino acid pool and the total body protein pool. Notice that the largest and

Table 16.4 *Distribution of radioactivity, as a percentage of that ingested, six hours after feeding rats on a control, histidine-imbalanced or corrected diet containing histidine-U-^{14}C*[1]

	Basal	Imbalance	Corrected
Liver	17.2	19.5	18.6
Kidney	2.5	2.9	2.2
Intestine	24.9	25.7	19.5
Blood plasma	6.7	7.9	5.4
Muscle plus carcass	41.3	37.6	42.2
CO_2, urine, faeces	6.0	5.1	11.4
Total	98.6	98.7	99.3

[1] Calculated from results of Soliman and King, 1969. Six male, 50 g, Sprague-Dawley rats were fed the basal diet twice a day for 1 h to train them to become meal-eaters. After 10 days, following a 12 h starvation period, basal, imbalanced or corrected diets labelled with histidine-U-^{14}C (about 3 µCi/g diet) were fed to two rats for 1 h. After a six hour interval the rats were killed.

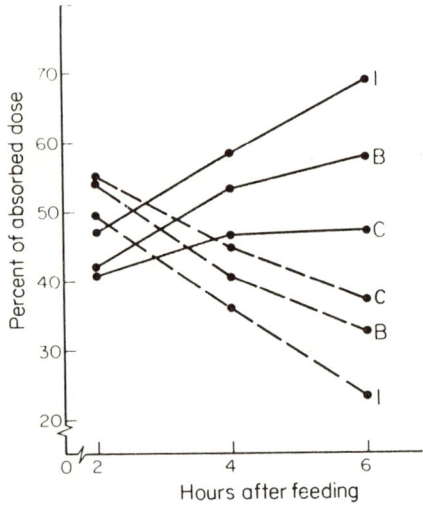

Figure 16.6 *Percentage of absorbed histidine-U-^{14}C dose retained in all the TCA-protein precipitates (——) and TCA-soluble fractions (---) after feeding basal (B), histidine-imbalanced (I) and corrected (C) diets. Used with permission from Soliman and King (1969)*

earliest shift occurs in rats fed the imbalanced diets. This corresponds to the time in which there is a marked decrease in the concentration of the limiting amino acid in plasma and other tissues (*Figure 16.5*). The most consistent increase in incorporation of the limiting amino acid among the various tissues has been found in the liver. There is some evidence that an increased incorporation of the limiting amino acid into intestinal tissue occurs even earlier than that of the liver (Benevenga, Harper and Rogers, 1968; Soliman and King, 1969). Within the liver, the increased incorporation has been found in the protein fraction (*Table 16.5*). This increase in incorporation has been

Table 16.5 Effect of amino acid imbalance on the incorporation of the limiting amino acid into liver protein six to eight hours after feeding

Limiting amino acid	Units	Incorporation into liver protein			References
		Control	Imbalance	Corrected	
Thr-U-^{14}C	dpm/mg protein	926	989		Yoshida et al., 1966
His-U-^{14}C	dpm/mg protein	1638	2402		Yoshida et al., 1966
His-U-^{14}C	dpm/mg protein	3260	5150	4160	Benevenga, Harper and Rogers, 1968
His-U-^{14}C	per cent of dose	12.2	16.1	11.1	Soliman and King, 1969

Yoshida et al.: meal-fed for two hours daily. Fed 7 g of diet with 8.3 µc of isotope per rat. Benevenga et al.: meal-fed for 1.5 h daily. Fed 5 g of diet with 13 µCi isotope. Soliman and King: meal-fed for one hour twice daily. Fed 3.0 µCi isotope per g diet

as much as 50 per cent. The 'apparent' decrease in rats fed the corrected diet is a result of the dilution of the label with the histidine added to the diet to correct the imbalance. If these dilutions are taken into account the corrected group has the highest incorporation of the limiting amino acid into hepatic protein, as might be expected (Benevenga, Harper and Rogers, 1968).

Before examining the biochemical effect of an imbalanced mixture of amino acids in further detail it is worth remembering the dynamic nature of the homeostatic response of the liver to a balanced pattern of amino acids. Even when there is enough protein in the diet for maximal growth, there is a diurnal shift in the extent of hepatic polysome aggregation which is diet-induced (Wurtman, 1970; Symmons, Maguire and Rogers, 1972). This is illustrated in *Figure 16.7* for both *ad libitum*-fed rats and meal-fed rats. If animals are adapted to a basal diet and then fed an amino acid-imbalanced diet or a corrected diet, all the diets cause polysome re-aggregation but the corrected diet is most effective (*Table 16.6*). The polysome patterns after feeding the imbalanced diet were more similar to those of the control than those of the corrected, however there always tended to be a somewhat higher percentage of hepatic polysomes in rats fed the imbalanced diet than in rats fed the control diet. These animals were adapted to their corresponding control diets for at least a week prior to the time of the test meal, therefore each group was depleted of its respective limiting amino acid. If, however, rats were adapted to a threonine-limiting control diet and then were fed an isoleucine-imbalanced diet, the isoleucine-imbalanced diet was as effective in causing polysome aggregation as the corrected diet (Pronczuk, Rogers and Munro, 1970). This latter phenomenon has been studied in more detail by Ip and Harper (1974a) who measured both polysome patterns and ^{14}C-leucine incorporation into hepatic protein after feeding threonine-imbalanced diets with and without prior threonine depletion. As can be

290 *Amino acid imbalances*

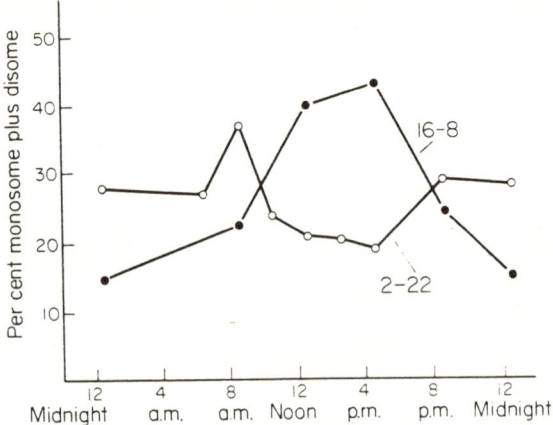

Figure 16.7 The time course of fluctuation of hepatic polysome profiles of rats meal-fed (two h/day, 2-22) control-fed (16 h/day, 16-8) a 15 per cent casein diet. Meal-fed animals were fed from 8.30 a.m. to 10.30 a.m.; control-fed animals were fed from 4.30 p.m. to 8.30 a.m. Each point represents the average of two to four animals. Used with permission from Symmons, Maguire and Rogers (1972)

Table 16.6 *Effect of feeding amino acid imbalanced diets on polysome patterns one or four hours after feeding*

	Before feeding	Isoleucine study (% monosome + disome)		Threonine study	
		1 h	4 h	1 h	4 h
Control	50	40	50	47	46
Imbalance	-	39	46	46	43
Corrected	-	27	41	31	35

Used with permission from Pronczuk, Rogers and Munro (1970).
Male albino 150 g rats were trained to eat during a daily two hour period using a 15 per cent casein diet for 10 days, followed by the appropriate amino acid control diet for 4 days prior to feeding the imbalanced diet. On the day of killing, the meal-fed rats were given 7 g of either control, imbalanced or corrected diet, which was consumed within 1 h.

seen in *Table 16.7*, the imbalanced mixture of amino acids was just as effective (two hours after feeding) in stimulating re-aggregation of hepatic polysomes and increasing the incorporation of ^{14}C-leucine into hepatic protein as the complete mixture of amino acids if the limiting amino acid was not pre-depleted. If threonine was previously depleted, however, the addition of the imbalanced mixture of amino acids did not cause any increase in either parameter over that found when the control diet was ingested. These effects are reminiscent of the effects seen by Munro and co-workers (Wunner, Bell and Munro, 1966) several years ago but place in perspective the role of the limiting amino acid in the diet in the protein synthetic response. It might be well to re-emphasise that dietary amino acids influence the rate of protein synthesis very quickly after each meal by modifying not only the flow

Table 16.7 *Effect of depleting rats of threonine on liver polysome aggregation and incorporation of ^{14}C-leucine into liver protein two hours after feeding rats a threonine-imbalanced diet*

	Polysome patterns % monosomes + disomes		Incorporation of ^{14}C-leucine (cpm/mg protein)	
	Pre-fed low protein diet			
Test meal	Balanced	Low thr	Balanced	Low thr
Control	37.0	40.5	401	344
Imbalance	28.0	38.7	565	369
Corrected	26.0	24.0	613	507

Used with permission from Ip and Harper (1974a).
Male 80 g, albino rats were trained to eat their food (9 per cent casein diet) in two hours daily. They were then fed either a threonine-limiting diet or a threonine-adequate diet as a three per cent agar gel. On the morning of the experiment the meal-fed rats were allowed to eat five g of either basal, threonine-imbalanced, or corrected diet and were killed two hours after feeding was initiated. Each animal was injected with 2.5 μCi of L-^{14}C (U)-leucine intraperitoneally 0.5 h before being killed

of substrates but also by modifying the protein synthetic machinery (see Munro, 1970). Ip and Harper (1974a) measured the concentration of free amino acids in the liver and found that the decrease in polysome aggregation and the decreased incorporation of leucine into protein was associated with a marked decrease in the concentration of liver threonine.

What is the mechanism of this response? In other words, how does the depletion of the limiting amino acid limit the extent of re-aggregation of polysomes when an amino acid mixture is ingested? The most obvious answer would appear to be that the limiting amino acid concentration must decrease to below that necessary to maintain tRNA charging at a fast enough rate to maintain the existing rate of protein synthesis. This would result in a decrease in the rate of elongation prior to the insertion of the last (the one closest to the carboxyl terminal end) limiting amino acid in each protein being made by a given polysome. This might cause a run-off of only a few or a large number of the ribosomes from an average mRNA and thereby markedly increase the monosome pool. If no other mechanism is invoked, however, a much smaller decrease in the proportion of polysomes would be predicted. Since the first part of the above mechanism fits the observed results, but the medium and heavy polysomes always decrease more than would be predicted, some other mechanism must be present.

Examining the first assumption above; from the liver threonine concentration measured at a time when polysome patterns are not fully aggregated because of threonine limitation, the threonine concentration has been found to be 112-578 × 10^{-6} M (Ip and Harper, 1974a; 1974b; Shenoy, 1973). The reported Km for rat liver threonine acyl synthetase is 4.3 × 10^{-6} M (Allende et al., 1966). Therefore, at the lowest liver concentration measured, the rate of charging of tRNA should be above 95 per cent of V_{max}. If the same calculation is made for other imbalances and deficiencies there always appears to be a high enough concentration of the limiting amino acid to have 75 per cent or higher of the V_{max} for that particular amino acyl synthetase.

Allen, Raines and Regen (1969) have measured the extent of tRNA charging in rats under various physiological states. It can be seen from *Table 16.8* that the extent of charging of tRNA's taken as a whole is not affected by food deprivation or the stomach-tubing of a complete or incomplete mixture of amino acids. It can be seen from the last

Table 16.8   Average per cent charging of 17 tRNAs from rat liver[1]

	Charging (%)	Range	Polysome[4] aggregation
Chow-fed	86	(75-97)	+
24-h food deprived	87	(78-95)	–
Complete AA mix[2]	88	(67-100)	+
Incomplete AA mix[3]	86	(72-94)	+

[1] Calculated from Allen, Raines and Regen, (1969)
[2] 24 h food deprived rats, stomach-tubed a complete amino acid mix
[3] The average of five-to-six experiments in which an amino acid mixture lacking one indispensable amino acid was stomach-tubed into 24 h deprived rats. Per cent of charging of tRNA for amino acid which was lacking from the mixture was excluded from the average
[4] Taken from Pronczuk et al. (1968) and Sidransky et al. (1967)

column that there appears to be no correlation between the proportion of charging of tRNA's as a whole and polysome aggregation. The results in *Table 16.9* illustrate that the extent of charging of the specific hepatic tRNA for each of five of the essential amino acids decreases after an amino acid mixture lacking each amino acid was stomach-tubed. Arginine and apparently lysine and leucine showed no decrease in the proportion charged when each was missing from the

Table 16.9   Average per cent charging of tRNA of indispensable amino acids in liver

tRNA class	Charging (%)		Polysome aggregation[3]
	When present in mix[1]	When absent from mix[2]	
Val	92	49	+
Ile	80	52	+
Arg	90	87	no data
Met	90	67	+
Trp	84	31	–
Phe	81	49	+
Leu	89	85[4]	+
Lys	88	86[4]	+

[1] Calculated from Allen, Raines and Regen, (1969). Average of six to seven experiments, one amino acid missing in each experiment but that tRNA excluded from the average
[2] Per cent charging of indicated tRNA from a single experiment from pooled livers
[3] Taken from Pronczuk et al. (1968) and Sidransky et al. (1967)
[4] Individual values from single livers

amino acid mixture. The experimental treatment of the rats was the same as that of Wunner, Bell and Munro (1966) in which it was found that only when they stomach-tubed an amino acid mixture lacking tryptophan (*Table 16.9*) did polysomes fail to aggregate. This would indicate that simply a decrease in the extent of charging of any tRNA with its respective amino acid does not automatically result in a lack of aggregation (or disaggregation). In the work of Allen, Raines and Regen, (1969) the concentration to which each amino acid in liver had to decrease before its respective tRNA was not fully charged was not determined.

We have attempted to determine the extent of tRNA charging under various physiological conditions, including the feeding of amino acid-imbalanced diets. We have used a modification of the method used by Allen, Raines and Regen (1969) and have determined the per cent of charging of tRNA's for leucine, isoleucine, phenylalanine and tryptophan. The effect of 48-hour starvation on the extent of charging of the four tRNA's is presented in *Table 16.10*. Among the four tRNA's, there was a decrease in charging only with tRNA tryptophan. Allen, Raines and Regen (1969) did not find a decrease in the extent of charging of any of the tRNAs after 24 hours of starvation. These results are not consistent with the hypothesis that tRNA charging may in some way give a signal for polysome aggregation since polysomes disaggregate before 24 hours of food deprivation in the rat (Symmons, Maguire and Rogers, 1972). The results may indicate that tryptophan is the amino acid which becomes limiting for liver protein synthesis

Table 16.10   Charging levels of tRNA in livers of fed or 48 h food-deprived rats

	Charging level (%)	
	Fed	Starved
Phe-tRNA	91 (81)	99 (86)
Ile-tRNA	79 (81)	78 (79)
Trp-tRNA	90 (78)	65 (78)
Leu-tRNA	91 (89)	83 (87)

[1] Shenoy (1973). The figures in parentheses are from Allen *et al.*, - 24 h starvation

after prolonged starvation; but shed little light on possible mechanisms relating to polysome aggregation. It might be well to point out that although the percentage of charging remains high, the total acceptor capacity does decrease during starvation (*Table 16.11*) even when expressed per gram of liver. Perhaps it should not be surprising to find that all tRNAs are as fully charged in livers of starved rats compared with fed rats since the concentrations of most of the essential amino acids do not decrease during starvation to near their Km values (*Table 16.12*). From the last column in *Table 16.12* it can be seen that the concentrations of the essential amino acids are generally 20-100 times that of each respective Km for the amino acid activating enzyme. The reported Km for methionyl acyl synthetase of rat liver

294 *Amino acid imbalances*

Table 16.11 Liver weights and acceptor capacities of Phe-tRNA from livers of 48 h food-deprived rats[1]

	Liver weight of fed control (%)	Acceptor capacity of fed control (%)
Experiment 1	51	38
Experiment 2	66	57

[1] Shenoy (1973)

Table 16.12 Range of concentrations of amino acids in livers of fed or food deprived rats

	Amino acid concentrations[1] (μmoles/kg liver)	Km of amino acyl syn.[2] μM
Arginine	Trace-40	1.2-3.0
Histidine	270-1170	–
Isoleucine	112-280	5.0
Methionine	65-180	500
Leucine	225-690	0.01-100
Lysine	410-1180	5-25
Phenylalanine	92-360	0.8-25
Threonine	220-1050	4.3
Tryptophan	80-150	1.0
Valine	267-650	100-190

[1] High and low values from three experiments in which pooled (from three to six rats) liver samples were analysed from rats fed stock diet or starved for 24 or 48 h.
[2] See *Table 16.2* for references

is 500 (Glenn, 1961), which is higher than the observed concentration of methionine in rat liver (*Table 16.12*). Since the Km values for the other amino acyl synthetases are considerably lower, the high value for methionine needs to be verified. Arginine presents a special problem in measurement. Because the concentration in liver is considerably lower than that in blood, the contamination of liver with blood can cause erroneously high values. A second major problem causing an error in the opposite direction is the rapid disappearance of arginine caused by the high activity of hepatic arginase. Although there is considerable uncertainty, the hepatic arginine concentration appears to be about 5-10 times its Km for the activating enzyme.

In an attempt to determine whether the tRNA for the limiting amino acid in the diet was fully charged and to determine the effect of an amino acid imbalance on the extent of charging of the tRNA of the limiting amino acid we have run two types of experiment, one in *ad libitum*-fed rats and the second in rats adapted to meal feeding. All the animals were first accustomed to a low amino acid diet in which isoleucine was limiting and then on the day of the experiment the rats were fed either the same isoleucine-limiting control diet, an isoleucine amino acid-imbalanced diet or a corrected diet (Shenoy, 1973). The results of these two experiments are shown in *Table 16.13*.

Table 16.13 Charging levels of tRNA of the limiting amino acid in livers of rats fed amino acid imbalanced diets

Diet	Per cent charging of Ile tRNA[1]	
	Ad lib-fed[2]	Meal-fed[3]
Basal	65 (47)[1]	68 (71)
Ile imbalance	90 (14)	65 (31)
Corrected	89 (47)	87 (112)

[1] Taken from Shenoy (1973), isoleucine concentration in the liver is given in parentheses in μM
[2] Six pooled livers from 200 g rats. During the latter half of the 12 h feeding period food was withheld for five hours and rats were then fed two and a half hours before killing. Food intake was respectively 2.3 g, 2.0 g and 2.6g.
[3] Six pooled livers from 165 g rats meal-fed 7 g of diet and killed four hours after feeding

The feeding of the low amino acid diet limiting in isoleucine resulted in a depression of the proportion of charging of tRNA-isoleucine. Charging was about 65 per cent whether the rats were fed *ad libitum* or meal-fed. There was no change in the extent of charging of the tRNA in meal-fed rats fed an isoleucine-imbalanced diet whereas there was a return to normal in the extent of charging of tRNA-isoleucine after feeding the *ad libitum*-fed rats the isoleucine-imbalanced diet (*Table 16.13*). Also, as expected, the feeding of the corrected diet (isoleucine-supplemented) restored the extent of charging of tRNA-isoleucine. The concentrations of isoleucine found in the liver of each experimental group are shown in parentheses in *Table 16.13*. It is not very satisfying to postulate that the concentration of isoleucine in the liver determines the extent of charging of tRNA-isoleucine. Indeed it would appear that something other than simply the concentration of isoleucine is influencing the extent of charging of the tRNA-isoleucine. Probably the most important variable is the rate of protein synthesis (i.e. ile-tRNA^{ile} utilisation). Whereas the Km and Vm of the activating enzymes are both constants (the quantity of the activating enzymes can change) it would appear that neither are normally limiting or normally regulating the rate of protein synthesis. Since many enzyme activities are modified by allosteric effectors it is possible that when more is known about the kinetic properties of individual tRNA species an explanation of the above results will be obvious. An alternate explanation may also become apparent when the rate of elongation can be measured in such a way as to allow an assessment of the rate of utilisation of ile-tRNA-isoleucine, compared with the maximum velocity *in vivo* of isoleucyl activating enzyme. Perhaps ile-tRNA^{ile} utilisation (where ile refers to isoleucine) simply exceeds synthesis under certain physiological conditions. This explanation would account for why, under various physiological conditions there is no correlation between the extent of tRNA charging and polysome aggregation. If a lack of charging of one tRNA does not contribute to polysome disaggregation or markedly reduce the rate of protein synthesis then it might be predicted that a lack of complete charging of a given tRNA when the amino acid for that tRNA is clearly limiting in the diet, could be shown. This might occur at more

296 *Amino acid imbalances*

than one degree of limitation of the amino acid in the diet. It would be expected, however, that the concentration of the limiting amino acid would decrease to closer to the apparent Km for that particular amino acid than apparently occurs.

Addition of amino acids to an *in vitro* protein synthetic system in which ^{14}C-phe-tRNA^{phe} (where phe is phenylalanine) is used as the labelled substrate has been shown to provide a small degree of stimulation of the ^{14}C-phe-tRNA^{phe} into protein, indicating one stimulatory effect of amino acids apart from their substrate function (*Table 16.14*). On the other hand, uncharged tRNA added to the *in vitro* system caused only a small depression of incorporation of ^{14}C-phe-tRNA^{phe} into protein (elongation). From these preliminary results it would be predicted that if this were the only effect of the decreased charging of a single tRNA, then the extent of depression of protein synthesis which would be predicted from the extent of depression of tRNA charging which we found after giving a diet limiting in a single essential amino acid (*Table 16.13*) would be about 10 per cent as compared with that found after giving an adequate amino acid mixture.

Table 16.14 Effect of amino acids and uncharged tRNA on the transfer of phenylalanine from Phe-tRNA into protein in cell free system from rat liver[1]

Additions to complete[2] system	Charging of tRNA[3] (%)	Protein synthesis (% of control)
None	100	100
5X amino acids[4]	100	113
1X uncharged tRNA[5]	50.1	84
2X uncharged tRNA[5]	33.2	75
3X uncharged tRNA[5]	26.6	66

[1] Shenoy (1973)
[2] Complete system includes ^{14}C-Phe-tRNA (1.5 O.D. units equivalent to 9000 dpm), 500 µg protein elongation factors, 250 µg ribosomal RNA, 0.4 µmoles of GTP, 12 µmoles MgCl$_2$, 160 µmoles NH$_4$Cl, 8 µmoles GSH, 100 µmoles of tris buffer, pH 7.4 in a total of 1.0 ml
[3] The first two groups were assumed to have tRNA 100 per cent charged whereas the extent of charging of the last three groups are measured values.
[4] Five times normal tissue concentrations (*Table 16.12*)
[5] 1.5 O.D. units, 3.0 O.D. units and 4.5 O.D. units of uncharged tRNA respectively

A new development important to the problem has recently been found relating the control of initiation of protein synthesis. Vaughan and Hansen (1973) have obtained evidence suggesting that uncharged tRNA, in general, decreases the rate of initiation in cultured human cells. Their results indicate that when the degree of charging of tRNA^{his} (his meaning histidine) or tRNA^{ile} is reduced by 20 per cent (by use of inhibitors) there is a four-fold or more reduction in the relative rate of translational initiation. They postulate that this reduction is a result of the formation of a regulatory repressor complex. If this type of a system were present in the liver then a decrease in the extent of charging of one tRNA would cause a breakup of polysomes with a concomitant increase in monosomes. However, this cannot be the major control component (i.e. via uncharged tRNA) of polysome aggregation since most of the conditions known to cause polysome disaggregation are not known to be associated with a decrease in charging of one or more tRNAs. The prospects for answers to just how the amino acid supply influences polysome aggregation and protein synthesis appear now to be in the elucidation of the control of initiation.

Before concluding, it is worth commenting on the metabolic effects of an excess of a single amino acid. The general effects and comparative toxicities, both acute and chronic, have been reported previously and have been reviewed (Gullino et al., 1955; Sassen, 1955; Sauberlich, 1961; Harper, Benevenga and Wohlhueter, 1970; Harper, 1973). Evidence is accumulating that administration of one of almost any of the essential amino acids will stimulate hepatic protein synthesis if the animal or the tissue preparation is in the proper physiological condition and the proper quantity is given (Sidransky et al., 1967; Sidransky, Verney and Sarma, 1971; Sarma et al., 1971; Park, Henderson and Swan, 1973; Pronczuk et al., 1968; Pronczuk, Rogers and Munro, 1970; Ip and Harper, 1974a,b; Hanking and Roberts, 1965a,b).

It should be pointed out that the stimulation which occurs is not uniform among all tissues. For example, it is known that high quantities of essential amino acids cause disaggregation of brain polysomes (Aoki and Siegel, 1970); cause inhibition of protein synthesis in the brain both *in vivo* and *in vitro* (MacInnes and Schlesinger, 1971, Peterson and McKean, 1969); and cause an inadequate brain development (Agrawal, Bone and Davison, 1970). The mechanism of how the stimulation of protein synthesis in the liver occurs is apparently different when the limiting amino acid is given as compared with giving an excess of an amino acid which is not limiting. Tryptophan (Pronczuk et al., 1968; Sidransky et al., 1967; 1971) was the first amino acid known to stimulate protein synthesis *in vivo* when administered to fasted or fed mice or rats. More recently Park, Henderson and Swan (1973) have shown that the administration of either methionine or phenylalanine (but not threonine or alanine) will also cause polysome aggregation in food-deprived rats. All of the essential amino acids have not been tested at high enough levels to know how extensive this effect is.

Although the mechanism of the above effect has not been elucidated it has been shown that the effect still occurs in adrenalectomised animals (Symmons, 1974; thus eliminating the indirect effect of specific amino acids in causing a release of corticosterone as shown by Munro, 1965) and that the amino acid acts to stimulate protein synthesis, even when it is not limiting in the diet (Sarma et al., 1971). Hanking and Roberts (1965b) have shown, using a liver slice preparation in which an increased rate of protein synthesis occurs as a result of high concentrations of phenylalanine, that the 'supranormal' concentrations of phenylalanine also resulted in an increased tissue uptake and flux of some amino acids while they had no effect or decreased that of others. From these results it would appear that these stimulatory effects might also be explained on the basis of an effect of amino acids on some part of the protein synthetic process, possibly initiation. Perhaps when the molecular explanation to the effects of specific amino acids on polysome aggregation and protein synthesis is found, we will then also have the explanation for the metabolic response of an animal to an amino acid imbalance.

Acknowledgements

1. The author is grateful to Dr. Alfred E. Harper for helpful suggestions and pertinent criticisms of the manuscript and to Miss Jane Bishop for her cheerful technical assistance and for her persistence in the literature search for the preparation of *Table 16.2*.

2. Unpublished results of the author included in this review are from research supported in part by Public Health Service Grant AM 11066.

References

AGRAWAL, H.C., BONE, A.H. and DAVISON, A.N. (1970). *Biochem. J.,* **117,** 325

AKI, K., OGAWA, K. and ICHIHARA, A. (1968). *Biochim. biophys. Acta.,* **159,** 276

ALFORD, M.A., BROTMAN, M., CHUDY, M.A. and FRASER, M.J. (1963). *Can. J. Biochem. Physiol.,* **41,** 1135

ALLEN, R.E., RAINES, P.L. and REGEN, D.M. (1969). *Biochim. biophys. Acta.,* **190,** 323

ALLENDE, C.C. and ALLENDE, J.E. (1964). *J. biol. Chem.,* **239,** 1102

ALLENDE, C.C., ALLENDE, J.E., GATICA, M., CELIS, J., MORA, G. and MATAMALA, M. (1966). *J. biol. Chem.,* **241,** 2245

ANDERSON, H.L., BENEVENGA, N.J. and HARPER, A.E. (1969). *J. Nutr.,* **97,** 463

AOKI, K. and SIEGEL, F.L. (1970). *Science,* **168,** 129

BENEVENGA, N.J., HARPER, A.E. and ROGERS, Q.R. (1968). *J. Nutr.,* **95,** 434

BOYD, J.W. (1961). *Biochem. J.,* **81,** 434

BOYKO, J. and FRASER, M.J. (1964). *Can. J. Biochem.,* **42,** 1677

CLARK, J.M. Jr. and EYZAGUIRRE, J.P. (1962). *J. biol. Chem.,* **237,** 3698

DAVIE, E.W., KONIGSBERGER, V.V. and LIPMANN, F. (1956). *Archs. Biochem. Biophys.,* **65,** 21

DEUTSCHER, M.P. (1967). *J. biol. Chem.,* **242,** 1132

DIPRISCO, G., BANAY-SCHWARTZ and STRECKER, H.J. (1968). *Biochem. biophys. Res. Commun.,* **33,** 606

ELLISON, J.S. and KING, K.W. (1968). *J. Nutr.,* **94,** 543

ELVEHJEM, C.A. and KREHL, W.H. (1947). *J. Am. Med. Assoc.,* **135,** 279

ELVEHJEM, C.A. and KREHL, W.H. (1955). *Borden's Rev. Nutr. Res.,* **16,** 69

FLECK, A., SHEPARD, J. and MUNRO, H.N. (1965). *Science,* **150,** 628
FLORENTINO, R.F. and PEARSON, W.N. (1962). *J. Nutr.,* **78,** 101
FRASER, M.J. and KLASS, D.B. (1963). *Can. J. Biochem. Physiol.,* **41,** 2123
FREEDLAND, R.A. and AVERY, E.H. (1964). *J. biol. Chem.,* **239,** 3357
GLENN, J.L. (1961). *Archs. Biochem. Biophys.,* **95,** 14
GOLDSTEIN, L., KNOX, W.E. and BEHRNEAN, E.J. (1962). *J. biol. Chem.,* **237,** 2855
GULLINO, P., WINITZ, M., BIRNBAUM, S.M., CORNFIELD, J., OTEY, M.C. and GREENSTEIN, J.P. (1955). *Archs. Biochem. Biophys.,* **58,** 253
HANKING, B.M. and ROBERTS, S. (1965a). *Nature,* **207,** 862
HANKING, B.M. and ROBERTS, S. (1965b). *Biochim. biophys. Acta.,* **104,** 427
HARPER, A.E. (1965). *Can. J. Biochem.,* **43,** 1589
HARPER, A.E. (1967). in *Handbook of Physiology,* **Section 6, vol.I,** 399, The American Physiological Society, Washington, D.C.
HARPER, A.E. (1970). *Amino Acid Balance and Food Intake Regulation* in *Parenteral Nutrition,* Ed. H.C. Meng and D.H. Law. p.181, C.C. Thomas; Springfield, Illinois
HARPER, A.E. (1973). in *Toxicants Occurring Naturally in Foods,* p.130 National Academy of Sciences, Washington, D.C.
HARPER, A.E., BENEVENGA, N.J. and WOHLHUETER, R.M. (1970). *Physiol. Rev.,* **50,** 428
HARPER, A.E. and ROGERS, Q.R. (1966). *Am. J. Physiol.,* **210,** 1234
HARTMAN, D.R. and KING, K.W. (1967). *J. Nutr.,* **92,** 455
HELE, P. and BARTH, P.T. (1966). *Biochim. biophys. Acta.,* **114,** 149
HIER, S.W., GRAHAM, C.E. and KLEIN, D. (1944). *Proc. Soc. exp. Biol. Med.,* **56,** 187
HILL, D.C. and OLSEN, E.M. (1963). *J. Nutr.,* **79,** 296
HOSHINO, J. and KROGER, H. (1969). *Hoppe-Seyler's Z. physiol. Chem.,* **350,** 595
IKEGAMI, H. and GRIFFIN, A.C. (1969). *Biochim. biophys. Acta.,* **178,** 166
IP, C.C.Y. and HARPER, A.E. (1974a). *J. Nutr.,* **104,** 252
IP, C.C.Y. and HARPER, A.E. (1974b). *Biochim. biophys. Acta.,* **331,** 251
KAPLAN, J.H. and PITOT, H.C. (1970). in *Mammalian Protein Metabolism,* **Vol.IV,** p.387. Ed. H.N. Munro, Academic Press, New York
KATUNUMA, N., HUZINO, A. and TOMINO, I. (1967). in *Adv. Enzymol. Reg.,* **5,** Ed. G. Weber. p.554. Pergamon Press, Inc., New York
KATUNUMA, N., MATSUDA, Y. and TOMINO, I. (1964). *J. Biochem. (Tokyo),* **56,** 499
KAUFMAN, S. (1970). in *Methods in Enzymology.* **Vol.XVIIA,** Eds. H. Tabor and C.W. Tabor. p.603. Academic Press, New York
KNOX, W.E., PIRAS, M.M. and TOKUYAMA, K. (1966). *J. biol. Chem.,* **241,** 297
KORZHOV, V.A. (1965). *Biochemistry/Biokhimiya,* **30,** 594 (English Translation)
KRAMAR, R., (1971). *Hoppe-Seyler's Z. physiol. Chem.,* **352,** 1267

KREBS, H.A. (1972). in *Adv. Enzymol. Reg.*, **10**, Ed. G. Weber, p.397 Pergamon Press Inc., New York
KUMTA, U.S. and HARPER, A.E. (1962). *Proc. Soc. expl. Biol. Med.*, **110**, 512
KUMTA, U.S., HARPER, A.E. and ELVEHJEM, C.A. (1958). *J. biol. Chem.*, **233**, 1505
LANKS, K.W., SCISCENTI, J., WEINSTEIN, I.B. and CANTOR, C.R. (1971). *J. biol. Chem.*, **246**, 3494
LEUNG, P.M-B., and ROGERS, Q.R. (1969). *Life Sci.*, **8**, 1
LEUNG, P.M-B., ROGERS, Q.R. and HARPER, A.E. (1968a). *J. Nutr.*, **95**, 474
LEUNG, P.M-B., ROGERS, Q.R. and HARPER, A.E. (1968b). *J. Nutr.*, **95**, 483
LEUNG, P.M-B., ROGERS, Q.R. and HARPER, A.E. (1968c). *J. Nutr.*, **96**, 303
LITWACK, G. (1966). *Biochim. biophys. Acta.*, **128**, 404
LOMBARDINI, J.B., TURINI, P., BIGGS, D.R. and SINGER, T.P. (1969). *Physiol. Chem. Phys.*, **1**, 1
MacINNES, J.W. and SCHLESINGER, K. (1971). *Brain Res.*, **29**, 101
McGEE, M., GREENGARD, O. and KNOX, W.E. (1971). *Enzyme*, **12**, 1
MEHANSHO, H., PENG, Y., VAVICH, M.G. and KEMMERER, A.R. (1973). *J. Nutr.*, **103**, 1512
MUNRO, H.N. (1965). *Proc. Nutr. Soc.*, **24**, 209
MUNRO, H.N. (1970). in *Mammalian Protein Metabolism*, **Vol.IV**, Ed. H.N. Munro. p.299. Academic Press, New York
NAKANO, Y., FUJIOKA, M. and WADA, H. (1968). *Biochim. biophys. Acta.*, **159**, 19
NAKANO, K., KISHI, T., KURITA, N. and ASHIDA, K. (1970). *J. Nutr.*, **100**, 827
NODA, K. (1970). *Agric. Biol. Chem.*, **34**, 710
NYHAN, W.L. Ed. (1974). *Heritable Disorders of Amino Acid Metabolism.* John Wiley and Sons Inc., New York
PAN, F. and TARVER, H. (1967). *Archs. Biochem. Biophys.*, **119**, 429
PANT, K.C., ROGERS, Q.R. and HARPER, A.E. (1972a). *J. Nutr.*, **102**, 117
PANT, K.C., ROGERS, Q.R. and HARPER, A.E. (1972b). *J. Nutr.*, **102**, 131
PARK, O.J., HENDERSON, L.M. and SWAN, P.B., (1973). *Proc. Soc. expl. Biol. Med.*, **142**, 1023
PENG, Y., TEWS, J.K. and HARPER, A.E. (1972). *Am. J. Physiol.*, **222**, 314
PERAINO, C. and PITOT, H.C. (1963). *Biochim. biophys. Acta.*, **73**, 222
PETERSON, N.A. and McKEAN, C.A. (1969). *J. Neurochem.*, **16**, 1211
PRONCZUK, A.W., BALIGA, B.S., TRIANT, J.W. and MUNRO, H.N. (1968). *Biochim. biophys. Acta.*, **157**, 204
PRONCZUK, A.W., ROGERS, Q.R. and MUNRO, H.N. (1970). *J. Nutr.*, **100**, 1249
ROGERS, Q.R. and LEUNG, P.M-B, (1973). *Fedn Proc. Fedn Am. Socs exp. Biol.*, **32**, 1709

ROUGE, M. (1969). *Biochim. biophys. Acta.*, **171**, 342
SANAHUJA, J.D. and HARPER, A.E. (1963). *Am. J. Physiol.*, **204**, 686
SARMA, D.S.R., VERNEY, E., BONGIORNO, M. and SIDRANSKY, H. (1971). *Nutr. Rep. Internat.*, **4**, 1
SASSEN, A. (1955). *Acta. Gastro-enterol. Belg.*, **18**, 944
SAUBERLICH, H.E. (1961). *J. Nutr.*, **75**, 61
SAYRE, F.W. and GREENBERG, D.M. (1956). *J. biol. Chem.*, **220**, 787
SCHIMKE, R.T. (1970a). in *Methods in Enzymology*, **Vol.XVIIA**, Eds. H. Tabor and C.W. Tabor. p.313. Academic Press, New York
SCHIMKE, R.T. (1970b). in *Methods in Enzymology*, **Vol.XVIIA**, Eds. H. Tabor and C.W. Tabor. p.313. Academic Press, New York
SCHWEET, R.S. (1958). *J. biol. Chem.*, **233**, 1104
SHENOY, S.T. (1973). Ph.D. Thesis, University of California, Davis, California
SIDRANSKY, H., BONGIORNO, M., SARMA, D.S.R. and VERNEY, E. (1967). *Biochem. biophys. Res. Commun.*, **27**, 242
SIDRANSKY, H., VERNEY, E. and SARMA, D.S.R. (1971). *Am. J. Clin. Nutr.*, **24**, 779
SMYTHE, C.V. (1942). *J. biol. Chem.*, **142**, 387
SOLIMAN, A-G.M. and KING, K.W. (1969). *J. Nutr.*, **98**, 255
SPOLTER, P.D. and BALDRIDGE, R.C. (1963). *J. biol. Chem.*, **238**, 2071
STRECKER, H.J. (1965). *J. biol. Chem.*, **240**, 1225
SUDA, M., NAKAGAWA, H. and KIMURA, H. (1971). in *Methods in Enzymology*, **XVII B**, Eds. H. Tabor and C.W. Tabor. p.454. Academic Press, New York
SWICK, R.W., BARNSTEIN, P.L. and STANGL, J.L. (1965). *J. biol. Chem.*, **240**, 3334
SYMMONS, R.A. (1974). Ph.D. Thesis, University of California, Davis, California
SYMMONS, R.A., MAGUIRE, E.J. and ROGERS, Q.R. (1972). *J. Nutr.*, **102**, 639
TEPPERMAN, J. and TEPPERMAN, H.M. (1958). *Am. J. Physiol.*, **193**, 55
VAUGHAN, M.H. and HANSEN, B.S. (1973). *J. biol. Chem.*, **248**, 7087
VESCIA, A. (1967). *Biochem. biophys. Res. Comm.*, **29**, 496
WEBSTER, G.C. (1961). *Biochim. biophys. Acta.*, **49**, 141
WEISS, J. and PITOT, H. (1969). cited in Kaplan, J.H. and Pitot, H. (1970)
WILSON, R.G., WORTHAM, J.S., BENTON, D.A. and HENDERSON, L.M. (1962). *J. Nutr.*, **77**, 142
WOHLHUETER, R.M. and HARPER, A.E. (1970). *J. biol. Chem.*, **245**, 2391
WUNNER, W.H., BELL, J. and MUNRO, H.N. (1966). *Biochem. J.*, **101**, 417
WURTMAN, R.J. (1970). in *Mammalian Protein Metabolism*, **Vol.IV**, Ed. H.N. Munro, p.445. Academic Press, New York
YOSHIDA, A., LEUNG, P.M-B., ROGERS, Q.R. and HARPER, A.E. (1966). *J. Nutr.*, **89**, 80
YOSHIDA, S. (1968). *J. Jap. Biochem. Soc. (Seikagahu)*, **40**, 16
YOSHIDA, T. (1967). *Vitamins* (Kyoto), **35**, 227
YOSHIDA, T. and KIKUCHI, G. (1972). *J. Biochem.*, **72**, 1503

V

PROTEIN NUTRITION OF NON-RUMINANTS

17

AMINO ACID REQUIREMENTS OF MEAT-PRODUCING POULTRY

C. CALET
Institut National de la Recherche Agronomique, Nouzilly, France

Measurement of Amino Acid Requirements

Before discussing the values obtained for amino acid requirements, the methods of measurement should be examined. This is of particular relevance since a new method of estimation from the free amino acids in the tissues has been proposed recently.

Although the proportion of amino acids present in plasma and muscle in the free form is small, these concentrations are not negligible (Munro, 1970) and are related to amino acid intake (Richardson, Cannon and Webb, 1965). Moreover, when protein anabolism is active or the protein catabolism reduced, the concentrations of free amino acids in tissues are low. For these reasons, a method based on tissue amino acids has promise. Zimmerman and Scott (1965) showed that a progressive increase of lysine in the diet produced a marked increase in the accumulation of free lysine in the blood at dietary intakes above that usually considered to be the requirement *(Figure 17.1)*. This has also been shown for arginine and valine. Larbier and Guillaume (1972) showed similar effects in the free amino acids of muscle. However, it is important to note that this effect appears only when the food intake is restricted. There is no such clear relationship when the animals are fed *ad libitum (Table 17.1)*. Zimmerman and Scott (1965) fed birds on a limited amount in 12 meals per day while in their restricted treatment Larbier, Guillaume and Blum (1971) allowed feeding for two hours per day only. Growth rates were low (about 15 g/day) and it is surprising that the requirements obtained by Larbier *et al.* (1971) were similar to those used in practice for maximum weight gain. From the data of Fonseca *et al.* (1970), it is possible to plot weight gain and free lysine in plasma against lysine in the diet *(Figure 17.2)*. In this instance the experimental conditions were similar to those in practice, the diet being maize-based and the growth rate normal. Under these conditions no clear requirement can be defined from plasma lysine concentration.

Several factors may influence free amino acid concentrations in blood and muscle. First, the free amino acid pool depends on the time which elapses between the end of the meal and blood sampling (Larbier, 1973) and this variation differs from one amino acid

Table 17.1 Muscle free lysine and methionine as related to dietary intake and weight gain during a ten day trial in chicks (Larbier et al., 1971)

Diets	Ad libitum				Meal-fed			
	C[1]	M[2]	L[3]	ML[4]	C	M	L	ML
Muscle free lysine content (mcg/g)	86.7	127.6	99.8	104.2	67.0	72.7	10.6	17.1
Muscle free methionine content (mcg/g)	9.7	12.7	8.3	7.7	15.9	traces	18.0	traces
Intake (g)								
Food	337	334	303	319	173	178	164	170
Lysine	4.2	4.1	2.7	2.8	2.1	2.2	1.4	1.5
Methionine	1.5	1.0	1.4	1.0	0.8	0.5	0.7	0.5
Weight gain (g)	145	144	129	129	91	90	76	76

[1] C: control diet
[2] M: methionine-supplemented diet
[3] L: lysine-supplemented diet
[4] ML: methionine + lysine-supplemented diet

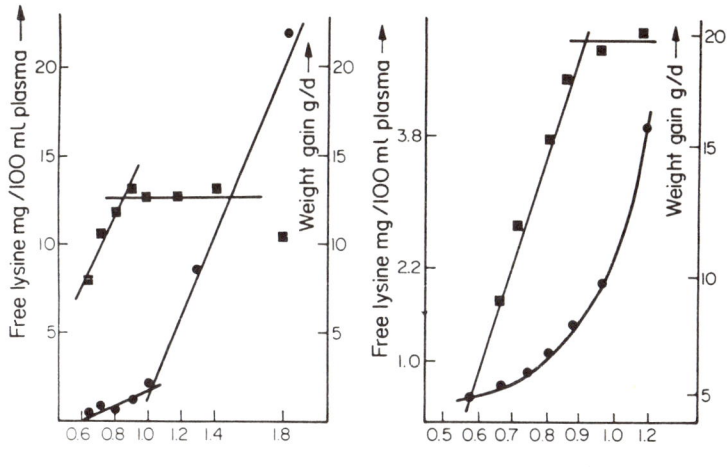

Figure 17.1

Figure 17.2

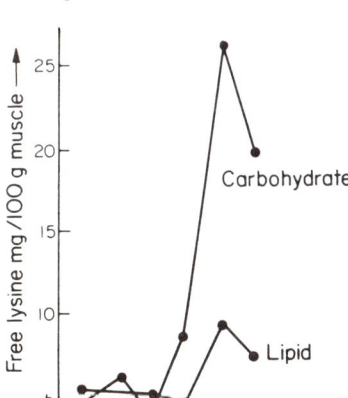

Figure 17.3

Figures 17.1, 17.2, 17.3 Effect of lysine content of the diet upon weight gain (■) and free lysine concentration of the plasma (or muscle) (●); variations according to experimental conditions: (1) crystalline amino acid diet (Zimmerman and Scott, 1965). (2) corn meal diet (Fonseca et al., 1970). (3) protein free sources of the diet: lipids vs carbohydrates (Larbier, 1974) to another (Larbier and Guillaume, 1972). Ohno, Tasaki and Okumara (1972) classified the amino acids into four groups in this respect. Smith (1966) had already pointed out that the plasma concentrations of one amino acid may depend on the proportions of the others in the diet. A possible explanation of this is a modification in gut absorption. A well-balanced diet is absorbed more rapidly than an unbalanced one and amino acids may be transported more rapidly to the sites of protein synthesis in the former case. Other factors have also been noted; temperature (May et al., 1972) and conditions of blood sampling (Hewitt and Lewis, 1972a). Finally, the nature of the non-protein constituents of the diet can modify the concentrations of free amino acids in muscle. Larbier (1974) found that the requirement of the broiler for lysine, determined from muscle lysine concentration, was 20 per cent higher when oil provided the dietary energy than when an equal amount of energy was provided by glucose (*Figure 17.3*).

So the hope of determining objectively amino acid requirements by such methods is not realised and experimental conditions must be carefully controlled to produce meaningful results. Thus we are obliged to turn again to the classical methods of weight gain, food conversion and nitrogen retention, knowing that each of these criteria can provide different values for requirements. Solberg (1971) was one of the most recent to note that lysine and methionine requirements are higher when estimated by the maximum efficiency of food conversion than when maximum weight gain is used. In general optimum nitrogen retention occurs at an even higher value, and can also be influenced by the utilisation of dietary energy (Calet et al., 1964).

Some Factors Affecting Amino Acid Requirements

Values obtained for amino acid requirements can vary widely. Austic and Nesheim (1970) record, for example, that arginine requirement can vary between 0.85 and 1.8 per cent under different conditions.

GENETIC
Among factors arising from the animals which contribute to variation, the strain is the most important. Doubtless, this is a reflection of different performances, since the higher-producing strains have the higher requirements. In this area we shall consider two special cases.

The first case concerns the animals selected by Nesheim (1966) and Hutt and Nesheim (1966) for different arginine requirements due to different renal arginase activities (Austic and Nesheim, 1970). Nesheim (1968) fed birds of a high arginine requirement (HA) and those of a low arginine requirement (LA) on a casein diet poor in arginine and showed that arginase activity in HA birds doubled while this activity remained unchanged, even after two weeks in LA birds. The lysine ketoglutarate reductase activity is also different in the two lines

(Nesheim, Austic and Wang, 1971), suggesting a difference in lysine requirement. Similarly, O'Dell, Amos and Savage (1965) found that broilers with high growth rates have low renal arginase activity.

The second example concerns dwarf birds in which the dw gene modifies energy metabolism. There is also an increase in the nitrogen requirement when expressed in relation to dietary energy (Guillaume, 1969). When the nitrogen content of the diet is increased, the difference in body weight between dwarf and normal birds is less (Guillaume, 1973). Dwarf birds also show concentrations of free amino acids in muscle which are 30 per cent less than those of normal birds (Guillaume, 1972), although both catabolic and anabolic activities are increased. Guillaume and Larbier (1974) gave a partial explanation of the lower concentrations of free amino acids in muscle, which suggests that some amino acids such as threonine participate more intensively in gluconeogenesis in the dwarf bird.

Data concerning the nitrogen requirements of dwarf birds are difficult to explain because these animals are not only genetically dwarf but also physically small and it is not certain that these phenomena are linked. Guillaume (1974) chose two female lines, one heavy and one light, and crossed them with normal or dwarf males. The offspring of each group were heavy or light, but half were normal and the other half dwarf. The chicks were grouped in families and fed on either a 15 or a 22 per cent protein diet. By a method of regression analysis which allows comparison between animals of the same family fed either the low or the high protein diet, it was shown that small birds tolerate better a diet of low protein; however, at constant body weight, dwarf birds have a higher nitrogen requirement than the normal birds.

AGE

Requirements also depend on age and it is generally considered that they decrease as the animal matures. Nevertheless this principle cannot be applied to all amino acids; as has been pointed out in the past, the decrease with age does not occur with the same intensity for each amino acid, and recent findings on lysine and methionine illustrate this point. The conclusions of these studies are not clear however; experimental conditions (natural or synthetic diet), method of expression (proportion of the diet, proportion of the protein) confusing the findings. With purified diets Graber, Scott and Baker (1971) found that methionine requirements did not exceed 0.63 and 0.70 per cent of diet at two and eight weeks respectively, or 4.14 and 4.48 per cent of the protein respectively. Chung, Griminger and Fisher (1973) came to the same conclusion finding the requirement for sulphur amino acids to be 0.61 per cent of diet. Boomgaardt and Baker (1973b) however, found that the requirement for sulphur amino acids decreased from 3.05 to 2.56 as a percentage of the protein. It must be noted that Graber *et al.* (1971) used methionine as the only sulphur amino acid while Boomgaardt and Baker (1973b) used a mixture of cystine and methionine. As pointed out by Graber and Baker

(1971) the requirement for sulphur amino acids is only 0.47 per cent when both sulphur amino acids are considered but it is 0.63 per cent for methionine alone. What practical conclusions may be drawn from such results? The values are lower than those used in practice because growth rates are 8-10 g/day in the starting period and 15-25 g/day during finishing; growth rates below those experienced in practice.

Also the recent studies of Boomgaardt and Baker (1973a,b,c) emphasise the complexities introduced by the use of synthetic diets since, even with an optimum ratio between methionine and cystine, the sulphur amino acid requirement is still too high (4 per cent protein).

For lysine, Boomgaardt and Baker (1973b) and Chung et al. (1973) agree that the requirement decreases with age although they do not agree on the values. Lysine requirement would be about 4.7 per cent protein.

DIETARY PROTEIN

Requirement classically varies with the protein and energy contents of the diet. However when expressed as a proportion of protein content, lysine requirement is independent of protein (Boomgaardt and Baker, 1973a) and of energy (Boomgaardt and Baker, 1973c) contents.

FOOD INTAKE

Requirements depend on food intake and on all factors which affect consumption, such as energy content of the diet (March and Biely, 1972) and temperature (Kubena et al., 1972). Also, as pointed out by March and Biely (1971) there is an interaction between protein requirement, energy requirement, environment and breed which is the main difficulty in defining requirements precisely. According to Hvidsten and Kolstad (1972) variations in food intake could explain most of the differences between values for requirements. Conversely dietary amino acids can also be an important factor in food intake regulation (Calet, Jouandet and Baratou, 1961; Calet et al., 1964) and therefore affect body composition. This latter point is important in broiler quality.

In the past the only goal in broiler production was maximum weight in minimum time at the highest efficiency. Today, there is increasing emphasis on the preferences of the consumer and thus taste, fattiness, wateryness, tenderness and overall acceptability must be taken into account. Many of these criteria are connected with the intramuscular fat. It would be of interest to determine the conditions which allow deposition of intramuscular fat without encouraging excessive accumulation of abdominal fat deposits. Depending on the criteria chosen, either fattening of the carcass or fattening of the muscle, the values of nitrogen requirement are different.

When the content of a relatively imbalanced protein in a diet is progressively increased, food consumption increases sharply. A maximum is reached around 15 per cent, after which there is some decrease (*Figure 17.4*) (Guillaume, Fendry and Imbach, 1965). This means

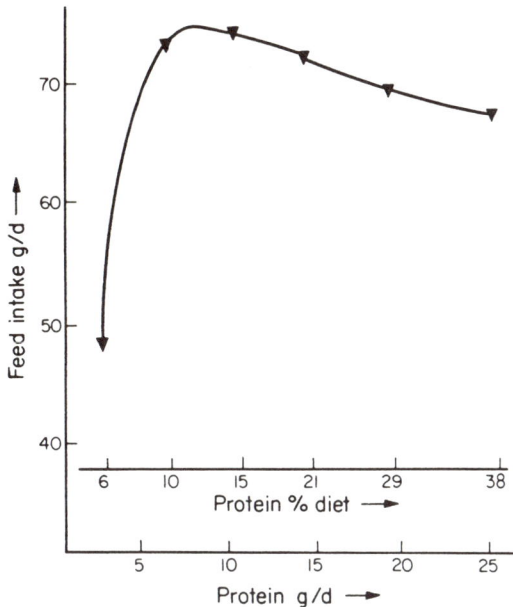

Figure 17.4 Influence of protein intake on food consumption

that a moderate deficit in all amino acids can provoke an increased consumption, as though the animal tries to compensate for the deficiency by increasing amino acid intake. As a consequence, the carcass becomes fatty (Combs, 1967). In contrast, an increase in the proportion of a well-balanced protein in the diet, above the requirement produces animals with less fat and better food conversion (Guillaume, 1966; Calet, 1971; Auckland and Fulton, 1973). However, the potential offered by this is limited. Increases in the protein content of a diet with a well-balanced nitrogen source, such as fish meal, cause decreases in the lipid content of the carcass but the efficiency of food conversion rapidly decreases. Above 23 per cent of such proteins in the diet, the weight gain tends to decrease.

The qualitative aspect of the nitrogen content of the diet must also be considered, i.e. the balance of amino acids. According to Sugahara, Baker and Scott (1969), a decrease of all the amino acids does not modify food consumption. The latter is decreased by a deficiency of one amino acid and the decrease depends on which amino acid is deficient: tryptophan and phenylalanine-tyrosine being most potent in this respect. The effects of amino acid deficits on the appetite have been clearly demonstrated but the effects of an excess have been largely ignored. Guillaume and Summers (1970) showed the advantage of an amino acid excess either in the free form or as an unbalanced mixture, such as a protein of low quality (*Table 17.2*). Even though the weight gain was not improved, supplementation with a low quality protein improved food conversion and gave less fatty carcasses.

Table 17.2 Amino acid excess on conversion efficiency (Guillaume and Summers, 1970)

Experiment 1 Additions to soyabean meal	5-8 weeks			
	Protein content (%)	Weight gain (g/day)	Conversion efficiency g food/g gain	Body fat (%)
Peanut meal	21	54.2	2.90	8.1
Lysine + Methionine	21	59.3	2.75	7.9
Peanut meal	25	59.2	2.70	6.0

Experiment 2 Dietary protein	0-4 weeks		
	Weight gain (g/day)	Conversion efficiency g food/g gain	Energy content of carcass (MJ)
Soyabean 20 per cent	20.5	1.74	2.95
Soyabean 20 per cent + Feathermeal 5 per cent	20.2	1.71	2.67

D'Mello and Lewis (1971) studied excess lysine in the diet and found that the primary effect is a decreased growth rate, however food intake started to decrease after six days of feeding the unbalanced diet. Although it was not measured, there is strong evidence to predict fatty animals. Leucine excess probably causes similar effects. Other amino acids also have a specific role in body composition. Velu, Baker and Scott (1972) fed birds on a well-balanced pattern of crystalline amino acids at 60 per cent of the proportions in an optimal diet. In these conditions, the lipid content of the chickens was 21 per cent. When only isoleucine was deficient by 40 per cent in an otherwise adequate diet leaner birds were obtained. But the same deficiency of lysine did not change the fat content of the body. Methionine deficiency seems to act in the opposite direction to isoleucine (Solberg, Buttery and Boorman, 1971) while a methionine supplementation increases linearly the protein content of the carcass (Boomgaardt and Baker, 1973c).

As a rule, the effect of deficiency or excess can be explained by their effects on food intake. However the variations in body composition are not always thus explicable. An increase in the methionine content of the diet affects body composition without any effect on food intake (Boomgaardt and Baker, 1973c; Hewitt and Lewis, 1972c). However, cystine causes an increase in food intake (Graber and Baker, 1971). Also, each amino acid, having its own metabolic pathway, will induce characteristic effects. Therefore the same deficit of different amino acids will cause different effects, which shades the concept of 'chemical score'.

Another approach has been published recently (Ivanov, 1972). So far total body fat has been considered, but this author looked at the effects of amino acids on muscle composition and found that methionine favoured deposition of intramuscular fat and consequently improved the flavour of the meat.

The difficulty of defining the requirement of one amino acid has been demonstrated since it is a function of the physiological state, the slope of the growth curve (Parks, 1971) and the quality of meat products.

Some Examples of Relationships Between Amino Acids

The requirement of one amino acid is essentially a function of the concentration of the other nutrients in the diet. We shall consider some interactions between amino acids. The general law between efficiency of protein utilisation and the balance of the amino acids of this protein is already well known as the theory of the limiting amino acid. However, certain amino acids have special relationships; it is these which will be examined.

SULPHUR AMINO ACIDS
The relationship between cystine and methionine in higher animals is well known. In birds, it is considered classically that a minimum of cystine has to be present in the diet for the growth of feathers. Doubtless methionine is transformed to cystine but methionine replaces cystine with an efficiency of only 79 per cent. Graber and Baker (1971), Graber *et al.* (1971) and Boomgaardt and Baker (1973c) estimated the requirement for methionine alone to be about 0.63-0.70 per cent although it is only 0.52 per cent in a mixture of methionine and cystine. The first authors considered, however, that these values were much closer when expressed on the basis of total sulphur amino acid intake; 475 and 495 mg/day respectively, and they became equal when compared on molecular basis (one mole of methionine or cysteine equivalent to 0.5 mole of cystine). These last observations emphasise the anomalies created by using proportion of the diet to express requirement and suggest that millimoles per 100 g of diet or even better millimoles per day should be used.

Such data lead us to revise the concept of sulphur amino acid requirement. A minimum content of cystine in the diet is advantageous in the sense that it spares methionine, however it should be dispensable, even in birds. Nevertheless, before this generalisation can be accepted, it must be remembered that the cystine-need as a proportion of total sulphur amino acid requirement might change with the age of the chicken, as shown by Graber *et al.* (1971): 56, 67 and 85 per cent respectively at two weeks, seven weeks and maturity.

GLYCINE–SERINE AND METHIONINE
There are conflicting reports concerning glycine, suggesting that this amino acid is or is not indispensable and values for requirement vary from 0.98-1.6 per cent of diet. These values are also a function of other amino acids in the diet. Fontaine and Reytens (1969) added 0.2 or 0.4 per cent glycine to a diet containing 0.77 per cent glycine

and observed a small decrease in growth. Nevertheless, if only 0.05 per cent methionine was added, the weight gain improved by 23 per cent. So the effects of the apparent excess of glycine appeared only if the diet was deficient in methionine. Baker and Sugahara (1970) using a diet devoid of glycine, serine and choline observed that growth rate was not improved by glycine or choline alone. However, simultaneous supplementation with choline and glycine induced a marked increase in weight gain. Similar results were obtained with either choline and serine or choline and sarcosine. Thus it appears that serine can entirely replace glycine, and that glycine would be indispensable only when the diet is deficient in serine or choline which can both give rise to glycine. In other words, serine spares glycine. In contrast threonine cannot replace glycine despite the similarity of their chemical formulae (D'Mello, 1973).

LYSINE-ARGININE

Nesheim's work (1966) drew attention to the bad effects of excess lysine in the diet. Allen and Baker (1972) and Allen et al. (1972) as well as many others have shown that this is due to a higher arginine requirement. An elegant demonstration of this interaction was given by D'Mello and Lewis (1970a). They used a diet slightly deficient in methionine, tryptophan, histidine and threonine, to which they added lysine (*Table 17.3*). Methionine supplementation caused a small growth improvement whereas addition of lysine depressed growth. The classical explanation would suggest that an excess of lysine would aggravate the deficiency of methionine. Methionine should therefore correct the effect of lysine; but it did not. However, arginine, which was not deficient in the basal diet, cured the depressive effect of lysine excess. At the same time these results were confirmed by Miller and Kifer (1970) with fish meal.

The reasons for the lysine-arginine antagonism are numerous. Nesheim and co-workers (1970; 1973) have shown that an excess of lysine stimulates renal arginase activity and hence, apparently, arginine breakdown. This is very clear in a strain having a high arginine requirement (Nesheim, 1968); the arginase activity was 4000 μmoles urea formed per hour with a well-balanced diet. With an excess of lysine in the diet, the activity was 5000 in a strain having a low arginine (LA) requirement and 8400 in a strain having a high arginine (HA) requirement. Another proof of this phenomenon is shown by the fact that a-aminoisobutyric acid, which depresses arginase activity, corrects the growth depression due to an excess of lysine. In favour of Nesheim's thesis, Stutz, Savage and O'Dell (1971) found that potassium salts with a metabolisable anion spare arginine and decrease its requirement. This potassium increases intracellular potassium, which decreases the free lysine concentration. This decrease is not due to increased catabolism but results from an increased availability of lysine for protein anabolism (Stutz, Savage and O'Dell, 1972).

Another explanation is given by Boorman (1971) who underlines the competition between arginine and lysine at the level of renal tubular

Table 17.3 Specificity of the lysine-arginine interaction (D'Mello and Lewis, 1970a)

	Gain (g/day)	Conversion efficiency (g gain/g food)
Basal diet	16.3	0.57
Basal + methionine 0.3 per cent	17.1	0.59
Basal + lysine 0.6 per cent	12.0	0.48
Basal + lysine + methionine	11.5	0.47
Basal + lysine + arginine	15.4	0.55

re-absorption. When the lysine perfusion was raised from one to four μmoles/min/kg, the re-absorption rate of arginine decreased from 97 to 66 per cent, while that of ornithine decreased from 89 to 5.3 per cent. Arginine perfusion depressed lysine re-absorption from 97 to 84.1 per cent. A similar competition has been found at the level of intestinal absorption in pig (Buraczewski et al., 1970). While with a well balanced diet the intestinal absorption rate of amino acids is high, around 92-95 per cent, an excess of lysine lowers these values to 50 per cent and an excess of arginine is worse; in this case, absorption rate of lysine is about 28 per cent.

These theoretical considerations are not without practical significance since there may be cases when the lysine-arginine ratio is too high (some fish meals, yeasts, etc.). In spite of its nutritional advantages, yeast cannot be used in chick diets at a content higher than 15 per cent without risks. This can be corrected with peanut meal rich in arginine or potassium supplementation. Mongin and Sauveur (unpublished results) improved the growth rate of chicks fed on a diet containing Torula yeast by supplementation with K_2CO_3 or arginine.

Although lysine excess leads to arginine breakdown, an excess of arginine does not lead to a loss of lysine (Wang, Crosby and Nesheim, 1973). An excess of arginine increases the requirement for methyl groups and consequently methionine (Keshavarz and Fuller, 1971a).

The relationship between arginine and lysine has been extensively studied but it is not an isolated example. Similar antagonisms have been found between leucine and isoleucine, leucine and valine (Allen and Baker, 1972; D'Mello and Lewis, 1970b) and threonine and tryptophan (D'Mello and Lewis, 1970c). Relationships also exist between arginine-glycine and methionine (Keshavarz and Fuller, 1971b) and between phenylalanine and tyrosine, this last amino acid having the ability to replace up to 42.5 per cent of the requirement for the aromatic amino acids (Sasse and Baker, 1972). There are also other specific interactions between indispensable amino acids to be discovered if the requirements of these were known with greater precision.

Before completing this section, it is of interest to examine the place of non-specific nitrogen sources. As pointed out by Calet (1967), as broiler performances are improved by selection, apparently some unimportant amino acids may become indispensable or may spare indispensable amino acids. A unanimous opinion is not established on this subject. Shannon et al. (1970) compared in an experimental diet the effects of diammonium citrate or glutamic acid with a supplement of proline and glycine. Weight gains increased with each supplementation. The authors showed also that nitrogen from diammonium citrate could enter into tissue synthesis since ammonium nitrogen is transferred to a-ketoglutaric acid. One may speculate whether any improvement caused by a diet containing excess glutamic acid is related to its proline content since the latter is known to be indispensable. However, Bhargava, et al. (1971) pointed out the synergism between proline and glutamic acid. The weight gain of chickens was improved by 30 per cent if the diet contained 1.5 per cent proline and 6.5 per cent glutamic acid compared with a diet containing eight per cent glutamic acid alone. Thus the proline requirement was one per cent in a diet containing five per cent glutamic acid but only 0.3 per cent when the glutamic acid content was doubled.

Towards Protein Saving by a Better Understanding of Amino Acid Requirements

The recent crisis in the supply of soyabeans caused problems for animal food manufacturers and emphasised the risk of protein deficiency, without easy remedy, even in the developed countries. Moreover the meat needs of the populations increase more and more and experts have calculated that by the year 2000, 200 million tons of proteins will have to be found to feed domestic animals. Before new protein sources or new strains of cereals or more economic animals are developed, wastage can be minimised by better adjustment of the nitrogen supplied to the requirements of the animals. Also, it may be advantageous, on economic grounds to decrease the amino acid contents of diets, knowing that performances will be depressed. We will consider these aspects below.

From Comb's work (1967), and as reported by Calet (1973), the lysine requirement of the chick up to two weeks of age is estimated as 1.1 per cent of a diet containing 12.54 MJ/kg ME, i.e. about 13 per cent less lysine than NRC recommendations. More recently Mafwila (1972) could not improve growth rate by increasing lysine content above 90 per cent of NRC recommendations. With 12.54 MJ/kg dietary energy, Petersen (1970) found an optimal growth between 0-7 weeks with a minimum crude protein content of 18.8 per cent. More notably, Packham and Payne (1973) found that growth rate is optimal or slightly decreased between seven and twenty days of age with a diet containing 16 per cent protein, 0.9 per cent lysine, 0.92 per cent arginine, 0.91 per cent leucine, 0.56 per cent valine and 0.51 per cent threonine. Their results are not surprising since Hewitt and Lewis (1972b) have shown that with a judicious adjustment of the amino acids, they could consistently use amino acid contents below accepted recommendations (*Table 17.4*).

There are not yet complete data for the period up to four weeks of age, only a few studies have been carried out over this period independently of the former period. Nevertheless, it is known that requirements

Table 17.4 Suggested requirements for amino acids (% diet)

	NRC	Dean and Scott (1965)	Hewitt and Lewis (1972)
Arginine	1.20	1.10	0.85
Glycine	1.00	1.60	0.61
Histidine	0.40	0.30	0.40
Isoleucine	0.75	0.80	0.61
Leucine	1.40	1.20	1.34
Lysine	0.15	1.12	0.85
Methionine + cystine	0.75	0.80	0.79
Phenylalanine + tyrosine	1.30	1.31	1.27
Threonine	0.70	0.65	0.53
Tryptophan	0.20	0.225	0.17
Valine	0.85	0.82	0.79

decrease with age and coefficients have been calculated to estimate the amino acid requirements for the older animal from the requirements of the first weeks. Between five and seven weeks, the values are 85 per cent of those of the period up to two weeks. Experience proves that this coefficient is too high, and even in 1967 Combs gave an estimate of 75 per cent and the values in the following table are in good agreement (*Table 17.5*)

Table 17.5 Effects of a lysine deficiency on cockerels at different ages

Age (weeks)	0 to 2½	2½ to 5	5 to 8
Lysine deficit (per cent NRC)	24	25	21
Weight gain deficit (per cent of control)	30	19	9
Excess in food conversion (per cent of control)	19	10	6

It is clear that margins of safety used during finishing in broiler nutrition are wider than in the starting period. During finishing, 0.95 per cent lysine is enough to ensure optimum growth and food conversion. If lysine, methionine and threonine are kept at 0.90, 0.43 and 0.51 per cent respectively, the broiler does not require more than 14 per cent protein in the diet between 28 and 63 days of age (Packam and Payne, 1973). If we take into consideration the fact that food consumption is more important during finishing, a better adjustment of the amino acid contents could save at least 2200 tonnes of soyabean meal per year in France's flock of 330 million broilers.

An alternative is to allow a slight decrease in broiler performance. It is, perhaps, more economic to aim for lower growth rates using slightly deficient diets. It is known that well below the requirement, the growth response is greater per unit increment in input than in the region of the requirement, this is the law of the diminishing returns. Recently Parks (1971) gave a mathematical interpretation of it. From Comb's (1967) and Petersen's (1970) results, weight gain can be related to lysine in the diet. With a diet of 12.54 MJ/kg, Guillaume (see

Calet, 1973) found the following equation:

$$y = -2404 x^2 + 4665 x - 1094$$

when x = lysine content of the diet (per cent)
y = body weight at seven weeks (g)

On the ascending part of the curve, we may establish that a deficit of five per cent in the lysine content decreases body weight by four per cent and two per cent during the starting and finishing period respectively. A deficit of 10 per cent produces values of 9.5 per cent and four per cent respectively.

In conclusion, a calculation might be done relating feedstuff costs and end-product returns to establish which conditions would give the optimal profit; the necessary inputs might not coincide with those for maximum performance.

Nitrogen Requirements of Other Species of Poultry

Nitrogen requirements of the other poultry species (turkey, quail, goose, duck, guinea-fowl and game birds) have not been so extensively studied.

In turkey the protein requirement decreases from 29 to 16 or 12 per cent during the five or seven months of the animal life for broiler turkey or heavy turkey respectively. However up to eight weeks, a supplement of four per cent protein improves weight gain (Summers and Moran, 1971). These authors found that the benefit is not due only to increased protein content but also to the increases of lysine and arginine. It should be noted that if an economy of protein is to be made, lysine must not exceed 1.55 to 0.85 or 0.6 per cent of the diet for the periods previously defined. According to Kummero, Jones and Loadholt (1971) the requirement for sulphur amino acids up to three weeks is about 1.17 per cent per MJ.

Quail requirements have been reviewed by Guillaume (1970). A diet of 22 per cent protein gives the same results at six weeks as a diet of higher protein, although early growth is improved on the latter. Identical results were obtained by growing in two phases and by a better adjustment of the needs; 26 per cent to two weeks, 18 per cent between two and six weeks.

The duck has been studied recently by Dean (1972) who points out the remarkable ability of the duck to compensate for a deficiency when young. It is possible to feed the animal with a relatively low-protein diet. By lowering the protein content of the diet from 28 to 22 per cent, weight gain is depressed by 30 per cent at 14 days, nine per cent at 28 days and is normal at slaughter-weight. However, if the dietary protein is too low, the duck becomes fat (36 per cent fat with a 28 per cent protein diet compared with 39.5 per cent fat with 16 per cent). According to Lühmann and Vogt (1972) the best results should be obtained with 22 per cent during the starting period and a

16 per cent protein diet thereafter. Auckland (1973) confirmed the
effect of a low-protein diet on body composition but was not able
to decrease fattening by feeding a restricted amount of a low-protein
diet. Restriction did not modify the composition of the fat-free dry
matter. During the first three weeks the ducklings received a 24 or 20
per cent protein diet and their body weight was 739 or 577 g; their
food conversion 2.03 or 2.34 and their fat content 34 or 41 per cent,
respectively. Under restricted feeding, food conversion was unchanged
and body weight was proportional to the extent of restriction. Leclercq
(1974) considers that lysine and sulphur amino acid requirements are
0.8 and 0.6 per cent of the diet respectively.

Little is known about the requirement of growing goose which eats
an appreciable amount of roughage containing a certain quantity of
nitrogen (Calet, 1970). Withdrawal of grass from the food of the
goose causes an increase in the protein required from the concentrate
diet from 17 to 22 per cent to achieve the same weight gain but
cannibalism is frequent and impossible to avoid. According to Koci
and Koci (1967), the lysine and sulphur amino acid requirements of
the young goose are 1.15 and 0.75 per cent respectively at 35 days
of age.

Guinea fowl have recently become important in France and there
were no data available on nutritional requirements. Growth rate is
low and from the work of Blum, Guillaume and Leclercq (1974) it
appears that nitrogen requirement, mainly during the finishing period,
is lower than would be expected. Three periods have to be considered
(0-4 weeks, 4-8 weeks and 8-12 weeks) for which protein and lysine
recommendations are 24, 19 and 16 per cent and 1.3, 0.95 and 0.72
per cent respectively. Slightly higher values led to an improvement of
weight gain but this was not justified by the cost of the food. In
contrast to the chicken, food intake in the guinea fowl is not so well
controlled by the characteristics of the diet and a decrease of the
protein content of the diet leads to a decrease in growth rate.

For game birds only the data of Vohra (1973) are available. Diets
with protein contents higher than 25 per cent are of no extra benefit
to male pheasants while the female could require a higher content
after 12 weeks. The nitrogen requirement for the partridge is 20 per
cent and higher contents do not improve weight gain up to seven
weeks. However, with a low-energy diet (9.70 MJ/kg) growth rate is
unaffected by a diet of 15 per cent protein.

The recent work concerning these other species shows that we may
not extrapolate to these species the results found with the chicken.
As such species are of interest it is important to undertake work in
these areas.

Conclusion

In conclusion, it seems important to recall the idea of Hewitt and
Lewis (1972b,c). Until that time requirements were assessed by considering one amino acid at a time by using a well-balanced experimental diet

in which the concentration of one amino acid was the only variable. The requirement was therefore defined as the minimum level of the amino acid in the diet which provoked the maximum performance.

However, all the recent work underlines how the utilisation of one amino acid is dependent on the concentrations of the others and even more specifically on the content of a certain one. All the amino acids should be considered simultaneously. This seems difficult but when investigators take into consideration only two amino acids, they demonstrate the antagonisms and synergisms already mentioned, and they produce a better estimate of the requirement. This method has demonstrated that the requirement can be 20 per cent lower than the usual recommendation.

While the resources of soyabean may be limited again in the world market and while the deficit in animal protein becomes more and more important, what precautions may be taken? Evidently we may find new and unconventional protein sources, create new strains of animals with low nitrogen requirements, but these solutions are not for tomorrow. A more precise and immediate approach would be to eliminate all form of wastage by fitting requirements more exactly to needs in an economically meaningful way.

References

ALLEN, N.K. and BAKER, D.H. (1972a). *Poult. Sci.*, **51**, 902
ALLEN, N.K. and BAKER, D.H. (1972b). *Poult. Sci.*, **51**, 1292
ALLEN, N.K., BAKER, D.H., SCOTT, H.M. and NORTON, H.W. (1972). *J. Nutr.*, **102**, 171
AUCKLAND, J.N. (1973). *J. Sci. Fd Agric.*, **24**, 719
AUCKLAND, J.N. and FULTON, R.B. (1973). *J. Sci. Fd Agric.*, **24**, 709
AUSTIC, R.E. and NESHEIM, M.C. (1970). *J. Nutr.*, **100**, 855
BAKER, D.H. and SUGAHARA, M. (1970). *Poult. Sci.*, **49**, 756
BHARGAVA, K.K., SHEN, T.F., BIRD, H.R. and SUNDE, M.L. (1971). *Poult. Sci.*, **50**, 726
BLUM, J.C., GUILLAUME, J. and LECLERCQ, B. (1974). *C.R.XVème World's Poult. Congr. New Orleans.* 461
BOOMGAARDT, J. and BAKER, D.H. (1973a). *Poult. Sci.*, **52**, 586
BOOMGAARDT, J. and BAKER, D.H. (1973b). *Poult. Sci.*, **52**, 592
BOOMGAARDT, J. and BAKER, D.H. (1973c). *J. Anim. Sci.*, **36**, 307
BOORMAN, K.N. (1971). *Comp. Biochem. Physiol.*, **39A**, 29
BURACZEWSKI, S., CHAMBERLAIN, A.G., HORSZCZARUK, F. and ZEBROWSKA, T. (1970). *Proc. Nutr. Soc.*, **29**, 51A
CALET, C. (1967). *Alim. Vie,* January, 1967, 31
CALET, C. (1970). *XIVe Congreso mundial de Avicultura, Madrid,* **Tome 1**, 315
CALET, C. (1971). *Xème Congrès International de Zootechnie, Versailles, July, 1971*
CALET, C. (1973). *C.R. Journées d'information sur l'alimentation azotée des animaux, November 1973,* 51 (I.N.R.A. Ed.)

CALET, C. GUILLAUME, J., DELPECH, P. and JACQUOT, R. (1964). *C.R. Acad. Sci. Paris*, **258**, 3104„,Série D
CALET, C., JOUANDET, C. and BARATOU, J. (1961). *Annls. Biol. anim. Biochim. Biophys.*, **1**, 5
CHUNG, E., GRIMINGER, P. and FISHER, H. (1973). *J. Nutr.*, **103**, 117
COMBS, G.F. (1967). *Proc. Maryland Nutr. Conf.*, 51
DEAN, W.F. (1972). *Proc. Cornell Nutr. Conf.*, 77
DEAN, W.F. and SCOTT, H.M. (1965). *Poult. Sci.*, **44**, 803-808
D'MELLO, J.P.F. (1973). *Nutr. Metab.*, **15**, 357
D'MELLO, J.P.F. and LEWIS, D. (1970a). *Br. Poult. Sci.*, **11**, 299
D'MELLO, J.P.F. and LEWIS, D. (1970b). *Br. Poult. Sci.*, **11**, 313
D'MELLO, J.P.F. and LEWIS, D. (1970c). *Br. Poult. Sci.*, **11**, 367
D'MELLO, J.P.F. and LEWIS, D. (1971). *Br. Poult. Sci.*, **12**, 345
FONSECA, J.B., ROGLER, J.C., FEATHERSTON, W.R. and CLINE, T.R. (1970). *Poult. Sci.*, **49**, 1519
FONTAINE, G. and REYTENS, N. (1969). *Rev. Agric.*, **4**, 551
GRABER, G. and BAKER, D.H. (1971). *J. Anim. Sci.*, **33**, 1005
GRABER, G., SCOTT, H.M. and BAKER, D.H. (1971). *Poult. Sci.*, **50**, 854
GUILLAUME, J. (1966). *Annls. Biol. anim. Biochim. Biophys.*, **6**, 411
GUILLAUME, J. (1969). *Annls. Biol. anim. Biochim. Biophys.*, **9**, 369
GUILLAUME, J. (1970). *Ann. Zootech.*, **19**, 5
GUILLAUME, J. (1972). *Ann. Génét. Select. anim.*, **4**, 233
GUILLAUME, J. (1973). *4th Europ. Poult. Conf., London*, 557
GUILLAUME, J. (1974). *Contribution à l'étude des effets métaboliques du gène dw.* Ph D. Thesis
GUILLAUME, J., FENDRY, M. and IMBACH, B. (1965). *Annls. Biol. anim. Biochim. Biophys.*, **5**, 293
GUILLAUME, J. and LARBIER, M. (1974). *C.R. Acad. Sci. Paris, Série D*, **27**, 1593
GUILLAUME, J. and SUMMERS, J.D. (1970). *Can. J. Anim. Sci.*, **50**, 355
HEWITT, D. and LEWIS, D. (1972a). *Br. Poult. Sci.*, **13**, 387
HEWITT, D. and LEWIS, D. (1972b). *Br. Poult. Sci.*, **13**, 449
HEWITT, D. and LEWIS, D. (1972c). *Br. Poult. Sci.*, **13**, 465
HUTT, F.B. and NESHEIM, M.C. (1966). *Can. J. Genet. Cytol.*, **8**, 251
HVIDSEN, H. and KOLSTAD, N. (1972). *Arch. Geflügelk.*, **36**, 94
IVANOV et al., (1964). cited by IVANOV, N. (1972). *Colloque O.N.U., Geneva, December 1972*, 13
KESHAVARZ, K. and FULLER, H.L. (1971a). *J. Nutr.*, **101**, 217
KESHAVARZ, K. and FULLER, H.L. (1971b). *J. Nutr.*, **101**, 855
KOCI, E. and KOCI, S. (1967). *Journées d'Etude sur l'Elevage et la Production de l'Oie, Paris*, Proc. in *Alim. Vie*, (1969), **57**, 119
KUBENA, L.F., DEATON, J.W., REECE, F.N., MAY, J.D. and VARDAMAN, T.H. (1972). *Poult. Sci.*, **51**, 1391
KUMMERO, V.E,, JONES, J.E. and LOADHOLT, C.B. (1971). *Poult. Sci.*, **50**, 752
LARBIER, M. (1973). *Symposium on New Developments in the Provision of Amino Acids in the Diets of Pigs and Poultry.* O.N.U., F.A.O., Geneva, December 1972
LARBIER, M. (1974). personal communication

LARBIER, M. and GUILLAUME, J. (1972). *Annls. Biol. anim. Biochim. Biophys.*, **12**, 637
LARBIER, M., GUILLAUME, J. and BLUM, J.C. (1971). *Nutr. Rep. Intern.*, **3**, 273
LECLERCQ, B. (1974). *L'Elevage*, **25**, 81
LUHMANN, M. and VOGT, H. (1972). *Arch. Geflügelk.*, **36**, 25
MAFWILA, J. (1972). PhD Thesis. University of Bonn
MARCH, B.E. and BIELY, J. (1971). *Poult. Sci.*, **50**, 1036
MARCH, B.E. and BIELY, J. (1972). *Poult. Sci.*, **51**, 665
MAY, J.D., KUBENA, L.F., REECE, F.N. and DEATON, J.W. (1972). *Poult. Sci.*, **51**, 1937
MILLER, D. and KIFER, R.R. (1970). *Poult. Sci.*, **49**, 999
MUNRO, H.N. (1970). *Mammalian Protein Metabolism.* Academic Press, New York
NESHEIM, M.C., (1966). *Proc. Cornell Nutr. Conf.*, 16
NESHEIM, M.C. (1968). *J. Nutr.*, **95**, 79
NESHEIM, M.C. (1971). *Proc. Cornell Nutr. Conf.*, 75
NESHEIM, M.C., AUSTIC, R.E. and WANG, S.H. (1971). *Fed. Proc.*, **30**, 121
O'DELL, B.L., AMOS, W.H. and SAVAGE, J.E. (1965). *Proc. Soc. exp. Biol. Med.*, **118**, 102
OHNO, T., TASAKI, I. and OKUMURA, J.I. (1972). *Nutr. Rep. Intern.*, **6**, 339
PACKHAM, R.G. and PAYNE, C.G. (1973). *World's Poult. Sci. J.*, **29**, 286. Spring Conf. U.K. Branch of W.P.S.A.
PARKS, J.R. (1971). *Am. J. Physiol.*, **221**, 1845
PETERSEN, V.E. (1970). *XIVe Congreso Mundial de Avicultura*, Madrid, **Tome II**, 1025
RICHARDSON, L.R., CANNON, M.L. and WEBB, B.D. (1965). *Poult. Sci.*, **44**, 248
SASSE, C.E. and BAKER, D.H. (1972). *Poult. Sci.*, **51**, 1531
SHANNON, D.W.F., BLAIR, R., McNAB, J.M. and LEE, D.J.W. (1970). *Proc. Nutr. Soc.*, **29**, 23A
SMITH, R.E. (1966). *J. Nutr.*, **89**, 271
SOLBERG, J. (1971). *Acta Agric. scand.*, **21**, 193
SOLBERG, J., BUTTERY, P.J. and BOORMAN, K.N. (1971). *Br. Poult. Sci.*, **12**, 297
STUTZ, M.W., SAVAGE, J.E. and O'DELL, B.L. (1971). *J. Nutr.*, **101**, 377
STUTZ, M.W., SAVAGE, J.E. and O'DELL, B.L. (1972). *Poult. Sci.*, **51**, 1283
SUGAHARA, M., BAKER, D.H. and SCOTT, H.M. (1969). *J. Nutr.*, **97**, 29
SUMMERS, J.D. and MORAN, E.T. Jr. (1971). *Poult. Sci.*, **50**, 858
VELU, J.G., BAKER, D.G. and SCOTT, H.M. (1972). *Poult. Sci.*, **51**, 938
VOHRA, P. (1973). *Feedstuffs*, **45**, 34, 26
WANG, S.H., CROSBY, L.O. and NESHEIM, M.C. (1973). *J. Nutr.*, **103**, 384
ZIMMERMAN, R.A. and SCOTT, H.M. (1965). *J. Nutr.*, **87**, 13

18

PROTEIN IN THE DIETS OF THE PULLET AND LAYING BIRD

C. FISHER
Poultry Research Centre, Edinburgh, Scotland
Present address: *Unilever Ltd., Colworth House, Sharnbrook, Bedford*

Introduction

From the viewpoint of practical nutrition, protein metabolism in the laying fowl differs little from that in mammals. The uricotelic mode of nitrogen excretion explains the essential nature of dietary arginine (Boorman and Lewis, 1971), and this has been confirmed directly in the adult fowl (Johnson and Fisher, 1956; Adkins, Harper and Sunde, 1962). Contrary to the situation in the growing bird, glycine is not essential for egg production (Johnson and Fisher, 1956) although the importance of glycine-serine interconversion and its relevance in this context has apparently not been investigated.

Amino acid imbalance has received little attention in the adult fowl. The effects of excess leucine on isoleucine and valine requirements in low-protein diets have been demonstrated (Bray, 1970; Muller and Balloun, 1972) but other interactions have not been investigated. However, it is probable that imbalances are not a problem in practical diets for the laying hen rations (Zimmerman *et al.,* 1969) and amino acids will be considered individually in the remainder of this paper.

The control of protein synthesis in the adult hen is attracting increasing attention. In particular the potential importance of discontinuous protein biosynthesis and its effect on the utilisation of absorbed amino acids is recognised. Yolk growth, which accounts for about 44 per cent of egg protein, is a continuous process in the normal laying hen (Warren and Conrad, 1939). The formation of the remaining protein in albumen, shell and shell membranes is at least partially discontinuous (Gilbert, 1971). The effect of this discontinuity on theoretical models of amino acid requirements has been investigated (Hurwitz and Bornstein, 1973).

However, most experimental evidence about protein requirements reflects the commercial practice of feeding a single diet *ad libitum* over a period of time. In these circumstances attention is directed mainly to measurements of average input (protein and amino acids) and average output (growth and egg production) and the relationship between them. This emphasis on input-output relationships is also merited by the fact that feeding hens is an important economic activity and it is this aspect of protein nutrition which will be emphasised here.

In this paper the term protein requirement is used to mean the requirement both for protein and amino acids. When it is required to distinguish between protein *per se* and its constituent amino acids, the term crude protein (N × 6.25) will be used.

General Considerations of Protein Requirements

The consideration of input-output relationships is greatly facilitated if requirements are defined as daily intakes rather than as percentages of the food.

$$\text{Percentage requirement, } P = 100 \times \frac{\text{(Intake requirement, } I\text{)}}{\text{(Food intake, } F\text{)}}$$

The intake requirement, I, will depend on the needs for output (growth, egg production, etc.) and for maintenance. Thus:

$$I = f'(O, W)$$

Where O = output of various forms (e.g. growth, egg production)
W = bodyweight, to which maintenance requirements are proportional.

This relationship is usually expressed in the factorial form:

$$I = a'O' + a''O'' + \ldots + bW^n$$

It is important to notice that food intake is also related to the same variables. Thus P can be represented as:

$$P = f'(O, W)/f''(O, W)$$

Since all the coefficients in this expression are positive there is an auto-correlation between I and F with the result that variations in P are of smaller magnitude than those in I. In some circumstances P is virtually independent of O and W because of these effects (Griminger and Scott, 1959; Fisher, 1967).

A nutritional requirement is a single point on a response curve relating output to input. Once an appropriate response curve has been defined a requirement is readily determined, and it is towards the definition of such curves that experimentation should be directed. The consideration of response curves has not received much attention in poultry nutrition but in a recent paper a new type of curve has been suggested (Fisher, Morris and Jennings, 1973) which forms the basis of the present discussion.

The model for this response curve is summarised in *Figure 18.1*, in which egg production is used as the form of output, but the principle can also be applied to the growing bird. The essential feature of the curve is to consider the response of an individual hen as a simple factorial model and then to derive the flock response as the integrated average of a large number of individual responses. The individual responses are assumed to vary only because of normally distributed and correlated variations in output (egg production, E' g/hen/day, in this example) and in bodyweight (W, kg). The resulting flock response curve is defined by seven parameters: average maximum output ($\bar{E}_{max}$),

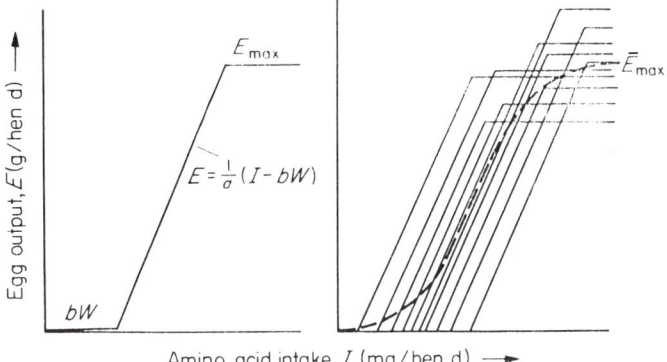

Figure 18.1 The basis of the model proposed by Fisher et al. (1973) to describe the relationship between egg output and amino acid intake. The left hand figure refers to an individual bird whereas the right hand figure shows responses for several individuals, varying in W and E_{max}, averaged to give a flock response curve (- -)
(see text for meaning of the symbols)

variation in maximum output (defined by $\sigma_{E\max}$), average bodyweight (W), variation in bodyweight (σ_W), the correlation between output and bodyweight (r_{EW}) and two constants representing the amounts of amino acid required per unit of output (a mg/gE) and per unit bodyweight (b mg/kgW). To date only single forms of output have been considered and maintenance requirements have been assumed to be directly proportional to W, although neither of these are essential limitations of the model. The equation for the curve and methods of fitting it to experimental data have been described by Curnow (1973); a computer simulation method can also be used (Fisher et al., 1973).

It can be shown (Fisher et al., 1973) that the optimum amino acid intake which yields maximum profit (i.e. the point on the curve relating cost of intake to value of output at which the slope is one) is given by:

$$I', \text{mg/d} = a\overline{E}_{\max} + b\overline{W} + x \left(\sqrt{a^2 \sigma_E^2 + b^2 \sigma_W^2 - 2ab\sigma_E\sigma_W} \right)$$

where x = the deviation from the mean of a standard normal distribution which is exceeded with probability ak in one tail.

k = cost per mg amino acid/value per g egg.

The value of x in standard units is obtained from tables and multiplied by the standard deviation of an individual's amino acid requirement as shown. If r_{EW} is zero this simplifies to $\sqrt{a^2 \sigma_E^2 = b^2 \sigma_W^2}$.

Two methods of fitting the model to experimental data are available, both requiring fixed estimates of some of the parameters to be made from either internal evidence or general experience of poultry populations. The method of Curnow (1973) estimates values of $\sigma_{E\max}$, $W\sigma_W$ and r_{EW} to be given and yields least-squares estimates of of $E_{\max}$, a and b. A computer simulation method described by Fisher et al. (1973) requires all the parameters except a and b to be fixed. The two methods give the same results (Fisher et al., 1973) and both are referred to below.

The Protein and Amino Acid Requirements of the Growing Pullet

The nutritional requirements of the growing pullet have been widely investigated but are still the subject of uncertainty. This is due partly to the variable results obtained in experiments but is also exacerbated by the complex nature of the responses. These involve growth, mortality and food-consumption during rearing, effects on sexual maturity and on subsequent egg production characteristics. It is a conventional practice to feed adequate diets until approximately eight weeks of age and this discussion is concerned mainly with the period from eight weeks to maturity. Only birds for consumable egg production are considered.

AMINO ACID REQUIREMENTS

Since a majority of the available results about the amino acid requirements of the growing pullet are unpublished they are presented in some detail in *Table 18.1*. Only methionine and lysine are referred to and the quoted dietary content of each amino acid was kept constant for the whole feeding period.

Rate of growth decreased if dietary lysine content fell below 0.4 per cent, whereas in all cases, maximum growth was attained at or below dietary contents of 0.5 per cent lysine. The three breeds in experiment three, which differed widely in mean bodyweight, had similar requirements when these were expressed as percentages, which agrees with results obtained in the early stages of growth (Griminger and Scott, 1959). Where growth-rate was low because of lysine deficiency a delay in sexual maturity usually occurred. The effects on egg production were small and probably non-significant but there was a tendency for the birds with a poor growth rate to show slightly increased egg production.

Maximum growth was achieved with dietary methionine contents of 0.25 per cent and 0.28 per cent in the two experiments shown. The effects on maturity and egg production were similar to those for lysine. The results of Kim and McGinnis (1972) in which egg production was measured for a fixed period after maturity, show very clearly the enhancing effect of growth restriction on rate of lay. The diets that were just adequate for growth contained 0.44 and 0.46 per cent total sulphur amino acids although it is probable that methionine was first-limiting.

The requirements indicated by these results are lower than those usually recommended. The National Research Council (1971) suggested requirements of 0.9 and 0.66 per cent lysine and 0.32 and 0.24 per cent methionine for the periods 6 to 14 and 14 to 20 weeks of age respectively. It appears that a revision of these figures downwards would be justified.

Lee, Gulliver and Morris (1971a) have related pullet performance to lysine intake and their figures also indicate higher requirements than the experiments in *Table 18.1*. However, their results are also based

Table 18.1 The lysine and methionine requirements of growing pullets

Experiment[1]	Treatment period (wks)	Dietary amino acid content (%)	Bodyweight at 20 weeks (kg)	Food intake for period (kg)	Age at maturity (days)	Egg production (per 100 hen/day)	Mean egg weight (g)
LYSINE							
1	10-20	0.36	1.79	6.97	161.5	64.2	59.1
		0.46	1.88	7.20	161.5	61.5	60.1
		0.56	1.89	6.91	163.5	62.5	59.2
		0.66	1.87	6.95	160.0	62.0	59.5
2	9-20	0.35	1.97	7.90	153.0	71.0	58.4
		0.43	2.06	7.72	-	-	-
		0.49	2.06	7.66	-	-	-
		0.55	2.01	7.73	150.5	69.4	58.2
3 Strain A	8-20	0.30	1.26	6.48	178	67.9[2]	60.7
		0.40	1.40	6.98	176	65.8	60.5
		0.50	1.45	6.98	177	67.6	61.0
		0.63	1.46	7.06	174	65.2	60.8
3 Strain B	8-20	0.30	1.33	6.76	179	55.7	61.3
		0.40	1.71	8.35	176	53.9	62.3
		0.50	1.82	8.23	173	55.7	62.2
		0.63	1.80	8.11	170	53.9	61.1
3 Strain C	8-20	0.30	1.70	8.22	180	63.4	62.8
		0.40	2.03	9.58	174	61.9	63.5
		0.50	2.12	9.29	171	61.3	63.0
		0.63	2.08	9.00	171	65.2	62.8
METHIONINE							
1	10-20	0.21	1.93	7.18	161	62.8	59.3
		0.25	1.99	7.21	159.5	63.7	60.0
		0.29	1.95	7.00	162	62.6	60.2
		0.33	1.97	7.43	162	61.3	59.5
4	8-20	0.18	1.07	4.50	177.8	79.6[3]	58.6
		0.23	1.13	4.58	176.5	75.8	58.3
		0.28	1.20	4.67	173.5	77.6	57.8
		0.33	1.20	4.74	172.8	77.6	58.7
		0.38	1.21	4.68	171.3	77.9	58.2

[1] Experiments 1 and 2, Fisher, unpublished
Experiment 3, University of Reading, unpublished
Experiment 4, Kim and McGinnis, (1972)
[2] Hen-housed egg production
[3] Hen-day egg production during 20 weeks after 50 per cent lay

on experiments in which constant proportions of lysine were fed from approximately eight to twenty weeks, during which time lysine requirements expressed as intakes will vary considerably. Thus the average results include periods when a given intake is limiting and others when it is adequate. The averaging of such results leads to a low estimate of lysine utilisation and an over-estimation of requirements when they are calculated from average growth rates for the whole period.

PROTEIN REQUIREMENTS

There is evidence that pullets can be reared satisfactorily when given only whole wheat as a source of protein (Foster, 1971) although lysine supplementation of such a diet might be required (Bragg, Biely and Goudie, 1971). Protein contents higher than this (i.e. from about 10 to 16 per cent) do not have any large or consistent effects on subsequent egg production. If growth is restricted, maturity will be delayed (Bullock, Morris and Fox, 1963; Lee, 1969; Blair, Bolton and Morley Jones, 1970; Foster, 1971; Auckland and Fulton, 1973) but in only a small minority of cases has subsequent egg production (Foster, 1971; Lillie and Denton, 1969) or egg weight (Blair *et al.*, 1970) been less although this might depend on breed (Wright *et al.*, 1968). However the economic benefits of using low-protein diets are very small (Wells, 1971) and are easily negated by very small depressions in egg production (Foster, 1971).

When protein content restricts growth by up to 20 per cent before eight weeks of age there is no effect on subsequent egg production (Peterson, Sauter and Lampman, 1966; Palafox, 1965; Summers, Pepper and Moran, 1969) although again exceptions occur (Lillie and Denton, 1967) and the economic savings are very small. A severe early growth depression will however permanently impair productive capacity (Summers *et al.*, 1969).

It has been suggested that deficiencies of specific amino acids such as lysine (Singsen *et al.*, 1965) or isoleucine (Naylor, Payne and Packham, 1972) can be used to control growth and maturity in pullets. However, the only comparative evidence (Lee, 1969) suggests that several amino acids have similar effects and that these are comparable with an overall protein deficiency.

Growth restriction is of greater importance in the rearing of broiler breeding pullets but there is evidence that the restriction of energy intake is to be preferred to low protein diets for this purpose (Lee, Gulliver and Morris, 1971b).

PROTEIN AND AMINO ACID REQUIREMENTS FOR EGG PRODUCTION

Protein in the diet meets the needs for essential amino acids and total nitrogen. Requirements for these are discussed here but space does not allow a review of the many studies of the crude protein requirements of the laying hen. Fisher (1970) found that approximately

80 per cent of the variation in response to crude protein intake could be accommodated within the sort of model suggested previously.

The results of experimental studies and the recommendations of the Agricultural Research Council (1963) and the National Research Council (1971) for the essential amino acid requirements of laying hens are shown in *Table 18.2*. The experimental methods and conditions differ greatly and offer many explanations for the variable results. Generally these values allow satisfactory egg production in practice but do not permit a more accurate definition of input-output relationships. Using the afore-mentioned model a start can be made in overcoming this deficiency.

The main difficulty in applying the model is that egg production in relation to nutrient intake is not always normally distributed; normal distribution is an essential assumption of the method. The distribution of egg production varies with flock age (Overfield, 1969) but for a period at 'peak' lay (i.e. once maturity is complete and while high egg production is maintained, usually between 30 and 40 weeks of age approximately) it approximates sufficiently closely to normal to allow deviations to be ignored. The results herein refer only to birds at this stage of lay; the question of requirements in older flocks is discussed below. Changes in bodyweight have been ignored since they are small at this age and appear to be entirely due to fat deposition (Weiss, 1958; Spandorf, 1963).

SULPHUR AMINO ACID REQUIREMENTS

The relationship between 'available' methionine intake and egg output in two strains of laying hen is shown in *Figure 18.2*. The response model was fitted to the data for each strain both separately and with common values of a and b.

The responses of the two strains clearly reflect differences in egg production. Similar values of a and b were obtained in all the analyses and when the residual sums of squares were compared after fitting the separate or combined curves the difference was non-significant ($p > 0.05$). The curves are also very similar (*Figure 18.2*). Carpenter and Anantharaman (1968) found 3.26 g 'available' methionine per 16 g nitrogen in egg. Assuming eggs contain 11.25 per cent protein, the methionine content is 3.67 mg/g and using the combined value of a of 4.358 the net efficiency of methionine utilisation for egg production is 84 per cent.

The estimates of a are clearly consistent with egg composition and, within wider limits of uncertainty, this is also true of the maintenance coefficient b. Leveille and Fisher (1960) estimated the methionine requirement for maintenance of nitrogen-balance in adult roosters to be 15 mg/kg/day and for maintenance of 'optimum' nitrogen-balance to be 71 mg/kg/day. Kandatsu and Ishibashi (1966) estimated the latter figure as 25 mg/kg/day.

In addition to the errors with which a and b are estimated (these cannot be accurately determined; Curnow, 1973), there is also

Table 18.2 *Summary of experimental evidence about the amino acid requirements of the laying hen and the recommendations of the Agricultural Research Council (ARC, 1963) and National Research Council (NRC, 1971). In addition to the calculated requirements of Johnson and Fisher (1958) shown here, tables of calculated requirements have also been given by Moran et al. (1967), Scott, Nesheim and Young (1969) and Hurwitz and Bornstein (1973)*

Reference	No. and range of dietary contents used (n/%)	Requirement (%)	Intake at requirement (mg/day)	Egg production g/day or %	Notes
Arginine					
ARC, (1963)	–	?	–	–	
NRC, (1971)	–	0.8	–	–	
Johnson and Fisher, (1958)	–	0.4	–	–	Calculated, egg-ratio method
Adkins et al. (1962)	9/0.27 to 0.7	0.6 to 0.7	376 to 1049	c 60%	Three experiments
Histidine					
ARC, (1963)	–	?	–	–	
NRC, (1971)	–	?	–	–	
Johnson and Fisher, (1958)	–	0.18	–	–	Calculated, egg-ratio method
Isoleucine					
ARC, (1963)	–	0.50	–	–	
NRC, (1971)	–	0.50	–	–	
Johnson and Fisher, (1958)	–	0.50	–	–	Calculated, egg-ratio method
Miller et al. (1954)	5/0.33 to 0.53	0.53	–	–	Two experiments
Combs, (1962)	2/0.53 and 0.64	0.64	612	44 g	
Bray, (1969)	9/0.26 to 0.62	0.40	472	45 g	
Leucine					
ARC, (1963)	–	0.70	–	–	
NRC, (1971)	–	1.20	–	–	
Johnson and Fisher, (1958)	–	0.68	–	–	Calculated, egg-ratio method
Cravens, (1948)	–	1.35	–	–	
Machlin, (1955)	3/1.0 to 1.6	1.00	–	–	
Anderson and Draper, (1956)	2/1.14 and 1.44	1.44	–	–	

Table 18.2 continued

Lysine				
ARC, (1963)	-	0.50	-	
NRC, (1971)	-	0.50	-	
Johnson and Fisher, (1958)	4/0.4 to 0.55	0.50	-	N-balance method
Ingram et al. (1951a)	4/0.35 to 0.60	0.52	-	Abstract only
Ingram and Little, (1958)	-	0.463 to 0.488	-	
Bray, (1969)	9/0.31 to 0.67	0.493	522	45 g
March and Biely, (1972)	various	-	800 to 850	45 g Supplementation study
Phenylalanine				
ARC, (1963)	-	0.35	-	
NRC, (1971)	-	?	-	
Johnson and Fisher, (1958)	-	0.42	-	Calculated, egg-ratio method
Sulphur amino acids				
CYSTINE (maximum utilised)				
ARC, (1963)	-	0.33	-	
NRC, (1971)	-	0.25	-	
Johnson and Fisher, (1958)	-	0.18	-	Calculated, egg-ratio method
METHIONINE (with cystine)				
ARC, (1963)	-	0.22	-	
NRC, (1971)	-	0.28	-	
Johnson and Fisher, (1958)	-	0.24	-	Calculated, egg-ratio method
Ingram and Little, (1958)	-	0.25	-	Abstract only
Ingram et al. (1951b)	4/0.22 to 0.70	0.38	-	
Leong and McGinnis, (1952)	4/0.18 to 0.33	0.28	-	41.7 g
Harms and Waldroup, (1963)	2/0.26 and 0.37	0.37	-	38.5 g
Spandorf, (1963)	5/0.17 to 0.28	0.25	352	41.5 g
Combs, (1964)	7/0.18 to 0.36	0.30	257	46.5 g
Harms et al. (1967)	5/0.21 to 0.35	0.31	337	40 to 43 g Two breeds
Bray, (1965)	8/0.16 to 0.33	0.22	308	40.6 g
Moran, (1969)	5/0.20 to 0.32	0.29	224	46.3 g
Harms and Damron, (1969)	5/0.19 to 0.35	0.27	363	42.6 g
Harms et al. (1969)	5/0.19 to 0.35	0.31	257	42 to 43 g Two basal diets
Fisher and Morris, (1970)	5/0.19 to 0.35	0.27	290 to 295	c40 g Two basal diets
Damron and Harms, (1973)	5/0.19 to 0.35	0.31	275	39.3 g
Reid and Weber, (1973)	3/0.24 to 0.33	0.29	288	41.4 g 22°C only
METHIONINE AND CYSTINE			260	
ARC, (1963)	-	0.55	-	
NRC, (1971)	-	0.53	-	
Harms and Damron, (1969)	5/0.34 to 0.66	0.58	541	42.2 g Corn-soya-fish diet

Table 18.2 continued

Threonine				
ARC, (1963)	-	0.40	-	
NRC, (1971)	-	0.40	-	
Johnson and Fisher, (1958)	-	0.36	-	Calculated, egg-ratio method
Ingram et al. (1951b)	8/0.27 to 0.62	0.42	374	Three experiments
			46.5 %	
Moran and Summers, (1968)	4/0.33 to 0.48	0.43	-	Abstract only
Tryptophan				
ARC, (1963)	-	0.13	-	
NRC, (1971)	-	0.11	-	
Johnson and Fisher, (1958)	-	0.12	-	Calculated, egg-ratio method
Ingram et al. (1951a)	4/0.12 to 0.27	0.12 to 0.15	-	Abstract only
Ingram and Little, (1958)	-	0.142	-	
Taylor et al. (1967)	4/0.12 to 0.16	0.14	167	c 60%
Bray, (1969)	10/0.075 to 0.31	0.11	117	79.3%
				46 g
Tyrosine				
ARC, (1963)	-	0.35	-	
NRC, (1971)	-	?	-	
Johnson and Fisher, (1958)	-	0.30	-	Calculated, egg-ratio method

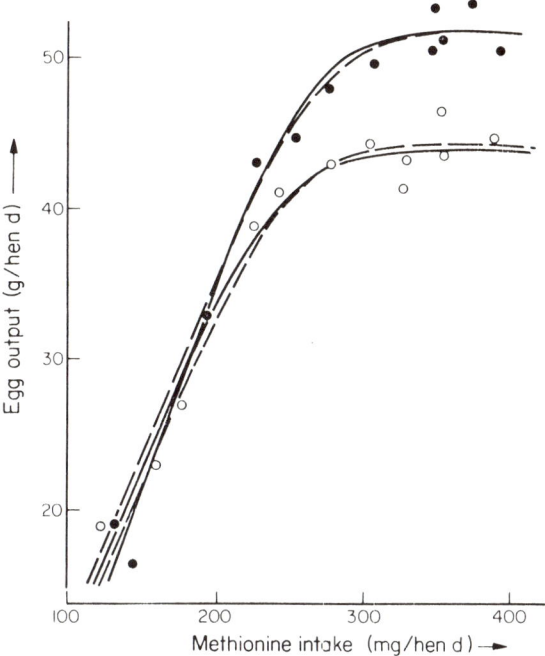

Figure 18.2 The relationship between egg output and methionine intake in two strains of laying hen. The curves were fitted by the method of Curnow (1973) as described in the text and yielded the following estimates of the coefficients a and b:-

Strain A ●	a = 3.89, b = 35.04
Strain B ○	a = 4.22, b = 23.75
Combined values for A and B	a = 4.36, b = 24.66

Dietary methionine levels were determined as 'available' methionine using the method of Ford (1962). Taken from Fisher (1970)

uncertainty, especially for the sulphur amino acids, about the figures used to describe foodstuffs. For practical purposes it is reasonable to assume values of $a = 4$ mg/g E and $b = 25$ mg/kg W and to use these to calculate methionine intake requirements under different economic circumstances (*Table 18.3*). Similar calculations can also be made with various partition equations obtained by regression analysis (*Table 18.4*). The coefficients obtained in such analyses tend to be larger than the previous estimates of a and b since they include, in part, the decrease in amino acid utilisation associated with increasing intake. The values obtained by regression analysis depend greatly on which part of the response curve is used.

These suggested methionine intake requirements for young laying pullets are of similar magnitude to published values (*Table 18.2*). In order to make a more detailed comparison with published experiments, requirements have been calculated in suitable cases and are shown with the experimental observations in *Figure 18.3*. In view of the uncertainties involved, the calculated requirements provide an excellent summary

Table 18.3 Calculated methionine requirements as affected by bodyweight (W), egg output (E) and the ratio between cost of methionine and value of egg (k). The method of calculation is described in the text

W (kg)	E (g/day)	Methionine requirements[1] (mg/day) k^2		
		0.004	0.006	0.008
1.5	45	295	289	285
1.5	50	324	317	312
2.0	45	308	302	298
2.0	50	337	330	332

[1] The following values were assumed in all cases: $a = 4$, $b = 25$, $r_{EW} = 0$, $\sigma_E = 0.2E$, $\sigma_W = 0.12W$

[2] The three values of k correspond to an egg value of 40 p/kg and methionine costs of 160, 240 and 320 p/kg

Table 18.4 Partition equations for the calculation of methionine requirements as determined by regression methods

Model:	Methionine requirement (mg/day) = $a_1 E + a_2 W + a_3 \Delta W$
where	E = egg output (g/day)
	W = bodyweight (kg)
	ΔW = change in bodyweight (g/day)

Author	a_1	a_2	a_3
Combs, (1960)	5.0	50	6.2
Shank, (1968)	5.39	37	4.5
Tolan and Morris, (1969)	4.3	69	12.14

of the experimental results and it is reasonable to conclude that there is close concurrence between the results described here and those from elsewhere.

There is no direct estimate of the amount of cystine that the laying hen can utilise, although in several experiments responses to methionine alone are described in terms of total sulphur amino acids (Reid and Weber, 1973; Novacek and Carlson, 1969). The National Research Council (1971) suggested that 47 per cent of the total sulphur amino acids could be cystine whilst the comparable recommendation of the Agricultural Research Council (1963) was 60 per cent. The only experimental evidence on this question is provided by Harms and Damron (1969) who studied responses to methionine in basal rations differing in cystine content. The results (*Figure 18.4*) indicate that approximately 50 per cent of the sulphur amino acid requirement can be met by cystine.

Damron and Harms (1973) obtained numerically large but non-significant responses in egg production by adding sodium sulphate to a methionine-deficient diet, but other work has not confirmed this effect (Jensen, 1972).

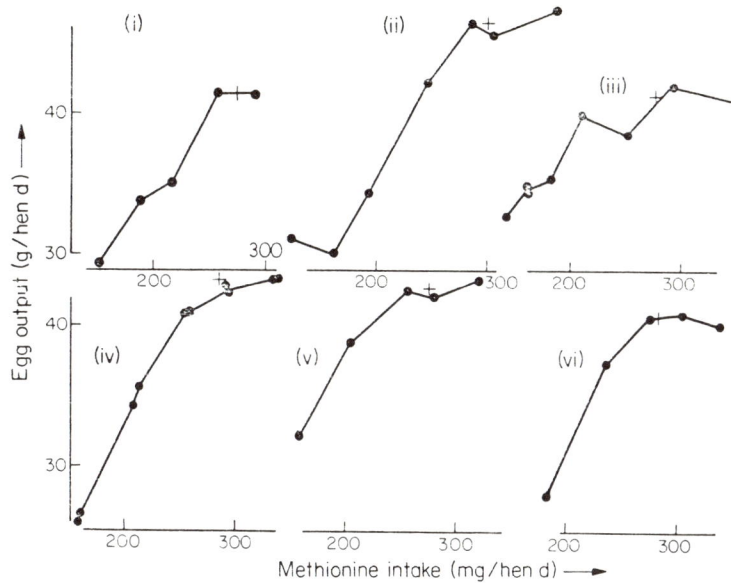

Figure 18.3 Comparison of calculated methionine requirements (+) and published experimental results (●——●). The calculated requirements were based on the following parameters: E_{max}, calculated from data at the rate shown in the figure W; calculated from data, $\sigma_{E_{max}} = 0.20E$, $\sigma_W = 0.12W$, $r_{EW} = 0$, $a = 4$, $b = 25$, egg value 40p/kg, methionine cost = £3/kg. Experimental data from (i) Spandorf (1962); (ii) Combs (1964); (iii) Bray (1965); (iv) Harms et al. (1969); (v) Harms and Damron (1969); (vi) Fisher and Morris (1970)

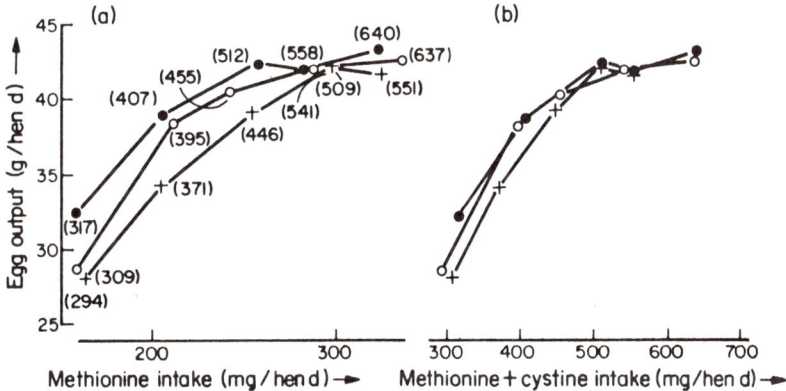

Figure 18.4 The relationship between methionine intake (a) and methionine plus cystine intake (b) and egg output. Results of an experiment reported by Harms and Damron (1969) in which methionine was added to basal diets composed principally of maize and soyabean meal (●——●), maize, soyabean and fish meal (○——○) and maize, soyabean meal and synthetic methionine (+——+). The figures in parentheses in (a) are methionine plus cystine intakes for each treatment

LYSINE REQUIREMENTS

A study of responses to lysine, using the methods described here, has been published by Pilbrow and Morris (1974). The response of egg output to lysine intake in each of eight breeds is shown in *Figure 18.5*. Response curves were fitted by iterative computer simulation assuming, initially, individual values for a and b, and in a second analysis, common values of these parameters for all breeds. Although

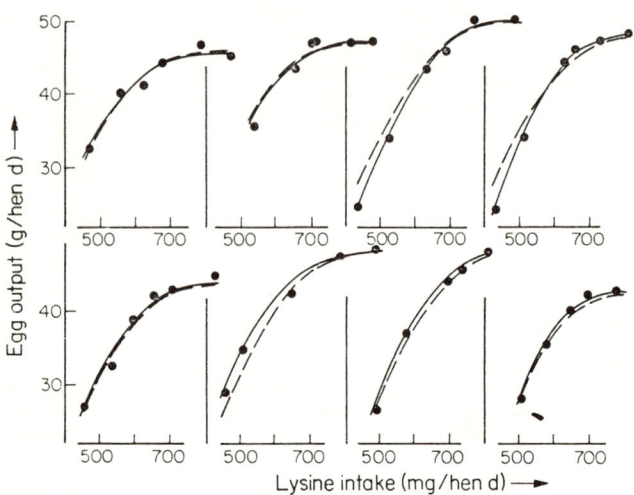

Figure 18.5 The response of egg output to lysine intake in eight breeds of laying hen between 36 and 47 weeks of age. (Pilbrow and Morris, 1974). The curves were fitted by an iterative computer simulation procedure (Fisher et al., 1973) as described in the text, assuming either breed specific (———) or common (— — —) values for a and b. The 'best fit' common values were a = 9.5, b = 90

the response curves obtained in the two analyses were very similar in all cases (*Figure 18.5*), the assumption of common values for a and b gave a significantly poorer fit to the data than the separate values. There was wide variation in the estimates of a and b for individual breeds. However, no meaning can be attached to these differences, since the estimates of the two parameters are correlated, in spite of their theoretical independence. This difficulty is largely overcome when data for different breeds are pooled since the estimate of b, the maintenance coefficient, then reflects both within- and between-breed differences in W.

The significantly poorer fit obtained with common values of a and b indicates breed differences in net lysine utilisation. These were not associated with egg composition (Pilbrow and Morris, 1974) and the calculated requirements for a bird with $E = 44$ g, $W = 2$ kg, ranged from 548 to 614 mg/d. If the maintenance requirement is assumed constant at 90 mg lysine/kg this indicates net efficiencies of lysine utilisation for egg production ranging from 80.2 to 94.6 per cent. However these differences are not well established and the common values of $a = 9.5$ mg/g and $b = 90$ mg/kg give response curves which are in close agreement with the data *(Figure 18.5)*.

These values of a and b again concur with expectation. Egg contains 7.9 mg lysine/g (for basis of calculation see *Figure 18.10*) indicating a net efficiency of lysine utilisation for egg production of 83 per cent. Direct estimates of maintenance requirements in adult roosters range from 29 to 60 mg lysine/kg W (Leveille and Fisher, 1959; Kandatsu and Ishibashi, 1966). Using these values a range of economically optimum lysine requirements can be calculated *(Table 18.5)*. These

Table 18.5 Calculated lysine requirements as affected by bodyweight ($\bar{W}$), egg output ($\bar{E}$) and the ratio between cost of lysine and value of egg (k). The method of calculation is described in the text

$\bar{W}$ (kg)	$\bar{E}$ (g/day)	Lysine requirements[1] (mg/day) k^2		
		0.0075	0.0100	0.0125
1.5	45	690	677	666
1.5	50	751	736	724
2.0	45	737	723	712
2.0	50	798	783	771

[1] The following values were assumed in all cases: $a = 9.5$, $b = 90$, $r_{EW} = 0$, $\sigma_E = 0.2\bar{E}$, $\sigma_W = 0.12\bar{W}$

[2] The three values of k correspond to an egg value of 40p/kg and lysine costs of 300, 400 and 500 p/kg

requirements are rather higher than usually recommended *(Table 18.2)* but there is reasonable agreement between the calculations and published response curves *(Figure 18.6)* except in the case of Bray (1969). The possible reasons for this discrepancy have been discussed by Pilbrow and Morris (1974) who concluded that uncertainty about lysine analyses provides the most likely explanation.

TRYPTOPHAN REQUIREMENTS

Unpublished evidence from the University of Reading (kindly supplied by Dr. T.R. Morris) indicates that values of $a = 2.25$ mg/g and $b = 10.25$ mg/kg can be used to calculate tryptophan requirements *(Table 18.6)*. In comparison with an egg content of 1.84 mg tryptophan/g the value for a indicates a net utilisation of 82 per cent for egg production. Maintenance requirements of roosters vary from 10 to 19 mg/kg W (Leveille and Fisher, 1960; Kandatsu and Ishibashi, 1966). There is only a single published response for comparison (Bray, 1969) but the agreement obtained is good *(Figure 18.6)*.

Table 18.6 *Calculated tryptophan requirements as affected by body-weight ($\bar{W}$), egg output ($\bar{E}$) and the ratio between cost of tryptophan and value of egg (k). The method of calculation is described in the text*

$\bar{W}$ (kg)	$\bar{E}$ (g/day)	Tryptophan requirements[1] (mg/day) k^2		
		0.05	0.075	0.1
1.5	45	141	136	132
1.5	50	155	150	145
2.0	45	146	141	137
2.0	50	160	155	150

[1] The following values were assumed in all cases: $a = 2.25$, $b = 10.25$, $r_{EW} = 0$, $\sigma_E = 0.2\bar{E}$, $\sigma_W = 0.12\bar{W}$

[2] The three values of k correspond to an egg value of 40p/kg and tryptophan costs of 2000, 3000 and 4000p/kg

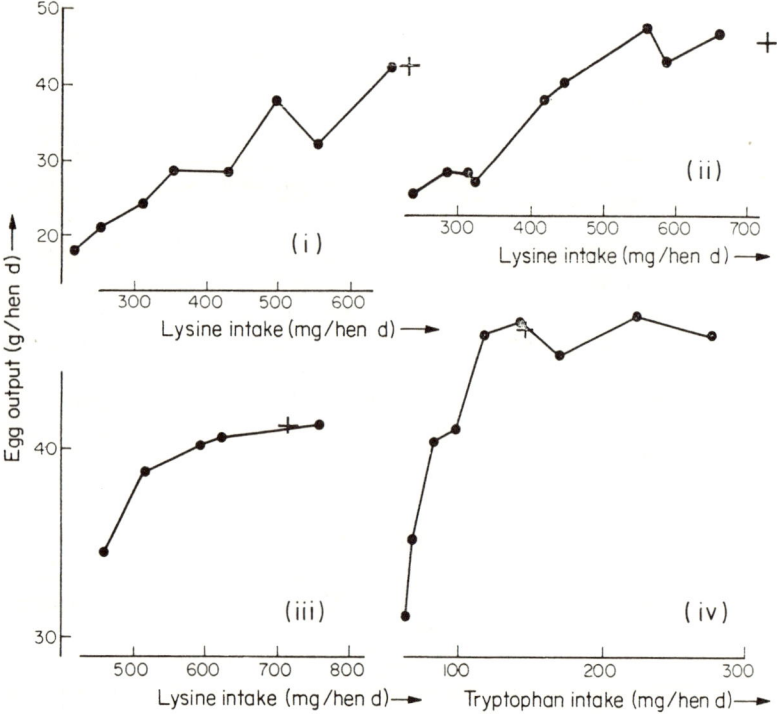

Figure 18.6 *Comparison of calculated lysine and tryptophan requirements (+) and published experimental results (●——●). The calculated requirements were based on the parameters given in* Figure 18.3 *except as follows: for lysine $a = 9.5$, $b = 90$, lysine cost = £4/kg; for tryptophan $a = 2.25$, $b = 10.25$, tryptophan cost = £30/kg. Experimental data from (i) Thomas (1966); (ii) Bray (1969); (iii) Fisher (1970); (iv) Bray (1969).*

NITROGEN REQUIREMENTS

The utilisation of nitrogen from non-protein sources (NPN) by the fowl is well established (Blair, 1972) and evidence of the economic use of such materials for laying hens (Young et al., 1965) stimulated much experimental work.

Requirements for nitrogen can be established only with adequate proportions of essential amino acids and uncertainty about these creates the greatest difficulty in evaluating experimental results. When NPN is added to an amino acid-deficient diet, egg production is always depressed (Moran, Summers and Pepper, 1967; Manoukas and Young, 1969). Manoukas and Young (1969) have suggested that a minimum ratio of 0.7 to 1.0 between essential and non-essential amino acid nitrogen should be maintained although the relationship between the first-limiting amino acid and nitrogen is probably more important. There are also indications that amino acid balance influences the response to NPN (Young et al., (1965). Finally Renner (1969) has suggested that utilisation of ammonium ions from NPN will depend partly on the metabolic fate of the accompanying anion. Because of these factors it is impossible to draw conclusions from experiments in which there is no response to NPN but several studies showing significant responses can be considered.

There is reasonably clear evidence that diammonium phosphate, ammonium sulphate and diammonium citrate can be utilised by the laying hen (Young et al., 1965; Chavez, Thomas and Reid, 1966; Ijaz, Fox and Morris, 1967; Pilbrow, 1970; Reid et al., 1972). Single and mixed non-essential amino acids can also be used (Young et al., 1965; Manoukas and Young, 1969; Fernandez, Salman and McGinnis, 1973). Monosodium glutamate is not utilised (Fernandez et al., 1973) but the information about urea is inconclusive. Fernandez et al. (1973) and Davis and Martindale (1973) both obtained responses whilst Chavez et al. (1966) and Blair and Lee (1973) failed to do so. Michie (1971) and Kamezi and Balloun (1973) successfully incorporated urea into layers' diets but did not include unsupplemented treatments in their experiments. Shannon, Blair and D'Mello (1969) showed that the nitrogen from diammonium phosphate and citrate was almost completely absorbed in colostomised hens but utilisation is clearly more than a matter of simple absorption since Fernandez et al. (1973) obtained evidence of enhanced nitrogen-retention from diammonium phosphate which was accompanied by a significant depression in egg production.

The results of experiments showing statistically significant responses to nitrogen sources other than essential amino acids are summarised in *Table 18.7*. The crude protein intakes at which responses occurred varied widely from 11.6 to 19.4 g/day and in many cases the NPN-supplemented diets supported egg production not significantly different from those supported by the higher-protein control diets. The relationship between egg output and crude protein intake is shown in *Figure 18.7*, but expressing the results in this way does little to reconcile the differences amongst experiments. However, apart from the responses obtained at very high intakes by Reid et al. (1972) the data of

Table 18.7 *Summary of experiments in which laying hens have shown significant responses to non-protein nitrogen or non-essential amino acids*

Reference	Basal crude protein content (%)	Nitrogen supplement	Basal protein intake (g/day)	Egg production	
				Basal diet (%)	Supplemented diet (%)
Chavez et al. (1966)					
Experiment 1	12.8	Diammonium citrate	12.8	64.7	68.8
Experiment 2	12.8	Diammonium phosphate	11.1	56.1	61.0
Ijaz et al. (1967)	12.0	Diammonium phosphate	13.7	56.1	62.6
Manoukas and Young, (1969)					
Experiment 1	12.5	NEAA[1] mixture	11.6	72.1	80.8
Experiment 2	12.5	NEAA mixture	12.0	68.1	74.0
Pilbrow, (1970)	11.5	Diammonium phosphate	15.3	(43.3 g/day)	(46.6 g/day)
Reid et al. (1972)					
Experiment 1	13.5	Ammonium sulphate and Diammonium citrate	14.5	60.9	66-69
Experiment 2	12.0	Diammonium phosphate	14.4	67.0	69.4
	14.0	Diammonium phosphate	16.9	66.1	70.1
	16.0	Diammonium phosphate	19.4	68.6	71.6
Experiment 4	13.3	Diammonium citrate	15.6	69.5	74.0
Fernandez et al. (1973)	12.5	Urea	14.1	71.1	74.7

[1] Non-essential amino acid

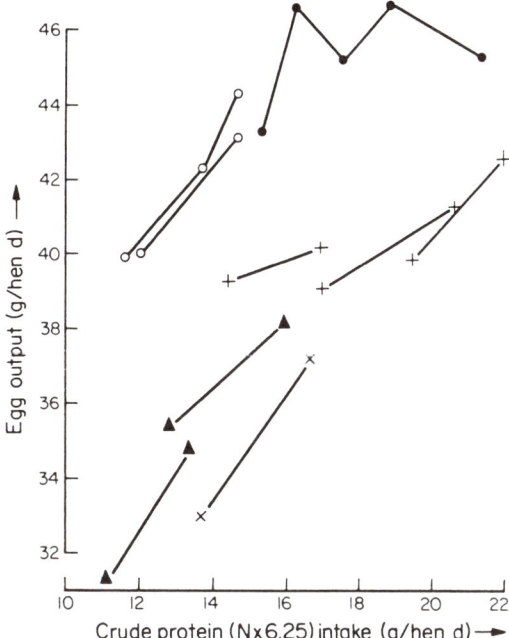

Figure 18.7 The relationship between egg output and crude protein intake from experiments in which non-protein nitrogen was added to low-protein basal diets. Data from Pilbrow (1970) ●——● ; *Manoukas and Young (1969)* ○——○ ; *Reid et al. (1972)* +——+ ; *Ijaz et al. (1967)* ×——× ; *Chavez et al. (1969)* ▲——▲

Pilbrow (1970) suggest that about 16 g crude protein/day is required by high-producing birds and the other data are consistent with this conclusion. There appears to be no simple relationship between crude protein requirement and egg production.

Factors Affecting Protein and Amino Acid Requirements

Protein and amino acid requirements are influenced by various dietary and non-dietary factors, but those of dietary origin (e.g. amino acid-vitamin interactions) are not peculiar to the laying hen and will not be discussed here.

The approach to protein requirements used in this paper indicates that non-dietary factors can influence requirements expressed as percentages in three ways: by changing the relationship between intake requirements and food intake, by modifying outputs but not the relationship between input and output, and by changing nutrient utilisation with or without a change of output (Fisher, 1967). When considering any one factor, combinations of these effects may be involved.

FACTORS AFFECTING REQUIREMENTS BY MODIFYING FOOD INTAKE

Protein requirements expressed as percentages have been shown to differ because of variations in food intake due to breed (Sharpe and Morris, 1965; Harms and Waldroup, 1962); environmental and seasonal temperature (Bray and Gesell, 1961; Bray and Morrissey, 1962; Reid and Weber, 1973) and dietary energy content (Smith, Payne and Lewis, 1963). Although such factors are of practical importance they are not problems of changes in protein utilisation.

FACTORS AFFECTING REQUIREMENTS BY MODIFYING OUTPUTS

Although the subject has not been exhaustively investigated it appears that factors which influence egg production only within a narrow range do not change protein utilisation and intake requirements can be anticipated from outputs (Fisher, 1967). This conclusion applies to comparisons of different housing systems (MacIntyre and Aitken, 1959), lighting patterns applied before maturity (Bray, Jennings and Morris, 1965) and environmental temperature (Bray and Gesell, 1961).

Whether all observed breed differences can be accounted for in terms of different outputs is not clear. An analysis of several experiments indicated that most differences could be explained in this way (Fisher, 1967) although a notable exception is one of four strains studied by Moreng *et al.* (1964) which showed a very high efficiency of protein utilisation. The analysis of responses to lysine by Pilbrow and Morris (1974) which has been described above also leads to the same conclusion.

FACTORS AFFECTING REQUIREMENTS BY MODIFYING PROTEIN UTILISATION

Although many variations in protein requirements can be accounted for in terms of food intake or output there are a few exceptions, the most notable being flock age. Some lighting patterns also influence protein utilisation (Bray, 1968, 1971) and there remains the possibility of residual genetic effects.

In all cases in which responses to protein have been measured independently in flocks at different stages of the first laying year, nutrient utilisation has been lower in the older birds (*Figure 18.8*) provided only that rate of lay is lower in the older birds. The small differences in bodyweight over such a period do not provide a satisfactory explanation for this observation. When responses at similar stages of the first and second years of lay are compared, the older birds are again less efficient if they are laying at a lower rate (Fisher, 1970, data for methionine) but not when rate of lay is the same in the two groups (T.R. Morris, unpublished data for tryptophan).

This apparent relationship between mean rate of lay in a flock and nutrient utilisation directs attention to the same relationship amongst individuals. Eggs are typically laid in cycles or clutches separated by a single non-laying day. A lower average rate of lay is a result of

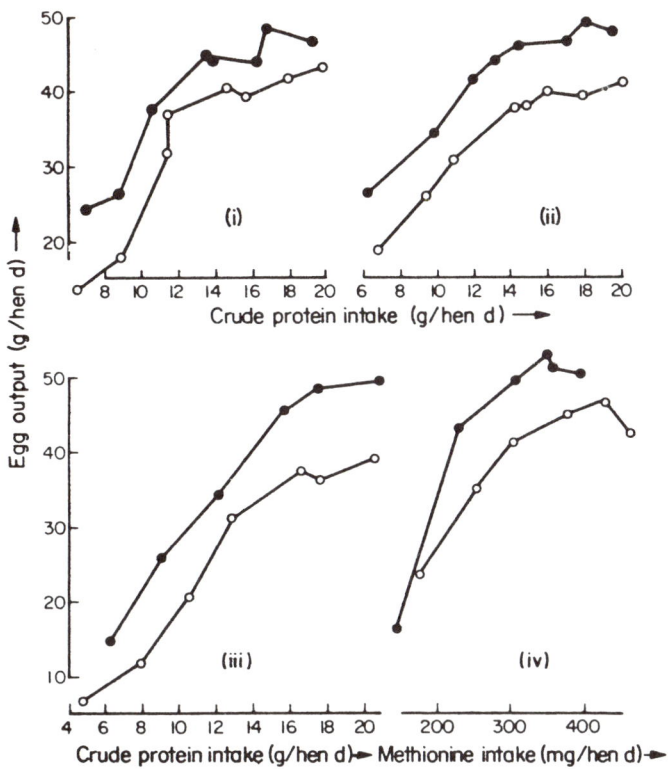

Figure 18.8 The relationship between egg output and protein or amino acid intake in young (●——●) and older (○——○) laying pullets. Data from (i) Bray (1968), 26 to 34 and 40 to 48 weeks of age; (ii) Bray (1971), 26 to 34 and 40 to 48 weeks of age, (iii) Jennings, Fisher and Morris (1972), 37 to 38 and 75 to 76 weeks of age, (iv) Fisher (1970), 38 to 40 and 68 to 70 weeks of age.

either shorter clutches, longer non-laying periods or both. When the non-laying periods are sufficiently long to cause intermittent variations in the rate of protein synthesis it is to be expected that the utilisation of ingested protein will decline, since intake is not adjusted under *ad libitum* feeding to correspond with the needs for synthesis. Some data for calculated methionine utilisation and rate of lay for individual birds of three ages are shown in *Figure 18.9*. Although the method of calculating methionine utilisation is comparatively crude the effect of rate of lay is shown quite clearly, with a rapid decline in utilisation occurring, as expected, below 50 per cent lay. The regression lines shown in the figure suggest that this effect does not provide a complete explanation for the effects of age on methionine utilisation since there is a significant age effect which is distinct from that of rate of lay. Differences in feather growth provide the most likely explanation for this observation (Fisher, 1970).

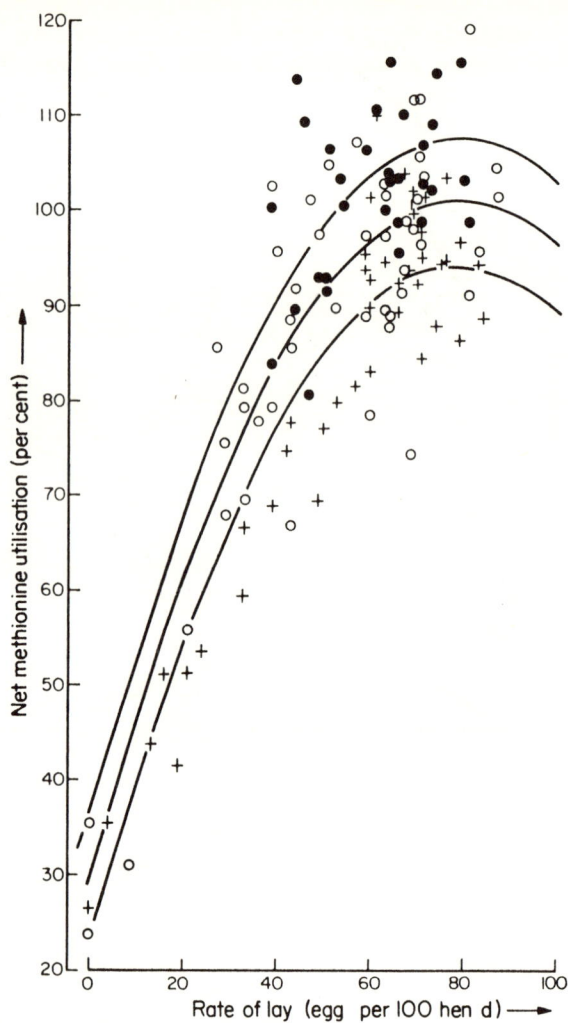

Figure 18.9 The relationship between net methionine utilisation and rate of lay in individual laying pullets of three ages:- 30 to 40 weeks (•), 45 to 55 weeks (o), 60 to 70 weeks (+). All birds received a diet containing 0.156 per cent methionine which was shown to be the first-limiting nutrient. Net methionine utilisation = methionine utilised/methionine intake where methionine utilised = $4E + 25W + 3.0 \Delta W$. The curves are parallel quadratic regressions. From Fisher (1970)

It is interesting that adverse lighting is the only other situation in which protein utilisation is decreased (Bray, 1968, 1971) and that an increase in the number of pausing birds also occurs in this case (Morris, Fox and Jennings, 1964).

These observations suggest a general model to explain the effect of non-dietary factors on protein intake requirements. If the factor has no effect on the incidence of pausing then protein utilisation will be unaffected and requirements will be proportional to output. On the other hand factors which increase the incidence of pausing will decrease average utilisation and requirements will not be proportional to output. Within limits, all of the available evidence is consistent with this hypothesis.

The uncertainty about breed effects can be resolved by the reasonable assumption that breeds differ mainly in clutch length in regularly laying birds but some strains may also have a higher incidence of pausing birds. Whether there is any residual genetic variation in net protein utilisation is not clear from existing evidence. Tolan and Morris (1969) found that after correction for output and maintenance the methionine requirements of individual birds still varied phenotypically by more than ± 100 mg/day around a mean of 335 mg. There is no direct evidence for a genetic component of such variation. The range in methionine utilisation shown in *Figure 18.9*, where the calculations were made in a different way, is much narrower than this.

Protein Utilisation in the Laying Hen

A normal laying hen consumes about 6.4 kg crude protein per annum and produces about 1.6 kg egg protein, a gross utilisation of dietary protein of 25 per cent. Very highly producing birds use crude protein for egg production with a net efficiency of about 33 per cent (Morris, 1972). The net utilisation of amino acids for egg production has been estimated above to lie between 80 and 85 per cent for methionine, lysine and tryptophan. Morris (1972) has suggested that the net utilisation of lysine, isoleucine and tryptophan is 100 per cent but, at least for lysine, there is some doubt about the basis of these calculations (Pilbrow and Morris, 1974).

The poor utilisation of dietary crude protein is largely due to poor amino acid balance. With reasonable assumptions, the gross utilisation of the first-limiting amino acid lies between 45 and 50 per cent at a maximum in long-term studies (*Figure 18.10*). This agrees with the maximum nitrogen retention of about 45 per cent of intake observed by Shapiro and Fisher (1965) when whole-egg protein is fed.

Figure 18.10 also shows that maximum gross utilisation is achieved at lower dietary protein contents than those producing maximum output and lower than those usually used. This is an inevitable consequence of a diminishing response curve which does not pass through the origin. Changed economic circumstances or a change of commercial objective to maximum protein utilisation would increase gross amino acid utilisation by up to 5 per cent.

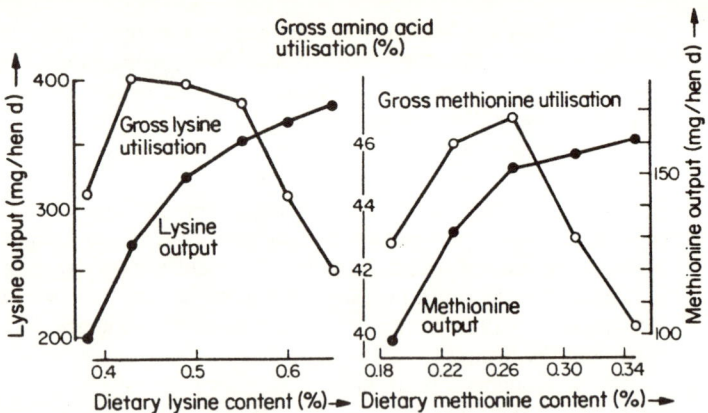

Figure 18.10. The gross utilisation of methionine and lysine for egg production in long term experiments. Based on data from Harms, Moreno and Damron (1969) and Pilbrow and Morris (1974). Gross utilisation was calculated as amino acid output in eggs/amino acid intake. Amino acid output was calculated from egg output data with the assumptions:- 10 per cent shell in egg output, 0 per cent and 2 per cent nitrogen in shell and content respectively, 7.02 and 3.12 g lysine and methionine/16 gN in egg contents respectively

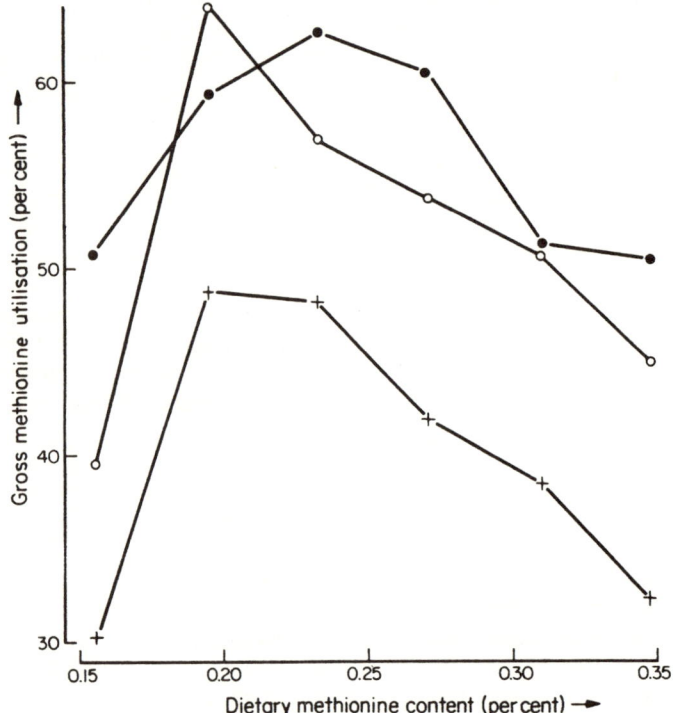

Figure 18.11 Gross methionine utilisation for egg production in birds aged 38 to 40 (●——●), 53 to 55 (○——○) and 68 to 70 (+——+) weeks of age. Methionine utilisation calculated as in Figure 18.10. Data from Fisher (1970)

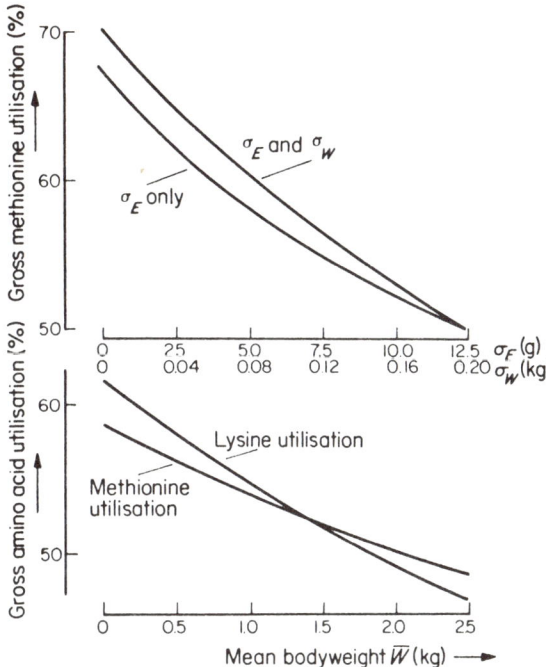

Figure 18.12 The effect of changing the variability of egg output (σ_E) and of bodyweight (σ_W) on the gross utilisation of methionine for egg production (upper figure) and the effect of mean bodyweight (W) on the gross utilisation of methionine and lysine for egg production (lower figure). Amino acid utilisation calculated as in Figure 18.10 assuming E = 50 g/day and with inputs determined by the method described in this paper when x = 2

The decline in net amino acid utilisation with age has already been shown (*Figure 18.9*). The gross utilisation of methionine for egg production at three ages is shown in *Figure 18.11* and it appears that improvements of about 10 per cent utilisation could be achieved in the later stages of lay if the age effect could be avoided. This requires the proportion of poor layers to be reduced, although in the case of methionine this would offset only about half the decline in efficiency (Fisher, 1970).

The model for response curves that has been described above associates the decline in protein utilisation with increasing intake to differences between individuals. In *Figure 18.12* the effect of flock variability on the calculated gross utilisation of methionine for egg production is illustrated. In comparison with the present position utilisation could be improved from about 50 per cent to 70 per cent if variability were completely eliminated. Over the more practical range of possibility, however, the effect is more limited and is almost entirely due to changes in the variability of egg output. Also shown in *Figure 18.12* is the effect of reducing bodyweight on the gross utilisation of methionine for egg production. Here again the scope for improvement in practical terms is very limited.

From this brief discussion it appears that the only prospect of substantially improving gross protein utilisation in the laying hen lies in adjusting amino acid balance more closely to meet requirements. A small further improvement will occur if economic objectives come to favour lower outputs (*Figure 18.10*) but after that only major and difficult changes in patterns of lay and population structure will be effective. Apart from a gradual reduction in bodyweight it seems uncertain that the effort required to make such changes will ever be worthwhile. In purely practical terms efficiency can be improved if the provision of safety margins to cover, amongst other things, uncertainty about requirements in individual flocks, can be avoided. The calculation of requirements by the methods outlined in this paper goes some way to achieving this objective.

Acknowledgements

I am most grateful to Dr. T.R. Morris for allowing me to refer to unpublished experimental results. The work described in this paper was carried out collaboratively with Dr. Morris, Mr. R.C. Jennings and Professor R.N. Curnow of the University of Reading

References

ADKINS, J.S., HARPER, A.E. and SUNDE, M.L. (1962). *Poult. Sci.*, **41**, 657
AGRICULTURAL RESEARCH COUNCIL (1963). *The Nutrient Requirements of Farm Livestock. No.1. Poultry. Summary of Recommendations.* Agricultural Research Council, London
ANDERSON, J.O. and DRAPER, C.I. (1956). *Poult. Sci.*, **35**, 562
AUCKLAND, J.N. and FULTON, R.B. (1973). *Br. Poult. Sci.*, **14**, 49
BLAIR, R. (1972). *Wld. Poult. Sci. J.*, **29**, 189
BLAIR, R., BOLTON, W. and MORLEY JONES, R. (1970). *Br. Poult. Sci.*, **11**, 249
BLAIR, R. and LEE, D.J.W. (1973). *Br. Poult. Sci.*, **14**, 9
BOORMAN, K.N. and LEWIS, D. (1971). in *Physiology and Biochemistry of the Domestic Fowl*, Ed. D.J. Bell and B.M. Freeman. p.339. Academic Press, London and New York
BRAGG, D.B., BIELY, J. and GOUDIE, C. (1971). *Poult. Sci.*, **50**, 761
BRAY, D.J. (1965). *Poult. Sci.*, **44**, 1173
BRAY, D.J. (1968). *Poult. Sci.*, **47**, 1005
BRAY, D.J. (1969). *Poult. Sci.*, **48**, 674
BRAY, D.J. (1970). *Poult. Sci.*, **49**, 1334
BRAY, D.J. (1971). *Proc. 14th World's Poult. Congr.*, **2**, 659
BRAY, D.J. and GESELL, J.A. (1961). *Poult. Sci.*, **40**, 1328
BRAY, D.J., JENNINGS, R.C. and MORRIS, T.R. (1965). *Br. Poult. Sci.*, **6**, 311

BRAY, D.J. and MORRISSEY, D.J. (1962). *Poult. Sci.*, **41**, 1078
BULLOCK, D.W., MORRIS, T.R. and FOX, S. (1963). *Br. Poult. Sci.*, **4**, 227
CARPENTER, K.J. and ANANTHARAMAN, K. (1968). *Br. J. Nutr.*, **22**, 183
CHAVEZ, R., THOMAS, J.M. and REID, B.L. (1969). *Poult. Sci.*, **45**, 547
COMBS, G.F. (1960). *Proc. Md Nutr. Conf. Fd Mfrs*, **28**
COMBS, G.F. (1962). *Proc. Md Nutr. Conf. Fd Mfrs*, **65**
COMBS, G.F. (1964). *Proc. Md Nutr. Conf. Fd Mfrs*, **45**
COMBS, G.F. (1968). *Proc. Md Nutr. Conf. Fd Mfrs*, **86**
CRAVENS, W.W. (1948). *Poult. Sci.*, **27**, 562
CURNOW, R.N. (1973). *Biometrics*, **29**, 1
DAMRON, B.L. and HARMS, R.II. (1973). *Poult. Sci.*, **52**, 400
DAVIS, R.H. and MARTINDALE, C.H. (1973). *Br. Poult. Sci.*, **14**, 153
FERNANDEZ, R., SALMAN, A.J. and McGINNIS, J. (1973). *Poult. Sci.*, **52**, 1231
FISHER, C. (1967). in *Protein Utilisation by Poultry*, Eds. R.A. Morton and E.C. Amoroso. p.174. Oliver and Boyd, Edinburgh
FISHER, C. (1970). *Ph.D. thesis, University of Reading*
FISHER, C. and MORRIS, T.R. (1970). *Br. Poult. Sci.*, **11**, 67
FISHER, C., MORRIS, T.R. and JENNINGS, R.C. (1973). *Br. Poult. Sci.*, **14**, 469
FORD, J.E. (1962). *Br. J. Nutr.*, **16**, 409
FOSTER, W.H. (1971). *Br. Poult. Sci.*, **12**, 333
GILBERT, A.B. (1971). in *Physiology and Biochemistry of the Domestic Fowl*. Eds. D.J. Bell and B.M. Freeman. p.129. Academic Press, London and New York
GRIMINGER, P. and SCOTT, H.M. (1959). *J. Nutr.*, **68**, 429
HARMS, R.H. and DAMRON, B.L. (1969). *Poult. Sci.*, **48**, 144
HARMS, R.H., DAMRON, B.L. and WALDROUP, P.W. (1967). *Poult. Sci.*, **46**, 181
HARMS, R.H., MORENO, R.S. and DAMRON, B.L. (1969). *Poult. Sci.*, **48**, 1652
HARMS, R.H. and WALDROUP, P.W. (1962). *Poult. Sci.*, **41**, 1985
HARMS, R.H. and WALDROUP, P.W. (1963). *Br. Poult. Sci.*, **4**, 267
HURWITZ, S. and BORNSTEIN, S. (1973). *Poult. Sci.*, **52**, 1124
IJAZ, M.A., FOX, S. and MORRIS, T.R. (1967). *Wld. Poult. Sci. J.*, **23**, 253
INGRAM, G.R., CRAVENS, W.W., ELVEHJEM, C.A. and HALPIN, J.G. (1951a). *Poult. Sci.*, **30**, 426
INGRAM, G.R., CRAVENS, W.W., ELVEHJEM, C.A. and HALPIN, J.G. (1951b). *Poult. Sci.*, **30**, 431
INGRAM, G.R. and LITTLE, P.R. (1958). *Poult. Sci.*, **37**, 1214
JENNINGS, R.C., FISHER, C. and MORRIS, T.R. (1972). *Br. Poult. Sci.*, **13**, 279
JENSEN, L.S. (1972). *Poult. Sci.*, **51**, 1822
JOHNSON, D. and FISHER, H. (1956). *J. Nutr.*, **60**, 275
JOHNSON, D. and FISHER, H. (1958). *Br. J. Nutr.*, **12**, 276
KAMEZI, R. and BALLOUN, S.L. (1973). *Poult. Sci.*, **52**, 44
KANDATSU, M. and ISHIBASHI, T. (1966). *Proc. 13th World's Poult. Congr.*, 173

KIM, S.M. and McGINNIS, J. (1972). *Poult. Sci.*, **51**, 1735
LEE, P.J.W. (1969). *Ph.D. thesis, University of Reading*
LEE, P.J.W., GULLIVER, A.J. and MORRIS, T.R. (1971a). *Br. Poult. Sci.*, **12**, 413
LEE, P.J.W., GULLIVER, A.J. and MORRIS, T.R. (1971b). *Br. Poult. Sci.*, **12**, 499
LEONG, K.C. and McGINNIS, J. (1962). *Poult. Sci.*, **31**, 692
LEVEILLE, G.A. and FISHER, H. (1959). *J. Nutr.*, **69**, 289
LEVEILLE, G.A. and FISHER, H. (1960). *J. Nutr.*, **72**, 8
LILLIE, R.J. and DENTON, C.A. (1967). *Poult. Sci.*, **46**, 1550
MACINTYRE, T.M. and AITKEN, J.R. (1959). *Can. J. Anim. Sci.*, **39**, 175
MACHLIN, L.J. (1955). *Poult. Sci.*, **34**, 984
MANOUKAS, A.G. and YOUNG, R.J. (1969). *Poult. Sci.*, **48**, 2037
MARCH, B.E. and BIELY, J. (1972). *Poult. Sci.*, **51**, 547
MICHIE, W. (1971). in *Research Investigations and Field Trials 1969-70.* p.71. North of Scotland College of Agriculture, Aberdeen
MILLER, E.C., SUNDE, M.L., BIRD, H.R. and ELVEHJEM, C.A. (1954). *Poult. Sci.*, **33**, 1201
MORAN, E.T. (1969). *Feedstuffs, Minneap.*, **41**, 26
MORAN, E.T. and SUMMERS, J.D. (1968). *Poult. Sci.*, **47**, 1698
MORAN, E.T., SUMMERS, J.D. and PEPPER, W.F. (1967). *Poult. Sci.*, **46**, 1134
MORENG, R.E., ENOS, H.L., WHITTETT, W.A. and MILLER, B.F. (1964). *Poult. Sci.*, **43**, 630
MORRIS, T.R. (1972). in *Egg Formation and Production*, Eds. B.M. Freeman and P.E. Lake. p.139. British Poultry Science, Edinburgh
MORRIS, T.R., FOX, S. and JENNINGS, R.C. (1964). *Br. Poult. Sci.*, **5**, 133
MULLER, R.D. and BALLOUN, S.L. (1972). *Poult. Sci.*, **51**, 1843
NATIONAL RESEARCH COUNCIL (1971). *Nutrient requirements of poultry*, NRC Publication 1345. National Academy of Sciences, Washington DC
NAYLOR, R.W., PAYNE, C.G. and PACKHAM, R.G. (1972). *Proc. Austral. Poult. Sci. Convent*, **83**
NOVACEK E.J. and CARLSON, C.W. (1969). *Poult. Sci.*, **48**, 1490
OVERFIELD, N.D. (1969). *N.A.A.S. Q. Rev.*, **86**, 84
PALAFOX, A.L. (1965). *Poult. Sci.*, **44**, 1405
PETERSEN, C.F., SAUTER, E.A. and LAMPMAN, C.E. (1966). *Poult. Sci.*, **45**, 1115
PILBROW, P.J. (1970). *Ph.D. Thesis, University of Reading*
PILBROW, P.J. and MORRIS, T.R. (1974). *Br. Poult. Sci.*, **15**, 51
REID, B.L. and WEBER, C.W. (1973). *Poult. Sci.*, **52**, 1335
REID, B.L., SVACHA, A.J., DORFLINGER, R.L. and WEBER, C.W. (1972). *Poult. Sci.*, **51**, 1234
RENNER, R. (1969). *J. Nutr.*, **98**, 297
SCOTT, M.L., NESHEIM, M.C. and YOUNG, R.J. (1969). *Nutrition of the Chicken.* M.L. Scott and Associates, New York
SHANK, F. (1968). *Ph.D. Thesis, University of Maryland*
SHANNON, D.W.F., BLAIR, R. and D'MELLO, J.P.F. (1969). *Br. Poult. Sci.*, **10**, 381

SHAPIRO, R. and FISHER, H. (1965). *Poult. Sci.*, **44**, 198
SHARPE, E. and MORRIS, T.R. (1965). *Br. Poult. Sci.*, **6**, 7
SINGSEN, E.P., NAGEL, J., PATRICK, S.G. and MATTERSON, L.D. (1965). *Poult. Sci.*, **44**, 1467
SMITH, G.H., PAYNE, C.G. and LEWIS, D. (1963). *Univ. Nott. Sch. Agric. Rep.*, 83
SPANDORF, A.H. (1963). *Proc. Md Nutr. Conf. Fd Mfrs*, 28
SUMMERS, J.D., PEPPER, W.F. and MORAN, E.T., (1969). *Poult. Sci.*, **48**, 1351
TAYLOR, B.R., PAYNE, C.G. and LEWIS, D. (1967). in *Protein Utilisation by Poultry*, Eds. R.A. Morton and E.C. Amoroso. p.137. Oliver and Boyd, Edinburgh
THOMAS, O.P. (1966). *Ph.D. thesis, University of Maryland*
TOLAN, A. and MORRIS, T.R. (1969). *Wld. Poult. Sci. J.*, **25**, 146
WARREN, D.C. and CONRAD, R.M. (1939). *J. agric. Res.*, **58**, 875
WEISS, H.S. (1958). *Poult. Sci.*, **37**, 848
WELLS, R.G. (1971). *Br. Poult. Sci.*, **12**, 313
WRIGHT, C.F., DAMRON, B.L., WALDROUP, P.W. and HARMS, R.H. (1968). *Poult. Sci.*, **47**, 635
YOUNG, R.J., GRIFFITH, M., DESAI, I.D. and SCOTT, M.L. (1965). *Poult. Sci.*, **44**, 1428
ZIMMERMAN, R.A., SNETSINGER, D.C., CARDIN, D.W. and GREENE, D.E. (1969). *Poult. Sci.*, **48**, 1894

19

PROTEIN NUTRITION OF THE BREEDING SOW

F.W.H. ELSLEY
School of Agriculture, University of Edinburgh, Scotland

Introduction

During recent years a number of reviews have been published which aim to evaluate the present state of knowledge of the protein requirements of sows. These reviews include ARC (1967), Lodge (1972), Elsley (1970; 1972a; 1972b) and Vanshoubroek and Spaendonck (1973). Since their compilation there have been a number of interesting papers concerned with the protein and amino acid needs of sows and some of these will be discussed later in this chapter. However they do not demand a major re-assessment of the statements made by Elsley (1972a,b).

It is appropriate however that the methods used for determining the needs of sows for protein are re-assessed, since the overall impression of the individual publications is of a wide range of attitudes towards this aspect of experimentation.

It can be established that investigations on the protein nutrition of sows are worthy of careful and comprehensive study. This view is demonstrated by calculations of the overall efficiency with which dietary protein supplied to the sow supports the growth of piglets. Piglet growth in this context represents the level of protein deposited within the piglet up to the stage that it is capable of an independent existence on conventional diets containing cereal and protein fractions. It is assumed that pigs attain this capability when they are ten kg liveweight.

An example of this calculation is given in *Table 19.1*. It must be stressed that such calculations depend on the validity and reliability of the data used. In this example, the feed intakes are those recommended by the ARC (1967); the estimate of nitrogen deposition by the sow is derived from unpublished data and the composition of pigs is calculated from the regression equations reported by Whitelaw *et al.* (1966). The piglet weights have been adjusted by the use of regression equations relating creep feed intake to liveweight gain of the pigs. The levels of performance of the sows, the interval between weaning and service are taken from experiments undertaken by the author which incorporated treatments very similar to those recommended by ARC (1967). It can be demonstrated that the overall calculations of efficiency, or inefficiency, would be altered, but not dramatically so, by the use of data from other sources.

Table 19.1 Protein utilisation by the breeding sow (kg protein)

Input	Standard (ARC, 1967)	Achieved (Elsley et al., 1969)
Input		
Pregnancy (114 days, 2 kg feed/day)	25.6	16.2
Lactation (42 days, max. 5.5 kg feed/day)	39.9	25.7
Weaning to conception (10 days, 2.6 kg feed/day)	4.2	4.2
Total sow feed input	69.7	46.1
Output		
Litter at 42 days - 95 kg	13.8	13.0
Increase in protein in sow	3.0	1.0
Total output	16.8	14.0
Overall efficiency (per cent)	24.1	30.4

It is clear that the gross efficiency with which dietary protein is used is very low (*Table 19.1*). Since the publication of the ARC recommendations in 1967 there have been strenuous efforts to improve the efficiency with which dietary protein is utilised by the sow. It is of interest, that in one experiment undertaken by Elsley, MacPherson and MacDonald (1971) the overall efficiency with which dietary protein was converted was improved to a figure of 30.4 by adjustment of protein concentrations in the diet and by the careful management of other aspects of the sow's nutrition.

On the basis of present nutritional knowledge it is apparent that the overall efficiency of the system outlined in *Table 19.1* can be improved by four main procedures:

(1) The adoption of realistic standards of the protein needs of the sow which do not permit the wasteful supply.
(2) Alteration of amino acid balance which results in a reduction in the concentration of protein in the diet.
(3) Changes in the adjustment of non-protein fractions of the diet, particularly energy, which improve the overall efficiency with which protein is utilised.
(4) The most important aspect of efficiency of the use of dietary protein is the concept of optional economic efficiency as opposed to maximal biological efficiency. It follows that the application of economic weightings to the input and output data may frequently result in the selection of a nutrient concentration that does not result in the maximum biological productivity.

To make use of these techniques to improve the efficiency of protein utilisation demands that a number of nutritional situations should be adequately defined. These can be described as follows:

(a) The biological effect of a change in the relative balance of one nutrient to another must be capable of description.
(b) Interactions between the input of nutrients at different phases of the reproductive cycle must be identified and quantified.
(c) Experiments must be designed so that the extent to which the sows respond to incremental changes in nutrients under investigation are capable of mathematical description.
(d) The scientist must produce a simple monitoring system which will give the farmer advance warning of any impending nutritionally induced crisis, particularly if the farmer has selected submaximal levels of nutrition in an attempt to maximise profitability.

It can be argued that experiments which do not contribute to points (a), (b), (c) and (d) have a limited value unless they supply information on the nutritive specification of unusual feeds. It can also be argued that on a world basis, many experiments are being undertaken or designed which are neither fundamental to our basic knowledge of nutrition nor which add to knowledge of the aspects listed in (a) to (d).

If these factors encompass the important objectives of experimental programmes relating to the protein needs of the sow, there only remains one major aspect of experimental technique that needs to be resolved. This is the criteria which the experimenter or developer should select to monitor the biological response to the nutritional treatments imposed. It is proposed that this paper will concentrate on this often controversial point of experimentation. The investigations mentioned in this chapter have been selected as convenient examples of various approaches that have been made and should not be considered to be the only or even the most important published work in any specified area of study.

Factorial Calculations of Protein Needs

A number of important factorial estimates of the protein requirements of pregnant and lactating sows have been made, including those of Moustgaard (1962), Gütte and Lenkeit (1960) and Vanschoubroek and Van Spaendonck (1973). The limitations of the factorial approach have been outlined already (Elsley, 1972a). The progress recently achieved by Whittemore and Fawcett (1974) has emphasised that models which are based on data verified by a number of workers can be relatively accurate in the prediction of field trial data. In the case of growing pigs, the models are relatively simple because they can take note of the homeostatic mechanisms controlling growth and form a clearly definable relationship between feed intake and fat deposition. In respect of sows there remains the difficulty of obtaining all the data required for the calculations within one experimental programme or within experimental programmes which allow extrapolation from one study to another. The need for indirect methods of estimating body composition also raises difficulty with respect to the accuracy of the data which have to relate in absolute terms to one another.

356 The breeding sow

Table 19.2 Digestible crude protein needs in pregnancy (g/day)

DLG (1965) DBR	286+
ARC (1967) UK	222
NRC (1959) USA	327
NRC (1964) USA	320
NRC (1968) USA	224
CVB (1970) Netherlands	290
Vanschoubroek and Van Spaendonck (1972)	151
Speer (1971)	185
Elsley (1971)	185

On the basis of the factorial approach, or modifications to it, the protein needs of sows outlined in *Table 19.2* were determined. The range of values quoted highlights the need to establish more clearly the basis on which these protein requirements were calculated.

Nitrogen Balance

The validity of the factorial approach is heavily dependent on the absolute accuracy of studies aimed at determining the nitrogen balance of sows. In some instances (ARC 1967) the estimates of requirements were exclusively based on nitrogen balance data.

The problems in undertaking nitrogen balance studies are widely appreciated, although the overall accuracy of the technique is considerably increased by the use of urinary catheters and selection of balance periods of at least ten days duration. When these technical difficulties have been eliminated there still remain aspects of interpretation that require careful examination.

PREGNANCY

Following the early calculation of Moustgaard (1962) considerable emphasis was placed on the need for high protein intakes by the pregnant sow to ensure maximum productivity. Such calculations omitted to take into account the undoubted improvement in the efficiency with which protein was utilised by the pregnant sow (Rombauts, 1959; Salmon-Legagneur, 1965; Heap and Lodge, 1967). Elsley *et al.* (1966) found that the nitrogen retention was nine per cent higher for pregnant as opposed to non-pregnant sows while Baker *et al.* (1966) found that the biological value of a diet was 99 per cent for pregnant and 69 per cent for non-pregnant gilts. The increase in nitrogen retention as gestation proceeds, also noted by Gütte and Lenkeit (1960), may allow the sow to satisfy her increasing needs for deposition within the uterus and the mammary gland without the need for increased dietary nitrogen. (Elsley *et al.*, 1966; Miller *et al.*, 1969).

The major differences between the factorial calculation of Moustgaard (1962) or Gütte and Lenkeit (1960) and the more recent careful calculations of Vanschoubroek and Van Spaendonck (1973) really represented the different attitudes to growth of maternal tissue. The sow mated at 105-120 kg live weight does not normally attain mature weight, even under generous feeding regimes and allowing for pregnancy anabolism, until the end of the third or fourth reproductive cycle. This means that the sow has the capacity to utilise protein and energy surplus to the needs of pregnancy, in an attempt to fulfil her target growth size. Estimates of protein requirements which are based on maximum nitrogen retention inevitably lead to the selection of protein intakes that are higher than those required for pregnancy *per se*. It also follows, in those circumstances where a high proportion of dietary protein is being used for normal growth processes, that the optimal amino acid balance will closely resemble that required for the later stages of finishing of the pig.

The effect upon nitrogen retention of increasing dietary protein concentration was clearly shown by Gütte and Lenkeit (1960) and Salmon-Legagneur (1965) and by many other workers. Elsley, Anderson and MacPherson (1966) found that increases in the dietary energy also markedly increased the protein retention particularly in the later stages of pregnancy. Such studies indicated, but failed to establish beyond doubt, that sows fed so that nitrogen retention was in the region of six g/day were retaining sufficient nitrogen throughout pregnancy to allow normal development of the uterus and its contents, and of the mammary gland.

Elsley *et al.* (1966) calculated that a total of 500 g nitrogen retained during pregnancy allowed for the total nitrogen deposited, a figure that is in agreement with the calculation of Moustgaard (1962) based on the data of De Villiers *et al.* (1958). In the experiments of Elsley, McDonald and MacPherson (1966) sows receiving 5.3 Mcal digestible energy/day and 28 g nitrogen/day retained in excess of 5.0 g nitrogen/day and analysis of over 100 pigs at birth failed to demonstrate any differences in body composition from groups of piglets produced by sows retaining much higher levels of daily nitrogen.

Following the early work of Rippel *et al.*, (1965a; 1965b) amino acid needs of sows have been calculated by the Illinois group based on nitrogen retention data. In most cases these have been short term balance studies, in the case of Rippel undertaken between the 102nd and 109th day of gestation. In parallel a series of papers were produced which identified the amino acids of adult pigs for maintenance (Baker *et al.*, 1966a,b,c). These studies allowed Rippel *et al.* (1965c) to compile a table of requirements for amino acids.

Subsequently Allee and Baker (1970) investigated the thesis of Rippel and his group that in maize-based diets, lysine and tryptophan were limiting and demonstrated that increased nitrogen retention followed the addition of 0.25 per cent L-lysine HCL and 0.041 per cent L-tryptophan. The addition of non-essential amino acids in the form of L-glutamic acid and glycine gave no further improvement.

Several groups have attempted to mimic the results of nitrogen balance experiments but using feedstuffs as opposed to synthetic amino acids. For example, Hesby *et al.* (1970b) found that sows receiving Opaque-2 maize exhibited higher nitrogen retention than sows receiving conventional maize.

LACTATION

The pattern of nitrogen metabolism during lactation has been studied in detail by Lenkeit and co-workers, and has been summarised by Gütte and Lenkeit (1960). These were substantially supported by the studies of Lodge (1959), Salmon-Legagneur (1965) and Elsley and MacPherson (1966). All these studies emphasised the inevitability of a period of negative nitrogen balance immediately following parturition, due to excessive production of urea and the importance of level of energy intake and milk yield upon the adequacy of the dietary protein intake. For example, Elsley and MacPherson (1966) found that the levels suggested by ARC (1967) for protein intake, supported positive nitrogen balance for the last 36 days of a 42 day lactation period on a feeding regime which allowed a 5.0 kg loss of body weight over the lactation period.

The selection of a protein intake which supports zero nitrogen balance during lactation is very dependent on the quality of protein and the level of energy with which it was associated. MacPherson and Elsley (1970) found that 760 g crude protein would allow slight positive balance whilst still supporting adequate milk production. Zero nitrogen balance could be achieved at lower dietary protein intakes, but this necessarily led to reduced milk yields. In the studies of Lewis and Speer (1973) positive nitrogen balance was achieved in sows receiving ten per cent protein diets which only allowed 560 g protein per day but which were comparable with the data of MacPherson and Elsley (1970) when account was taken of the intakes of essential amino acids. Mahan et al. (1971) found that 900 g crude protein was required, but this was in association with low energy intakes whilst Ganguli et al. (1971) found that 420 g crude protein/day was sufficient but in this case in association with low milk yields.

An overall efficiency of protein utilisation of 65 per cent for body synthesis and milk production was found by the German workers (Gütte and Lenkeit, 1960) which is good agreement with the 68 per cent found by the Rowett group (MacPherson and Elsley, 1970). The gross efficiency of conversion of dietary crude protein into milk protein within these studies has varied from 30 to 45 per cent but on average is higher than the 33 per cent efficiency of utilisation of dietary protein used by ARC (1967) to estimate protein requirements for lactating sows based on calculated yields of milk protein.

As in pregnancy, the most important role of nitrogen balance has been in the study of amino acid balance where diets were iso-energetic and iso-nitrogenous. In these instances, the relative advantage of one amino acid complex can easily be demonstrated in spite of the fact that measurements of milk yield are inefficient and that the juxtaposition of sows and piglets raises a range of technical problems associated with the accurate collection of faeces and urine. The problems are compounded since Gütte and Lenkeit (1960) and Elsley (1967) reported that the nitrogen retention of sows during the first few days of lactation is inversely related to the nitrogen retained in late pregnancy. This means that nitrogen balance studies on lactating sows should be undertaken on sows receiving similar amounts of energy and protein. These errors could be overcome by within-sow comparisons of diets in a single lactation but this technique is complicated by the fact that sows require several days to reach nitrogen equilibrium after a change in the level of dietary protein.

In spite of these inherent technical difficulties, nitrogen balance experiments have contributed to the knowledge of amino acid needs. As an example Kracht (1964) conducted a series of interesting trials which highlighted the improvement in biological value of the protein measured by nitrogen balance data of sows receiving barley diets supplemented by L-lysine monohydrochloride. The addition of up to 20 g L-lysine monohydrochloride increased the biological value of barley diets from 56 to 71.5 per cent which compares favourably with 73 per cent found by Kracht (1964) for a diet based on barley and fish meal. The number of sows used in these trials and the variable feed intakes associated with some of the treatments make the absolute values suspect, but do not cast doubt on the premise advanced by Kracht.

OVERALL ASSESSMENT OF NITROGEN BALANCE STUDIES

This brief study of nitrogen balance indicates that it is a useful technique to monitor feedingstuffs or diets differing only in amino acid balance. However, the experiments are most difficult to assess when the treatments differ in both energy and protein concentrations and when long term studies involving gradual but important changes in body composition take place. It is certainly important to relate nitrogen balance at one stage to the preceding nutritional history of the animal. Finally, any assessments of dietary protein requirements based on nitrogen balance should be carefully assessed against the results of trials in which other criteria have been measured.

BLOOD METABOLITES AS MEASURES OF NUTRITIONAL ADEQUACY

It has been demonstrated by many workers that alterations in blood metabolites, particularly alterations in maternal plasma amino acids, can be used to demonstrate severe protein under-nutrition. This technique has been used by many workers following the work of Lucas *et al.* (1969) who assessed the adequacy of a 20 per cent protein maize-soya-bean diet diluted with maize starch to produce diets containing 8, 12, 16 or 20 per cent protein. On the basis of plasma protein profiles of the lactating sows they concluded that the sulphur containing amino acids were the most limiting under the conditions of their experiment. The technique has also been used to select the optimum level of a particular amino acid in the diet. For example, the studies undertaken by Chen, D'Mello and Elsley (unpublished data) were based on the supply of low protein intakes during pregnancy of 190 g digestible protein, in a diet formulated to contain only 10 per cent protein and 0.42 per cent lysine in the diet. The sows were then subjected to a range of levels of lysine during lactation and the plasma amino acid concentrations determined at several stages of the lactation

360 *The breeding sow*

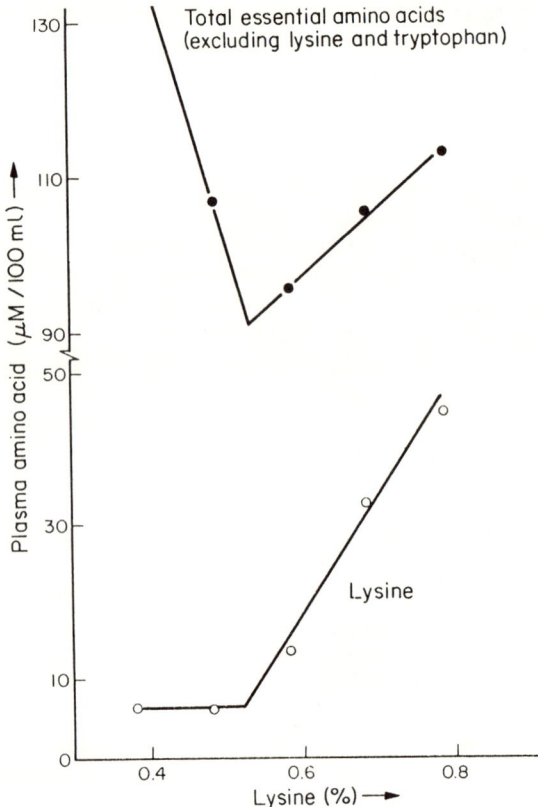

Figure 19.1a The effect on plasma amino acid concentration of increases in percentage lysine in the diet of lactating sows

(*Figures 19.1a,b*). This experiment suggests that 0.56 per cent lysine is required to ensure satisfactory blood profiles.

An alternative approach to the use of plasma amino acids has been the further development of techniques of determining blood urea by the Nottingham group of workers. The studies of Sohail, Cole and Lewis (1974) have indicated that blood urea concentration can be easily measured and used to indicate the adequacy of a nutritional regime.

These blood metabolites are subject to alteration by a range of external factors, such as timing of removal of the blood sample relative to feeding time, a range of analytical difficulties and the considerable financial costs of the methods. These difficulties could be overcome by careful research techniques but there remains the doubt as to whether they are more sensitive than simpler and apparently less sophisticated techniques.

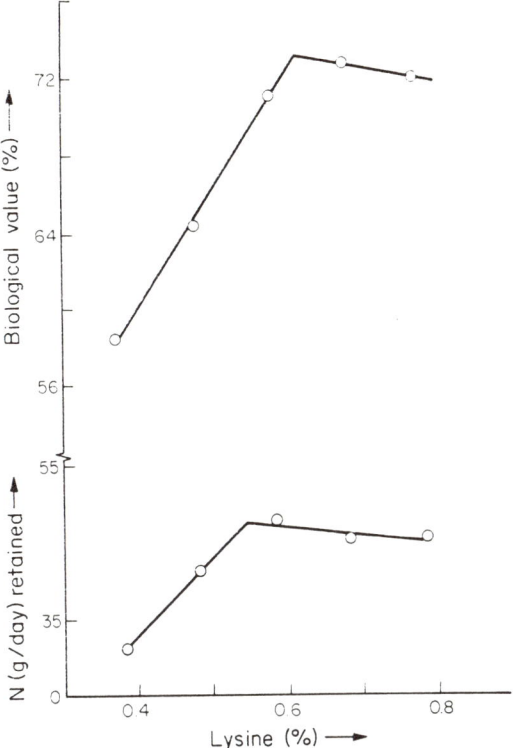

Figure 19.1b The effect on nitrogen retention and biological value of increases in percentage lysine in the diet of lactating sows

SOW PRODUCTIVITY

It is important that the results from factorial calculations or indirect measures based on nitrogen balance or blood metabolites should be assessed against results gained in field trials. It is frequently assumed that such trials although expensive are basically simple to undertake. Nothing could be further from the truth. Quite apart from the difficulties of assembling sufficient animals to obtain the replication which allows the detection of significant differences in sow productivity, there is the fundamental problem of deciding which aspects of sow productivity should be used in the assessment process. In addition, in some instances, the features of sow productivity are inversely correlated. For example, sow weight gain in pregnancy is inversely related to weight gain in lactation whilst the inter-relationship between milk productivity and creep feed consumption by the piglets is widely recognised.

PREGNANCY

Throughout the last two decades there has been a large number of experimental reports aimed at identifying the critical protein standards that should be adopted in pregnancy, based on the results of field trials (*Table 19.2*). These reports were reviewed by Elsley (1972a,b) and although the more recent published experiments have added to our knowledge they have not substantially altered the conclusions reached. These experiments all stress that the protein needs of sows determined by field trials are much lower than comparable results gained from metabolism studies. This is due to the difficulty of isolating maternal and foetal deposition of protein in the metabolic studies.

The estimates in *Table 19.2* depend largely on the traits of birth-weight and post-natal growth. The information is only sparsely available for the effect of protein intake in pregnancy on breeding regularity, longevity and overall sow health. It is important however to recognise that the studies suggested by Speer (1971) and Elsley (1972a,b) still include a degree of nutritional insurance.

This is demonstrated by the fact that where extreme dietary protein deprivations are imposed, reproductive performance is reduced but pregnancy and lactation still continue to a remarkable extent. For example, in some experiments, treatments have been imposed to a severity never attainable in commercial conditions. Pond *et al.* (1968), and Pond, Barnes and Krook (1969) supplied synthetic diets which only allowed pigs a daily intake of 9.0 g crude protein and this treatment reduced litter size and liveweight gain of the sow and had a marked effect on the subsequent lactation. The salient feature of the experiment however was that the performance of the pigs was little affected compared with that of sows receiving higher protein intakes; results which were supported by De Geeter *et al.* (1970a,b); De Geeter (1971). Similarly, the use of more conventional feedstuffs by Clawson *et al.* (1963) in a trial which allowed an intake of 136 g crude protein/day and by Elsley, MacPherson and MacDonald (1971) which allowed an intake of 140 g crude protein/day in pregnancy both allowed satisfactory performance which did not differ significantly from sows receiving higher intakes of protein.

Svajger *et al.* (1970) also found that low protein intakes reduced the regularity of breeding by producing anoestrous animals, prolonged oestrous cycles and a reduction in ovulation rates; but these effects were detected with diets containing 2 per cent protein (36 g protein/day) in pregnancy and 5 per cent in lactation, as compared with sows receiving diets containing 17 per cent crude protein in both pregnancy and lactation.

LACTATION

The study of the nutritional needs of lactating sows poses a range of technical problems which have not yet been satisfactorily solved. These include the basic difficulties of measuring milk yield and composition; the extent to which the sow is drawing on body reserves of nutrients, largely energy; the nutritional adequacy of the preceding and subsequent stages in the reproductive cycle; the length of lactation period and the length of the breeding life which the sow is expected to fulfil. These have been discussed by Elsley (1972b) and space does not allow their complete re-appraisal. Three points however require emphasis:

Firstly, the development of an accurate and simple technique for measuring milk yield of sows still eludes research workers, and even radio-isotope procedures outlined by McDougall and Fowler (1974) fail to convince that the techniques can be used in more than a limited range of experimental circumstances. Secondly, milk production is closely associated with dietary energy or the mobilisation of body lipids and in turn protein needs are linked to the level of milk production. If the conditions under which the experiment is conducted preclude high milk yields then the applicability of the results is restricted. Thirdly, and this is the crucial aspect, careful thought is required to establish the level of milk production that is necessary to maintain a specified level of piglet growth. This in turn will allow the development of the most attractive nutritional and economic regime for this level of production. The concept that an optimal milk production target is required rather than a maximal output has not been sufficiently recognised, the implications of such a philosophy upon nutritive standards are, of course, very considerable.

An examination of the results from trials that have been published provides a confusing picture. For example, a series of trials undertaken and summarised by Salmon-Legagneur (1965) indicated that sows receiving 700 g crude protein per day performed in a similar manner to sows receiving 893 g crude protein per day and in particular the protein fractions of the milk were unaltered. MacPherson, Elsley and Smart (1969) found that the performance of sows receiving diets as low as 14.0 per cent crude protein containing 0.60 per cent lysine did not differ significantly from sows receiving 19 per cent crude protein and 0.97 per cent lysine. O'Grady (1971) found that sows receiving diets containing 11.7 or 15.8 per cent crude protein during performed equally well regardless of whether the lysine content at each protein level was three or four per cent of the protein. Neilsen (1968) also reported that the reduction of the protein concentration from 16 to 13 per cent did not affect the performance of the piglets, had little effect on the live weight of the sows but did lead to a small reduction in yield of milk and milk protein. The results for the diets were similar when supplied at high or lower levels of intake.

On the other hand Mahan, Becker and Jenson (1971a,b) found that there was an improvement in the growth of the litter up to 16 per cent protein for lactating gilts and of 18 per cent for second litter sows although the sows only received 3.5 kg as gilts and 4.1 kg of feed daily in the second lactation. Recently Greenhalgh et al. (1974) have reported that there was an increase in performance of piglets following the supply of a diet containing 17 per cent crude protein as compared with a diet containing 12 per cent protein during lactation. In many of these experiments, protein concentration has been confounded with the quality of the protein. In some experiments this effect of protein quality has been eliminated by the use of maize starch as a diluent of basic diets in order to achieve a range of protein intakes of identical quality. Holden et al. (1968) fed diets containing 8, 12, 16 or 20 per cent crude protein to both gestating and lactating sows. Pregnant sows received 1.82 kg/day of feed while the lactating sows received the diet *ad libitum*. Although piglets from sows receiving the eight per cent protein diet grew more slowly than the pigs from other groups, the differences were small and no major differences were detected between sows receiving 12 per cent and higher protein intakes. In the experiments of Elsley, MacPherson and MacDonald (1971) pregnant sows received 7, 10 or 13 per cent crude protein and in lactation 12.5 or 16.0 per cent crude protein. There were no differences in quality of the protein since dilution of basal diets by maize starch was used to produce the range of experimental diets. The higher protein intakes in lactation led to slight increases in piglet growth which could be associated with the higher yields of milk and of milk protein produced.

None of the experiments so far mentioned has been of sufficient size to allow confident conclusions to be drawn in respect of the effect of protein intake or amino acid concentration upon the long term productivity of the sow. In addition each of the experiments indicates that the nutritional treatment imposed has influenced some measure of sow productivity without emphasising which of the measures is of prime importance or whether one parameter can be used as an indication of general nutritional adequacy.

The recent publications of Lewis and Speer (1973; 1974) are therefore of interest since they include a comparison of a number of criteria in their assessment of the amino acid needs of sows. The agreement between the estimates is as welcome as it is surprising. It should be emphasised, however, that in the first experiment comparisons were made between diets differing only in lysine concentration and in the second differing only in tryptophan. One would anticipate that short term experiments studying these aspects would be expected to reveal similar results but that a more confused picture would emerge if the treatments compared, say, different dietary protein intakes associated with differing energy intakes. The criteria quoted in *Table 19.3* do not include any measure relating to long term productivity.

Table 19.3 Lysine and tryptophan needs using different criteria

Lysine needs using different criteria (per cent of diet)	
Milk protein	0.56
Milk solids	0.49
Nitrogen balance	0.54
Nitrogen retention	0.56
Piglet gain	0.55
PFAA-lysine	0.53
Plasma urea	0.53
	(Lewis and Speer, 1973)
Tryptophan needs using different criteria (per cent of diet)	
Milk protein	0.074
Milk yield	0.072
Nitrogen retention	0.082
Piglet gain	0.069
Plasma tryptophan	0.063
Plasma urea	0.072
	(Lewis and Speer, 1974)

SIMPLE MONITORING SYSTEMS FOR PROTEIN ADEQUACY

In the majority of the field trials, results indicate that the nutritional treatments imposed may not affect piglet growth but that usually they are reflected in changes in the live weights of the sows. In these experiments where, for example, more severe treatments have been imposed and where the birthweight of the piglets, the milk yield or the growth of the piglets have been affected, then these differences have been closely associated with changes in live weight of the sows. Furthermore in experiments where low protein intake has even led to problems in breeding regularity then large losses in the weights of the sows over the experimental period have been demonstrated. These results indicate that an apparently simple method of assessing the protein needs of sows may be to select protein intakes which allow a specified pattern of live weight change within which little effect on other aspects of sow productivity of commercial importance are likely to be detected.

The establishment of this optimum live weight change is a difficult process but Elsley (1971) has stressed that a gain of 12-15 kg per reproductive cycle may preclude any symptoms of gross bad performance of sows. It is not suggested that this single criterion obviates the need for other measurements but merely that the recording of other performance traits or metabolic data must provide additional information of practical value if they are to be justified.

The overall conclusion is that the absence of a useful specification of the optimum protein needs of sows, or the description of the response function of protein intake and performance is probably due as much to the range of criteria that have been used to assess the needs of the sow as to the variation in response attributable to environmental or genetic variables. The resolution of the problems of interpretation of the variables attributable to experimental technique is of considerable potential value to the industry and yields further evidence for the thought that there is much to be undertaken in nutritional research that does not result in the mere extension to the flood of published literature.

References

ALLEE, G.L. and BAKER, D.H. (1970). *J. Anim. Sci.*, **30**, 748

ARC (1967). *The Nutrient Requirements of Farm Livestock, No.3, Pigs*, HMSO, London

BAKER, D.H., BECKER, D.E., NORTON, H.W., JENSEN, A.H. and HARMON, B.G. (1966a). *J. Nutr.*, **88**, 382

BAKER, D.H., BECKER, D.E., NORTON, H.W., JENSEN, A.H. and HARMON, B.G. (1966b), *J. Nutr.*, **88**, 391

BAKER, D.H., BECKER, D.E., NORTON, H.W., JENSEN, A.H. and HARMON, B.G. (1966c). *J. Nutr.*, **89**, 441

CLAWSON, A.J., RICHARDS, H.L., MATRONE, G. and BARRICK, E.R. (1963). *J. Anim. Sci.*, **22**, 662

DEGEETER, M.J. (1971). *Diss Abstr. Internat.* (B), **32** (4) (1939-40)

DEGEETER, M.J., HAYS, V.W., CROMWELL, G.L. and KRATZER, D.D. (1970b), *J. Anim. Sci.*, **31**, 1020 (Abstr.)

DEGEETER, M.J., HAYS, V.W., KRATZER, D.D. and CROMWELL, G.L. (1970a), *J. Anim. Sci.*, **31**, 199 (Abstr.)

DEVILLIERS, V., SORENSEN, P.H., JAKOBSEN, P.E. and MOUSTGAARD, J. (1958). *Copenhagen Vet-og Landbohjsk Inst Sterilitetsforsk Aarsberetn*, p.139

ELSLEY, F.W.H. (1967). *Proceedings of the Symposium on the Nutrition of Sows*, University of Nottingham. Ed. M. Bannerman

ELSLEY, F.W.H. (1970). *Rep. Rowett Inst.*, **26**, 108

ELSLEY, F.W.H. (1971). in *Lactation*, Ed. I.R. Falconer. Butterworths, London

ELSLEY, F.W.H. (1972a). in *Handbuch der Tierernährung. Bd 11*, Ed. W. Lenkeit, K. Brierem and E. Craseman. p.330. Paul Parey, Hamburg and Berlin

ELSLEY, F.W.H. (1972b). in *Proc. Symp. Improvement of Sow Production*, Ed. A.S. Jones, V.R. Fowler and J.C.R. Yeats. *Occ. Publ. Rowett Res. Inst. No.3.*

ELSLEY, F.W.H., ANDERSON, D.M., McDONALD, I., MacPHERSON, R.M. and SMART, R. (1966). *Anim. Prod.*, **8**, 391

ELSLEY, F.W.H., ANDERSON, D.M. and MacPHERSON, R.M. (1966). *Proceedings of the 177th Meeting of the Nutr. Soc.*, Dundee

ELSLEY, F.W.H., McDONALD, I. and MacPHERSON, R.M. (1966). *Anim. Prod.*, **8**, 353 (Abstr.)

ELSLEY, F.W.H. and MacPHERSON, R.M. (1966). *Anim. Prod.*, **6**, 259 (Abstr.)

ELSLEY, F.W.H., MacPHERSON, R.M. and McDONALD, I. (1971). *Proc. Inter. Congr. Anim. Prod.*, Versailles

GANGULI, M.C., SPEER, V.C., EWAN, R.C. and ZIMMERMAN, D.R. (1971). *J. Anim. Sci.*, **33** (2), 394

GREENHALGH, J.F.D., ELSLEY, F.W.H., GRUBB, D.A., LIGHTFOOT, A.L., SAUL, D.W., SMITH, P., WALKER, N., WILLIAMS, D. and YEO, M.L. (1974). *Proc. Br. Soc. Anim. Prod.*, **3**, 110 (Abstr.)

GUTTE, J.O. and LENKEIT, W. (with FARRIES, E., KIRCHHOFF, W., SOEHNGEN, F.K., STREUTER-PETERMÖLLER, A. and WARNECKE, W.) (1960). *Z. Tierphysiol., Tierernährung Futtermittelk.*, **15**, 165

HEAP, F.C. and LODGE, G.A. (1967). *Anim. Prod.*, **9**, 237

HESBY, J.H., CONRAD, J.H., PLUMLEE, M.P. and HARRINGTON, R.B. (1970). *J. Anim. Sci.*, **31**, 481

HESBY, J.H., CONRAD, J.H., PLUMLEE, M.P. and MARTIN, T.G. (1970). *J. Anim. Sci.*, **31**, 474

HOLDEN, P.J., LUCAS, E.W., SPEER, V.C. and HAYS, V.W. (1968). *J. Anim. Sci.*, **37**, 1578

KRACHT, W. (1964). *Arch. Tierernähr*, **14**, 39

LEWIS, A.J. and SPEER, V.C. (1973). *J. Anim. Sci.*, **37**, 104

LEWIS, A.J. and SPEER, V.C. (1974). *J. Anim. Sci.*, **39**(5), 978 (Abstr.)

LODGE, G.A. (1959). *J. Agric. Sci., Camb.*, **53**, 172

LODGE, G.A. (1969). in *Nutrition of Animals of Agricultural Importance, Part 2.* Ed. Sir David Cuthbertson. p.1053. Pergamon Press, Oxford, London and Edinburgh

LODGE, G.A. (1972). in *Pig Production*, Ed. D.J.A. Cole. Butterworths, London

LUCAS, E.W., HOLDEN, P.J., SPEER, V.C. and HAYS, V.W. (1969). *J. Anim. Sci.*, **29**, 429

McDOUGALL, I.A. and FOWLER, V.R. (1974). *Proc. Br. Soc. Anim. Prod.*, **3**, 112

MacPHERSON, R.M. and ELSLEY, F.W.H. (1970). unpublished data

MacPHERSON, R.M., ELSLEY, F.W.H. and SMART, R.I. (1969). *J. Anim. Sci.*, **29**, 429

MAHAN, D.C., BECKER, D.E., HARMON, B.G. and JENSEN, A.H. (1971). *J. Anim. Sci.*, **32**, 482

MAHAN, D.C., BECKER, D.E., NORTON, H.W. and JENSEN, A.H. (1971). *J. Anim. Sci.*, **33**(1), 35

MAHAN, D.C., BECKER, D.E. and JENSEN, A.H. (1971a). *J. Anim. Sci.*, **32**(3), 470

MAHAN, D.C., BECKER, D.E. and JENSEN, A.H. (1971b). *J. Anim. Sci.*, **32**(3), 476

MILLER, G.M., BECKER, D.E., JENSEN, A.H., HARMON, B.G. and NORTON, H.W. (1969). *J. Anim. Sci.*, **28**, 204

MOUSTGAARD, J. (1962). in *Nutrition of pigs and poultry*. Ed. J.T. Morgan and D. Lewis, p.189. Butterworths, London

NIELSEN, H.E. (1968). *Int. Conv. 'Technical and Economic Aspects of Pig Feeding'*, Reggion Emilia, Italy

O'GRADY, J.F. (1971). *Ir. J. Agric. Res.*, **10**(1), 17

POND, W.G., BARNES, R.H. and KROOK, L. (1969). *New Yorks Food and Life Sciences*, **2**(4), 12

POND, W.G., DUNN, J.A., WELLINGTON, G.H., STOUFFER, J.R. and VAN VLECK, L.D. (1968). *J. Anim. Sci.*, **27**, 1583

RIPPEL, R.H., HARMON, B.G., JENSEN, A.H., NORTON, H.W. and BECKER, D.E. (1965a), *J. Anim. Sci.*, **24**, 209, 373, 378

RIPPEL, R.H., HARMON, B.G., JENSEN, A.H., NORTON, H.W. and BECKER, D.E. (1965b). *J. Anim. Sci.*, **24**, 373, 378

RIPPEL, R.H., RASNUSSEN, O.G., JENSEN, A.H., NORTON, H.W. and BECKER, D.E. (1965c). *J. Anim. Sci.*, **24**, 203

ROMBAUTS, P. (1959). *C.R. Acad. Sci.*, **248**, 2410

SALMON-LEGAGNEUR, E. (1965). *Annls Zootech.*, **14**, 5

SOHAIL, M.A., COLE, D.J.A. and LEWIS, D. (1974). *Proc. Br. Soc. Anim. Prod.*, **3**, 110

SPEER, V.C. (1971). *Feedstuffs, Minneap.*, **43**(41), 29

SVAJGER, A.J., HAYS, V.W., CROMWELL, G.L. and DUTT, R.H. (1970). *J. Anim. Sci.*, **31**, 212

SVAJGER, A.J., HAMMELL, D.L., DEGEETER, M.J., HAYS, V.W., CROMWELL, G.L. and DUTT, R.H. (1972). *J. Reprod. Fert*, **30**, 455

VANSCHOUBROEK, F. and SPAENDONCK, R. VAN (1973). *Z. Tierphysiol. Tiernähr. Futtermittelk.*, **31**(1), 1

VANSCHOUBROEK, F. and SPAENDONCK, R. VAN (1973). *Z. Tierphysiol. Tiernähr Futtermittelk.*, **31**(2), 71

WHITELAW, A.W.W., ELSLEY, F.W.H., JONES, A.S. and BOYNE, A.W. (1966). *J. agric. Sci.*, **66**, 203

WHITTEMORE, C.W. and FAWCETT, R.F. (1974). *Anim. Prod.*, **19**, 221

20

THE AMINO ACID REQUIREMENTS OF GROWING PIGS (1)

S. POPPE
Animal Husbandry Section, University of Rostock, East Germany

Increasing the utilisation of available feed is an urgent problem in animal feeding. The utilisation of protein in the feed is becoming increasingly important. Protein quality, which is primarily determined by the amino acid content of the diet concerned, plays a decisive role in the feeding of pigs.

Assessment of crude protein in the nutrition of monogastric animals is being replaced to an increasing extent by causal considerations. This is particularly true with regard to amino acids, because the protein requirement of these animals corresponds, in the final analysis, to a relatively precisely definable amino acid requirement.

Extensive investigations to determine the amino acid requirement of growing pigs have been carried out in the European socialist countries during the last 15 years. A large number of experiments have been performed in the Soviet Union by Tomme *et al.* (1967; 1971; 1971a); Tkatschow *et al.* (1971); Borz (1963); Owsjanikow (1963); Sceglov (1966; 1969) and Gradusow (1967). Various methods have been used for determining the amino acid requirement. The requirements for the lysine and sulphur-containing amino acids differ considerably, depending on the animals and methods used. An example of the variation in estimates of requirements is given in *Tables 20.1* and *20.2*.

Several authors have used the nitrogen balance as a criterion for determining requirements. A set of rules for performing such investigations was drawn up by Poppe and Wiesemüller (1969) in order to obtain more comparable values by the use of a uniform method.

Use of the nitrogen balance to determine amino acid requirements should always be supported by growth and slaughter experiments. Growth trials alone are unsuitable because growth and energy requirement per unit live weight gain allow no assessment of the quantitative characteristics of the carcass and its chemical composition. Investigations (Wiesemüller, Poppe and Sieg, 1974) have shown that quantitative correlations exist between energy consumption, energy expenditure and daily deposition of meat.

If a sufficient supply of essential nutrients is provided, pigs can meet their potential for daily gain of lean meat. This potential can also be realised if the energy consumption lies a little below the *ad libitum* intake when the diet is of adequate nutrient density. This leads to the

Table 20.1 The lysine requirement of growing pigs (per cent of air dry feed)

Live weight (kg)	20	30	40	50	60	70	80	90
Navratil and Simecek (1963, 1966)		0.86		0.75				
Awrutina et al. (1966)	0.98		0.88			0.68	0.56	
Flicek (1966)	0.84			0.68			0.48	
Poppe and Wiesemüller (1968)			0.80			0.65		0.55
Platikanov et al. (1969; 1970)			0.77		0.69			
Dinu and Georgescu (1970)		0.67				0.59		0.55
Borz and Shurba (1963)		1.10				0.56		
Owsjanikow (1963)	0.69	1.01			0.37			
Tomme et al. (1971)		0.68				0.59	0.60	
Tkatschow et al. (1971)		0.90			0.75			
Müller and Gebhardt (1973)		0.65	1.22			0.29		0.52
DDR-Normen (1973)			0.70			0.55		
UdSSR-Normen (1969)		0.81			0.64	0.59		

Table 20.2 The requirement for methionine + cystine of growing pigs (per cent of air dry feed)

Live weight (kg)	20	30	40	50	60	70	80	90
Navratil and Simecek (1963)		0.49		0.48				
Awrutina (1966)	0.60		0.54			0.43	0.44	0.33
Flicek (1966)	0.55		0.49					
Wiesemüller and Poppe (1968)		0.50			0.45			0.40
Platikanov et al. (1969, 1970)			0.43		0.40		0.37	
Dinu and Georgescu (1970)			0.30			0.27		
Borz and Shurba (1963)	0.40	0.38				0.31		
Owsjanikow (1963)		0.62			0.36	0.49		
Tomme et al. (1971)			0.51			0.48		
Tkatshow et al. (1971)			0.60		0.50			0.40
Müller and Gebhardt (1973)		0.40	0.74				0.23	0.45
DDR-Normen (1973)			0.50					0.40
UdSSR-Normen (1969)		0.59			0.43		0.42	0.42

formation of less fat and, due to the slightly protracted duration of the finishing phase, more lean is produced. If energy consumption is further reduced, daily protein synthesis also decreases. The carcass quality thus deteriorates and the energy expenditure is increased.

The investigations outlined here fully confirm the values obtained for the amino acid requirement by means of the nitrogen balance. As found by Rerat and Lougnon (1966), the values found for amino acid requirement are greatly dependent on the criterion used for their determination.

Two examples serve to illustrate the effect of the chosen criterion on the level of the amino acid requirement.

Young Growing Pigs

Piglet growth during the rearing phase (7-35 kg live weight, 30-90 days of age) has been used as a basis for the estimation of amino acid requirements.

The amino acid norms currently in use are related to the maximum performance of the piglets and do not give sufficient consideration to the subsequent phases in the life of the animals. Thus, it becomes necessary to check amino acid requirements, and consequently the rearing intensity of piglets. The literature contains reports on experiments showing, in the case of early weaned piglets reared on a relatively lysine-deficient diet until the finishing stage is reached, that there are not adverse effects on carcass characteristics. Information regarding so-called compensatory growth (increased growth rate and improvement of feed utilisation during the finishing phase of animals previously on a low-level feed plane during their early development) varies considerably and does not permit definite conclusions to be drawn. Such experiments have been performed in the GDR only on piglets with a live weight of 12-14 kg and not with animals aged three or four weeks.

The introduction of a single mixed feed for piglets has considerable advantages for the feed industry and a large number of pig farms. In order to check the results obtained by such a single phase feeding technique, Knobloch (1972) first reared 163 piglets on rations which differed in their crude protein and amino acid contents; *Table 20.3* shows the dietary treatments used. Treatment 1 provided the same diet over the whole period (28 to 88 days of age). Treatment 2 provided a higher protein diet up to 60 days of age and then the same level as Treatment 1. Treatment 3 involved three diets of decreasing protein level through the rearing period. After the rearing phase, 96 of these animals were tested in a growth and slaughter trial.

The fact that the piglets take in less of the feed mixtures with a low crude protein and lysine content must be stressed. The slightly lower energy consumption, and particularly the considerably lower consumption of lysine and the other amino acids, caused smaller increases in live weight and a higher feed utilisation up to 60 days of age in the case of Treatment 1. After this all piglets received the same diet. The daily live weight increase from 60 to 88 days was almost identical in the animals on all three treatments, the piglets on Treatment 1 tending to

Table 20.3 The effect of different protein and lysine levels (in Dry Matter) on the development of piglets

	Treatment 1		Treatment 2		Treatment 3	
Number of animals	53		56		54	
	Crude protein (%)	Lysine (%)	Crude protein (%)	Lysine (%)	Crude protein (%)	Lysine (%)
Age of piglets (days)						
28-42	19.2	0.84	23.0	1.28	25.8	1.52
42-60	19.2	0.84	23.0	1.28	23.0	1.28
60-88	19.2	0.84	19.2	0.84	19.2	0.84
Live weight (kg) at						
28 days		6.77		6.91		6.86
42 days		9.31		10.72		11.41
60 days		16.10		19.00		19.50
88 days		33.40		36.90		37.00

have a lower energy utilisation. However, these animals were 3-4 kg lighter at 88 days than the animals on Treatments 2 and 3. From the economic standpoint this performance of the young animals is not as important as the age at which a fattening pig reaches its prescribed final weight (or a sow can be inseminated). An economic advantage is gained if the required production can be achieved in the same period of time in the case of animals subject to less intensive rearing during the initial phase as for animals reared in the traditional manner.

The effect of different feeding intensities on the subsequent growth and carcass characteristics of early weaned piglets is shown in *Table 20.4*. Thirty two animals reared on each of the three dietary treatments were used. From 88 to 150 days of age, the feed used contained 0.72 per cent lysine and 0.48 per cent methionine + cystine. During the subsequent finishing phase, the feed contained only 0.62 per cent lysine and 0.40 per cent methionine + cystine.

Table 20.4 shows that animals on Treatment 1 are clearly superior during the first part of the finishing phase. The gains for animals on Treatment 1 during the first fattening phase were similar to the rearing phase, whereas those recorded for the intensively reared animals on Treatments 2 and 3 decreased. This initial deterioration in daily gain has been observed in some commercial enterprises. After intensively reared piglets have achieved very high daily gains on some rearing methods the gains achieved in the finishing accommodation have been unsatisfactory for long periods; the cause not being solely the change of environment.

As a result of the good initial finishing performance of the pigs on Treatment 1, they achieved the highest daily gains and best energy utilisation throughout the whole finishing period, although the differences obtained by comparison with those on Treatments 2 and 3 were not significant. Thus, slower growth at the piglet stage was compensated to such a degree that the animals on Treatment 1 reached the prescribed live weight at the same age as the animals on the other rations. The

Table 20.4 Growth and carcass characteristics of pigs reared at different intensities during early life

	Treatment 1		Treatment 2		Treatment 3	
Age (days)	Daily[1] gain	Efficiency	Daily gain	Efficiency	Daily gain	Efficiency
89-150	617	2.04	540	2.32	572	2.22
150-slaughter	698	2.60	675	2.62	715	2.51
89-slaughter	660	2.35	612	2.50	646	2.38

Growth	Age (days)	Liveweight (kg)	Age (days)	Liveweight (kg)	Age (days)	Liveweight (kg)
	88	33.4	88	36.9	88	37.0
	150	71.6	150	69.8	150	71.1
	216	117.8	220	116.8	215	117.7

Carcass characteristics[2]	Treatment 1	Treatment 2	Treatment 3
Prime cuts (per cent)	46.8	47.2	46.0
Fat (per cent)	25.6	25.5	26.8
Mean thickness of fat layer (cm)	3.52	3.77	3.76
'Eye' muscle area (cm²)	33.2	33.7	31.3
Proportion meat/fat	0.55	0.54	0.58

[1] Daily gain in g and efficiency as kEFs/kg gain
[2] Energy feed units

characteristics of the carcasses of pigs on all three rations did not differ but pigs on Treatment 1 tend to have the lowest fat thickness.

It has been possible to reproduce these positive results in a large-scale experiment performed at Losten. The animals in the experimental group (single-phase) were given a weaned piglet diet and those in the control group (two-phase) were given pre-starter feed *ad libitum* after 10 days of age. Whereas the weaned piglet feed was supplied without interruption until 95 days of age, the control piglets were given piglet pre-starter (23% crude protein) to 28 days of age and subsequently received a piglet rearing feed (19.2 per cent crude protein) until 95 days of age.

The most important results are shown in *Table 20.5*, which shows that both groups of animals developed relatively slowly up to 60 days of age. However, they had reached a satisfactory live weight by 95 days of age.

Breeding Boars

The second example characterises the problems associated with determining the amino acid requirements during the rearing of breeding boars.

The starting point for determining the amino acid requirements can be either general performance characteristics, i.e. live weight gain and feed utilisation, or breeding performance, i.e. the production of sperm.

The development of sexual potency of boars is affected to a large extent by various environmental factors, particularly nutritional and housing conditions during rearing. In the case of the individual animal, the genetic potential for sperm production can only be fully realised if measures which have a favourable effect on the development of the sexual control system and sexual organs are introduced when the animals

Table 20.5 Comparison of the single phase and the two-phase feeding methods at Losten

	Experimental group (single phase)	Control group (two-phase)
Number of animals used	555	645
Losses (35-95 days of age)	3.20%	3.20%
Mean live weight (kg) at:		
35 days of age	11.45	11.89
60 days of age	14.50	14.30
95 days of age	39.50	37.50
Feed consumption (kg/piglet) up to:		
35 days of age	3.95	no data
35-95 days of age	66.00	no data
Feed utilisation (kg feed/kg gain)		
35-95 days of age	2.36	no data

reach maturity, i.e. during a period of intense organ growth. At Rostock, Kleemann (1972) investigated the question of possible nutritional control of subsequent sperm production during the period of maturity. In these experiments, young boars were restricted-fed because precise determination of the nitrogen balance is difficult with *ad libitum* feeding. The amount of dry feed provided daily was determined using the known norms obtained with castrated animals as a guideline. The dry weight of food, energy, lysine, methionine and cystine consumption during the different live weight ranges is shown in *Table 20.6*.

However, it must be stated that, from about 70 kg live weight, difficulties occurred with regard to the food intake, so that the daily dry matter intake could only be increased to a slight extent. A food intake of 3.05 kg (2.67 kg dry matter) per animal could not be reached until the live weight was 110 kg. Since, at higher live weights, the maximum dry feed intake was achieved on average for all animals, feeding can, with certain restrictions, be regarded as having been semi-*ad libitum*. The lower food intake of boars compared with castrated male animals and gilts during the rearing phase has been described by several workers (Charette, 1961; Teague *et al.*, 1964; Blair and English, 1965; Tribble *et al.*, 1965; Witt and Schröder, 1969; Farries and Kallweit, 1970). The reduced food intake is associated with improved utilisation of the nutrients for growth, this being reflected in a marked improvement in feed and energy utilisation.

In order to achieve a fast growth rate during the fattening phase, certain minimum requirements must be set for the energy and amino acid concentration in the feed taking account of the low food intake of boars. If the increased nutrient requirement of boars is not taken into consideration, castrated males will have a better growth rate than boars during the fattening phase as a result of their higher food intake on *ad libitum* feeding (Walstra and Kroeske, 1968).

Experiments have shown that an increase of seven g/animal/day of lysine in conjunction with additional methionine does not result in a

Table 20.6 Dry matter, energy and amino acid intake of young boars (g/animal/day)

Group	Live weight (kg)	Dry matter	EFs	Lysine	Methionine + cystine
1	40	1739	1230	18.9	8.6
2	40	1739	1232	18.9	11.6
3	40	1739	1236	24.9	11.6
1	60	2176	1510	23.3	12.1
2	60	2176	1513	23.3	17.1
3	60	2176	1518	30.3	17.1
1	80	2401	1666	24.0	13.3
2	80	2401	1672	24.0	21.3
3	80	2401	1676	31.0	21.3
1	100	2581	1797	24.5	13.9
2	100	2581	1804	24.5	23.9
3	100	2581	1808	31.6	23.9
1	120	2721	1863	29.2	15.9
2	120	2721	1872	29.2	27.9
3	120	2721	1878	37.4	27.9

significantly better growth of young boars between 40 and 110 kg live weight. Thus, the animals in the control group (Group 1) were fed according to established requirements. The assessment of the methionine + cystine requirements is more difficult. As in the case of lysine, it will be higher than that required for maximum lean growth in castrated animals due to the greater potential of the boar.

Table 20.7 shows the results obtained for sperm production during the period covered by the experiment together with the standard values calculated by Hühn (1969) for corresponding ages and live weights. Since sperm production is closely related to the amount of methionine + cystine available daily in the diet the later use of the young boar for insemination or impregnation cannot be ignored when determining the requirement for sulphur-containing amino acids. The concentration of methionine + cystine in the diet should be high in order to obtain satisfactory sperm production when using the boar for insemination.

After critical evaluation of the available results, provisional amino acid requirements for rearing young boars can be suggested (*Table 20.8*). It is not possible to meet this requirement for methionine + cystine from normal feed ingredients if lysine restriction is enforced, because such a narrow ratio between lysine and sulphur-containing amino acids is not found. Thus, it is necessary to use synthetic DL-methionine when rearing young boars. If, on the other hand, lysine is disregarded when formulating the ration, the requirement for methionine + cystine is easily met, e.g. by using a combination of fishmeal and shredded sesame extract, without the need for synthetic methionine. A certain excess of lysine is then unavoidable, but this need not have any serious negative effect.

In conclusion, it can be said that the amino acid supply, and particularly the supply of methionine + cystine, during the development of young boars is of great importance for sperm production later in life.

Table 20.7a. Volume of ejaculate from young boars aged 7-8 months (ml)

Group	Number of animals	Mean	± Standard error
1	60	139 ± 5.0	38.9
2	60	218 ± 9.8	75.8
3	72	158 ± 5.3	44.8
Hühn (1969)		135 - 165	

Table 20.7b Concentration of sperm from young boars aged 7-8 months $(mill./m^3)$

1	60	0.25 ± 0.01	0.10
2	60	0.23 ± 0.01	0.08
3	72	0.31 ± 0.01	0.11
Hühn (1969)		0.19 - 0.24	

Table 20.7c Number of sperm per ejaculation from young boars aged 7-8 months (10^9)

1	60	32.3 ± 1.4	10.7
2	60	47.5 ± 1.3	9.7
3	72	46.2 ± 1.5	13.0
Hühn (1969)		29.0 - 39.5	

Table 20.8 Provisional lysine and methionine + cystine requirements for rearing young boars, on the basis of a high level of sperm production during later use

Live weight (kg)	Lysine	Methionine + cystine
	(g/animal/day)	
40	18	12
60	21	16.5
80	23	20-22
100	24.5	22-24

For this reason all efforts at improving sperm production and fertility of boars should be concentrated in the rearing period when organs are in a state of development and react more readily to nutritional effects.

The Amino Acid Requirement

Although this paper deals with the amino acid requirement of growing finishing pigs, in the strictest sense it concerns only the amino acid requirements of castrated males upon which the experiments described were performed. Half the fattening pigs, however, are gilts. Gilts deposit about 12 per cent more protein than castrated males and will therefore have a higher amino acid requirement. However, a difference of 10 per cent can still lie within the range of normal biological distribution. If the fattening of boars is introduced in the future because considerably more lean and less fat can be produced, the currently valid amino acid requirements will no longer be applicable. Boars deposit about 30 per cent more protein than castrated males. According to our initial investigations, their amino acid requirement is about 30 per cent higher than our current standards.

The determination of the amino acid requirement is an important factor in economic feeding. The other condition, which is of equal importance is that this requirement should be met by the available feedstuffs.

Apart from the energetic evaluation of the foodstuffs, which can be expected to improve considerably as a result of the introduction of the GDR feed evaluation system in the German Democratic Republic, the utilisation of the protein in the feed is becoming increasingly important. The increasing shortage of protein foodstuffs on a worldwide scale is also making the assessment of the nutritional value of proteins in foodstuffs an urgent problem. Popow (1960) drew the following conclusions in one of his last papers concerning protein evaluation of feedstuffs:

(1) None of the existing methods for assessing the protein quality of the feed (biological, chemical etc.) is suitable as a parameter for the quality of protein when actually feeding farm animals. The main drawback of such methods is that their characterisation of the 'protein nutritional value' is not associated with figures showing the protein requirement of the animals.
(2) The final assessment of the nutritional value of proteins for animals which are completely dependent on the protein in the food supply (pigs and poultry) must be based on the calculation of the essential amino acids which are available to (or digestible by) the animals. The nutritional value of protein in a feedstuff together with the protein requirement of the monogastric animal should be characterised by the amount of digestible crude protein and the amount of essential amino acids contained in it (at least of the critical amino acids lysine, methionine and tryptophan).

Popow (1960) also stated that it is time to move in this direction in the investigation of feedstuffs.

In the past few years it has been possible, by means of highly automated amino acid analysers, to further unravel the natural laws governing the absorption of amino acids as one factor which decisively influences their utilisation. Therefore it is necessary to investigate more closely the natural laws governing the utilisation of amino acids and the factors which affect it. The animal must be supplied with amino acids by means of food before high rates of protein synthesis can occur. The availability of amino acids for the synthesis of protein can, according to Herrmann (1970), be affected by the following factors:

(1) digestibility,
(2) possible chemical modification which permits absorption but hinders intermediate utilisation,
(3) differences in the time of absorption, transport or activation of the amino acids.

The first of these factors, i.e. digestibility, has the greatest effect. The digestibility of the amino acids in various feedstuffs for various animal species have been investigated in order to obtain guidelines which permit precise prediction of the performance of pigs and poultry with respect to the amino acid supply.

When amino acid requirement is mentioned, it is the crude or total amino acid which is implied. In the case of other nutrients (at least the digestible ones) and in the case of energy even, the net values are used to characterise the corresponding requirements. In the last few years, extensive investigations have been performed concerning the true digestibility of the amino acids in different feedstuffs (Herrmann, 1970; Hackl, 1971; Meier, 1970; Kristen, 1968). Methodical questions (Schurig, 1973) and factors affecting the true digestibility of the amino acids (Breite, 1973; Rossbach, 1974; Heinz, 1974; Schadereit, 1973) have also been investigated during the same period.

In order to provide an impression of the importance of such investigations, the effect of the technical drying process on the true digestibility of a few amino acids is considered.

Table 20.9 *The true digestibility (per cent) of nitrogen, lysine, methionine and cystine in potatoes of varying dry matter*

Form	Dry matter (%)	True digestibility of			
		Nitrogen	Lysine	Methionine	Cystine
0[1]	16.6	75.1	78.5	67.4	67.5
1	87.8	74.3	68.6	61.1	63.5
2	92.9	57.9	52.1	53.1	56.2
3	96.5	31.9	3.6	7.8	22.9

[1] Form 0 is steamed potatoes
Forms 1, 2 and 3 are potatoes dried to different dry matter levels

True protein digestibility is highest in this form for steamed potatoes. The more intensive the drying process (the dry matter content can, with certain reservations be regarded as a scale), the lower is the true digestibility of the proteins and amino acids. The amino acids lysine and methionine are most adversely affected (*Table 20.9*).

These results and other investigations show that considerable errors can occur in the ability of particular ingredients and complete diets to supply amino acids if the crude amino acids are used as a basis. *Table 20.10* provides some details of the true digestibility of amino acids in various feedstuffs which illustrate this problem. There are considerable differences between the various feed constituents and between the amino acids, so that it would appear necessary to take true digestibility of the amino acids into consideration when assessing the capability of individual feed constituents and whole rations to supply amino acids. At the same time, corresponding particulars must

Table 20.10 The level and true digestibility of amino acids in some ingredients used in pig nutrition (dry matter basis)

	Crude protein (g/kg)	Lysine			Methionine			Cystine	
		Total (g/kg)	True digestibility (%)	True digesti-bility (g/kg crude protein)	Total (g/kg)	True digestibility (%)	True digesti-bility (g/kg crude protein)	Total (g/kg)	True digestibility (%)	True digesti-bility (g/kg crude protein)
Fish meal	604	61.6	92.2	56.8	19.9	85.8	17.1	5.4	70.7	3.8
Powdered skimmed milk	339	20.7	94.7	19.6	8.8	94.1	8.3	3.0	97.4	2.9
Shredded soya extract	453	29.9	97.0	29.0	5.9	96.2	5.7	7.2	97.6	7.0
Shredded groundnut extract	441	15.9	84.4	13.4	4.0	90.0	3.6	8.8	89.3	7.9
Shredded sesame extract	441	9.7	92.7	9.0	11.0	97.1	10.7	6.6	97.7	6.5
Feed yeast Prowona	394	28.0	83.1	23.3	2.4	74.9	1.8	4.7	68.5	3.2
Field beans	284	17.6	96.7	17.0	2.8	83.0	2.3	3.7	93.5	3.5
Wheat	141	4.1	89.4	3.7	1.8	98.9	1.8	3.2	95.0	3.0
Wheat	165	5.0	89.5	4.5	3.1	99.8	3.1	3.0	93.9	2.8
Barley	118	4.6	78.3	3.6	1.4	97.4	1.4	2.7	89.9	2.4
Barley	156	5.1	71.1	3.6	1.6	99.5	1.6	3.0	90.3	2.7
Potatoes, steamed	86	5.3	81.5	4.3	2.0	78.8	1.6	1.4	85.5	1.2
Potatoes, steamed and ensiled	79	5.1	77.2	3.9	1.6	70.7	1.1	1.1	76.2	0.8
Potatoes, dried	89	3.5	68.7	2.4	1.7	73.1	1.2	1.4	73.2	1.0

be provided regarding the requirements of the animals. Our initial investigations on the requirements of growing pigs (castrated males) for true digestible lysine and methionine + cystine have led to the conclusions given in *Table 20.11*.

Table 20.11 The lysine and methionine + cystine requirement of castrated males (g/animal/day)

Live weight (kg)	True digestible lysine	True digestible methionine + cystine
30	9.4	5.6
40	11.0	6.7
50	11.8	7.9
60	12.4	8.9
70	12.9	9.3
80	13.2	9.4
90	13.6	9.4

It is our opinion that more rational feeding of fattening pigs is possible on this basis. Better formulation of mixed feed for the supply of amino acids is possible, particularly under the conditions associated with the production of pigs for slaughter on a large scale commercial basis.

References

AWRUTINA, A. (1966). *Aminosäurenernährung bei Schweinen und Geflügel*, Leningrad
BLAIR, R. and ENGLISH, P.R. (1965). *J. agric. Sci.*, **64**, 169
BORZ, I.L. and SHURBA, W.A. (1963). *Aminosäurenernährung der Schweine und des Geflügels. Moskau*, **S.97**
BREITE, S. (1973). *Diss. Univ. Rostock*, **144 S**
CHARETTE, L.A. (1961). *Can. J. Anim. Sci.*, **41**, 30
DINU, I. and GEORGESCU, D. (1970). *Rev. zootechn. si. Med. Vet.*, **20**, 10, 43
FARRIES, E. and KALLWEIT, E. (1970). *Sonderheft, Z. Landw. Forschg.*, **13**
FLICEK, V. (1966). *Z. Cislo, Rok.*, IV *Rocnik*
GRADUSOW, J.N. (1967). *Synthetische stickstoffhaltige Präparate in der Tierzucht, Moskau, Verlag Kolos*
HACKL, W. (1971). *Diss. Univ. Rostock*, **157 S**
HEINZ, T. (1974). *Diss. Univ. Rostock*, **154 S**
HERRMANN, U. (1970). *Diss. Univ. Rostock*, **139 S**
HÜHN, U. (1969). *Forschungsbericht Sektion Tierproduktion.* Friedrich-Schiller-Universität, Jena
KLEEMANN, F. (1972). *Diss. Univ. Rostock*, **145 S**
KNOBLOCH, F. (1972). *Diss. Univ. Rostock*, **168 S**
KRISTEN, H. (1968). *Habil-Schr., Univ. Rostock*, **122 S**
MEIER, H. (1970). *Diss. Univ. Rostock*, **120 S**

MÜLLER, H. and GEBHARDT, G. (1973). *Wiss. Zeitschr. der Karl Marx Universität, Leipzig*, **22**, 264
NAVRATIL, B. and SIMECEK, K. (1963). *Zivocisna vyroba VI*, **8 (XXXVI)**, 575
NAVRATIL, B. and SIMECEK, K. (1966). *Zivocisna vyroba, Prag, Rochik* **II, XXXIX**
OWSJANIKOW, A.I. (1963). *Aminosäurenernährung der Schweine und des Geflügels, Moskau*, **S.43**
PLATIKANOV, N., ANGELOV, A., KANEV, M. and PLANAJOTOV, P. (1969). *Wiss. Z. Univ. Rostock, Math. Nat. Reihe*, **XVIII**, 19
PLATIKANOV, N., ANGELOV, A., KANEV, M., PLAJANOTOV, P. and ANGELOWA, L. (1970). *Sitzungsberichte der DAL Berlin*, **XIV, H.3**, 45
POPPE, S. and WIESEMÜLLER, W. (1968). *Arch. Tierernähr.*, **18**, 5
POPPE, S. and WIESEMÜLLER, W. (1969). *Int. Z. d. Landw.*, **H.6**, 1
POPOW, I.S. (1960). *Ausgewählte Werke Verlag 'Ähre' Moskau*, 1966
RERAT, A. and LOUGNON, J. (1966). *A.E.C. Nr.*, **6**, 341
ROSSBACH, F. (1974). *Diss. Univ. Rostock*, **151 S**
SCEGLOV, V.V. and FICEV, A.U. (1966). *Aminosäurenernährung der Schweine Minsk* (russ.) *Verlag Uroshai*
SCEGLOV, V.V. and FICEV, A.U. (1969). *Wiss. Z. Univ. Rostock, Math. Nat. Reihe*, **XVIII, H, Teil II**, 267
SCHADEREIT, R. (1973). *Diss. Rostock*, **130 S**
SCHURIG, J. (1973). *Diss. Univ. Rostock*, **109 S**
TEAGUE, H.S., PLIMPTON, R.F., CAHILL, A., GRIFO, A, and KUNKLE, L.E. (1964). *J. Anim. Sci.*, **63**, 341
TKATSCHOW, I.F. and GRIGORJEW, W.W. (1971). *Informationen der Landw. Wissenschaft.*, **11**, 18
TOMME, M.F., FILIPOWITSCH, E.G. and PTAK, I.R. (1971). *Vorträge* **WASCHNIL, 8**, 28
TOMME, M.F. and KROCHINA, W.A. (1967). *Arbeiten Wish XXX, Moskau*, 62
TOMME, M.F., MACHAJEV, E.A., KROCHINA, W.A., FILIPOWITSCH, E.G., RJADSCHIKOW, W. and SKORKIN, G.K. (1971a). *Vorträge Waschnil*, **12**, 20
TRIBBLE, L.F., AMICK, G.L., LASLEY, N. and ZOBRISKY, S.E. (1965). *Research Bulletin 881*, University of Missouri, College of Agriculture
WALSTRA, P. and KROESKE, D. (1968). *Wld Rev. Anim. Prod.*, **4**, 59
WIESEMÜLLER, W. and POPPE, S. (1968). *Arch. Tierernähr.*, **18**, H.5
WIESEMÜLLER, W., POPPE, S. and SIEG, G. (1974). *Untersuchungen zum Aminosäurenbedarf von Mastschweinen (Börge) im Druck*
WITT, M. and SCHRÖDER, J. (1969). *Fleischwirtschaft*, **49**, 353
WÜNSCHE, J., BOCK, H.D. and KREIENBRING, F. (1973). *Überarbeitete Aminosäuren Gehaltstabellen für die Mischfutterindustrie und Landwirtschaft*
UdSSR-Normen (1969). *Normen und Rationen in der Fütterung der landwirtschaftlichen Nutztiere Moskau*, **S. 121**

21

THE AMINO ACID REQUIREMENTS OF GROWING PIGS (2)

T. HOMB
Agricultural University of Norway

Introduction

Some twenty years ago it was clearly established that the growing pig, like the growing rat, needs ten amino acids to be provided in its diet (Mertz, Beeson and Jackson, 1952). The establishment of quantitative requirements for each essential amino acid were the objectives of extensive studies from the 1940's onwards. However, the findings are still to some degree controversial, due to variation in animal populations, feed level, diet composition, required carcass quality etc. Nevertheless, the huge amount of published material provides a useful background for analysing the reasons behind the divergent results.

Methodology

The methods used in establishing the requirements for amino acids and protein may be divided into two categories, namely the factorial method and growth studies (which are sometimes supplemented with biochemical criteria).

THE FACTORIAL METHOD

The general principle of the factorial method is based on the sum of all expenditures for maintenance and productive purposes, taking into account the fact that animals are not able to utilise all the dietary amino acid provided.

A serious attempt to determine the amino acid requirements of adult, non-pregnant gilts has been published by the Illinois group (Baker *et al.*, 1966). They chose a daily nitrogen deposition of 1 g as the criterion for maintenance requirement (taking nitrogen in integuments into consideration). Of the amino acids essential for growth three were shown non-essential for maintenance: arginine, histidine, and leucine. A requirement of 25 mg/kg $W^{0.75}$ lysine and 26 mg methionine/kg $W^{0.75}$ was suggested. A mixture of the crystalline L-amino acids essential for growth, plus L-glutamic acid, made up the experimental ration. By omitting one or more of the amino acids in the synthetic diet it was possible to reach valid conclusions for each amino acid.

Another attempt was made by Wiesemüller, Poppe and Kristen (1969), who also used nitrogen balance data as criteria for maintenance, in this case with young pigs weighing 20-35 kg liveweight. Their results suggested maintenance requirements for a 30 kg pig of 1.8 g lysine and 0.9 g methionine. Metabolic faecal nitrogen (MFN) is also involved in the factorial method. Eggum (1973a) has reviewed his own and other results on the amino acid content of MFN of pigs. On average the lysine content is about 5.0 per cent and methionine + cystine 3.6 per cent (of N × 6.25). A 30 kg pig will, according to Eggum, have a daily expenditure of 0.21 g lysine and 0.15 g methionine + cystine. The different sources agree fairly well on this point, and compared to the expenditure for growth these amounts account for only a small part of the pig's total requirement.

Amino acid deposition in the body is the most important single factor in this type of estimate. Of the numerous data on the amount of protein deposited in growing pigs the work of Just Nielsen (1971) is chosen as an example. The Danish pigs had an average daily retention of 15-16 g nitrogen from 20 to 90 kg live weight, a little more for females than for castrated males. These figures were based on slaughter experiments. Nitrogen balance figures disclosed an increasing trend in daily nitrogen retention from 20 to around 50 kg live weight after which the values were nearly constant.

Buraczewski (1973) has reviewed the literature on the amino acid content of the protein (N × 6.25) in the empty body of pigs. Although the animals varied in age and weight, the amino acid content was surprisingly consistent. His own results showed an increase in lysine from 6.0 per cent to 6.9 per cent as the live weight increased from 10 to 100 kg, with a rise in the sulphur-containing amino acids from 2.6 to 3.0 per cent.

Investigations based on the factorial method seem to give increasingly more reliable results. One important reason for this has been improved analytical techniques. However, as Wiesemüller and Poppe have stated, the results from balance experiments should be verified in growth studies, including slaughter investigations.

GROWTH STUDIES

Of the criteria considered in growth studies the following could be listed: rate of growth, feed conversion ratio, lean content of carcasses, nitrogen balance, serum or plasma-free amino acids, blood urea, and urinary-urea. The conclusions, however, might vary with the methods and criteria used. The simplest way to establish requirement figures for a specified amino acid is to carry out a growth study with graded levels of the amino acid in question and to measure the growth rate with *ad libitum* feeding. If the feed conversion ratio is taken into account also, the possibility of a compromise exists, although these two criteria are very often correlated. A large number of data of this type are reported in the literature especially from the USA (Homb, 1963). The NRC recommendations (NRC, 1968) represent 'minimal levels for normal growth and performance'. This might explain the lower levels compared with most European standards. As an example of the European work, Clausen (1963) has defined the necessity to meet the need of individually restricted-fed pigs in good houses, considering at the same time growth rate, feed conversion ratio and carcass leanness. As in other investigations carcass leanness was the factor most difficult to satisfy, one of the reasons being that increasing levels of protein and amino acids in the diet lead to a lower energy utilisation (Kielanowski, 1972). The genotype of the animal will undoubtedly affect results. It is reasonable to believe that the high Danish amino acid recommendations may be explained by the pig population existing there. American balance experiments clearly show differences in daily nitrogen retention between fat-type and meat-type pigs (McConnell, Barth and Griffin, 1971), and this will affect the protein and amino acid needs of the animals. Homb (1972) also found considerable individual variability in protein deposition, especially for pigs weighing more than 90 kg live weight. To evaluate the different factors against each other is a difficult task.

While growth studies usually take a long time, balance trials are generally of shorter duration. Blood analysis or urine analysis may be done in experiments of still shorter duration and are much cheaper (Cline, 1973). The significance of this type of experiment as a basis for determining the requirements of amino acids has been argued. In recent Canadian investigations a new technique has been introduced in which a short interval feeding regime is an important feature. It has the effect of reducing the wide fluctuations in plasma amino acid levels. Long-term studies have the advantage that any compensatory growth can be observed. There are many indications of compensatory protein deposition in pigs, some of the first being Robinson (1964) and Lewis (1968) at the University of Nottingham. It seems logical that this also has some implication upon amino acid nutrition.

Total and Available Amino Acids

Until now it has been common to base requirements upon the total amount of each essential amino acid in the diet. Several attempts have been made to find appropriate values for the availability of the amino acids. This is best done by animal experiments, and the methods used have recently been reviewed by Eggum (1973a,b). The digestibility of the amino acids might differ from one to the other, and it is very tedious to elucidate figures for the availability of each amino acid in all types of feed. Materials like dried skim milk and fish meal show a high degree of variability, one of the reasons being different heat-treatments. Until reliable figures for the availability of each amino acid are obtained, Eggum (1973a,b) recommends the correction of total amino acids with values for true digestibility of the protein in the feed in question. However, lysine should be considered separately, because of the low true digestibility in grains, which is of utmost importance in pig feeding. Cheaper biological methods that can give information about the availability of methionine (*Streptococcus zymogenes*) and lysine (*Tetrahymena pyriformes*) have been developed (Carpenter, 1973).

In the case of lysine it appears that a chemical method (dinitrofluorobenzene [FDNB] reactive lysine) discloses heat damage of the protein in feedstuffs. A recent review by Carpenter (1973) provided a fairly close correlation between the FDNB method and results from chick feeding tests. The feeds tested in this case comprised fish meals and meat meals. The method is more questionable when different batches of grain are compared. Carpenter also admits that the FDNB method includes a technical fault, which even for animal feed sources leads to a discrepancy between chemical and biological data.

A question of practical importance in pig feeding is how to evaluate protein and amino acid requirements when the protein content of grain varies. Firstly, the content of lysine as a percentage of the protein usually decreases with increasing protein content (Eggum, 1973b). Secondly, a high protein barley will contribute a greater percentage of total protein than a low protein barley. Thirdly, the availability of amino acids will be lower in a mixture with high protein barley. In the Norwegian studies behind the recommended allowances (0.80 per cent lysine, 0.55-0.60 per cent methionine + cystine of the air-dry diet) for pigs 20-55 kg live weight; barley provided 53 per cent of total protein, the rest originating from herring meal and soya bean meal. The following provision was given: 'We are aware that if barley with a higher protein content is used, the situation will be changed...'. The situation from the early 1960's has already changed, and research in progress indicates that the allowances for lysine should be revised because of the higher protein levels (12-13 per cent) in the barley grown today compared with a few years ago (9-10 per cent). Madsen and Mortensen (1974) have proposed the use of apparent digestibility coefficients for protein obtained in pig experiments as criteria of availability of essential amino acids. On this basis they have determined the content of digestible lysine and threonine in a mixture consisting of barley meal (varying in protein) and soya bean meal, when assuming a constant protein percentage in the mixture (16 per cent). When 18 per cent soya bean meal was included, the mixture contained 0.65 per cent digestible lysine, whereas 10 per cent soya bean meal plus protein-rich barley give only 0.50 per cent digestible lysine. This clearly shows the weakness of basing the requirement figures on the total content of protein and amino acids.

The same type of calculation could be done by using figures for true digestibility of protein, as proposed by Eggum (1973a). His data showed high true digestibility of protein in fish meal (95.6 per cent) and soya bean meal (91.7 per cent), but lower figures for protein in barley and oats (72-73 per cent).

In the following, total amino acid content will be used as the most important parameter, even when the results mentioned indicate that available amino acids undoubtedly will be of greater interest in the future.

Expression of Requirements

There are several alternatives:
(1) As intake (g/day) of each essential amino acid.
(2) As a percentage of air-dry feed or dry matter (DM).
(3) As a certain part of the feed energy.
(4) As a percentage of the crude protein.

The fourth alternative must be excluded as less useful, because it is too complicated for practical purposes, although it is interesting from a theoretical point of view and in the estimation of protein quality.

The third method, which is widely used in calculating broiler rations, does not play the same role in pig feeding, because supplementation with fat is rare. Usually the energy concentration of pig diets varies little.

While the NRC (1968) prefers to give the figures solely as percentages of the diet, the ARC (1967) expresses the need partly as a ratio with digestible energy (g/Mcal ME) and partly as a percentage of the ration. The American recommendations apply only to a certain energy density of the diet. Conclusions made by Clausen (1963) as well as by Rerat (1971) are in favour of stating the requirement of each amino acid in daily amounts. Declarations of trade feed mixtures usually include crude protein content and often the level of some of the essential amino acids. As an example of official feed regulations, the standard feed mixtures for pigs in Norway must have a minimum percentage of lysine and sulphur-containing amino acids, while the protein percentage may vary within certain limits.

Variations in Amino Acid Requirements Due to Age and Sex

All the evidence shows that the protein requirement as a percentage of dietary dry matter decreases with increasing age, and this is also thought to apply to each essential amino acid. However, the need for sulphur-containing amino acids does not show the same decline as, for instance, lysine. This concept which was stated by Mitchell (1950) is explained by the increasing importance of integuments and decreasing requirement for growth as the animals become older. This point was later clarified for pigs by the Cornell group (Kroening, Pond and Loosli, 1965) who found a relatively low need of sulphur-containing amino acids in baby pigs, and by Wiesemüller and Poppe for pigs weighing 20-90 kg live weight (Weisemüller and Poppe, 1968).

Newly born pigs respond more drastically to protein quality than do older pigs, and milk protein is in most cases found to be superior to other sources. An exceptionally high degree of protein utilisation has been observed in newborn pigs receiving sows' milk (Wöhlbier, 1928). Research on amino acid nutrition has mainly been centred on pigs between 20 and 90 kg live weight, and less data are available for baby pigs and heavier pigs. One of the reasons for the relatively sparse data for heavier pigs is that the requirement can be met easily by practical rations. Because the daily protein and amino acid requirements increase less than the energy requirement, most results show declining protein utilisation with age. By producing diets in accordance with the protein need at all stages it is questioned whether growing pigs from 30 to 90 kg do show decreasing protein utilisation (Homb, 1972a,b) (*Table 21.1*).

Table 21.1 Protein utilisation in pigs of different weights (Homb, 1972)

Live weight (kg)	34	52	72	93	104
Dry matter (kg/day)	1.1	1.6	2.1	2.6	2.7
Dietary crude protein (per cent of DM)	18.4	17.3	15.1	13.3	13.1
Net protein utilisation (per cent)	51	49	52	50	43

The pigs reported in *Table 21.1* received net energy and protein according to their requirements (based on the factorial method) at all stages, and the protein quality of the diets remained unchanged throughout. Except for the last period, which gave a distinct drop in nitrogen deposition for these pigs, the true digestibility and biological value of the protein did not show any tendency to decline. This is in contrast to investigations where protein content has been held constant for too long a period, or protein quality has been diminished due to increasing percentage of grain in the ration.

Available information from balance trials as well as growth trials indicates variations between boars, castrates and females in protein deposition and therefore in protein and amino acid requirements. In balance trials Kielanowski (1972) obtained the following daily protein deposition from 30-80 kg live weight: boars 132 g, gilts 111 g, castrated gilts 98 g, and castrated boars 97 g. The small amount of research work carried out to determine protein requirements of growing boars indicates that 16-17 per cent digestible crude protein (20-21 per cent crude protein) is necessary to give maximum growth rate and lowest feed conversion ratio in young boars of 20-60 kg live weight (Hanssen, 1972; Nielsen, 1969)). Based on average values for amino acids, the diets used by Hanssen at our institute (unpublished) contained 1.0-1.2 per cent lysine and 0.8-0.9 per cent methionine + cystine in the growing period (20-50 kg live weight) followed by 0.7-0.8 per cent lysine and 0.6-0.7 per cent methionine + cystine in the period to 100 kg live weight. However, in feeding boars it is neither necessary nor desirable to reach maximum daily gain, and Hanssen, therefore recommends somewhat lower levels of protein and amino acids in practice.

The difference in leanness between the carcasses of gilts and barrows is also an indication that the amino acid requirements of gilts is a little higher than for barrows. Feeding gilts and barrows separately has led to slightly better results than when the sexes are mixed. (Frölich and Thomke, 1969). Until now no distinction in protein and amino acid requirement has been made for practical purposes.

The relationship between energy on the one hand and protein and amino acids on the other is a controversial matter. As this question is discussed in other chapters it will be only mentioned here.

The Requirements for Individual Amino Acids

In this chapter it is sought to extract from the impressive amount of published material, requirement figures suitable for most practical conditions. Because lysine and sulphur-containing amino acids are most likely to be limiting, these are discussed in more detail than the others. Protein requirement is also discussed in connection with lysine.

LYSINE (AND PROTEIN)
For understandable reasons only limited data are available for very young piglets. If sow's milk serves as a guideline, its dry matter contains about 2.2 per cent lysine (or about 40 g Mcal ME). Young piglets fed milk substitute have, under practical conditions, equally good gain on a diet containing 24 per cent crude protein and 1.3 per cent lysine in DM. However there are no direct comparisons between these two feeding regimes and level of feeding in each case will be of importance. It would seem that 2.2 per cent lysine in DM is unnecessarily high. The ARC's (1967) recommendations of requirements for young piglets applies to the weight range 9-20 kg live weight, and for such pigs the average value 1.1 per cent lysine and 20 per cent crude protein in the dry matter is chosen as the standard. Nielsen (1973) has found that baby pigs from three to ten weeks of age respond positively in growth rate to higher protein content. Although 26 per cent and 31 per cent crude protein of air-dry feed gave slightly better growth rates in the period mentioned, 21 per cent crude protein resulted in as good a weight gain and performance in the subsequent period and in carcass characteristics as the higher protein diets. The diet with 31 per cent crude protein contained 2.4 per cent lysine, while the diets with 26 and 21 per cent contained 1.8 and 1.4 per cent lysine respectively. A diet containing 17 per cent crude protein and 0.9 per cent lysine gave lower growth rates during the piglet period than the 21 per cent diet. However, the daily weight gain from 20 to 90 kg live weight was good enough. Thus, the Danish results indicated somewhat higher requirements for the period 9-20 kg live weight than recommended by the ARC.

Wiesemüller, Poppe and Knobloch (1973) reported higher growth rates in pigs from 7 to 19 kg live weight by increasing lysine from 0.84 to 1.28 or 1.52 per cent. Taking into account the subsequent period of growth their conclusion was that an air-dry feed mixture with 18-19 per cent crude protein and 0.88-0.97 per cent lysine gave a satisfactory weight gain. Ground wheat is recommended as the basal feed for early weaned piglets.

For the live weight range 20-90 kg recommendations vary considerably, the lowest figures originate from the NRC (1968) and are 0.7 per cent lysine for 20-35 kg live weight pigs and 0.50 per cent for finishing pigs (in air-dry feed). The highest figures come from Denmark. Clausen's (1963) data for daily lysine intake have been recalculated here as a percentage of air-dry feed, with the following results: 20-30 kg live weight, 1.1 per cent lysine and 16-17 crude protein; 50 kg liveweight, 0.85 per cent lysine and 14-15 per cent crude protein; 80 kg live weight, 0.70 per cent lysine and 13 per cent crude protein. The background of Clausen's figures is a diet of skim milk and barley meal, supplemented with lysine and methionine. This diet has given the best results, taking into account the lean-fat ratio of the carcasses. East German experiments have also given relatively high figures for the lysine requirement. Levels decreasing from 0.90 to 0.56 per cent of the air-dry diet for the period from 20 to 90 kg live weight have been recommended (Poppe and Wiesemüller, 1968).

In between the lowest and highest recommendations are those of the ARC (1967), Rerat and Lougnon (1971) and many others. The two sources mentioned agree well, and the Norwegian results are in fair agreement with these (Homb and Matre, 1967).

As suggested earlier, compensatory growth might enable pigs to do almost as well with a suboptimal supply of protein and amino acids for a period of time, provided that this is accounted for in the subsequent growth period. This has been shown by Frölich and Thomke (1969) in Sweden. Their findings are in good agreement with the Norwegian results, which indicate 0.70 per cent lysine in the air-dry diet is satisfactory for the period 20-90 kg live weight.

As a summary of this discussion the requirements for protein and lysine under normal conditions for growing-finishing pigs and replacement gilts are given in *Table 21.2*.

Table 21.2 Crude protein and lysine requirements of growing-finishing pigs and replacement gilts (per cent of air-dry diet)

Live weight (kg)	Crude protein (per cent)	Lysine
Young piglets	20-30	1.1 - 1.6
20	16-18	0.8 - 1.0
50	13-15	0.65-0.75
90	12-13	0.55-0.65
120	11-12	0.45-0.55

The requirements in *Table 21.2* are given as intervals for the following reasons. As stated earlier, gilts need slightly more than barrows, and boars (from 20-30 kg live weight) have a 15-20 per cent higher requirement than the figures show. Performance tests of boars and gilts might have been planned in such a way that higher than necessary levels were used. The use of high-protein barley requires a higher lysine level in the diet than low-protein barley.

The price of protein-rich feeds may affect the choice of protein and lysine level. High protein prices might, therefore, justify a reduction of protein, while a high premium on lean carcasses might have the opposite effect.

In Norway an attempt has also been made to state the requirement of available lysine for pigs of 20-60 kg live weight, using Carpenter's

FNDB method. The recommendations were 0.65-0.70 per cent available lysine, which in this case corresponded to 0.80 per cent total lysine (Homb and Matre, 1967; 1971).

METHIONINE AND CYSTINE

It is assumed that cystine may meet about one half of the need for the sulphur-containing amino acids. Kroening, Pond and Loosli (1965) found that young pigs needed 0.5 per cent methionine + cystine when the protein content of the diet was 12 per cent, 0.6 per cent by using 18 per cent crude protein, and 0.7 per cent when the ration contained 25 per cent crude protein. This indicates that practical rations for early weaned pigs should have 0.6-0.7 per cent methionine + cystine. On the basis of American and British results the ARC (1967) suggest that 0.5-0.6 per cent methionine + cystine in the air-dry diet is sufficient for pigs up to 45-50 kg live weight. Norwegian figures (0.55-0.6 per cent) agree fairly well with these, as do East German recommendations based on balance trials (Wiesemüller and Poppe, 1973) for pigs of 20-50 kg live weight (0.52-0.50 per cent). Danish recommendations are higher; a recalculation of Clausen's (1963) data indicates the levels of 0.70 per cent methionine + cystine for the 20 kg live weight pig and 0.50 per cent for a pig weighing 50 kg live weight.

Somewhat contradictory recommendations have appeared, from 0.30 per cent methionine + cystine (Lewis, 1966) to 0.53-0.57 per cent (Clausen, 1963). The ARC (1967) is of the opinion that 'there is insufficient evidence on which to specify requirements lower than those given for pigs up to 45-50 kg live weight.' (The same view applies also to some other essential amino acids.) Wiesemüller and Poppe (1968) have calculated requirements to be 0.50-0.39 per cent methionine + cystine for pigs from 50 to 90 kg live weight. Standards for the Norwegian trade mixtures are 0.60 per cent for 20-60 kg live weight pigs and 0.48 per cent for pigs of 60-100 kg live weight. The requirements for methionine + cystine are summarised in *Table 21.3* below.

Table 21.3 Requirements for methionine + cystine (per cent of air-dry diet)

Liveweight (kg)	Methionine + cystine
Young piglets	0.65-0.70
20-50	0.55-0.65
50-100	0.45-0.55

TRYPTOPHAN

Gallo and Pond (1966) suggested a tryptophan level of 0.18-0.22 per cent of a diet containing 20 per cent crude protein to be the minimum requirement for optimal performance of young pigs from 21 to 45 days of age. Other American results have shown slightly lower figures for pigs of ten kg live weight (0.15 per cent, Rerat, 1971). For growing-finishing pigs, decreasing levels from 0.13 to 0.09 per cent are recommended by the NRC (1968), while the ARC (1967) suggested 0.13-0.17 per cent tryptophan for pigs up to 45-50 kg live weight. Clausen's (1963) data have been recalculated to 0.18 per cent for a pig weighing 20 kg live weight, falling to 0.13 per cent for the 90 kg live weight pig (all figures given as the percentages of air-dry diet). Commercial feed mixtures in Norway for pigs from 20-60 kg live weight contain 0.15-0.17 per cent tryptophan (Homb and Matre, 1971). All the recommendations have assumed an adequate supply of niacin. The conclusions concerning tryptophan requirement are given in *Table 21.4*.

Table 21.4 Requirements for tryptophan (per cent of air-dry diet)

Live weight (kg)	Tryptophan
Young piglets	0.18-0.20
20-50	0.13-0.15
50-100	0.10-0.12

THREONINE

The few observations on the threonine requirement for young piglets have suggested a level of 0.6-0.78 per cent of the air-dry diet, the highest figure for two day old pigs (ARC, 1967), and the lowest for pigs of 10 kg live weight (Rerat, 1971). Remarkably consistent figures for the requirement have been suggested for pigs in the live weight range 20-50 kg: 0.45 per cent (NRC, 1968), 0.45-0.5 per cent (ARC, 1967), 0.48 per cent (Rerat, 1971). Usually, it is easy to satisfy this need with normal feed ingredients. The Norwegian feed mixtures for pigs of this weight have a content of about 0.6 per cent (Homb and Matre, 1971), and the Danish barley/skim milk diet has a gradually decreasing level from 0.72 to 0.58 per cent threonine as the pigs grow from 20 to 50 kg live weight.

The requirements for threonine are given in *Table 21.5* but there is little information for finishing-pigs.

Table 21.5 Threonine requirements (per cent of air-dry diet)

Live weight (kg)	Threonine
Young piglets	0.6 -0.75
20-50	0.45-0.5

ISOLEUCINE

American and British experiments indicate the requirements set out in *Table 21.6*.

Table 21.6 Isoleucine requirements (per cent of air-dry diet)

Live weight (kg)	Isoleucine
Young piglets	0.7-0.8
20-50	0.5-0.65
Finishing pigs	0.35 (NRC)

The common rations in most countries contain sufficient of this amino acid. Both Danish and Norwegian feeding systems include a surplus of isoleucine compared with the recommendations.

OTHER ESSENTIAL AMINO ACIDS

Under practical conditions it is unlikely that the remaining five essential amino acids are limiting, and this is probably the reason why little research work has been conducted to clarify the requirements for these. The American recommendations are given in Table 21.7.

Table 21.7 Requirements for other essential amino acids (per cent of air-dry diet)

| | Live weight | |
	5-10 (kg)	20-35 (kg)
Arginine	-	0.20
Histidine	0.27	0.18
Leucine	0.90	0.60
Phenylalanine + tyrosine	-	0.50
Valine	0.65	0.50

References

AGRICULTURAL RESEARCH COUNCIL, (1967). *The nutrient requirements of farm livestock, No.3. Pigs.* Technical reviews and summaries. Agricultural Research Council, London
BAKER, D.H., BECKER, D.E., NORTON, H.W., JENSEN, A.H. and HARMON, B.G. (1966). *J. Nutr.*, **89**, 441
BURACZEWSKI, S. (1973). *Brno. Proc.*, **C-1**
CARPENTER, K.J. (1973). *International Association of fish meal manufacturers. Symposium on the use of fish meal in animal feeding.* 31
CLAUSEN, H. (1963). *Landøkonomisk Laboratoriums Efterårsmøde.* 209
CLINE, T.R. (1973). *Proc. 21st Annual Pfizer Research Conference,* 40
EGGUM, B.O. (1973a). *Beretning nr. 406 fra Forsøgslaboratoriet.* 173
EGGUM, B.O. (1973b). *International Association of fish meal manufacturers. Symposium on the use of fish meal in animal feeding.* 14
FRÖLICH, A. and THOMKE, S. (1969). *Svinskötsel,* **59**, (1), 24
GALLO, J.T. and POND, W.G. (1966). *J. Anim. Sci.*, **35**, 774
HANSSEN, J.T. (1972). *Husdyrforsøksmøtet,* 78. Rådet for husdyrforsøk, Oslo
HOMB, T. (1963). *Meld. Norges landbr. høgsk.*, **42**, 20

HOMB, T. (1972a). *Festkrift til Knut Breirem*, **61**. Norges landbrukshøgskole
HOMB, T. (1972b). *Z. Tierphysiol., Tierernähr. Futtermittelk.*, **29**, 123
HOMB, T. and MATRE, T. (1967). *Z. Tierphysiol., Tierernähr. Futtermittelk.*, **23**, 129
HOMB, T. and MATRE, T. (1971). *Z. Tierphysiol., Tierernähr. Futtermittelk.*, **28**, 86
JUST NIELSEN, A. (1971). *Royal Veterinary and Agricultural University, Copenhagen. Yearbook*, 81
KIELANOWSKI, J. (1972). in *Handbuch der Tierernährung*, **Bd. 2**, 528, Paul Parey, Hamburg
KRIDER, J.L. and CARROL, W.E. (1971). *Swine Production*. McGraw Hill, New York
KROENING, G.H., POND, W.G. and LOOSLI, J.K. (1965). *J. Anim. Sci.*, **24**, 519
LEWIS, D. (1966). in *Recent advances in animal nutrition*, 188. J. and A. Churchill, London
LEWIS, D. (1968). *Proceedings EAAP, study commission animal nutrition*, Dublin
MADSEN, A. and MORTENSEN, H.P. (1974). *Tolvmandsbladet*, **46**, 3
McCONNELL, J.C., BARTH, K.M. and GRIFFIN, S.A. (1971). *J. Anim. Sci.*, **32**, 654
MERTZ, E.T., BEESON, W.M. and JACKSON, H.D. (1952). *Arch. Biochem.*, **38**, 121
MITCHELL, H.H. (1950). in *Protein and amino acid requirements of mammals*, Ed. Albanese. p.1. Academic Press, New York
MITCHELL, J.R. Jr., BECKER, D.E., JENSEN, A.H., HARMON, B.G. and NORTON, H.W. (1968). *J. Anim. Sci.*, **27**, 1327
MUNRO, H.N. (1964). in *Mammalian protein metabolism*, **Vol.1**, Eds. Munro and Allison. p.381. Academic Press, New York and London
NATIONAL RESEARCH COUNCIL. (1968). *Nutrient requirements of swine*. National Academy of Sciences, Washington D.C. Publication 1599
NIELSEN, H.E. (1969). **375**. *beretning fra Forsøgslaboratoriet*, Copenhagen
NIELSEN, H.E. (1973). *Beretning nr. 405 fra Forsøgslaboratoriet*, Copenhagen
POPPE, S. and WIESEMÜLLER, W. (1968). *Arch. Tierernähr.*, **18**, 392
ROBINSON, D.W. (1964). *Animal Prod.*, **6**, 227
RERAT, A. (1971). *Nutr. Abstr. Rev.*, **42**, 13
WIESEMÜLLER, W. and POPPE, S. (1968). *Arch. Tierernähr.*, **18**, 405
WIESEMÜLLER, W. and POPPE, S. (1973). *Symposium on Amino Acids*, Brno, **C 3**
WIESEMÜLLER, W., POPPE, S. and KRISTEN, H. (1969). *Arch. Tierernähr.*, **19**, 273
WIESEMÜLLER, W., POPPE, S. and KNOBLOCH, F. (1973). *Tierzucht*, **27**, 370
WOHLBIER, W. (1928). *Biochem. Z.*, **202**, 29

VI

PROTEIN NUTRITION OF RUMINANTS

UTILISATION OF NITROGEN IN RUMINANTS

J.R. MERCER AND E.F. ANNISON
Unilever Research Laboratory, Sharnbrook, Bedford
Present address: Department of Animal Husbandry, University of Sydney, Australia

Introduction

The unique aspects of nitrogen metabolism in ruminants impinge mainly on the supply of amino acids to the small intestine rather than the amino acid requirements of tissues, which appear to be similar to those of non-ruminants (Lewis and Mitchell, 1974). The key feature, the relationship between the nitrogen metabolism of the microbial population in the rumen with that of the host, will be discussed with emphasis on the synthesis and utilisation of amino acids by rumen bacteria, the factors affecting the efficiency of synthesis of microbial protein, the extent to which food protein escapes microbial degradation, and to recent data which has been obtained on nitrogen cycling using isotope dilution procedures based on ^{14}C and ^{15}N labelled substrates.

There are three main sources of amino acid supply: proteins, peptides and amino acids of dietary origin which escape ruminal degradation, microbial protein produced in the rumen, and endogenous nitrogen sources consisting largely of material secreted into the alimentary tract together with any tissue shed by the epithelium of the mouth and alimentary tract. These facets of nitrogen metabolism are closely interrelated since the peptides, amino acids and ammonia produced by the microbial degradation of dietary protein in the rumen serve as nitrogen sources for the microbial population. Excess ammonia (NH_3) is absorbed into the portal system, and part of the urea formed in the liver is returned to the rumen in saliva to supplement the rumen NH_3 pool. Quantitative data on nitrogen cycling has emerged from recent studies with $^{15}NH_3$ (Pilgrim *et al.*, 1970; Mathison and Milligan, 1971; Nolan and Leng, 1972; Nolan, Norton and Leng, 1973).

Uptake of Amino Acids by Rumen Bacteria

It is now generally accepted that rumen bacteria obtain most of their amino acid requirements by *de novo* synthesis from NH_3 and various carbon sources produced during the fermentation of carbohydrates (Allison, 1969). Although many rumen bacteria have an absolute requirement for NH_3 (Bryant, 1963) there is evidence that some species assimilate exogenous amino acids (Bryant and Robinson, 1963) in spite of the high turnover and degradation rates (Lewis, 1955; Wright and Hungate, 1967b) and the normally low concentrations and small transient increases in concentration after feeding (Annison, 1956; Leibholz, 1965; Wright and Hungate, 1967a). Amino acids supplied as peptides are more efficiently utilised than free amino acids by some species (Pittman and Bryant, 1964; Wright, 1967) although the mechanism of utilisation is not clear. Allison (1970) suggested that the bacterial cell wall may have a non-specific binding capacity for peptides which are subsequently split to amino acids prior to transport into the cell.

Growth requirements for cysteine (Bryant, 1961) and methionine (Pittman and Bryant, 1964) by certain strains of rumen bacteria have been identified but the importance of these requirements is perhaps overestimated unless consideration is given to the role of other bacteria in satisfying them. This symbiosis between different species of rumen bacteria is well illustrated by the results of Pittman, Lakshmanan and Bryant (1967) and Zalin (1971). Pittman, Lakshmanan and Bryant (1967) showed that *Bacteroides ruminicola* cannot accumulate ^{14}C from ^{14}C-labelled proline or from ^{14}C-labelled glutamic acid and only incorporated small amounts of ^{14}C from ^{14}C-labelled valine. Zalin (1971) on the other hand has observed high rates of uptake of 17 amino acids and the incorporation of significant amounts of ^{14}C from ^{14}C-labelled amino acids into microbial protein by mixed cultures of rumen bacteria taken from a sheep fed lucerne hay cubes. An average of 6.5 per cent of the radioactivity of the essential amino acid fraction was incorporated, the highest values being for methionine (9.8 per cent) and isoleucine (8.5 per cent). Zalin concluded however, that the rates of uptake of the pre-formed amino acids did not determine the rate of microbial protein synthesis since an inverse relationship between rate of uptake and rate of synthesis of the amino acid from ^{14}C-glucose could not be demonstrated. *In vivo* experiments in sheep fed lucerne diets using ^{15}N-labelled substrates have shown that 20-40 per cent of the microbial protein is synthesised directly from the rumen amino acid pool (Pilgrim, *et al.*, 1970; Nolan and Leng, 1972). The tendency of heterotrophic bacteria to utilise increased amounts of amino acids and peptides when these substrates are available (Roberts *et al.*, 1955; Warner, 1956) is in line with the reduction in the microbial utilisation of ammonia observed when increased levels of organic nitrogen are fed (Mathison and Milligan, 1971).

Synthesis of Amino Acids by Rumen Bacteria

The utilisation of NH_3 for microbial protein synthesis in the rumen is of great practical interest in view of the relative shortage and high cost of protein, but information on the mechanisms of amino acid synthesis by rumen microbes is surprisingly scanty.

The fixation of NH_3 is achieved via a number of systems, the most important of which is probably glutamate dehydrogenase in both NAD^+ and $NADP^+$ linked forms (Palmquist and Baldwin, 1966; Somerville, (1968)

Glutamine synthetase (Chalupa et al., 1970a) is also probably of some importance and carbamyl phosphokinase (Chalupa et al., 1970a) and asparagine synthetase (Burchall, Reichett and Wolin, 1964) have also been found in bacterial extracts. The evolutionary significance of the possession of such diverse NH_3 fixing systems may lie in the adaptation of rumen bacteria to the utilisation of NH_3 under widely different conditions in the rumen. An example of this type of adaptation was shown by Palmquist and Baldwin (1966) who found that increasing amounts of concentrate in the ration tended to increase the proportion of the $NADP^+$-linked glutamate dehydrogenase at the expense of the NAD^+-linked enzyme. Kinetic studies on the purified ammonia assimilatory enzymes from both pure and mixed cultures of rumen bacteria might well yield the valuable basic information necessary for the closer control over and the more efficient utilisation of the ammonia produced in the rumen.

Several enzymes for the distribution of amino nitrogen in rumen microorganisms have been detected but not extensively studied. Aspartate and alanine amino transferase activity have both been detected in extracts of pure and mixed cultures of bacteria (Joyner and Baldwin, 1966; Chalupa et al., 1970a) and Tsubota and Hoshino (1969) have demonstrated the existence of a number of branched chain amino acid transferases involved in transamination reactions resulting in the formation of glutamic acid in both bacterial and protozoal extracts. Studies with ^{15}N-labelled urea and ammonium salts have generally confirmed the high amino transferase activity of rumen microbes (Boggs, 1959; Little, 1972) although Little found some evidence for the direct synthesis of all amino acids from NH_3. When rumen fluid was incubated with $^{15}NH_4Cl$ in the presence of the amino transferase inhibitor 4-amino-3-isoxazolidine (cycloserine) (Webb, 1966), the first sample taken after only 5 min incubation revealed that all the amino acids in the bacterial protein were significantly enriched with ^{15}N, but no single amino acid showed preferential enrichment. The total enrichment was about 42 per cent of that found when cycloserine was absent. The significance of these results is questionable, however, since the effectiveness of the inhibitor in the systems studied was not examined. We have found that the incubation of extracts of rumen bacteria with cycloserine reduces the activities of both aspartate and alanine amino transferase by only about 50 per cent (Mercer and Annison, unpublished).

The information at present available suggests that both the NH_3 fixing enzymes and the amino transferase system are responsible for the synthesis of amino acids by rumen bacteria, and although dietary effects on the activities of some of the enzymes have been shown (Chalupa et al., 1970b; Salem, Devlin and Marquardt, 1973), there is insufficient basic information to permit the influence of dietary conditions on microbial amino acid synthesis to be effectively predicted.

Some control of bacterial amino acid biosynthesis may occur later in the pathways than the initial fixation of ammonia or the subsequent operation of the amino transferase system. An interesting possibility in this regard is the sequence of reactions initiated by the enzyme aspartokinase (ATP: L-aspartate-4-phosphotransferase. E.C. 2.7.2.4), for the synthesis by separate pathways of lysine, methionine, threonine and isoleucine. The existence of multiple metabolic feedback patterns involving one or more of these amino acids, singly or together, which control the activity of the aspartokinase in different micro-organisms has recently been revealed (Datta, 1969; Dungan and Datta, 1973). If such control mechanisms exist in rumen bacteria the amino acid composition of soluble food proteins might be expected to influence the yield of microbial protein which is known to fluctuate widely in response to dietary changes (Thomas, 1973).

Microbial Protein Synthesis

Bauchop and Elsden (1960) in their much cited work demonstrated that for several anaerobic bacteria, grown in batch culture under optimum conditions, there was a fairly precise relationship between the bacterial cell yield and the ATP made available by the fermentation of an energy source. The yield of dry bacterial cells was about 10.5 g per mole of ATP produced by substrate fermentation. The yields of dry matter/mole ATP (molar growth yield or Y_{ATP}) were subsequently found to be close to the value of 10.5 for many anaerobic bacteria, including some rumen organisms (Thomas, 1973). This apparent constancy of yield led many workers to assume that bacterial protein yield could be predicted from the amount of hexose (or equivalent) fermented in the rumen, and the latter was often calculated from rates of VFA or methane production using the known stoichiometry of carbohydrate fermentation (Hungate, 1966). In fact, much subsequent work has shown that Y_{ATP} may vary widely (Hutton and Annison, 1972) and some of the possible reasons for this variation have been discussed by Thomas (1973).

This uncertainty concerning cell yields, when considered with the paucity of information concerning the digestibility of bacterial protein in the small intestine only emphasises the caution that must be used when making factorial calculations concerning the supply of amino acids to ruminants (Hutton and Annison, 1972; Armstrong and Annison, 1973).

There is growing evidence that maximum microbial protein synthesis occurs at relatively low rumen NH_3 concentrations even when energy is not limiting. Henderson, Hobson and Summers (1969) showed that in continuous culture NH_3 became limiting for the growth of *Bacteroides amylophilus* when the concentration in the uninoculated medium fell below 8.8 mg/100 ml and the cell crop of pure cultures of ruminal bacteria was proportional to the NH_3 concentration in the medium between 1 and 7 mg/100 ml. Allison (1970) has shown that the growth of *B. amylophilus* is limited at NH_3 concentrations lower than 6.4 mg/100 ml. The *in vitro* results of Satter and Slyter (1972) and of Nikolic, Jovanovic and Filipovic (1974) also suggest that the optimal level of rumen NH_3 is between 6 and 8 mg/100 ml.

We have examined this problem *in vivo* by feeding sheep, at hourly intervals, high energy/low protein diets at gradually increasing nitrogen intakes (Mercer and Annison, unpublished). Rumen NH_3 levels stayed relatively constant at a concentration of about 8 mg/100 ml until the nitrogen intake exceeded 9.2 g/day when there was a sharp rise in rumen NH_3 concentration for each further increment of nitrogen. The plasma urea concentration followed a similar pattern of change, being 13 mg/100 ml up to 9.2 g nitrogen intake/day but rising sharply to 23 mg/100 ml when the intake was increased to 9.3 g nitrogen/day. The results of Hume, Moir and Somers (1970) are particularly interesting since they showed that in sheep fed purified diets there was no increase in rumen tungstic acid precipitable nitrogen above a rumen NH_3 level of 8.8 mg/100 ml. However, the rumen NH_3 concentration which maximised the flow of tungstic acid precipitable nitrogen through the omasum was approximately 13 mg/100 ml and may be considerably higher (Miller, 1973).

This relationship between rumen NH_3 concentration and microbial protein synthesis should be studied over a wide range of feeding regimes. These considerations emphasise that the rumen NH_3 level is likely to prove the most effective practical guide to the usefulness or otherwise of dietary NPN, as emphasised by Loosli and McDonald (1968).

The dilution rate (Hobson and Summers, 1967; Roberts and Miller, 1969), the molar percentage of propionic acid in rumen fluid (Jackson, Rook and Towers, 1971) and the sulphur content of the diet (Hume and Bird, 1970) have all been shown to influence the microbial synthesis of protein. The presence or absence of protozoa also appears to influence the yield since the high rumen butyrate molar percentages associated with high rumen ciliate populations in animals fed barley diets at a restricted level (Eadie *et al.*, 1970) also seem to result in lower rates of microbial protein synthesis (Jackson, Rook and Towers, 1971).

Dietary Protein and Amino Acid Supply

The contribution of dietary protein to the amino acid supply of the ruminant is dependent on the extent to which it is degraded in the rumen, and on the amino acid composition and digestibility of that fraction which reaches the small intestine. Two major factors which influence the breakdown of dietary protein in the rumen are its solubility, and the level of food intake. Proteolytic activity in the rumen is normally high, and is not influenced by the protein content of the diet (Blackburn and Hobson, 1960).

Henderickx and Martin (1963) showed that the extent of breakdown of protein in the rumen was closely related to its solubility. Physical or chemical treatments which reduce solubility confer a degree of protection from microbial attack but the benefit to the animal stemming from these procedures will depend on the essential amino acid profile of the material which reaches the small intestine, and its digestibility in that organ.

Chalmers, Cuthbertson and Synge (1954) showed that the heat treatment of casein to reduce its solubility greatly reduced NH_3 production in the rumen, and increased the nutritive value of casein as assessed by nitrogen retention. Many subsequent investigations have confirmed that heat treatment may improve the nutritive value of feed proteins (Whitelaw, Preston and Dawson, 1961; Tagari, Ascarelli and Bondi, 1962; Chalmers, Jayasinghe and Marshall, 1964; Sherrod and Tillman, 1962; Danke et al., 1966). The concept of chemical protection originated with the use of vegetable tannins (Leroy, Zelter and Francois 1964; Tagari et al., 1965) but the most striking advance in this field has been the discovery that the formaldehyde treatment of casein confers substantial protection without loss of biological value (Ferguson, Hemsley and Reis, 1967; Reis and Tunks, 1969). Favourable effects of formaldehyde on the nutritive value of other feed proteins have not been reported, largely because the proteins used in ruminant feeds are usually vegetable proteins of modest biological value and of low solubility following processing, and furthermore, in the growing animal or the lactating cow, high feed intakes probably increase the extent to which feed protein escapes ruminal breakdown (Annison, 1972). The influence of food processing, which may include screw pressing, solvent extraction, drying, grinding and heat pelleting using direct steam injection is often overlooked when the solubility of protein in ruminant diets is considered.

The measurement of the proportion of dietary protein which escapes degradation in the rumen requires the use of techniques which distinguish food protein from microbial and endogenous protein in digesta obtained from animals prepared with simple or re-entrant duodenal fistulae located 6-12 cm from the pyloric sphincter. Re-entrant preparations permit the total collection of digesta (Coelho da Silva et al., 1972) a procedure which has been greatly facilitated by the development of automatic sampling techniques (Evans, Axford and Offer, 1971; Corse, 1974). Diaminopimelic acid (DAP) was used by Weller, Gray and Pilgrim (1958) and Hutton, Bailey and Annison (1971) to estimate the rate of synthesis of bacterial protein (but not protozoal protein). Smith and McAllan (1970) used the ratio of ribonucleic acid (RNA) to total nitrogen in rumen fluid and rumen bacteria to estimate the extent of conversion of dietary nitrogen to bacterial and protozoal nitrogen, but the analytical procedure is tedious. The use of ^{35}S as a bacterial marker was first suggested by Henderickx (1961) and used subsequently by Walker and Nader (1968) and Roberts and Miller (1969). In these studies however, no attempt was made to separate ^{35}S-labelled cystine from ^{35}S-labelled methionine or from inorganic ^{35}S. Recently the relative specific radioactivities of ^{35}S-cystine in rumen microbial and duodenal preparations has been used to determine the contributions of food and microbial protein to duodenal digesta (Leibholz, 1972). Hume (1974) has reported the use of the ratio of ^{35}S in the microbial (M) and the unfractionated duodenal digesta (D) to determine the extent of degradation of different dietary protein supplements in the rumen (*Table 22.1*). The technique is based on the premise of Harrison, Beever and Thomson (1972) that in the absence of food protein the ratio of the specific radioactivities of sulphur amino acids in the M and D fractions will be unity. The improvement

Table 22.1 The passage of undegraded food protein and protein supplements from the rumen and the effect of different levels of intake on the ruminal degradation of protein supplements

Animal Protein supplement/feed	Lamb Sunflower seed meal	Lamb Sunflower seed meal	Wether Lupin seed meal	Wether Soyabean meal	Wether Ground- nut meal	Wether Ground- nut meal	Wether Peruvian fish meal
Dry matter intake g/day	850	1700	800	800	631	800	635
Nitrogen intake g/day	15.2	30.4	14.8	15.6	14.8	14.3	15.2
Nitrogen from supplement/feed g/day	12.0[6]	24.0[6]	14.6	15.4	5.2	14.1	5.2
NAN[8] in duodenum	9.1[6]	17.5[6]	14.5	18.2	15.2	16.2	15.4
NAN[8] from supplement/feed in duodenum g/day	3.3[6]	4.5[6]	5.7	8.7	1.1	6.2	3.7
Supplementary/feed nitrogen undegraded in rumen (per cent)	28	19	35	61	21	37	70
Reference	1	1	2	2	3	2	3

Table 22.1 continued

	Wether Fish meal	Lactating ewe Peruvian fish meal	Wether Unwilted grass silage	Wether Fresh grass	Wether Frozen grass	Wether Dried chopped grass	Wether Dried ground grass	Wether Formalin silage	Wether Formalin silage dried
Dry matter intake g/day	800	2238	1043	750	750	750	1058	1023	1018
Nitrogen intake g/day	17.1	46.7	20.3	20.0	19.4	19.2	28.3	24.0	28.5
Nitrogen from supplement/feed g/day	16.9	20.6	–	–	–	–	–	–	–
NAN[8] in duodenum	20.6	64.7[7]	25.1	15.7	18.2	23.7	31.7	31.4	33.3
NAN[8] from supplement/feed in duodenum g/day	12.2	21.9	4.5	9.1	10.1	13.6	20.8	22.1	24.2
Supplementary/feed nitrogen undegraded in rumen (per cent)	71	≃ 100	22	48	52	71	73	93	85
Reference	2	4	5	5	5	5	5	5	5

[1] Miller, 1973; [2] Hume, 1974; [3] Mercer, Allen and Miller, unpublished; [4] Mercer, 1972; [5] Beever, Thomson and Harrison, 1974; [6] Abomasal nitrogen; [7] Total nitrogen; [8] Non-ammonia nitrogen

of the ^{35}S incorporation technique has recently been reported by Beever et al. (1974) who isolated the ^{35}S-labelled methionine in the M and D fractions and elegantly confirmed the validity of the technique. Miller (1973) measured the influence of level of feed intake on the ruminal breakdown of several proteins in the sheep. Highly soluble proteins were extensively degraded irrespective of the level of food intake, but the extent of breakdown of less soluble proteins was markedly reduced by the high flow rates from the rumen associated with high levels of feeding (*Table 22.1*). Data supporting these conclusions were reported by Ørskov and Fraser (1973), who showed that the extent of breakdown of soya protein in sheep was significantly reduced at high levels of intake. A similar effect was observed in lactating ewes by Mercer (1972) who fed a barley diet with and without Peruvian fish meal. The animals were fed at hourly intervals at near *ad libitum* levels and the difference in the quantity of nitrogen and amino acids which passed to the duodenum on the two diets suggested that all of the fish meal protein escaped degradation. The earlier experiments of Whitelaw and Preston (1963) showed the importance of protein solubility on nitrogen retention in calves and also emphasised the difficulty of characterising proteins according to their solubilities, since the choice of solubilising solution had a marked effect on the solubility. There is an obvious need for the adoption of a standard procedure to measure both the rate and extent of protein solubility to allow meaningful comparisons to be made of data from different laboratories.

In recent years a wealth of data has appeared on the supply of amino acids to the duodenum and their absorption from the small intestine (Clarke, Ellinger and Phillipson, 1966; Ørskov, Fraser and McDonald, 1971; Neudoerffer et al., 1971; Coelho da Silva et al., 1972; Klooster and Boekholt, 1972; Hogan, 1973).

Two recent advances in our knowledge of protein digestion and absorption of amino acids in the small intestine of ruminants are of particular interest, viz., that the digestibility of bacterial protein may vary according to the diet and that there is a preferential absorption of essential amino acids from the small intestine.

The biological value of rumen bacterial and protozoal protein is known to vary widely (Hungate, 1966) and Bergen, Purser and Cline (1967) suggested that the changes in the rumen microbial population caused by dietary changes were responsible for the accompanying changes in nitrogen utilisation, since they found that the release of amino acids by *in vitro* pepsin-pancreatin digestion from the proteins of different rumen bacterial species varied widely. Burris, Bradley and Boling (1974) have applied the same technique to the bacterial protein obtained from steers fed a protein supplement of either soya bean meal, fish meal or linseed meal. They showed that although the amino acid composition of the bacterial preparations was reasonably constant there were considerable differences in the release of amino acids both within and between bacterial preparations. With the exception of threonine, the essential amino acids were released more readily than the non-essential amino acids and the jugular blood plasma free essential amino acid concentrations in the donor animals paralleled the *in vitro* enzymatic release rate of these amino acids.

The preferential absorption of essential amino acids relative to non-essential amino acids has been shown in both sheep (Ben-Ghedalia *et al.*, 1974) and cows (Klooster and Boekholt, 1972). About 70 per cent of the amino acids that leave the abomasum are removed during passage through the small intestine (Clarke *et al.*, 1966; Coelho da Silva *et al.*, 1972; Klooster and Boekholt, 1972; Hogan, 1973). Ben-Ghedalia *et al.* (1974) have shown that the most active area of intestinal absorption of amino acids in sheep occurs 7-15 m from the pylorus and that maximum activities of the proteolytic enzymes trypsin, chymotrypsin and carboxypeptidase occur in this section of the intestine. Furthermore, they showed that of the essential and non-essential amino acids passing through the pylorus, 61 and 43 per cent respectively were absorbed by the time the digesta reached a site in the intestine 15 m from the pylorus. The results of Clarke *et al.* (1966), Klooster and Boekholt (1972) and Coelho da Silva *et al.* (1972) support these findings. The preferential absorption of essential amino acids across the intestinal epithelium was recognised earlier in the rat (Levin, 1970).

The relationship between the amino acid supply in the feed and the absorption of amino acids from the small intestine has often been demonstrated in monogastrics by observing the postprandial changes which occur in the plasma free amino acid concentrations (Dent and Schilling, 1949; Longenecker and Hause, 1959). Such a relationship has not been demonstrated in ruminants (Oltjen *et al.*, 1967) but highly significant positive linear correlations between the concentrations of the essential amino acids in hydrolysed duodenal digesta and in the blood plasma of sheep have been observed (Mercer, 1972).

Utilisation of Dietary NPN

The ability of rumen micro-organisms to utilise NH_3 for protein synthesis permits the replacement of dietary protein with ammonium salts, or with NPN sources which give rise to NH_3 in the rumen. All aspects of this subject have been fully reviewed in recent years (Loosli and McDonald, 1968; Chalupa, 1972; 1973).

Virtanen (1966) showed that lactating cows could be maintained indefinitely on protein-free diets, but only modest levels of production were achieved in a few selected animals. Two important findings were that a long period of adaptation to high NPN diets was essential, and that when 20 per cent of the dietary nitrogen was supplied as protein, feed intakes rose and normal levels of milk production were achieved (Virtanen, 1967). In animal production systems based on urea and molasses, small amounts of fish meal greatly increased growth rates (Preston, 1972). There is evidence from plasma amino acid data that diets in which NPN is the sole nitrogen source may lead to amino acid deficiency (Virtanen, 1966). Supplementary protein presumably corrects this situation either directly, since a proportion will escape ruminal degradation or indirectly, by permitting the increased growth of rumen micro-organisms which require peptides or amino acids as sources of nitrogen, and thereby increasing microbial protein production.

As stressed earlier, there is much evidence that maximum utilisation of NH_3 for microbial protein synthesis in the rumen occurs at relatively low concentrations (5-8 mg NH_3-nitrogen/100 ml). Supplementary dietary NPN is only likely to prove worthwhile when preliminary monitoring has revealed that this level is not reached when the unsupplemented diet is fed.

When conditions in the rumen are favourable for NPN feeding, in order to avoid high and wasteful rumen NH_3 levels the feeding regime must be designed to balance the rate of release of NH_3 against the rate at which energy for microbial growth is made available by microbial fermentation. Unfortunately, ammonium salts and urea, the cheapest and most readily available sources of NPN both generate NH_3 rapidly in the rumen, and must be fed in small amounts at frequent intervals, or together with readily available sources of energy such as molasses or suitably processed cereals. An alternative approach is to use NPN sources which liberate NH_3 in the rumen at relatively slow rates. Comparison of the rates of release of NH_3 in the rumen from several nitrogen sources would suggest that uric acid has the most appropriate characteristics (Oltjen et al., 1968). Uric acid occurs in large amounts in poultry droppings, and the efficient use of dried poultry waste in certain ruminant diets has been convincingly demonstrated (Oliphant, 1974).

There is increasing recognition that considerable improvements in the utilisation of urea in cereal based diets may be achieved by processing the cereal to increase its rate of fermentation by rumen micro-organisms. Steam flaking increases the digestibility in the rumen of sorghum and maize (Armstrong, 1972), and the degree of gelatinisation of the starch grains might well be correlated with the susceptibility of the cereal to ruminal fermentation (Armstrong, 1972). The reported success of the commercial product Starea, a mixture of cereal and urea treated with steam at high temperature and pressure (Bartley et al., 1968) must be attributed to the increased availability of the energy of the cereal, since the urea is not chemically or physically bound to the cereal. Shiehzadeh and Harbers (1974) have recently shown that gelatinised sorghum grain and potato starch substantially improve the utilisation of urea in high roughage rations fed to lambs.

Under normal feeding conditions the Krebs-Henseleit pathway copes with the transient increases in the absorption of NH_3 from the rumen (Chalupa et al., 1970b). Urea however, is not the sole product of NH_3 detoxification. McLaren et al. (1961) demonstrated the synthesis of glutamic acid in rumen mucosal homogenates and Hoshino, Sarumaru and Morimoto (1966) showed that the ruminal mucosa splits and synthesises glutamine. The importance of glutamine in the detoxification of absorbed NH_3 has been demonstrated in rat (Duda and Handler, 1958), cat (Berl et al., 1962) and sheep (Bartik, Rosival and Zwick, 1971) tissues. The latter workers showed that intra-ruminal doses of urea resulted in rapid increases in plasma glutamine and asparagine concentrations followed by a more gradual increase in the concentration of plasma urea. The possible involvement of glutamic acid in NH_3 detoxification has also been demonstrated in lactating sheep (Mercer, 1972). Higher entry rates of glutamic acid occurred when a barley diet was supplemented with urea than when the unsupplemented diet was fed.

The economic advantages of replacing protein with NPN in ruminant diets will ensure continued activity in this field, but success will depend on devising feeding regimes which optimise rumen NH_3 utilisation by correctly balancing NH_3 availability against the supply of energy from microbial fermentation.

Quantitative Aspects of Nitrogen Metabolism in the Rumen

Three groups of workers have used ^{15}N labelled NH_3 to study the kinetics of NH_3 utilisation and transfer in forage fed mature sheep (Pilgrim, Gray and Belling, 1969; Pilgrim et al., 1970; Mathison and Milligan, 1971; Nolan, Norton and Leng, 1973). Classical isotope dilution procedures were employed, but the difficulties faced by these investigators should not be overlooked. The measurement of the degree of enrichment of ^{15}N in nitrogenous materials is complex and time consuming, and of low sensitivity and precision relative to the assay of radioisotopes. Comprehensive studies are inevitably restricted to only a few animals but notwithstanding the problems inherent in ^{15}N usage, the data produced by the different groups are in good agreement and have yielded valuable quantitative data (*Table 22.2*).

The experimental procedures of the three groups of investigators had much in common. In all cases mature sheep were continuously fed i.e. fed at intervals of 1 h or less over 24 h, in order to achieve the steady state conditions necessary for isotope dilution measurements. ^{15}N labelled ammonium salts were continuously infused into the rumen for long periods (up to 216 h) to ensure the maximum incorporation of label into microbial protein. Mathison and Milligan (1971) used sheep prepared with both rumen and abomasal fistulae, and the analysis of abomasal digesta provided data on microbial protein synthesis. In spite of the length of the infusion periods used by these workers, the time-course curves of the build-up of ^{15}N in bacterial and protozoal protein showed that plateau values had not been obtained, suggesting that the calculated contribution of rumen NH_3 to microbial protein was a minimum value. The data on which the curves were based were inevitably too scanty to permit mathematical projection to plateau values.

In these studies, about 60 per cent of the dietary nitrogen was digested in the rumen. Of this available nitrogen about 30 per cent was utilised as amino acids or peptides without entering the rumen NH_3 pool. Some 30 per cent of the nitrogen which entered the NH_3 pool was recycled within the rumen through the pathways ruminal $NH_3 \rightarrow$ microbial protein $\rightarrow$ amino acids $\rightarrow NH_3$, implying substantial degradation of microbial cells within the rumen. Lysis of micro-organisms may result from bacteriophage action (Hoogenraad et al., 1967), predation of

Table 22.2 *Quantitative data on nitrogen metabolism in the sheep rumen derived from isotope dilution experiments with ^{15}nitrogen-labelled substrates*

Diet	Lucerne hay pellets	Lucerne chaff	Wheaten hay	Lucerne chaff	Hay	Barley	Lucerne chaff
Dietary nitrogen (g/day)	28.4	27.2	12.5	22.9	16.8	10.6	23.4
Rumen NH$_3$ (mg nitrogen/l)	155	201	-	-	143	58	224
NH$_3$ production (per cent dietary nitrogen)	17	25	-	55	60	84	71
Urea inflow (g/day)	-	-	-	-	6	3	1.2
Bacterial nitrogen from NH$_3$ (per cent)	-	-	77	63	50	57	80
Protozoal nitrogen from NH$_3$ (per cent)	-	-	53	38	40	55	-
NH$_3$ absorption (g/day)	3	4.3	5.6	8.8	7.2	1.6	2
Undigested food nitrogen (g/day)	-	-	-	-	3.5	3	9.5
Reference	Pilgrim, Gray and Belling (1969)		Pilgrim, Gray, Weller and Belling (1970)		Mathison and Milligan (1971)		Nolan and Leng (1972)

bacteria by protozoa (Coleman, 1967) or death from other causes (Hungate, 1966).

Ammonia production in the rumen in the three isotope dilution studies ranged from 17-84 per cent of the dietary nitrogen intake (*Table 22.2*) but the NH_3 irreversibly lost was approximately constant at 8-17 g/day in all three studies. At this production rate Nolan and Leng (1972) calculated that the loss of NH_3 from the rumen could be completely accounted for by the movement of fluid from the rumen, suggesting that little NH_3 was absorbed directly from the rumen. This unexpected finding is at odds with most current views, but apart from the observations of McDonald (1948) on the levels of NH_3 in ruminal vein blood, there is no direct information on this topic.

Urea Metabolism

The ability of the ruminant to extensively recycle nitrogen by the microbial utilisation of NH_3 in the digestive tract has led to numerous studies on the transfer of urea from blood to the tract (Nolan, Norton and Leng, 1973; Harrop and Phillipson, 1974). There is evidence that the extent of this process may be regulated by the nitrogen intake of the animal, since ruminants are known to conserve nitrogen when dietary supplies are low by reducing the urinary output of urea (Schmidt-Nielsen *et al.,* 1958), a process which must permit a greater proportion of blood urea to enter the alimentary tract. Furthermore, many studies have shown that there is a limit to the passage of urea into the rumen at high levels of blood urea (Gärtner, Decker and Hill, 1961; Weston and Hogan, 1967; Vercoe, 1969; Thornton, 1970). The prevailing levels of NH_3 in the rumen (Varady *et al.,* 1967), the numbers of ureolytic bacteria in the ruminal epithelium (Houpt and Houpt, 1968) and the concentration of CO_2 in the rumen (Thorlacius, Dobson and Sellers, 1971) have all been suggested as factors that regulate the entry of urea into the rumen. Many of the *in vivo* and *in vitro* techniques that have been used in these studies are open to question, however, since they have often involved unphysiologically high levels of blood urea, or the use of isolated rumen pouches (Nolan, Norton and Leng, 1973). These approaches have been largely superseded by isotope dilution techniques based on ^{14}C and ^{15}N labelled urea which provide the most effective method of studying urea biokinetics in the intact, normal animal. (Cocimano and Leng, 1967; Nolan and Leng, 1972). Procedures based on the two isotopically labelled materials provide complementary data. The $^{14}CO_2$ liberated by the degradation of ^{14}C-urea (largely in the alimentary tract) equilibrates with the large body pool of CO_2 and ^{14}C-recycling into urea is negligible. Nevertheless, the entry rate measured using ^{14}C urea includes re-cycled urea, and the difference between this value and the renal urea excretion rate is a measure of the quantity of urea degraded in the alimentary tract (Cocimano and Leng, 1967). When ^{15}N-urea is used, however, that fraction of the $^{15}NH_3$ liberated in the alimentary tract and possibly in the tissues which escapes utilisation returns to the blood pool after re-synthesis into urea in the liver. The plasma urea ^{15}N enrichment data provides a measure of the extent of re-cycling. This approach, first used by Regoeczi et al. (1965) in the rabbit, has been brilliantly exploited by Nolan and Leng (1972) in the sheep. These workers have combined isotope dilution studies of ^{14}C-urea in plasma with similar studies on $^{15}NH_3$ metabolism in the rumen. From the combined data a model of nitrogen metabolism in sheep was constructed incorporating quantitative data on the movement of nitrogen between the various nitrogenous pools in the body. This work has revealed that only about 1.2 g of the urea entering the digestive tract (6.3 g) contributed to the rumen NH_3 pool. Furthermore, the authors were able to show that the amount of urea appearing as ruminal NH_3 could all be accounted for by a total salivary flow of 8 l/day assuming that the concentrations of urea in mixed salivary secretions was 60 per cent of that in plasma (Somers, 1961). This was a most unexpected finding in view of the reported passage of much greater quantities of urea to the rumen (Juhasz, 1965; Egan, 1965; Egan and Moir, 1965; Weston and Hogan, 1967; Houpt and Houpt, 1968). Nolan and Leng (1972) considered the possibility that the conversion of urea into NH_3 in rumen epithelium (Houpt, 1970) might account for the small apparent transfer of urea from blood, since the NH_3 might be re-absorbed without entering the rumen, but arteriovenous difference studies of urea uptake across the rumen did not support this possibility.

The data of Nolan and Leng (1972) suggest that the digestive tract distal to the rumen is the major site of urea degradation in the sheep, but the relative importance of the small intestine, caecum and large intestine is not known (Nolan, Norton and Leng, 1973).

Urea probably enters the small intestine in digestive secretions, and in anaesthetised sheep the concentrations in the digesta of the duodenum and ileum were shown to be similar to that in blood (Nolan, Norton and Leng, 1973). Quantitative studies on the nitrogenous materials reaching the terminal ileum have shown that the amounts of urea are low. Appreciable quantities of NH_3 however, were detected and this almost certainly arises from the hydrolysis of urea entering the small intestine in digestive secretion, or by diffusion from blood (Clarke, Ellinger and Phillipson, 1966; Coehlo da Silva et al., 1972).

Thornton et al. (1970) elegantly demonstrated the transfer of urea from blood to the hind-gut by infusing glucose into the terminal ileum. This procedure increased the output of faecal nitrogen and reduced urinary urea output leading Thornton et al. (1970) to conclude that the transfer of urea from the blood to either of the fermentative areas of the gut is the preferred pathway of urea excretion in ruminants. In more direct studies using isotope dilution techniques Nolan, Norton and Leng (1973) showed that about a third of the NH_3 produced in the caecum was derived from blood urea, and suggested that most of this nitrogen entered the caecum in digesta from the small intestine.

More effort clearly needs to be directed towards nitrogen transactions in the hind-gut if we are to define overall nitrogen metabolism in the ruminant. The examination of blood draining specific sites in the tract coupled with isotope dilution procedures based on ^{14}C and ^{15}N should prove fruitful, particularly if carried out on animals of known nitrogen and energy balance.

Nitrogen Metabolism in the Whole Animal

The application of a range of new techniques to the study of nitrogen metabolism in ruminants has yielded important qualitative and quantitative information on nitrogen transactions within the rumen, and in the whole animal. These new findings are depicted graphically in *Figure 22.1*, and it is of some interest to compare this updated 'nitrogen cycle' with earlier versions (Annison and Lewis, 1959; Preston, 1970).

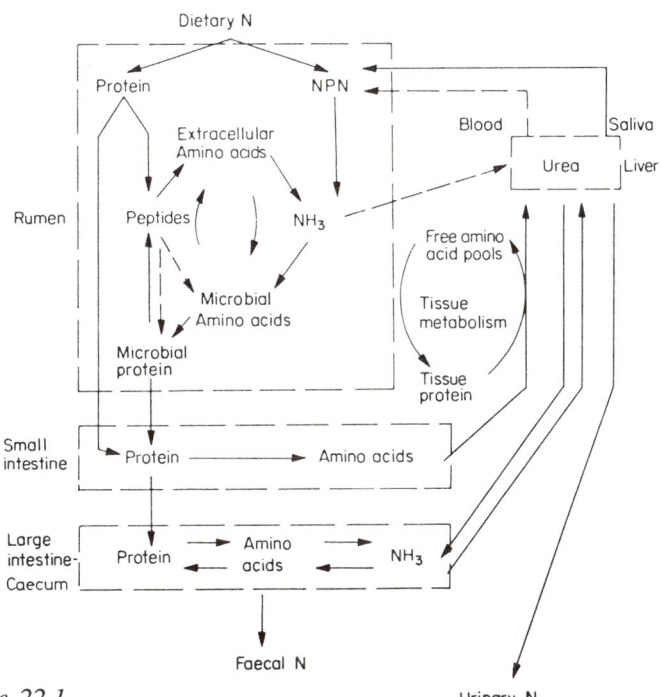

Figure 22.1

Major differences include the extent of ammonia re-cycling within the rumen, the low levels of NH_3 transfer from the rumen into blood, and vice-versa, and the important role of the caecum and large intestine in urea metabolism. These advances largely stem from the brilliant isotope dilution studies of Leng and his colleagues (Cocimano and Leng, 1967; Nolan and Leng, 1972, Nolan, Norton and Leng, 1973).

Valuable quantitative data on the extent of degradation of food protein in the rumen, and microbial protein synthesis have been obtained by using animals prepared with re-entrant fistulae to permit complete collection of digesta. The combination of this technique with comprehensive isotope dilution studies in animals on complete nitrogen balance should yield quantitative data which can be incorporated into computer simulation models of nitrogen metabolism (Morris et al., 1974). Such models permit the integration of new and existing information, and evaluate the relevance and sensitivity of the parameters of the system.

References

ALLISON, M.J. (1969). *J. Anim. Sci.*, **29**, 797

ALLISON, M.J. (1970). in *Physiology of Digestion and Metabolism in the Ruminant.* p.456. Ed. A.T. Phillipson. Oriel Press, Newcastle upon Tyne

ANNISON, E.F. (1956). *Biochem. J.*, **64**, 705

ANNISON, E.F. (1972). in *Nutrition Conference for Feed Manufacturers: 6.* p.2. Ed. H. Swan and D. Lewis. Churchill Livingstone, Edinburgh and London

ANNISON, E.F. and LEWIS, D. (1959). *Metabolism in the Rumen*, 1st Ed., p.93. Methuen, London

ARMSTRONG, D.G. (1972). in *Cereal Processing and Digestion*, p.9, U.S. Feed Grains Council, London

ARMSTRONG, D.G. and ANNISON, E.F. (1973). *Proc. Nutr. Soc.*, **32**, 107

BARTIK, M., ROSIVAL, I. and ZWICK, K. (1971). *Acta vet. Brno*, **40**, 285

BARTLEY, E.E., DEYOE, C.W., PFOST, H.B., ANSTAETT, F.R., BOREN, F.W., HELMER, L.G., MEYER, R.M., PERRY, N.B., SUNG, A.C. and STILES, D.A. (1968). *Feedstuffs*, **40**, 19

BAUCHOP, T. and ELSDEN, S.R. (1960). *J. gen. Microbiol.*, **23**, 457

BEEVER, D.E., HARRISON, D.G., THOMSON, D.J., CAMMELL, S.B. and OSBOURN, D.F. (1974). *Br. J. Nutr.*, **32**, 99

BEEVER, D.E., THOMSON, D.J. and HARRISON, D.G. (1974). in *Proceedings of Fourth International Symposium of Ruminant Physiology.* Sydney, Australia

BEN-GHEDALIA, D., TAGARI, H., BONDI, A. and TADMOR, A. (1974). *Br. J. Nutr.*, **31**, 125

BERGEN, W.G., PURSER, D.B. and CLINE, J.H. (1967). *J. Nutr.*, **92**, 357

BERL, S., TAKAGAKI, G., CLARKE, D.D. and WAELSCH, H. (1962). *J. biol. Chem.*, **237**, 2562

BLACKBURN, T.H. and HOBSON, P.N. (1960). *Br. J. Nutr.*, **14**, 445

BOGGS, D.E. (1959). Ph.D. Thesis. Cornell University, Ithaca, New York

BRYANT, M.P. (1961). *U.S. Dept. Agric. Res. Service*, ARS 44-92 (Mimeo)

BRYANT, M.P. (1963). *J. Anim. Sci.*, **22**, 801

BRYANT, M.P. and ROBINSON, I.M. (1963). *J. Dairy Sci.*, **46**, 150

BURCHALL, J.J., REICHETT, E.C. and WOLIN, M.J. (1964). *J. biol. Chem.*, **239**, 1794

BURRIS, W.R., BRADLEY, N.W. and BOLING, J.A. (1974). *J. Anim. Sci.*, **38**, 200

CHALMERS, M.I., CUTHBERTSON, D.P. and SYNGE, R.L.M. (1954). *J. agric. Sci., Camb.*, **44**, 254
CHALMERS, M.I., JAYASINGHE, J.B. and MARSHALL, S.B.M. (1964). *J. agric. Sci., Camb.*, **63**, 283
CHALUPA, W. (1972). *Fedn Proc. Fedn Am. Socs exp. Biol.*, **31**, 1152
CHALUPA, W. (1973). *Proc. Nutr. Soc.*, **32**, 99
CHALUPA, W., CLARK, J., OPLIGER, P. and LAVKER, R. (1970a). *J. Nutr.*, **100**, 161
CHALUPA, W., CLARK, J., OPLIGER, P. and LAVKER, R. (1970b). *J. Nutr.*, **100**, 170
CLARKE, E.M.W., ELLINGER, G.M. and PHILLIPSON, A.T. (1966). *Proc. R. Soc. B.*, **166**, 63
COCIMANO, M.R. and LENG, R.A. (1967). *Br. J. Nutr.*, **21**, 353
COELHO DA SILVA, J.F., SEELEY, R.C., THOMSON, D.J., BEEVER, D.E. and ARMSTRONG, D.G. (1972). *Br. J. Nutr.*, **28**, 43
COLEMAN, G.S. (1967). *J. gen. Microbiol.*, **47**, 449
CORSE, D.A. (1974). *Proc. Nutr. Soc.*, **33**, 141
DANKE, R.J., SHERROD, L.B., NELSON, E.C. and TILLMAN, A.D. (1966). *J. Anim. Sci.*, **25**, 181
DATTA, P. (1969). *Science N.Y.*, **165**, 556
DENT, C.E. and SCHILLING, J.A. (1949). *Biochem. J.*, **44**, 318
DUDA, G.D. and HANDLER, P. (1958). *J. biol. Chem.*, **232**, 303
DUNGAN, S.M. and DATTA, P. (1973). *J. biol. Chem.*, **248**, 8534
EADIE, J.M., HYLDGAARD-JENSEN, J., MANN, S.O., REID, R.S. and WHITELAW, F.G. (1970). *Br. J. Nutr.*, **24**, 157
EGAN, A.R. (1965). *Aust. J. agric. Res.*, **16**, 463
EGAN, A.R. and MOIR, R.J. (1965). *Aust. J. agric. Res.*, **16**, 437
EVANS, R.A., AXFORD, R.F.E. and OFFER, N.W. (1971). *Proc. Nutr. Soc.*, **30**, 40A
FERGUSON, K.A., HEMSLEY, J.A. and REIS, P.J. (1967). *Aust. J. Sci.*, **30**, 215
GÄRTNER, K., DECKER, P. and HILL, H. (1961). *Pflügers Archiv fur die Gesamte Physiologie des Menschen und der Tiere*, **274**, 281
HARRISON, D.G., BEEVER, D.E. and THOMSON, D.J. (1972). *Proc. Nutr. Soc.*, **31**, 60A
HARROP, C.J.F. and PHILLIPSON, A.T. (1974). *J. agric. Sci. Camb.*, **82**, 399
HENDERICKX, H. (1961). *Archs int. Physiol. Biochem.*, **69**, 449
HENDERICKX, H. and MARTIN, J. (1963). *C.r. Rech., Inst. Encour. Rech. scient. Ind. Agric., Bruxelles*, **31**, 7
HENDERSON, C., HOBSON, P.N. and SUMMERS, R. (1969). in *Proc. IV Int. Symp. on the Continuous Culture of Microorganisms.* p.189. Czechoslovak Academy of Sciences, Prague
HOBSON, P.N. and SUMMERS, R. (1967). *J. gen. Microbiol.*, **47**, 53
HOGAN, J.P. (1973). *Aust. J. agric. Res.*, **24**, 587
HOOGENRAAD, N.J., HIRD, F.J.R., HOLMES, I. and MILLIS, N.F. (1967). *J. gen. Virol.*, **1**, 575
HOSHINO, S., SARUMARU, K. and MORIMOTO, K. (1966). *J. Dairy Sci.*, **49**, 1523

HOUPT, T.R. (1970). in *Physiology of Digestion and Metabolism in the Ruminant.* p.119. Ed. A.T. Phillipson. Oriel Press, Newcastle upon Tyne

HOUPT, T.R. and HOUPT, K.A. (1968). *Am. J. Physiol.,* **214,** 1296

HUME, I.D. (1974). *Aust. J. agric. Res.,* **25,** 155

HUME, I.D. and BIRD, P.R. (1970). *Aust. J. agric. Res.,* **21,** 315

HUME, I.D., MOIR, R.J. and SOMERS, M. (1970). *Aust. J. agric. Res.,* **21,** 283

HUNGATE, R.E. (1966). *The Rumen and its Microbes.* Academic Press, New York and London

HUTTON, K. and ANNISON, E.F. (1972). *Proc. Nutr. Soc.,* **31,** 151

HUTTON, K., BAILEY, F.J. and ANNISON, E.F. (1971). *Br. J. Nutr.,* **25,** 165

JACKSON, P., ROOK, J.A.F. and TOWERS, K.G. (1971). *J. Dairy Res.,* **38,** 33

JOYNER, A.E. and BALDWIN, R.L. (1966). *J. Bacteriol.,* **92,** 1321

JUHASZ, B. (1965). *Acta. vet. Acad. Sci. hung.,* **15,** 25

KLOOSTER, A.TH. VAN'T and BOEKHOLT, H.A. (1972). *Neth. J. agric. Sci.,* **20,** 272

LEIBHOLZ, J. (1965). *Aust. J. agric. Res.,* **16,** 973

LEIBHOLZ, J. (1972). *Aust. J. agric. Res.,* **23,** 1073

LEROY, F., ZELTER, S.Z. and FRANCOIS, A.C. (1964). *C.r. hebd. Seanc. Acad. Sci., Paris,* **259,** 1592

LEVIN, R.J. (1970). *Life Sciences,* **9,** 61

LEWIS, D. (1955). *Br. J. Nutr.,* **9,** 215

LEWIS, D. and MITCHELL, R.M. (1974). this volume

LITTLE, W. (1972). Ph.D. Thesis. University of Nottingham

LONGENECKER, J.B. and HAUSE, N.L. (1959). *Archs Biochem. Biophys.,* **84,** 46

LOOSLI, J.K. and McDONALD, I.W. (1968). in *Non-protein Nitrogen in the Nutrition of Ruminants.* FAO. Agricultural Studies No.75

McDONALD, I.W. (1948). *Biochem. J.,* **42,** 584

McLAREN, G.A., ANDERSON, G.C., MARTIN, W.G. and COOPER, W.K. (1961). *J. Anim. Sci.,* **20,** 942

MATHISON, G.W. and MILLIGAN, L.P. (1971). *Br. J. Nutr.,* **25,** 351

MERCER, J.R. (1972). Ph.D. Thesis. University of Cambridge

MILLER, E.L. (1973). *Proc. Nutr. Soc.,* **32,** 79

MORRIS, J.G., BALDWIN, R.L., MAENG, W.J. and MAEDA, B.T. (1974). in *Tracer Studies on Non-protein Nitrogen for Ruminants;* FAO and IAEA, Vienna

NEUDOERFFER, T.S., LEADBEATER, P.A., HORNEY, F.D. and BAYLEY, H.S. (1971). *Br. J. Nutr.,* **25,** 343

NIKOLIC, J.A., JOVANOVIC, M. and FILIPOVIC, R. (1974). in *Tracer Studies on Non-protein Nitrogen for Ruminants;* FAO and IAEA, Vienna

NOLAN, J.V. and LENG, R.A. (1972). *Br. J. Nutr.,* **27,** 177

NOLAN, J.V., NORTON, B.W. and LENG, R.A. (1973). *Proc. Nutr. Soc.,* **32,** 93

OLIPHANT, J.M. (1974). *Anim. Prod.,* **18,** 211

OLTJEN, R.R., KOZAK, A.S., PUTNAM, P.A. and LEHMANN, R.P. (1967). *J. Anim. Sci.,* **26,** 1415

OLTJEN, R.R., SLYTER, L.L., KOZAK, A.S. and WILLIAMS, E.E. (1968). *J. Nutr.*, **94**, 193
ØRSKOV, E.R. and FRASER, C. (1973). *Proc. Nutr. Soc.*, **32**, 68A
ØRSKOV, E.R., FRASER, C. and McDONALD, I. (1971). *Br. J. Nutr.*, **25**, 243
PALMQUIST, D.L. and BALDWIN, R.L. (1966). *Appl. Microbiol.*, **14**, 60
PILGRIM, A.F., GRAY, F.V. and BELLING, C.B. (1969). *Br. J. Nutr.*, **23**, 647
PILGRIM, A.F., GRAY, F.V., WELLER, R.A. and BELLING, C.B. (1970). *Br. J. Nutr.*, **24**, 589
PITTMAN, K.A. and BRYANT, M.P. (1964). *J. Bacteriol.*, **88**, 401
PITTMAN, K.A., LAKSHMANAN, S. and BRYANT, M.P.. (1967). *J. Bacteriol.*, **93**, 1499
PRESTON, R.L. (1970). *Fedn Proc. Fedn Am. Socs exp. Biol.*, **29**, 33
PRESTON, T.R. (1972). *Wld Anim. Rev.*, **1**, 24
REGOECZI, E., IRONS, A., KOJ, A. and McFARLANE, A.S. (1965). *Biochem. J.*, **95**, 521
REIS, P.J. and TUNKS, D.A. (1969). *Aust. J. agric. Res.*, **20**, 775
ROBERTS, R.B., ABELSON, P.H., COWIE, D.B., BOLTON, E.T. and BRITTEN, R.J. (1955). *Publs. Carnegie Instn.*, No.607, p.13
ROBERTS, S.A. and MILLER, E.L. (1969). *Proc. Nutr. Soc.*, **28**, 32A
SALEM, H.A., DEVLIN, T.J. and MARQUARDT, R.R. (1973). *Can. J. Anim. Sci.*, **53**, 503
SATTER, L.D. and SLYTER, L.L. (1972). *J. Anim. Sci.*, **35**, 273
SCHMIDT-NIELSEN, B., OSAKI, H., MURDAUGH, H.V. and O'DELL, R. (1958). *Am. J. Physiol.*, **194**, 221
SHERROD, L.B. and TILLMAN, A.D. (1962). *J. Anim. Sci.*, **21**, 901
SHIEHZADEH, S.A and HARBERS, L.H. (1974). *J. Anim. Sci.*, **38**, 206
SMITH, R.H. and McALLAN, A.B. (1970). *Br. J. Nutr.*, **24**, 545
SOMERS, M. (1961). *Aust. J. exp. Biol. med. Sci.*, **39**, 145
SOMERVILLE, H.J. (1968). *Biochem. J.*, **108**, 107
TAGARI, H., ASCARELLI, I. and BONDI, A. (1962). *Br. J. Nutr.*, **16**, 237
TAGARI, H., HENIS, Y., TAMIR, M. and VOLCANI, R. (1965). *Appl. Microbiol.*, **13**, 437
THOMAS, P.C. (1973). *Proc. Nutr. Soc.*, **32**, 85
THORLACIUS, S.O., DOBSON, A. and SELLERS, A.F. (1971). *Am. J. Physiol.*, **220**, 162
THORNTON, R.F. (1970). *Aust. J. agric. Res.*, **21**, 323
THORNTON, R.F., BIRD, P.R., SOMERS, M. and MOIR, R.J. (1970). *Aust. J. agric. Res.*, **21**, 345
TSUBOTA, H. and HOSHINO, S. (1969). *J. Dairy Sci.*, **52**, 2024
VARADY, J., BODA, K., HAVASSY, I., BAJO, M. and TOMAS, J. (1967). *Physiologia Bohemoslov*, **16**, 571
VERCOE, J.E. (1969). *Aust. J. agric. Res.*, **20**, 191
VIRTANEN, A.I. (1966). *Science, N.Y.*, **153**, 1603
VIRTANEN, A.I. (1967). *Agrochimica*, **11**, 289
WALKER, D.J. and NADER, C.J. (1968). *Appl. Microbiol.*, **16**, 1124
WARNER, A.C. (1956). *Biochem. J.*, **64**, 1
WEBB, J.L. (1966). *Enzyme and Metabolic Inhibitors.* Vol.II. p.359. Academic Press, New York and London

WELLER, R.A., GRAY, F.V. and PILGRIM, A.F. (1958). *Br. J. Nutr.*, **12**, 421
WESTON, R.H. and HOGAN, J.P. (1967). *Aust. J. biol. Sci.*, **20**, 967
WHITELAW, F.G. and PRESTON, T.R. (1963). *Anim. Prod.*, **5**, 131
WHITELAW, F.G., PRESTON, T.R. and DAWSON, G.S. (1961). *Anim. Prod.*, **3**, 127
WRIGHT, D.E. (1967). *Appl. Microbiol.*, **15**, 547
WRIGHT, D.E. and HUNGATE, R.E. (1967a). *Appl. Microbiol.*, **15**, 148
WRIGHT, D.E. and HUNGATE, R.E. (1967b). *Appl. Microbiol.*, **15**, 152
ZALIN, R.J. (1971). Ph.D. Thesis. University of London

AMINO ACID REQUIREMENTS OF RUMINANTS

D. LEWIS and R.M. MITCHELL
University of Nottingham

It is recognised that in defining the dietary protein component, attention should be given to the adequacy of supply of each individual essential amino acid and the fraction of amino nitrogen that can contribute to those amino acids that are individually dispensable but are required as a group. Such an attitude has hardly existed until recently in the case of ruminants and little attention has been given to anything beyond the total level of dietary nitrogen. It is likely that there has been little awareness of the amino acid requirements of ruminants for two reasons; no effective techniques have been available for determining them and it has not been possible predictably by dietary adjustment to alter amino acid supply to meet requirements. It is therefore desirable to overcome the two problems simultaneously by devising procedures to determine requirements and to protect potential supplementary amino acids against breakdown in the rumen.

Several early studies were concerned with the amino acid nutrition of ruminants but were not really directed towards the quantitative establishment of requirements. Thus Lofgreen, Loosli and Maynard (1947) observed that the supplementation of the diets of lambs with dried egg improved appetite and caused a greater nitrogen retention than similar diets containing linseed meal; it was implied that the dietary amino acid supply was improved. An attempt was made by Bigwood (1964) to assess whether the supply of each amino acid in the diet of the dairy cow was adequate for milk production. It was recognised that the dietary supply of amino acids did not represent what actually became available to the animal, beyond the duodenum. Experimental work was conducted to examine changes in the rumen and the dietary pattern in the dairy cow was then arbitrarily modified in the light of these changes. Amino acid needs were calculated from daily output in milk protein and a value for catabolised amino acids arrived at following a series of assumptions. It was concluded that amino acid supply limited milk production and that the first limiting amino acid was lysine. It must be emphasised that many of the assumptions can be seriously questioned.

One approach that has been adopted in an effort to recognise the adequacy of amino acid supply to the ruminant has been to determine the biological value of microbial products leaving the rumen. This has of course been determined using a non-ruminant as experimental animal, usually the rat. It must be emphasised that the material leaving the rumen includes unchanged dietary protein as well as bacterial and protozoal products in differing proportions. Nevertheless the determined biological values have been reasonably constant. Thus Johnson et al. (1944) found biological values of around 70 both for rumen protozoal and bacterial products. The data that has become available more recently has been summarised by Chalupa (1972) and an overall value of 75-80 would seem to be more reasonable. This must mean that some amino acid is only present at around 75 per cent of the relative level that would ideally meet the needs of the animal. Following the establishment of amino acid requirements it should be possible, in theory, to raise this value to 100, perhaps by supplementing the diet with protected amino acids, and so substantially improve the efficiency of utilisation of the dietary nitrogen fraction.

It was shown by Cuthbertson and Chalmers (1959) that the administration of casein to a sheep via a duodenal cannula led to a substantially greater nitrogen retention than when an equivalent amount was administered via a rumen cannula. However, efforts to establish amino acid requirements took on a new impetus following the findings of Reis and Schinkel (1963) that abomasal infusions of sulphur-containing amino acids to a sheep fed a diet of around 10 per cent protein led to a substantial increase in the rate of wool growth. The observations of Schelling and Hatfield (1967) are however rather perplexing in that positive responses in terms of nitrogen retention were found simultaneously to several individual amino acids though they can all hardly be regarded as being first limiting. The greatest response was found in the case of lysine. It has recently become increasingly apparent that, for a particular dietary situation, it is desirable to identify for the ruminant the sequence of amino acid limitation followed by the establishment of quantitative data for requirements.

It is usual to determine a dietary requirement for a particular nutrient by devising a basal diet specifically deficient in that nutrient and then adding incremental amounts within a series of treatments. A production objective is then usually recorded (for example liveweight gain) and the point at which a further response is not found becomes identified as the requirement. This is however hardly possible in the case of the amino acid needs of the ruminant since incremental supplements within the diet would probably not be paralleled by equivalent changes in the material passing along the duodenum. It is necessary to devise indirect procedures and two approaches have recently been developed. The first is based upon an interpretation of either plasma amino acid levels or amino acid catabolism in relation to graded postruminal administration. The second is based upon a factorial summing of the quantities of amino acids required for each constituent function.

Metabolic Interpretation

It has been recognised for some time in the case of non-ruminants that it is possible to relate plasma amino acid levels under controlled conditions to the adequacy or otherwise of dietary supply. It can be considered for example that supplementation of the diet with an amino acid in short supply would encourage protein synthesis; thus the plasma level of that amino acid would probably not increase unduly. Beyond the point at which the dietary supply was adequate the plasma level could be expected to increase at a rate determined by the effectiveness of pathways to deal with surpluses, for example catabolic or excretory routes. Within such a concept an inflexion point in the graph relating plasma level to dietary supply is taken to identify the point of dietary need. Since there is an increase in anabolic processes up to the point of meeting requirements the plasma levels of other essential amino acids might be expected to fall. Beyond the point of meeting requirements an increase in catabolism could perhaps be recognised by an increased respiratory loss of labelled carbon dioxide when a labelled dietary supplement was offered or an increase in plasma urea levels.

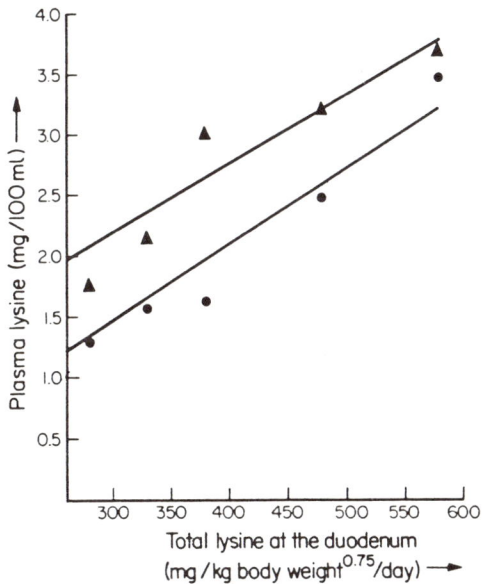

Figure 23.1 Plasma lysine concentration in relation to increasing passage of lysine at the duodenum

A particularly interesting approach was proposed by Mercer and Miller (1973) for the establishment of the methionine requirement of lambs. They used the urinary excretion of ^{35}S after an abomasal injection of $[^{35}S]$methionine as the index of recognising the inflexion in catabolic routes.

Using sheep fitted with duodenal re-entrant cannulae Wakeling, Lewis and Annison (1970) obtained data relating plasma methionine and lysine under standard conditions to the total methionine and lysine passing the duodenum. Since the plasma lysine increased linearly (*Figure 23.1*) it was considered not to be the first limiting amino acid. In the case of methionine (*Figure 23.2*) however, there was an inflexion in the curve and this was considered to identify the point of requirement. A similar pattern of relationship was shown by Brookes *et al.* (1973) in their effort to determine lysine requirement by a similar

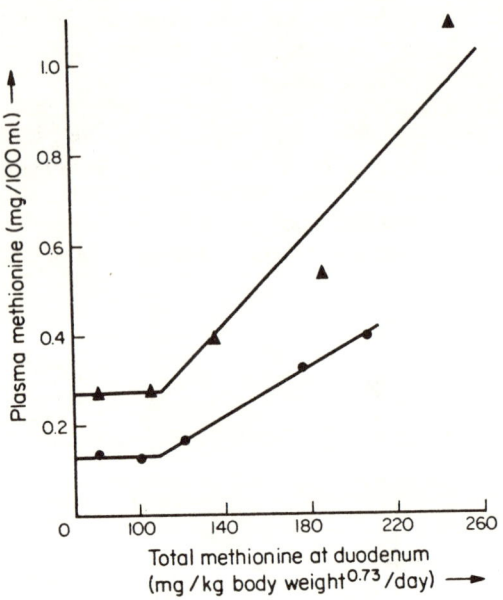

Figure 23.2 Plasma methionine concentration in relation to increasing passage of methionine at the duodenum

approach except that the point at which the supply was considered to be adequate was identified not by monitoring plasma amino acid levels but by observing an increase in $^{14}CO_2$ in expired gases when labelled lysine was administered. The studies of Wakeling, Lewis and Annison (1970) have been extended by Mitchell (1974) who was able to identify threonine as the second limiting amino acid (*Figure 23.3*) and also establish a requirement when the methionine need and a slight margin of excess had been met by supplementation. Similar data have been obtained for the young calf (Williams and Smith, 1974) and the findings that are now available have been summarised in *Table 23.1*. It is possible to relate these values of amino acid requirements to data that have been collected for the total passage of amino acids at the duodenum (*Table 23.2*). Though in several cases it can be seen that methionine is the first limiting amino acid this does not always seem to be the case. It is, however, as yet hardly possible to specify the

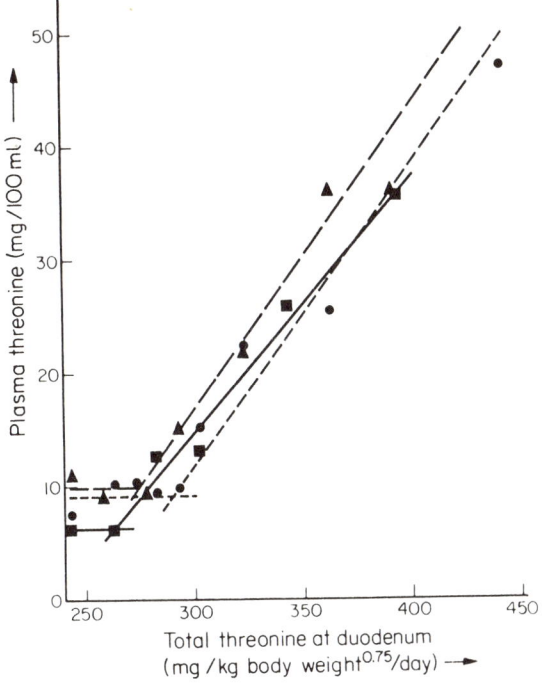

Figure 23.3 Plasma threonine concentration in relation to increasing passage of threonine at the duodenum

full pattern of amino acid needs of the ruminant but this approach is clearly capable of leading to such data becoming available.

Factorial Approach

A factorial approach has been used to establish the protein requirements of ruminants (Agricultural Research Council, 1965) and has recently been modified (Hutton and Annison, 1972; Armstrong and Annison, 1973) to assess the amino acid requirements of ruminants. The requirement for a particular essential amino acid is considered to be the sum of the needs for maintenance, wool growth, tissue protein deposition and foetal growth or milk production as appropriate; the value is also corrected for an assumed availability of the amino acid to the animal.

The need for maintenance is calculated from endogenous urinary excretion and as an example the value for a 45 kg sheep based on 0.09 g nitrogen/kg body weight $^{0.73}$/day is 1.45 g nitrogen per day. This quantity of endogenous urinary nitrogen is partitioned into the equivalent amounts of each amino acid on the basis of the values of Schweigert and Payne (1956) for the amino acid composition of lamb muscle protein. The retention of nitrogen for wool growth has been

Table 23.1 Amino acid requirements of ruminants

Age and liveweight	Diet and level of intake	Requirement in mg/kg $W^{0.75}$/day in terms of amino acid passing in the duodenum				
		Methionine	Threonine	Lysine	Glutamate	
Growing lambs 25 kg	Semi-purified; urea as nitrogen source 6.3 per cent $W^{0.75}$	63^1	63^1	90^1	250^1	Plasma amino acids; nitrogen retention (Nimrick et al, 1970)
Growing lambs 25 kg[2]	Barley/Straw 712 g/day	125-143	-	-	-	^{35}S excretion; plasma urea levels (Mercer and Miller, 1973)
Growing lambs 45 kg	Corn/Alfalfa 900 g/day	-	-	408	-	CO_2 excretion (Brookes et al, 1973)
Mature wethers 50 kg	Semi-purified; groundnut protein 4.2 per cent $W^{0.73}$	105-125	310-340	<545	-	Plasma amino acids (Wakeling, 1970)
Mature wethers 40 kg	as above	158	260-276	-	-	Plasma amino acids (Mitchell, 1974)
Mature wethers 40 kg	as above	137-197	-	-	-	CO_2 excretion (Mitchell, 1974)
Mature merinos 40 kg[3]	Wheaten/lucerne hay 800 g/day	207^1	-	-	-	Plasma amino acids (Reis et al, 1973)
Calves 140 kg	20 g nitrogen/kg DM	197	-	-	-	Plasma amino acids (Williams and Smith, 1974)

[1] Not including contribution from amino acids passing in duodenum
[2] Assumed 25 kg liveweight
[3] Assumed 40 kg liveweight

considered to be 1.5 g per day (Agricultural Research Council, 1965) which can be again partitioned to constituent amino acids (Biochemists' Handbook, 1961).

It was considered by the Agricultural Research Council (1965) that the gain in carcass nitrogen as a percentage of live weight gain for a sheep of over 40 kg was 2.4 per cent. Thus a sheep gaining weight at 0.2 kg per day would deposit 4.8 g of nitrogen in the carcass, a value that can be partitioned to the constituent amino acids again according to the data of Schweigert and Payne (1956). Comparable calculations can be carried out for foetal growth and milk production. Thus an estimate of amino acid requirements can be obtained by summing these values and applying a factor for availability as obtained by Coelho da Silva (1971) in determining the digestibilities of individual amino acids when a variety of diets were fed to sheep. There was not in the procedure developed by Armstrong and Annison (1973) any account taken of metabolic faecal nitrogen which can be related to dry matter intake.

Table 23.2 *Amino acid requirements and supply at the duodenum (mg/kg body weight $^{0.75}$/day)*

Requirement	Methionine 140	Threonine 300	Lysine 450
Coelho da Silva (1971)			
Chopped lucerne	131	422	610
Dried grass (early)	155	413	525
Dried grass (medium)	167	385	442
Wakeling, (1970)	81	258	279
Mitchell, (1974)	127	343	309
Oldham, (1973)			
Barley/Fish	65	270	459
Molasses/Soya	57	185	336

An attempt was made by Armstrong and Annison (1973) to relate net uptakes of amino acids from the small intestine to requirements either obtained by the factorial approach or by interpretation of metabolic data. They assumed the methionine requirement derived from metabolic data to be 115 mg per kg body weight$^{0.75}$. Uptake from the small intestine seemed adequate on diets of dried grass or dried grass and flaked maize but inadequate in the case of dried lucerne or wheat fibre and barley grain diets; a value of 140 mg has however been assumed in *Table 23.2*. In nearly all cases the threonine supply was not adequate. A somewhat different picture is obtained if the factorial approach is used to establish requirements; the threonine supply seems far in excess of that needed and only in the case of the wheat fibre - barley grain ration does the supply of sulphur amino acids appear to be inadequate.

Much further information is needed on the amino acid requirements of ruminants, this is essential before real progress can be made in improving the efficiency of nitrogen utilisation.

References

ARMSTRONG, D.G. and ANNISON, E.F. (1973). *Proc. Nutr. Soc.*, **32**, 107

AGRICULTURAL RESEARCH COUNCIL (1965). *The Nutrient Requirements of Farm Livestock.* No.2, *Ruminants.* Agricultural Research Council: London

BIGWOOD, E.J. (1964). in *The role of the gastrointestinal tract in protein metabolism.* Ed. H.N. Munro. Blackwell Scientific Publications, Oxford

BIOCHEMISTS' HANDBOOK (1961). Ed. C. Long. p.702. E. and F.N. Spon Ltd., London

BROOKES, I.M., OWENS, F.N., BROWN, R.E. and GARRIGUS, U.S. (1973). *J. Anim. Sci.*, **36**, 965

COELHO DA SILVA, J.F. (1971). *The digestion of nitrogenous constituents in forage and forage-cereal diets by adult sheep.* Ph.D. Thesis, University of Newcastle upon Tyne

CHALUPA, W. (1972). *Fedn Proc. Fedn Am. Socs exp. Biol.*, **31**, 1.152

CUTHBERTSON, D.P. and CHALMERS, M.I. (1959). *Biochem. J.*, **46**, XVII

JOHNSON, B.C., HAMILTON, T.S., ROBINSON, W.B. and CAREY, J.C. (1944). *J. Anim. Sci.*, **3**, 287

HUTTON, K. and ANNISON, E.F. (1972). *Proc. Nutr. Soc.*, **31**, 151

LOFGREEN, G.P., LOOSLI, J.K. and MAYNARD, L.A. (1947). *J. Anim. Sci.*, **6**, 343

MERCER, J.R., and MILLER, E.L. (1973). *Proc. Nutr. Soc.*, **32**, 86A

MITCHELL, R.M. (1974). *The amino acid requirements of sheep.* Ph.D. Thesis. University of Nottingham

NIMRICK, K., HATFIELD, E.E., KAMINSKI, J. and OWENS, F.N. (1970). *J. Nutr.*, **100**, 1301

OLDHAM, J.D. (1973). *Dietary carbohydrate and nitrogen interactions in the rumen.* Ph.D. Thesis. University of Nottingham

REIS, P.J. and SCHINCKEL, P.G. (1963). *Aust. J. Biol. Sci.*, **16**, 218

REIS, P.J., TUNKS, D.A. and SHARRY, L.E. (1973). *Aust. J. Biol. Sci.*, **26**, 637

SCHELLING, G.T. and HATFIELD, E.E. (1967). *J. Anim. Sci.*, **26**, 929

SCHWEIGERT, B.S. and PAYNE, B.J. (1956). American Meat Institute Foundation Bulletin No.3

WAKELING, A.E., LEWIS, D. and ANNISON, E.F. (1970). *Proc. Nutr. Soc.*, **29**, 60A

WAKELING, A.E. (1970). *The amino acid requirements of ruminants.* Ph.D. Thesis. University of Nottingham

WILLIAMS, A.P. and SMITH, R.H. (1974). *Proc. Nutr. Soc.*, **33**

FACTORS INFLUENCING THE SUPPLY OF NITROGEN AND AMINO ACIDS TO THE INTESTINE OF DAIRY COWS

H. HAGEMEISTER, W. KAUFMANN and E. PFEFFER
Institut für Milcherzeugung, Kiel, West Germany

Bacterial transformations play an important role in the ruminant digestion of feed protein. They may cause considerable shifts of the amount of protein available to the host animal. Detailed reviews concerning this subject have been prepared in recent years (Kurilow and Krotkowa, 1971; Hutton and Annison, 1972; Phillipson, 1972; Henderickx, 1973; Miller, 1973; Thomas, 1973; Nolan, Norton and Leng, 1973). From these, it may be concluded in summary that evaluation of dietary protein and resulting standards of requirements are influenced by:
(1) Breakdown of dietary protein in the forestomachs.
(2) Synthesis of bacterial protein in the forestomachs.
(3) Digestibility of protein in the intestine.
(4) Possible changes in the supply of amino acids.
(5) Possible changes in the proportions of amino acids absorbed.
Further systematic investigations are required to render an evaluation of the different factors and to give information about the actual supply of nitrogen to the animals under different feeding conditions. Extreme rations as well as more usual dietary components of dairy rations should be used.

With a few exceptions, research work published so far was done on sheep. We feel that further work using dairy cows is needed, not only because rates of passage through the digestive tract are different, but also because different contents of digestible protein and energy are required.

For this purpose a technique was developed (Dirksen, Kaufmann and Pfeffer, 1972), which allows long-term experiments with lactating dairy cows using typical dairy feeds and non-protein nitrogen (NPN) compounds.

The aims of these investigations were:
(1) To quantify the most important factors influencing the utilisation of protein.
(2) To find basic rules, which lead to a general survey of possible interactions, when different feed components are fed, including NPN.

Method of Measuring Protein Supply in the Dairy Cow

Two to four dairy cows were used, each fitted with a rumen fistula and duodenal re-entrant cannula. Each trial lasted 21 days with constant feed intake. The daily rations were fed in six equal portions and twice daily 25 g polyethylene was introduced into the rumen as a marker substance. Following a 10 day collection period for faeces, duodenal flow was measured continuously for 72 hours (Dirksen, Kaufmann and Pfeffer, 1972). The measured flow was corrected in order to give 100 per cent recovery of polyethylene flow (Dijkstra and Kemmink, 1970; van't Klooster et al., 1972).

Some of the rations were frozen forage, which had varying contents of crude protein due to different intensities of nitrogen-fertilisation and to inclusion of different amounts of clover. In addition, mixed rations were used consisting of hay (35-40 per cent), barley and oats (25-30 per cent), tapioca (0-30 per cent) and a protein feed to be tested. The proportion of this protein feed ranged up to 30 per cent according to the respective protein content, so that 45 to 50 per cent of the total protein in the ration was supplied by this source.

Bacterial protein flowing to the intestine was calculated from diaminopimelic acid (DAPA) synthesised in the rumen, using the method of Hutton, Bailey and Annison (1971); DAPA and nitrogen were also determined in the rumen in each trial. Factors for calculating bacterial nitrogen from DAPA showed a considerable variation (up to 20 per cent of the mean). For our calculations we used the mean of all measurements, amounting to 21.6 g bacterial nitrogen/g DAPA.

Non-bacterial protein in duodenal contents was determined as the difference between total and bacterial protein. This consists mainly of dietary protein passing undegraded from the forestomachs. Endogenous secretions contained in the digesta, protozoa (Ibrahim and Ingalls, 1972; Coehlo da Silva et al., 1972a,b) and possible absorption of amino acids from the forestomachs (Cook, Brown and Davis, 1965; Demaux et al., 1961; Leibholz, 1971) were not taken into consideration separately. In the same way fermentation processes within the large intestine were neglected when digestibilities in the intestine were calculated. These influences should be of minor importance for the main picture of protein digestion in the dairy cow, and they are probably fairly constant.

With reference to the work of Annison (1956) and Henderickx and Martin (1963) the protein components in the mixed rations were classified following an in vitro method developed by Melosch (1974), in which NH_3 formation is measured after eight hours incubation in buffered ruminal juice. Relative to urea, the following order was obtained: casein treated with formaldehyde (Ferguson, Hemsley and Reis, 1967) 10; coconut pellets 10; soyabean meal 30; fish meal 38; yeast 43; rape-seed meal 45; groundnut meal 61; horse bean meal 80; casein 87; urea 100. In Tables 24.1, 24.2 and 24.3 the feeds are presented in this order.

Table 24.1 Non-bacterial protein-nitrogen (undegraded dietary protein nitrogen) at the duodenum

Type of ration	Crude protein in the ration (percentage of DM)	Dietary nitrogen (g/day)	Non-bacterial protein-nitrogen at the duodenum	
			(g/day)	(g/100 g dietary nitrogen)
Forages				
Ryegrass, 200 kg nitrogen/ha	7.7	95	64	67
Ryegrass, 400 kg nitrogen/ha	11.2	133	75	56
Clover/grass	16.4	215	98	46
Mixed rations (45 per cent of nitrogen supplied by:)				
Casein, treated with formaldehyde	17.4	238	157	66
Coconut pellets	15.7	228	131	57
Soyabean oil meal	15.1	236	107	45
Fish meal	15.6	229	75	33
Yeast	12.5	197	83	42
Rapeseed oil meal	15.5	230	92	40
Peanut oil meal	15.7	233	105	45
Horsebean meal	15.2	231	107	46
Casein, untreated	17.5	238	60	25
Urea	16.1	230	53	23

Table 24.2 *Bacterial protein-nitrogen at the duodenum*

Type of ration	Digestible OM of the ration (kg/day)	OM digested in the forestomach (kg/day)	Bacterial protein-nitrogen at the duodenum	
			(g/day)	(g/100 g OM digested in the forestomachs)
Forages				
Ryegrass, 200 kg nitrogen/ha	4.20	2.98	77	2.60
Ryegrass, 400 kg nitrogen/ha	4.19	2.90	65	2.24
Clover/grass	4.82	3.00	76	2.55
Mixed rations				
(45 per cent of nitrogen supplied by:)				
Casein, treated with formaldehyde	5.75	3.36	104	3.10
Coconut pellets	6.10	4.30	85	1.98
Soyabean oil meal	6.02	4.05	134	3.30
Fish meal	6.12	4.30	153	3.56
Yeast	6.20	4.32	152	3.52
Rapeseed oil meal	6.44	4.24	140	3.30
Peanut oil meal	6.32	4.26	133	3.12
Horsebean meal	6.73	4.27	130	3.04
Casein, untreated	6.04	4.13	130	3.36
Urea	6.00	4.31	143	3.32

Table 24.3 Total and bacterial protein-nitrogen at the duodenum and digestibility of protein-nitrogen in the intestine

Type of ration	Total protein-nitrogen at the duodenum		Proportion of bacterial protein-nitrogen at the duodenum (per cent)	Digestibility of protein nitrogen in the intestine (per cent)	Total available protein-nitrogen (g/100 g dietary nitrogen)
	(g/day)	(percentage of dietary nitrogen)			
Forages					
Ryegrass, 200 kg nitrogen/ha	141	148	55	67	100
Ryegrass, 400 kg nitrogen/ha	140	105	46	67	70
Clover/grass	174	81	44	68	55
Mixed rations (45 per cent of nitrogen supplied by:)					
Casein, treated with formaldehyde	261	110	40	69	77
Coconut pellets	216	95	39	70	66
Soyabean oil meal	241	102	56	70	72
Fish meal	228	100	67	74	74
Yeast	235	119	65	71	85
Rapeseed oil meal	232	101	60	75	76
Peanut oil meal	239	102	56	78	80
Horsebean meal	237	103	55	75	77
Casein, untreated	199	84	70	76	64
Urea	196	85	73	72	61

Flow of Non-bacterial Protein Nitrogen (Undegraded Dietary Protein) to the Intestine

Table 24.1 shows the crude protein contents of three forages and ten mixed rations, the average daily intake of nitrogen and the amounts of non-bacterial nitrogen passing through the duodenum in absolute terms and relative to the intake of nitrogen. Raising the crude protein content of green forage from 7.7 to 16.4 per cent increased the amount of non-bacterial nitrogen passing through the duodenum. Relative to 100 g dietary nitrogen, however, there was a decrease from 67 g to 46 g.

With the exception of the yeast ration, the mixed rations had a relatively constant crude protein content between 15.2 and 17.5 per cent and produced similar nitrogen intakes. The amounts of non-bacterial nitrogen passing daily through the duodenum, in absolute terms and relative to nitrogen intake, decreased in the order given. These differences, however, were evident only for those nitrogen sources, which showed considerable differences in their *in vitro* rates of fermentation as well. With standard feedstuffs, except coconut meal, the amount of non-bacterial nitrogen passing to the intestine was relatively constant and amounted to 40-46 g per 100 g dietary nitrogen. When fish meal was fed, unexpectedly little non-bacterial nitrogen was found at the duodenum.

Thus, when *in vitro* rates of fermentation are near the mean, it seems very difficult to show a relationship to the amount of dietary protein passing into the gut undegraded. This was also noted by Gladjaewa (1972), who showed significant differences between the proteins of barley and peas, and by Miller (1973), who used fish meal and groundnut meal. Possible differences in the rates of degradation may be compensated by different times of retention in the rumen Ørskov and Fraser, 1973), which might be caused by different particle sizes.

In general it may be deduced, with the exceptions stated above, that 40 per cent of the ingested dietary protein passes to the intestine undegraded. To increase the reliability of this general statement, more information about the proportion of protozoa and of endogenous secretions included in non-bacterial nitrogen seems to be of particular importance, since the literature available is still limited in this respect (Hogan and Weston, 1968; Karilow and Krotkowa, 1971; Harrison and Hill, 1962; Coehlo da Silva *et al.,* 1972a; Smith and McAllan, 1973; Ternuth and Butler, 1973).

Flow of Bacterial Protein Synthesised in the Forestomachs

Table 24.2 shows the amounts of organic matter digested in the entire digestive tract and in the stomachs as well as the amounts of bacterial protein nitrogen (calculated from DAPA) per day and per 100 g organic matter disappearing from the stomachs. Independently of the type of ration fed, 60 to 70 per cent of the digestible organic matter disappeared from the stomachs. With the exception of rations containing green forage or coconut pellets, 3 to 3.5 g bacterial protein nitrogen were formed per 100 g organic matter metabolised in the stomachs. This shows a very close correlation between the energy supply to the bacteria and their synthetic yield. This is in agreement with published results (Conrad and Hibbs, 1968; Hogan and Weston, 1970; Hume, 1970; Hume, Moir and Somers, 1970; Walker and Nadler, 1970; Hutton and Annison, 1972; Thomas and Clapperton, 1972; Hagemeister and Pfeffer, 1973). Deviations may be explained by different methods used for determining bacterial protein nitrogen and different factors used for calculating bacterial protein from DAPA.

Clearly, this dependence of bacterial protein synthesis on energy supply can be valid only as long as the nitrogen requirement for this synthesis is covered. Hume, Moir and Somers (1970) and Thomas and Clapperton (1972) expect a decrease in bacterial protein synthesis when crude protein in the ration falls below 10 per cent. Nevertheless, the ruminant seems to be able to divert considerable amounts of endogenous urea via the rumino-hepatic cycle into the rumen, where it may provide the microbes with nitrogen for protein synthesis.

A deficiency of nitrogen for synthetic purposes could also occur with rations containing normal amounts of protein, if the rate of degradation of this protein in the forestomachs is extremely low. For coconut pellets a very low rate of degradation had been found *in vitro*. The low rate of synthesis per 100 g of organic matter metabolised in the forestomachs when forage was fed, may partly be explained by deficiency of nitrogen and in addition the depressing effect of the relatively high contents of cellulose cannot be excluded (Henderickx and Martin, 1963; Walker and Nader, 1970).

In typical dairy rations the protein content is sufficient to supply the bacteria in the forestomachs with adequate amounts of nitrogen. Therefore, in general, the amount of bacterial protein may be calculated from the amount of organic matter metablised in the forestomachs. If the proportion of digestion in the forestomachs is known, digestible energy may be used for this calculation as well (Sutton, 1971; Kaufmann and Hagemeister, 1973).

It has been found that, per 100 g of organic matter fermented in the forestomachs, an average of 20 g bacterial protein are synthesised. If 60 to 70 per cent of a ration's digestible organic matter is broken down in the rumen, 13 g bacterial protein on average should be synthesised per 100 g digestible organic matter. With a mean digestibility of 70 per cent, this means 9.1 g digestible bacterial protein per 100 g digestible organic matter. This 100 g digestible organic matter corresponds to roughly 90 units starch equivalent (SE), so that 100 units SE are accompanied by roughly 10 g of digestible bacterial protein. This means that the maintenance requirements of protein may be covered in total by bacterial protein, if we accept that the animals' requirements are met by the protein available to the gut. When yields of 25 kg milk per day are produced, the animal requires about 10 000 units SE and 1800 g digestible protein. Because only about 1000 g digestible bacterial protein will be supplied, the remainder, about 45 per cent of the requirement, will have to be supplied by undegraded dietary protein.

Table 24.4 Flow of amino acid-nitrogen to the intestine (g amino acid-nitrogen/day)

Type of ration	Total amino acid nitrogen	Essential amino acids									Non-essential amino acids						
		Arg	His	Ileu	Leu	Lys	Met	Phe	Thr	Val	Asp	Ala	Glu	Gly	Pro	Ser	Tyr	Cys
Forages																		
Ryegrass, 200 kg nitrogen/ha	76	2.3	1.1	3.5	5.2	4.7	0.9	2.8	3.8	4.7	7.6	6.7	7.6	13.7	3.1	3.9	2.0	1.7
Ryegrass, 400 kg nitrogen/ha	72	2.2	1.3	3.4	4.9	4.3	1.0	2.9	3.6	4.2	6.9	6.2	7.0	13.5	2.9	3.9	2.1	1.5
Clover/grass	102	3.5	2.0	4.5	7.0	6.3	1.2	3.9	4.8	6.4	9.7	8.7	8.8	22.1	3.5	4.6	3.0	1.6
Mixed rations (45 per cent of nitrogen supplied by:)																		
Casein, treated with formaldehyde	157	5.0	2.9	8.5	12.8	9.9	2.5	6.2	7.6	10.1	13.5	9.2	24.1	16.5	11.0	9.2	4.7	2.9
Soyabean oil meal	130	4.9	1.9	7.1	9.8	8.8	2.1	4.8	6.7	8.7	13.8	10.4	15.6	17.5	6.0	6.4	4.1	3.2
Fish meal	130	4.0	3.2	5.9	8.4	7.3	1.8	4.1	6.1	8.0	12.8	11.8	14.6	23.9	5.7	6.2	3.4	2.3
Yeast	138	5.8	3.0	6.4	8.3	10.8	2.0	4.0	8.8	8.4	14.5	11.7	14.6	19.1	5.7	7.8	3.9	2.9
Rapeseed oil meal	124	4.4	2.8	6.1	8.6	6.6	1.5	3.8	5.9	9.9	10.9	10.9	13.5	20.5	6.0	6.6	3.7	2.2
Peanut oil meal	125	4.4	2.9	6.1	8.4	7.2	1.5	4.3	6.2	9.6	11.5	10.7	13.6	20.2	5.7	6.2	4.0	2.2
Horsebean meal	139	4.6	2.6	6.5	9.1	8.9	1.6	5.1	6.5	7.8	13.6	10.9	18.5	23.5	6.3	6.5	4.0	2.5
Casein, untreated	111	3.7	2.2	5.6	7.7	7.0	1.5	4.0	5.9	7.0	11.4	9.0	13.4	16.9	4.9	5.7	2.9	2.3
Urea	92	2.9	1.4	4.5	6.0	6.2	1.5	3.6	4.2	6.0	8.9	6.6	11.4	15.6	4.3	3.9	2.4	2.1

Table 24.5 Disappearance of amino acids from the intestine (per cent)

Type of ration	Total amino acid nitrogen	Arginine	Leucine	Lysine	Threonine	Valine
Forages						
Ryegrass, 200 kg nitrogen/ha	72	74	73	74	70	69
Ryegrass, 400 kg nitrogen/ha	72	75	67	72	66	68
Clover/grass	74	85	70	73	67	67
Mixed rations (45 per cent of nitrogen supplied by:)						
Casein, treated with formaldehyde	69	70	69	66	68	67
Soyabean oil meal	76	81	74	76	70	74
Fish meal	79	82	79	74	77	77
Yeast	75	73	66	69	69	67
Rapeseed oil meal	76	81	74	71	71	78
Peanut oil meal	83	87	82	81	81	82
Horsebean meal	79	82	76	80	76	75
Casein, untreated	79	82	78	78	77	77
Urea	73	77	67	75	65	70

Digestibility of Protein in the Intestine

Table 24.3 summarises results presented in *Tables 24.1* and *24.2* concerning the amounts of nitrogen passing to the intestine. The lower energy supply in forage rations compared to mixed rations should be borne in mind. Forage rations showed a closer dependence of the protein supply on the energy intake than on the intake of dietary nitrogen.

In most cases *in vitro* fermentation of protein had little influence on the amount of protein passing to the intestine. With a few exceptions more than half of the protein in duodenal contents was of bacterial origin.

Table 24.3 also shows the digestibility of protein passing through the duodenum. Whereas great variations in the apparent digestibility of protein are usually accepted due to varying contents and types of protein in the rations, the difference between nitrogen passing through the duodenum and excreted in faeces varied very little, extremes being 67 and 78 per cent. This confirms the results of Hogan and Weston (1968); Weston and Hogan (1968); Offer, Evans and Axford (1972); Coehlo da Silva *et al.* (1972a,b) and van't Klooster and Boekholt (1972). No influence of the proportion of bacterial protein nitrogen on digestibility in the intestine can be seen, although Ørskov, Fraser and McDonald (1971) and Salter and Smith (1974) found lower digestibilities of bacterial protein compared to dietary protein.

The last column of *Table 24.3* indicates the amount of digestible protein-nitrogen per 100 g of dietary nitrogen. Due to the relatively constant digestibility in the intestine, these values reflect the amounts of nitrogen measured at the duodenum.

Amino Acids Flowing to and Disappearing from the Intestine

Investigations of the amino acid pattern of protein flowing to the intestine proved that the proportions of individual amino acids are relatively constant with different rations and that the amounts of amino acid nitrogen add up to 50-60 per cent of total nitrogen in the digesta (*Table 24.4*). This agrees with results of Potter, Little and Mitchell (1968), and Champredon, Pion and Thivend (1971). It has to be remembered that the amino acid composition of bacteria is largely independent of the ration fed (Purser and Buechler, 1966; Purser, 1970). With extreme rations, changes in the amino acid pattern of bacterial populations are possible (Harmeyer and Jeskova, 1971) as well as changes in the ratio of dietary to bacterial protein, as can be seen for the mixed ration including formaldehyde treated casein. It may be concluded that for most rations fed under practical conditions biological value of protein flowing to the intestine varies only very little.

Table 24.5 shows digestibility in the intestine of total α-amino acid nitrogen and of individual essential amino acids. It can be seen that here also only minor differences occur, which should be of no importance under the conditions of practical feeding of dairy cows. Because microbial transformations in the hind gut were not taken into consideration in our results, we cannot discuss differences in the disappearance of amino acids from the small intestine alone, as found by Coehlo da Silva et al. (1972a,b), van't Klooster and Boekholt (1972) and Tamminga (1973) in animals fitted with duodenal as well as ileal re-entrant cannulae.

Availability of Protein from Non-protein Nitrogen (NPN) Compounds

The dependence shown between breakdown of feed and bacterial synthesis in the forestomachs (10 g digestible bacterial protein per 100 units SE) allow an evaluation of NPN compounds in dairy feeding. Two aspects are possible: the first being that if parts of the dietary protein are replaced by NPN on an isonitrogenous basis, then this NPN will in a similar way serve as a source of nitrogen for bacterial synthesis. Bacterial synthesis will not exceed that from dietary protein, of which 40 per cent will pass into the intestine undegraded (*Table 24.1*). Thus, replacing dietary protein nitrogen by NPN will impair the protein supply of the host animal to an extent which is equal to part of the dietary protein, which would have passed to the intestine.

In *Table 24.6*, availabilities of protein are calculated for different milk yields and different proportions of NPN in the ration. The rate of fermentation of feed protein was taken to be 60 per cent, which means 40 per cent passed into the gut undegraded.

With increasing proportions of NPN in the ration, protein nitrogen available to the host animal falls by 40 per cent of the protein nitrogen replaced. At maintenance and with low yields the supply still exceeds the requirements, due to the net synthesis resulting from the operation of the rumino-hepatic cycle. With increasing yields this surplus decreases. With a yield of 30 kg milk per day even without NPN, that available at the intestine is less than nitrogen intake.

Therefore the calculated loss of available protein when NPN is included in the ration is of little importance, as long as protein available in the intestine exceeds the animals' requirements. Feeding NPN should be limited according to the milk yield and to the ratio of digestible protein to units of SE in the ration.

Secondly, NPN is utilised only if the energy supply allows a greater capacity for bacterial synthesis than can be met by ammonia resulting from the breakdown of dietary protein. With this view, a possible additional supply of nitrogen via the rumino-hepatic cycle is neglected. Utilisation of NPN can be expected only with wide ratios of digestible protein to units of SE.

In *Table 24.7* efficiencies of utilisation have been calculated. The values fall with increasing yields, because the narrow ratio of digestible protein to digestible energy results in sufficient nitrogen from degraded dietary protein. With increasing proportions of NPN in the ration, efficiency of utilisation increases because insufficient amounts of ammonia

Table 24.6 Calculated amounts of protein available in the gut (g/day) with varying proportions of NPN in the ration and with varying milk yields (digestible protein/units SE).

Yield	Percentage nitrogen as NPN	Digestible protein required (g/day)	Protein available in the gut	
			(g/day)	(percentage of requirement)
Maintenance (1:10)	0	300	420	140
	10		408	136
	20		396	132
	30		384	128
10 l/day (1:6.4)	0	900	935	104
	10		899	100
	20		863	96
	30		827	92
20 l/day (1:5.7)	0	1500	1450	97
	10		1390	93
	20		1330	89
	30		1270	85
30 l/day (1:5.4)	0	2100	1965	94
	10		1881	90
	20		1797	86
	30		1713	82

Rate of degradation in the forestomachs 60 per cent

Table 24.7 Calculated efficiency of utilisation of NPN with varying proportions in the ration and varying yields (digestible protein/units SE)

Yield	Percentage nitrogen as NPN	Percentage of NPN utilised
Maintenance (1:10)	10	100
	20	100
	30	100
10 l/day (1:6.4)	10	100
	20	80
	30	73
20 l/day (1:5.7)	10	27
	20	43
	30	50
30 l/day (1:5.4)	10	0
	20	28
	30	39

Rate of degradation in the forestomachs 60 per cent

are produced from dietary protein and the deficit has to be covered by NPN. It has been calculated that the maximum utilisation of bacterial synthesis (10 g digestible protein per 100 units SE) is connected with a low supply of protein to the animal.

A digestibility in the rumen of 60 per cent means that, per 100 g digestible protein, 40 g passes to the gut undegraded. If losses of protein are to be avoided, 60 g digestible bacterial protein must be synthesised, which requires an intake of 600 units SE. A ratio of digestible protein to SE of 1:6 is thus the limit for utilisation of NPN; with more narrow ratios the supply of ammonia from dietary protein is sufficient. If a ratio of 1:10 is to be fed, all digestible protein could originate from bacterial synthesis, which means that the requirements could be covered totally by NPN.

Table 24.8 *Calculated maximum use of NPN as percentage of dietary nitrogen with the aim of a supply according to feeding standards and an efficiency of utilisation of NPN roughly 100 per cent*

Digestible protein : SE[1]	NPN as percentage of total dietary nitrogen
1 : 6	none
1 : 7	- 25
1 : 8	- 50
1 : 9	- 75
1 : 10	-100

[1] Rate of degradation in the forestomachs 60 per cent

Calculations of the proportions of NPN in dietary nitrogen, which may be effectively utilised at ratios of digestible protein to SE between 1:6 and 1:10 are presented in *Table 24.8*. According to the feeding standards, dairy cows require 1:10 for maintenance, 1:6.4 for yields of 10 kg daily and 1:5.7 for yields of 20 kg daily. Therefore feeding NPN seems efficient only with yields up to 15 kg daily. In this range, requirements of bacteria exceed those of the host animal. Consequently NPN is not suited for high yielding dairy cows. These results are in good agreement with feeding trials published by Helmer and Bartley (1971), Huber *et al.* (1972) and Møller (1973), whereas Conrad and Hibbs (1968) found slightly better efficiencies of utilisation.

Finally it has to be mentioned that a lower rate of digestion in the rumen than the 60 per cent taken in this calculation does not alter the principle of evaluating NPN. However, it would enable the use of NPN at more narrow ratios of digestible protein to SE. Our measurements so far indicate that lower rates of fermentation can be expected only in exceptional cases. Possibly treatments of protein resulting in decreased fermentation in the forestomachs could improve the utilisation of NPN at narrow ratios of digestible protein to digestible energy.

References

ANNISON, E.F. (1956). *Biochem. J.* **64**, 705
CHAMPREDON, C., PION, R. and THIVEND, P. (1971). *Annls Biol. anim. Biochim. Biophys.*, **11**, 298
COELHO, DA SILVA, J.F., SEELEY, R.C., THOMSON, D.J., BEEVER, D.E. and ARMSTRONG, D.G. (1972a). *Br. J. Nutr.*, **28**, 43
COELHO DA SILVA, J.F., SEELEY, R.C., BEEVER, D.E., PRESCOTT, J.H.D. and ARMSTRONG, D.G. (1972b). *Br. J. Nutr.*, **28**, 357
CONRAD, H.R. and HIBBS, J.W. (1968). *J. Dairy Sci.*, **51**, 276
COOK, R.M., BROWN, R.E. and DAVIS, C.L. (1965). *J. Dairy Sci.*, **48**, 475
DEMAUX, G., LE BARS, H., MOLLÉ, J., RERAT, A. and SIMMONET, H. (1961). *Bull. Acad. vet. Fr.*, **34**, 85
DIJKSTRA, N.D. and KEMMINK, A. (1970). *Versl. Landbouwk. Onderz.*, **738**, 23
DIRKSEN, G., KAUFMANN, W. and PFEFFER, E. (1972). *Fortschr. Tierphysiol., Tierernähr., Futtermittelk.*, P. Parey, Hamburg
FERGUSON, K.A., HEMSLEY, J.A. and REIS, P.J. (1967). *Austr. J. Sci.*, **30**, 215
GLADJAEVA, O.F. (1971). *Doklady vses. Akad. sel'skochoz. Naukim V.I. Lenina, Moscow*, **10**, 31
HAGEMEISTER, H. and PFEFFER, E. (1973). *Z. Tierphysiol., Tierernähr., Futtermittelk.*, **31**, 275
HARMEYER, J. and JEZKOVA, D. (1971). *Z. Tierphysiol., Tierernähr., Futtermittelk.*, **28**, 256
HARRISON, F.A. and HILL, K.J. (1962). *J. Physiol.*, **162**, 225
HELMER, L.G. and BARTLEY, E.E. (1971). *J. Dairy Sci.*, **54**, 25
HENDERICKX, H.K. (1973). in *Giesecke, Henderichx. Biologie und Biochemie der mikrobiellen Verdauung.*, p.168. BLV, Munich
HENDERICKX, H.K. and MARTIN, J. (1963). *Verslagen over Navorsingen*, **31**, J.W.O.N.L., Brussels
HOGAN, J.P. and WESTON, R.H. (1968). *Proc. Aust. Soc. anim. Prod. (Armidale)*, **7**, 364
HOGAN, J.P. and WESTON, R.H. (1970). in *Physiology of digestion and metabolism in the ruminant.* Ed. A.T. Phillipson. p.474. Oriel Press, Newcastle upon Tyne
HUBER, J.T., ANDRUS, D.F., ERICKSON, R.E. and POLAN, C.E. (1972). *J. Dairy Sci.*, **55**, 708
HUME, I.D. (1970). *Aust. J. Agric. Res.*, **21**, 305
HUME, I.D., MOIR, R.J. and SOMERS, M. (1970). *Austr. J. agric. Res.*, **21**, 283
HUTTON, K., BAILEY, F.J. and ANNISON, E.F. (1971). *Br. J. Nutr.*, **25**, 165
HUTTON, K. and ANNISON, E.F. (1972). *Proc. Nutr. Soc.*, **31**, 151
IBRAHIM, E.A. and INGALLS, I.R. (1972). *J. Dairy Sci.*, **55**, 971
KAUFMANN, W. and HAGEMEISTER, H. (1973). *6th Symp. Energy Metabol. Europ. Ass. Anim. Prod.*, Hohenheim

KLOOSTER, A.TH. VAN'T and BOEKHOLT, H.A. (1972). *Neth. J. agric. Sci.*, **20**, 272
KLOOSTER, A.TH. VAN'T, KEMP, A., GEURINK, J.H. and ROGERS, P.A.M. (1972). *Neth. J. agric. Sci.*, **20**, 314
KURILOW, N.W. and KROTKOWA, A.P. (1971). *Fisiologija i biochimija pitschewarenija jwatschnich*, p.317. Moscow, Isdatelstwo Kolos
LEIBHOLZ, J. (1971). *Austr. J. agric. Res.*, **22**, 639
MELOSCH, V. (1974). *Z. Tierphysiol., Tierernähr. Futtermittelk.*, **33**, 195
MILLER, E.L. (1973). *Proc. Nutr. Soc.*, **32**, 79
MOLLER, P.D. (1973). *The influence of different carbohydrate sources on the utilisation of urea nitrogen by lactating dairy cows. 412. beretning fra forsogslaboratoriet*, Copenhagen
NOLAN, J.V., NORTON, B.W. and LENG, R.A. (1973). *Proc. Nutr. Soc.*, **32**, 93
OFFER, N.W., EVANS, R.A. and AXFORD, R.F.E. (1972). *Proc. Nutr. Soc.*, **31**, 104A
ØRSKOV, E.R. and FRASER, C. (1973). *Proc. Nutr. Soc.*, **32**, 68A
ØRSKOV, E.R. FRASER, C. and McDONALD, I. (1971). *Br. J. Nutr.*, **25**, 225
PHILLIPSON, A.T. (1972). *Proc. Nutr. Soc.*, **31**, 159
POTTER, G.D., LITTLE, C.O. and MITCHELL, G.E. (1968). *J. Anim. Sci.*, **27**, 1174
PURSER, D.B. (1970). *J. Anim. Sci.*, **30**, 988
PURSER, D.B. and BUECHLER, S.M. (1966). *J. Dairy Sci.*, **49**, 81
SALTER, D.N. and SMITH, R.H. (1971). *Proc. Nutr. Soc.*, **33**, 42A
SMITH, R.H. and McALLAN, A.B. (1973). *Proc. Nutr. Soc.*, **32**, 84A
SUTTON, J.D. (1971). *Proc. Nutr. Soc.*, **30**, 243
TAMMINGA, S. (1973). *Z. Tierphysiol., Tierernähr., Futtermittelk.*, **32**, 185
TERNUTH, J.H. and BUTLER, H.L. (1973). *Br. J. Nutr.*, **29**, 387
THOMAS, P.C. (1973). *Proc. Nutr. Soc.*, **32**, 85
THOMAS, P.C. and CLAPPERTON, J.C. (1972). *Proc. Nutr. Soc.*, **31**, 165
WALKER, D.J. and NADER, C.J. (1970). *Austr. J. agric. Res.*, **21**, 747
WESTON, R.H. and HOGAN, J.P. (1968). *Austr. J. agric. Res.*, **19**, 963

25

PROTEIN REQUIREMENTS IN RELATION TO THE LACTATION CYCLE

A.J.H. VAN ES
Institute for Livestock Feeding and Nutrition Research, Hoorn, Netherlands
Department of Animal Physiology, Agricultural University, Wageningen, Netherlands
H.A. BOEKHOLT
Department of Animal Physiology, Agricultural University, Wageningen, Netherlands

Introduction

During the lactation cycle, milk-protein production ranges from zero in the dry period to 1.75 kg for a cow producing 50 kg milk with 3.5 per cent protein. In addition, protein is being utilised for maintenance and this amounts to some 0.3 kg for a cow of 550 kg body weight. Remarkably some cows can increase the rate of their nitrogen metabolism sevenfold. Thus, it is important as a first approach to study the nitrogen intake of the animal and the subsequent supply of amino acids to the blood during the lactation cycle.

Blood amino acids in the ruminant arise from four sources: (1) absorbed microbial amino acids, (2) absorbed food amino acids, (3) absorbed residues of digestive enzymes, and (4) amino acids mobilised from the tissues. Recently more information on feed intake and thus on nitrogen intake of dairy cows has become available (Van Es, 1974). Also, knowledge on the absorption of nitrogen from the gastrointestinal tract of dairy cows has increased considerably (Van 't Klooster and Boekholt, 1972; Kaufmann and Hagemeister, 1973; Tamminga, 1973). The quantitative understanding of the total absorption of amino acids from the gastrointestinal tract during lactation is far from complete. Nevertheless, there is sufficient knowledge to allow for a study on the limits of amino acid supply by means of simulation. Such an approach in which information on feed intake potentials is combined with data on supply of amino acids from the intestines to the blood, also avoids the danger of discussing protein without paying sufficient attention to energy (Broster, 1973).

When the amino acid supply from the gastrointestinal tract is too low, milk protein production may fall and/or amino acids from the tissues may be mobilised. A special case forms the period (one or two weeks) just after parturition, during which an amount of protein is mobilised from the tissues due to a change in hormone levels, even if there is no need for additional amino acids (Lenkeit, 1972). Mobilisation of protein from the tissues for synthesis of milk protein may result in a negative nitrogen balance in the body. Nitrogen balance experiments with dairy cows may give information on the upper limit of the mobilisation of tissue protein and on the relationships between nitrogen balance, energy balance and change of body weight. Such experiments may yield information on the efficiency of the utilisation of nitrogen or apparently digested nitrogen for milk protein production. It should, however, be taken into account that during lactation, protein may be used for purposes other than milk protein synthesis, e.g. gluconeogenesis or energy supply.

Finally, we must ask the question - are protein standards derived from balance experiments and from feeding trials (usually performed after the peak of milk yield) suited for application under practical circumstances? There may be a need to increase these standards to eliminate the risk of reduced milk yields as a consequence of the normal variation in protein and energy supply under farm conditions. Increased levels may be required from the second week of lactation until peak milk production, because of a possible positive long-term effect of adequate protein supply on milk yield.

Simulation of the Supply of Amino Acids to the Blood During the Lactation Cycle

The simulation was carried out upon cows of 550 kg live weight, yielding 5, 10, 15, 20, 25 and 30 kg milk. The rations used were based on either hay, wilted grass silage, corn silage or urea-supplemented corn silage, and a concentrate mixture with or without urea. The level of amino acid reaching the blood was estimated by making assumptions based on the work of Kaufmann and Hagemeister (1973) and Tamminga (1975):

(1) Computation of the amount of concentrate dry matter required for maintenance and the production of 5 to 30 kg milk at a given forage intake (hay and wilted silage: 9, 8, 7 and 6 kg dry matter; corn silage: 13, 12, 11 and 10 kg dry matter). The criterion was the requirement of starch equivalent (SE) for maintenance (2.833 kg starch equivalent and for each kg of four per cent fat-corrected milk 0.286 kg starch equivalent).

At higher intake levels of mixed ration, each additional kg of concentrate or ground pelleted roughage reduces the voluntary intake of long roughage by about 0.5 kg. Cows producing 20-25 kg milk (if fed 10 kg of concentrate, 8.7 kg dry matter) consume on average not more than 1.5 kg dry matter per 100 kg live weight of good quality (long) hay or wilted silage. Taken together, these data determine the upper level of the intake of dry matter for most lactating cows fed practical rations. However, it should be recognised that considerable between-animal variation in intake potential occurs and that cows yielding more than 30 kg milk per day may have higher upper levels of intake. With regard to the reduction of the voluntary intake of long roughage by concentrate or ground roughage, corn silage was considered to consist of 70 per cent long roughage and 30 per cent concentrate.

(2) Of the dietary protein 40, 50 or 60 per cent was assumed to reach the duodenum unattacked by micro-organisms; 40, 50, 60 or 70

per cent of the apparently digestible organic matter of the ration was assumed to be digested prior to the duodenum and to be available for microbial protein synthesis in the forestomachs; 100 g of this available digestible organic matter was assumed to give a production of 15, 20 or 25 g of microbial protein (in the cases where the ration contained less nitrogen than that required for the assumed production of microbial protein, the microbial protein production was reduced correspondingly); 70 per cent of the unattacked and microbial proteins were assumed to be absorbed from the small intestine into the blood.
(3) For survey purposes, the amino acids absorbed into the blood were expressed as a percentage (S) of an assumed amino acid requirement, arbitrarily set at 331 g for maintenance and 57 g per kg of four per cent fat corrected milk.

Table 25.1 describes the rations and sets out assumptions regarding feeding value and nitrogen utilisation. To satisfy energy requirements for the cornsilage rations, a maximum forage intake (13 kg dry matter) was allowed up to a milk yield of 25 kg; for a yield of 30 kg, forage intake was reduced to 11 kg. Similar values for the grass silage rations were: maximum (9 kg dry matter) for up to 25 kg milk and 7 kg at 30 kg milk; for the hay rations: maximum (9 kg dry matter) for up to 20 kg, 8 kg at 25 kg and 6 kg at 30 kg of milk.

When the total dry matter intake contains 35 per cent long forage or less, digestive disturbances and feed refusal may occur. At the minimum level of forage intake and a production of 30 kg milk, the percentages of long forage are 34, 33 and 40 per cent for the grass silage, hay and cornsilage rations respectively. This means that for most cows it is not safe to increase the intake of dry matter still further to meet the energy requirements at production levels above 30 kg, because this will result in an even lower long-forage content of the rations. Increasing the net energy content of concentrate and forage may result in rations meeting the energy requirement for slightly higher production. Most cows above these higher production levels will draw on their reserves. If the value of S, the ratio of absorbed amino acids to the arbitrarily chosen amino acid requirement expressed in per cent units, was above 100 for a given combination of assumptions and milk yield, the entries H, G, C and U were made for the rations with hay, grass silage, cornsilage and urea-containing cornsilage respectively in *Table 25.2*. This was done to produce the condition of a microbial protein production of 20 g per 100 g digestible organic matter leaving the stomach prior to the duodenum. Under some conditions, and less often in the case of the silage rations, values above 100 are reached at milk yields above 25 kg. Part of the low values of S for the cornsilage rations can be explained by the low nitrogen intake (*Table 25.3*). Using, in the cornsilage- and urea-cornsilage ration, a concentrate mixture with 320 and 210 g crude protein per kg dry matter respectively instead of 170 g, improved the S-values (some values for a few extreme assumptions are given in *Table 25.2*). The improvement was not very great because in all cornsilage rations forage intake was assumed to be high so that concentrate intake, and thus its nitrogen-contribution, was relatively small, even in the case of elevated nitrogen content. The high intake

Table 25.1 Information regarding the assumptions made to compute the quantity of amino acids absorbed into the blood during the lactation cycle

	In the dry matter			$d_0{}^1$	Non-amino-acid nitrogen in total nitrogen	$I_T{}^2$
	Organic matter	Nitrogen × 6.25	Starch equivalent			
	(g/kg)	(g/kg)	(g/kg)	(%)	(%)	(kg)
Hay	900	140	425	65	10	9,8,7,6
Wilted grass silage	890	160	500	70	35	9,8,7,6
Cornsilage	920	100	600	75	20	13,12,11,10
Cornsilage with urea	920	150	600	75	47	13,12,11,10
Concentrate mixture	950	170	720	80	5	13,12,11,10
Concentrate mixture with urea	950	170	720	80	25	13,12,11,10

[1] d_0 = apparent digestibility of organic matter
[2] I_T = intake of dry matter

Food 'protein' leaving the forestomachs unattacked: 40, 50 and 60 per cent.
Digestible organic matter leaving the stomach prior to the duodenum as a part of the total digestible organic matter: 40, 50, 60 and 70 per cent
Microbial protein synthesised per 100 g digestible organic matter leaving the stomach prior to the duodenum: 15, 20 and 25 g.
Starch equivalent required: 1000 + 3.33 × (live weight, kg) + 286 × (four per cent fat-corrected milk, kg)g.

level of the cornsilage also explains the many S-values above 100 for milk yields up to 10 kg. Here, forage intake usually exceeded energy requirements, so that more nitrogen was available than when energy needs are just met.

In the case of the hay and grass silage rations, decreasing or increasing the microbial protein production of 20 g (per 100 g digestible organic matter leaving the stomach prior to the duodenum) by 5 g, changed those S-values which were below 100 by –9 to –15 and +9 to +13 respectively. These changes were similar for the conditions of maximal as well as minimal forage intake. Increasing the non-protein-nitrogen (NPN) content of the crude protein of the concentrate mixture from 5 to 25 per cent resulted in a decrease of the S-values by five to eight units in all cases.

With the cornsilage rations containing concentrate of 170 g crude protein per kg dry matter, shortage of nitrogen fed determined the effect on the S-values (below 100) of changing the microbial protein production of 20 g (per 100 g digestible organic matter leaving the stomach prior to the duodenum) by –5 or +5 g. In the cornsilage ration without urea, the effect of this change was negligible, in those with urea it was of similar size as for the hay ration. Increasing the NPN content of the crude protein of the concentrate mixture from

Table 25.2 Ratios of absorbed amino acids to an arbitrarily chosen amino acid requirement for four ration types; assumed production of microbial protein: 20 g per 100 g digestible organic matter leaving the stomach prior to the duodenum. A letter code entry is only made if the ratio is above 100 per cent

		Milk yield, kg (four per cent-fat-corrected)					
Z^1		5	10	15	20	25	30
		(40 per cent of food protein not attacked in forestomachs)					
50	h	HGCU²	HCU				
50	l	HCU					
60	h	HGCU	HGCU	H	H		
60	l	HGCU	HG	H			
70	h	HGCUcu	HGCUcu	HG	HG	H	H
70	l	HGCUcu	HGU	HG	HGc	Hc	Hc
		(50 per cent of food protein not attacked in forestomachs)					
50	h	HGCU	HGCU	H	H		
50	l	HGCU	H	H			
60	h	HGCU	HGCU	HG	HG	H	H
60	l	HGCU	HG	HG	HG	H	H
70	h	HGCU	HGCU	HGU	HG	HG	HG
70	l	HGCU	HGU	HG	HG	HG	HG
		(60 per cent of food protein not attacked in forestomachs)					
40	h	HGCUcu	HGCUcu	H	H		
40	l	HGCUcu	Hu	H			
50	h	HGCU	HGCU	HG	HG	H	H
50	l	HGCU	HG	HG	H	H	H
60	h	HGCU	HGCU	HG	HG	HG	HG
60	l	HGCU	HGU	HG	HG	HG	HG
70	h	HGCU	HGCU	HGU	HG	HG	HG
70	l	HGCU	HGU	HGU	HG	HG	HG

[1] Z = percentage of digestible organic matter leaving the stomach prior to the duodenum
h, l = highest and lowest assumption for forage intake
[2] H, G, C, U: ration with hay, grass silage, cornsilage and urea-containing cornsilage respectively
c, u: as C and U but with concentrates containing 320 and 210 g crude protein per kg dry matter respectively; for values of Z of 40 and 70 only.

5 to 25 per cent resulted in decrease of the S-values by zero to six units.

It is probable that a low rate of disappearance of digestible organic matter from the stomach prior to the duodenum combines with a high percentage of food protein which leaves the forestomachs unattacked. This would improve the amino acid supply to the blood considerably.

A higher rate of digestion of amino acid-nitrogen reaching the duodenum than the assumed 70 per cent (e.g. to 77 per cent) would increase all S-values by 10 units, also a considerable improvement.

The simulation served its purpose well. The influence on S for each ration due to its content of protein, NPN and digestible organic matter was clearly shown. The same holds true for the influence on S caused by the assumptions regarding breakdown of food protein and synthesis of protein by rumen microorganisms and regarding absorption of amino acids into the blood.

Table 25.3 *Daily quantities of digestible organic matter, nitrogen × 6.25, amino acid and non-amino acid nitrogen × 6.25 of the four rations used in the simulation*

Ration based on:	5 kg milk daily					30 kg milk daily				
	Conc.[1]	DOM	I_N	I_{AAN}	I_{NPN}	Conc.	DOM	I_N	I_{AAN}	I_{NPN}
	(kg/d)	(kg/d)	(× 6.25, kg/d)			(kg/d)	(kg/d)	(× 6.25, kg/d)		
Maximal forage:										
Hay	0.6	5.7	1.4	1.2	0.1	10.5	13.3	3.1	2.8	0.2
Grass silage	-	5.6	1.4	0.9	0.5	9.6	12.9	3.1	2.5	0.6
Cornsilage	-	9.0	1.3	1.0	0.3	5.0	12.8	2.2	1.8	0.3
Cornsilage (32 per cent)[2]	-	9.0	1.3	1.0	0.3	5.0	12.8	2.9	2.6	0.3
Urea-cornsilage	-	9.0	1.9	1.0	0.9	5.0	12.8	2.8	1.8	1.0
Urea-cornsilage (21 per cent)[2]	-	9.0	1.9	1.0	0.9	5.0	12.8	3.0	2.0	1.0
Minimal forage:										
Hay	2.4	5.3	1.2	1.1	0.1	12.3	12.9	2.9	2.7	0.2
Grass silage	1.8	5.1	1.3	0.9	0.4	11.7	12.6	2.9	2.5	0.4
Cornsilage	-	6.9	1.0	0.8	0.2	7.5	12.6	2.3	2.0	0.3
Cornsilage (32 per cent)[2]	-	6.9	1.0	0.8	0.2	7.5	12.6	3.4	3.1	0.3
Urea-cornsilage	-	6.9	1.5	0.8	0.7	7.5	12.6	2.8	2.0	0.8
Urea-cornsilage (21 per cent)[2]	-	6.9	1.5	0.8	0.7	7.5	12.6	3.1	2.3	0.8

[1] Conc. = concentrate dry matter
DOM = digestible organic matter
I_N, I_{AAN}, I_{NPN} = intake of total nitrogen, of amino acid nitrogen and of non-protein nitrogen respectively
[2] Concentrate mixture with an increased protein content of 21 or 32 per cent

Supply of Amino Acids to the Blood from Tissues

Lenkeit (1972) discussed the results of nitrogen balance trials carried out by Oslage and Farries (1966) which show a retention of 1 kg nitrogen from the beginning of the dry period to just after parturition. Paquay, de Baere and Lousse (1972) concluded from similar trials with dry non-pregnant cows, that the animal can draw from its reserves up to 2.7 kg nitrogen. However, this was over a long time period when the energy supply was severely reduced. Energy balances were not determined by these authors. Low energy supply in early lactation, according to Broster (1971), leads to low milk yields over the whole lactation period. This considerably restricts the use of nitrogen reserves to supply the blood with an additional quantity of amino acids.

In the last section it was shown that cows producing 30 kg milk or less can be fed to energy requirement using normal rations. Where appetite is high, many cows succeed in consuming sufficient amounts of energy for the production of 30-40 kg milk per day. Earlier it was found that, in a relative sense, the protein supply from the food to the blood decreases at higher milk yields, which might result in negative nitrogen balances or reductions in milk yield, despite positive or zero energy balances. Thus, it seemed worthwhile to study the relationship between nitrogen and energy balances during lactation and, moreover the relation of these balances with changes in body weight, as the latter are easier to measure than the former.

For this study, results of work with lactating cows from four sources were available: (1) Balance data from the Ruminant Nutrition Laboratory, Beltsville, USA, kindly provided by Dr. P.W. Moe in a computer listing. (2) Balance data from the Oskar-Kellner Institute, Rostock, GDR. (3) Data from balance trials performed before 1960. (4) Data from balance trials performed at the Department of Animal Physiology, Wageningen, the Netherlands. *Table 25.4* refers to general information on the material (*see also* van Es, 1975b).

In this material, the nitrogen balances were usually within the range from −20 to +20 g, i.e. within −5 and +5 per cent of the nitrogen intake. However, negative balances up to −40 g occurred in the data of Denisov and the Beltsville laboratory. In the other trials nearly all were between −20 and zero. Positive balances up to 40 g occurred in the data of Kleiber, Hashizume and Wageningen, up to about 60 in those of Beltsville and even higher in those of Denisov and Ritzman.

To obtain information on mobilisation of nitrogen from and deposition in the tissues it was thought useful to study the relationship between nitrogen balance and energy balance. For the larger sets of data from the sources mentioned above, the correlation between the two balances was: +0.50 (Beltsville), +0.46 (Rostock), +0.64 (Denisov) and +0.39 (Wageningen). Nearly all trials with a positive energy balance had positive nitrogen balances. Assuming 1 g nitrogen balance to consist of 6.25 g protein containing 149 kJ, 15-25 per cent of the positive energy balance was protein-energy. Percentages of a similar magnitude are found in older steers during fattening. The experiments

Table 25.4 General data on balance trials with lactating cows used for statistical treatment[4]

	n^2	W (kg)	I_T (kg/d)	M_E (MJ/d)	q (%)	R_E (MJ/d)	L_E (MJ/d)	D_N (g/d)	L_N (g/d)	R_N (g/d)
Beltsville	342	552	13.0	139	58	3	49	263	88	10
coefficient of variation		18	26	27	11	24[1]	52	38	47	22[1]
Rostock	24	555	13.4	151	61	6	55	247	92	2
coefficient of variation		4	11	11	6	8[1]	18	28	16	8[1]
Wageningen	390	529	14.5	149	56	0	57	249	90	8
coefficient of variation		7	15	15	8	12[1]	27	30	24	12[1]
Kellner	36	437	12.0	118	53	5	38	137	63	0
coefficient of variation		7	10	13	5	4[1]	22	40	22	7[1]
Møllgaard	14	448	10.9	117	60	2	44	207	72	10
coefficient of variation		15	17	19	6	7[1]	30	31	29	8[1]
Fries	7	362	7.6	79	55	3	23	148	40	−1
coefficient of variation		4	6	8	3	6[1]	39	7	38	6[1]
Ritzman	8	577	10.4	105	53	−5	30	188	52	23
coefficient of variation		13	8	15	10	15[1]	15	26	12	39[1]
Kleiber	14	438	8.1	86	60	−3	29	106	45	14
coefficient of variation		12	7	8	3	7[1]	19	18	16	11[1]
Denisov	76	542	15.5	161	58	4	57	300	102	21
coefficient of variation		7	15	27	8	22[1]	29	31	27	36[1]
Hashizume	20	547	13.8	139	55	16	41	146	72	31
coefficient of variation		7	10	17	6	7[1]	18	27	19	12[1]
Selected trials[3]	156	511	12.7	129	56	−7	53	160	89	0
coefficient of variation		16	23	29	10	19[1]	41	38	40	20[1]

[1] Standard deviation
[2] n: number of trials; W = body weight; I_T = dry-matter intake; M_E = metabolisable energy; $q = M_E$ as a percentage of intake of gross energy; R_E = energy retained in tissues; L_E = milk energy; D_N = apparently digested nitrogen; L_N = milk nitrogen; R_N = nitrogen retained in the tissues
[3] Trials in which digested crude protein was less than 80 per cent of standards, selected from the above material
[4] See also van Es (1975b)

with negative energy balance, however, did not show a corresponding protein-energy mobilisation of 15-25 per cent. In nearly all sets of data with energy balances from 0 to -20 MJ, positive nitrogen balances were found averaging +8 and +4 g at energy balances of -8 MJ and -16 MJ respectively. Only at energy balances below -20 MJ were most nitrogen balances negative, averaging -10 g at energy balances between -20 and -40 MJ with a high degree of variation. At even lower energy balance, some nitrogen balances were between -20 and -70 g. It is doubted if these values are reliable because cows having energy balances of that size seldom have a steady-state metabolism.

Systematic and non-systematic errors may obscure true nitrogen balance. The coefficients of variation due to errors of measurement and day-to-day variation of data on food-nitrogen, faecal-nitrogen, milk-nitrogen and urinary-nitrogen may be estimated at 1, 2, 2 and 2 respectively. This leads to a standard deviation of the nitrogen balance of a cow consuming 400 g nitrogen with a digestibility of 70 per cent and urinary- and milk-nitrogen productions of 160 g and 120 g respectively, of:

$$\sqrt{4^2 + 2.4^2 + 3.2^2 + 2.4^2} = \text{about 6 g}$$

This means that in most experiments, particularly those with negative energy balances from 0 to -20 MJ, the nitrogen balances did not significantly differ from zero.

In most of the work at Wageningen, body-weight changes of the cows during the balance trials were known over periods of 14-21 days. It is clear that the accuracy of these figures is low because the period was so short that the differences in gutfill easily masked true weight changes. The coefficients of correlation between weight change (ΔW, kg/d) and energy balance (R_E, MJ/d), weight change and nitrogen balance (R_N, g/d) and R_E and R_N were 0.31, 0.25 and 0.39 respectively. In the 198 trials from the same material with a positive R_E these correlations were 0.25, 0.23 and 0.56 respectively. In the 172 trials with a negative R_E they were 0.16, 0.21 and 0.21 respectively.

The regression equations of R_N on R_E suggest that for a positive R_E, R_N could be predicted by multiplying R_E by one and for a negative R_E by only a quarter, so that in the latter case little nitrogen is mobilised. In both cases a change of R_N could be predicted approximately from a change in body weight by multiplying the latter by 3 ± 1.

It is well known that nitrogen balances may have a considerable systematic error (van Es, 1975a) due to the fact that ammonia losses occur during the collection period. With lactating cows it is especially difficult to collect the excreta without any loss of nitrogen. At high levels of feed intake, the separate collection of urine and faeces during the collection period is not often thorough. Urine which does not go directly to the collection vessel but mixes with faeces easily loses ammonia when drying. Most (about 80 per cent) of this ammonia is absorbed by the condensed water of the air-conditioner in a respiration chamber. In the trials at Wageningen, an average amount of 3 g nitrogen (with extremes of 1 to 10 g) was lost per day in this manner. The losses were about 1.8 per cent of the urinary nitrogen.

The above mentioned nitrogen balances were not corrected for this nitrogen loss. The true nitrogen balances, thus, are lower. If we assume that the corrected nitrogen-balance data do not have other systematic errors, it may be concluded that in this material the cows with energy balances between 0 and -20 MJ hardly mobilised any nitrogen. This could, however, be due to the fact that most animals in these balance trials received an adequate supply of nitrogen because most of the trials were part of studies on energy metabolism in which nitrogen limitation is avoided. For this reason we also studied the relationship between nitrogen balance and energy balance for the 156 trials in which the quantity of digestible crude protein was below 80 per cent of the current protein standard for lactating cows in the Netherlands (equal to $0.13[3.33 W + 1000] + [L \times 63]$ g digestible crude protein for a cow of W kg bodyweight producing L kg of four per cent fat-corrected milk per day). It should be stated that this standard is expressed in digested protein measured in trials with sheep. Cows usually digest dietary protein about 10 per cent less efficiently than sheep, due to the higher feeding level of lactation and due to species differences. This means that in these selected 156 trials, the animals were fed about 90 per cent or less of the assumed protein standard. In the material selected in this way a similar relationship between nitrogen balance and energy balance was found as for the whole material. Obviously, cows also with a low nitrogen supply do not mobilise great quantities of nitrogen from their tissues unless the energy balances are below -20 MJ. It seems that reduction of milk protein synthesis is preferred over nitrogen mobilisation from the tissues. The conclusion is in agreement with the results of Paquay, de Baere and Lousse (1972) who needed long periods of underfeeding to obtain sizeable nitrogen mobilisation from the tissues. Because feeding below the energy requirements for longer periods results in lower milk yields during the whole lactation (Broster, 1971) relying on nitrogen mobilisation with regard to nitrogen supply to dairy cows seems inappropriate. An exception may be made for the first week of lactation when the hormone-induced nitrogen mobilisation helps to cover the nitrogen requirements of the cow which at that time usually eats less than it should to meet energy and nitrogen requirements.

The Conversion of Dietary Nitrogen into Milk Nitrogen

With regard to the efficiency of the conversion of dietary nitrogen into milk nitrogen, it is important to study the relationship between intake of apparently digestible nitrogen or of nitrogen and the sum of nitrogen in milk and nitrogen balance. Theoretically, by increasing the nitrogen content of a ration containing little nitrogen one would expect increases of the nitrogen balance and milk yield. At higher nitrogen contents of the ration no further increase of the nitrogen balance and the milk yield and thus of the milk-nitrogen-production would be expected, due to the fact that the surplus nitrogen is excreted in the urine (van Es, 1972). Plotting the data showed that this concept was true. At high intakes of nitrogen both high and low nitrogen deposits in milk and tissue were found. At lower intakes hardly any data were lying above a straight line running through a point somewhat to the right of the origin with a slope of about 36 degrees (tg 36.5° = 0.74): $y = 0.74x - 22$. In the first plot of our material up to 1970, the number of points near this line was rather small. In most trials performed up to that time the rations were composed in such a way that they would meet the animal's requirement for digestible protein as given by the Dutch standards. However, due to lower feed intake or higher milk yield than expected, some cows did eat less than they required according to the standards during the actual trial. The points for these trials were lying close to the line mentioned above. Afterwards, balance trials with low nitrogen supply were performed, so that the number of points near the line increased in number (Boekholt, 1972). Then the data of Beltsville became available and were plotted in a similar way. As in these experiments the rations were not often composed to meet assumed protein requirements and many had low nitrogen intakes. Again few points were to be found to the left of this line.

Points to the left of the line may occur due to random errors of the data, those to the right, however, can be due to random error as well as to excess nitrogen supply with regard to requirement. It was assumed that excess nitrogen supply did not play a part in rations supplying 80 per cent or less of the requirement for digestible protein mentioned previously. Of all the experiments given in *Table 25.4*, 156 fulfilled this condition. The equation of the regression line was:

$$L_N + R_N = 89.5 + (0.56 \pm 0.02)(D_N - 159.6) \quad RSD = 18.7 \quad (1)$$

(D_N, L_N, R_N: digested nitrogen, milk nitrogen and retained nitrogen in tissues [= N balance] respectively, g/day; RSD = residual standard deviation). Only 72 experiments fulfilled the condition of a nitrogen supply at or below 70 per cent of the requirement for digestible crude protein; the corresponding line was:

$$L_N + R_N = 72.1 + (0.58 \pm 0.04)(D_N - 125.2) \quad RSD = 16.2 \quad (2)$$

It did not differ statistically from the former and in both cases no important differences within the data due to their origin could be detected.

Multiple regression calculations were performed with the data of the 156 experiments which gave the first line. Of the added independent variables of M_E^* (intake of metabolisable energy per metabolic body weight), q (content of metabolisable energy in the gross energy), L_E^* (milk energy per metabolic body weight), W (body weight) and D_N/M_E (ratio of digested nitrogen to metabolisable energy) only the first reduced the residual standard deviation of the regression slightly, from 18.7 to 16.6. The coefficients of variation of the independent variables M_E^*, q, L_E^*, W and D_N/M_E were 19, 10, 17, 16 and 26 per cent respectively. Dividing the dependent and the first independent variable by W or $W^{3/4}$ did not result in any better fit. An improvement was obtained by using the sum of milk nitrogen and the positive nitrogen balance as dependent and the sum of digested nitrogen and the absolute value of the nitrogen balance (if negative) as first independent variable: the RSD decreased to 15.5 when using only this independent variable. Physiologically this model seems to be more correct. Again only M_E^* as second independent variable improved the fit. Dividing by W or $W^{3/4}$ had little effect. An even better improvement resulted from using the sum of milk nitrogen and nitrogen balance (if positive) as dependent variable and the sum of digested nitrogen and negative nitrogen balance (if the energy balance was positive) as independent variable:

$$N_L + R_N(\text{pos.})$$
$$= 96.1 + (0.61 \pm 0.02)(D_N - R_N(\text{if neg. and } R_E \text{ pos.}) - 161$$
$$\text{RSD} = 13 \quad (3)$$

Additional independent variables did not improve the RSD. According to the last model, in the case of negative energy balances, a negative nitrogen balance is not assumed to increase the animal's nitrogen supply.

Similar results were found when using the 72 trials in which the intake of digestible nitrogen was only 70 per cent of the Dutch standards.

Regressing the sum of milk nitrogen and nitrogen balance on nitrogen intake instead of on digested nitrogen in 101 of the above 156 trials where nitrogen-intake data were present, did change the regression coefficient from 0.62 ± 0.03 to 0.45 ± 0.03 and the intercept from -12 to -33. This could easily be explained by the digestibility of the nitrogen and metabolic faecal nitrogen. In this material it did not increase the RSD; obviously nitrogen intake and digested nitrogen were of equal value for the prediction of the sum of milk nitrogen and nitrogen balance.

It is remarkable that all regression lines showed an approximately 60 per cent conversion of digested nitrogen into milk nitrogen and nitrogen balance and intercepts very close to zero. It would mean that the maintenance requirement for digested nitrogen is very low, which is unrealistic. This result is probably due to a slight error in the estimation of the slope resulting (due to extrapolation at a considerable distance from the mean values of the dependent and independent variables) in an incorrect value for the intercept. Thus, instead of comparing estimates on digested-nitrogen requirements for milk production and maintenance, derived from slope and intercept of the regression lines separately with present protein standards, it is better to compare the sum of both estimates at the mean value of the dependent variable with the corresponding standard. According to the Dutch standard, a cow of 550 kg producing 17 kg four per cent fat-corrected milk per day (containing 90 g nitrogen, an amount equal to the mean value of the dependent variable in equation (1) if the nitrogen balance is put equal to zero) requires 230 g digested nitrogen. This standard is expressed in digested nitrogen obtained from the rations by sheep at the maintenance level. Lactating cows, as stated above, would obtain about 10 per cent less from the same rations. Expressed in digested nitrogen for lactating cows the standard is about 207 g. According to the regression equation (1) only 160 g, i.e. 20 per cent less, was needed. The other regression lines gave similar results.

The question remains whether the conclusion, that this material with low nitrogen supply and milk-nitrogen synthesis required 20 per cent less digested nitrogen than is needed according to standard, can be generalised. The lower requirement has probably to be considered to be a minimum requirement when there is no great shortage of energy. Slope and fit of the regression line increased by excluding negative nitrogen balances as a nitrogen source in cases where the energy balances were negative. Furthermore, the conclusion does not tell us anything about the effect of low nitrogen supply on milk production in later periods of the lactation cycle. Well conducted and long lasting feeding trials plus a better understanding of the regulation of milk synthesis are needed to answer such questions.

Little information was obtained on the amount of absorbed amino acids needed for maintenance and milk synthesis. Only the fact that the composition of the digested organic matter nor the milk-yield level markedly influenced its conversion into milk nitrogen and nitrogen balance, appears to be in contradiction with the results of the simulation. There it was found that several assumptions influenced the amount of absorbed amino acids markedly. This may be due to the small variation regarding composition of rations for dairy cows in the treated material as well as in practice. Another explanation is that the choice of values for some of these assumptions automatically determines the value of the other assumptions.

That only about 60 per cent of the digested nitrogen is converted into milk nitrogen, is partially due to the fact that some of the digested nitrogen is ammonia- or urea-nitrogen, and partially due to the utilisation of absorbed amino acids for energy purposes. It is not clear whether substrate is being used for energy purposes in intermediary metabolism regardless of origin or not. If not, the ratio of amino acid energy to other energy in the tissue fluids would be the determining factor. It would seem useful to pay more attention to this ratio.

Developments in Practical Dairy Husbandry which Increase the Chances of Sub-optimal Amino Acid Supply to the Blood

In 1973, concentrate protein became so expensive that the feed industry had to reduce the safety margins of the guaranteed protein content of their mixtures and, furthermore, to use feedstuffs of which the content of digestible protein varies more from batch to batch or is not known precisely. The result at the farm may be a greater fluctuation in the nitrogen supply, a supply which sometimes may even be below the animals' requirements despite correctly computed rations. The risk of selling concentrate mixtures with a protein content below guaranteed levels can be made smaller by nitrogen analysis of suspected feedstuffs and by not using high percentages of one batch of these feedstuffs but, instead, small percentages of several batches and of several of such feedstuffs.

The high farm labour cost has also led to measures which endanger the protein supply of some animals, especially those on high production level. Cows housed in groups are usually fed roughages *ad libitum,* so that it is not known how much the individual cows consume. Variation in forage intake between cows is considerable, thus in this system those animals which eat less than average may have a low total nitrogen intake. Giving the concentrate mixture in a separate milking stall may also result in low intakes of protein because the time spent there by slowly eating cows with high yields is often not long enough. The development of feeding boxes which can be opened by one or a few selected cows to obtain additional feed seems promising in this regard. If provided with a system which supplies only small quantities of concentrates at a time, it would be even more suited for optimal protein supply to high yielding cows because this would reduce the occurrence of unfavourable conditions in the rumen (e.g. low pH resulting in feed refusal or high ammonia levels giving inefficient nitrogen utilisation).

The use of only one concentrate mixture for all dairy cows is another measure which endangers the amino acid supply of high yielding cows in early lactation. The increase of feed intake in this period lags slightly behind the increase of milk yield. Moreover, the high levels of concentrates to be supplied to meet energy requirements enhances the chance of periods with reduced intake or abnormal rumen metabolism; it also lowers the intake of roughage, a fact which is not often realised by the farmer.

We may conclude that for obtaining maximal milk yield during the whole lactation period of high-yielding cows, special attention should be paid to the nitrogen intake of the animal in early lactation to ensure sufficient absorption of amino acids into the blood.

References

BOEKHOLT, H.A. (1972). *Z. Tierphysiol., Tierernähr. Futtermittelk.,* **30,** 145

BROSTER, W.H. (1971). *Dairy Sci. Abstr.,* **33,** 253

BROSTER, W.H. (1973). *Proc. Nutr. Soc.,* **32,** 115

ES, A.J.H. VAN. (1972). *Handbuch der Tierernährung,* II, p.51. Paul Parey, Hamburg and Berlin

ES, A.J.H. VAN. (1975a). *Proc. 9th Int. Congr. Nutr.,* Mexico, 1972, III, p.107, Karger, Basel

ES, A.J.H. VAN. (1975b). *Livestock Production Science,* **2,** 95

KAUFMANN, W. and HAGEMEISTER, H. (1973). *Kraftfutter,* **56,** 12, 609

LENKEIT, W. (1972). *Festskrift Breirem,* p.123. Mariendals Boktrykkeri, As

OSLAGE, H.J. and FARRIES, E. (1966). *Landbauforschung Völkenrode,* **16,** 53

PAQUAY, R., DE BAERE, R. and LOUSSE, A. (1972). *Br. J. Nutr.,* **27,** 27

PAQUAY, R., GODEAU, J.M., DE BAERE, R. and LOUSSE, A. (1973). *J. Dairy Res.,* **40,** 93

TAMMINGA, S. (1973). *Z. Tierphysiol., Tierernähr. Futtermittelk.,* **32,** 185

TAMMINGA, S. (1975). *Neth. J. agric. Sci.,* **23,** 89

VAN 'T KLOOSTER, A.TH. and BOEKHOLT, H.A. (1972). *Neth. J. agr. Sci.,* **20,** 272

FACTORS INFLUENCING PROTEIN AND NON-PROTEIN NITROGEN UTILISATION IN YOUNG RUMINANTS

E.R. ØRSKOV
Rowett Research Institute, Bucksburn, Aberdeen, Scotland

Growth in young animals is almost invariably associated with a high rate of protein deposition in relation to the intake of available energy. While the actual rate of protein deposition is influenced by the adequacy of dietary protein and energy, the limit depends on the genetic potential of the animal. Although in many situations it may not be economical to achieve the genetic potential for growth of an animal, a knowledge of the potential is of fundamental importance for the nutritionist to enable him to make accurate predictions of response in food utilisation resulting from the manipulation of dietary protein and energy intake.

Although the ruminant animal conforms in principle to the above generalisation, the presence of the forestomach, where microbial protein is synthesised, adds a further dimension to its nutrition. In the present paper the general principles of protein utilisation by the young ruminant are discussed and the contribution made by microbial protein in meeting the animal's potential for protein retention is assessed. Finally, an attempt is made to transform the more fundamental information into a form that can be used in practice to assess the utilisation of protein and non-protein nitrogen (NPN) by the young ruminant.

Protein Utilisation by the Young Ruminant

There are many factors which affect the capacity of the young ruminant to synthesise tissue protein. The most important of these factors are undoubtedly the genetic potential and the sex of the animal, the level of energy input and the stage of maturity.

The effect of available energy intake on the rate of protein retention has been discussed by Balch (1967), and Andrews and Ørskov (1970a,b) provided quantitative evidence of this principle with lambs, demonstrating clearly that the optimum protein concentration in the diet for early growth was related to the level of feeding, and that female animals at similar weights to males retained less nitrogen. Since the microbial protein synthesis in the rumen is related to the amount of carbohydrate fermented, Ørskov (1970) suggested that it would be appropriate to express the protein utilisation by the ruminant animal in similar terms. The principles of this approach are illustrated in *Figure 26.1,* which gives the relationships between the intake of digestible organic matter (DOM) and stage of maturity with growing lambs in a series of experiments carried out at the Rowett Research Institute (Andrews and

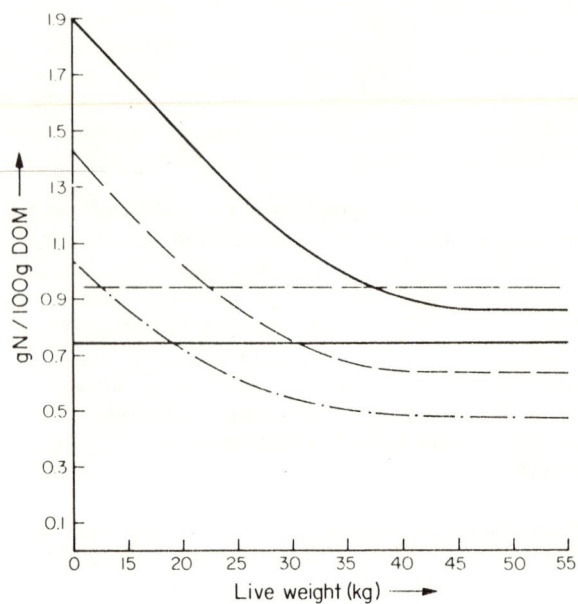

Figure 26.1 The effect of feeding level and stage of growth on nitrogen retention in lambs with near maximal intake (———) 85 per cent (– – –) and 70 per cent (– · —) of maximal intake. The horizontal lines show the potential rate of retention with microbial protein (———) or microbial and undegraded protein in barley based diets (– – –).

Ørskov, 1970a,b; Ørskov and Boyne, 1973). The effect of feeding level is shown by the three response curves, representing maximal intake with no physical restriction to voluntary intake and 85 and 70 per cent of maximal intake. These different levels of energy input can also be achieved by adding roughage to diets or using different qualities of roughage which can impose different levels of physical restriction on voluntary intake because of their slow rate of fermentation.

The exact position of the curves relating feeding level and nitrogen retention to stage of maturity will no doubt depend on the genotype studied so that it is possible only to relate these to body weight when the particular genotype is specified.

In *Figure 26.1* energy intake has been expressed in DOM simply because of the ease with which it can be determined. It can be converted to metabolisable energy (ME) by assuming that each g of DOM contains 4.5 kcal (19 KJ) and that 82 per cent of the digestible energy is metabolisable. It should be pointed out here that the relationships have been constructed using fully functioning ruminant animals, though in some cases part of the protein has been given so that the rumen was by-passed.

Figure 26.1 illustrates clearly that the attainable rate of nitrogen retention by the host animal varies over a wide range; it depends on stage of maturity and it varies greatly with feeding level. In addition, it is different for different genotypes and within genotypes there are differences between the two sexes. Previous nutrition is another factor on which little information is available, but unpublished work from this Institute indicates that, at the same live weight, animals that have previously received insufficient protein will retain more nitrogen than those that have received an adequate supply of protein.

Contribution of Microbial Protein to the Potential Protein Deposition

As referred to in the previous section, the production of microbial protein will, on the whole, be proportional to the organic matter fermented. These relationships will not be discussed in detail here. For a very comprehensive discussion the reader is referred to Hungate (1966). The anaerobic fermentation of carbohydrate imposes strict limitation on the amount of ATP available for the microbes and thereby the amount of cells produced when nitrogen is not limiting. With slowly fermenting substrates such as straw, the amount of microbial cells produced per unit of carbohydrate fermented (Hobson, 1971) will undoubtedly be lower owing to a greater maintenance cost of the fermenting microbial mass. There is, however, no biochemical reason upon which to expect a relationship between the quantity of microbial cells synthesised per unit of substrate fermented and the type of rumen fermentation. On the other hand, possible differences could occur if different groups of bacteria, which are particularly efficient in their utilisation of ATP, also produced distinct types of fermentation. Hume (1970), and Ishaque, Thomas and Rook (1971) produced some evidence to suggest a greater microbial nitrogen output with a greater proportion of propionic acid in the rumen. This has not been confirmed by other workers (Ørskov *et al.*, 1974).

The amount of cell protein produced per unit of organic matter fermented has now been assessed by several workers, the results of which have recently been summarised by Thomas (1973). In his review Thomas estimated that when the production of microbial protein was assessed by the use of a-diamino-pimelic acid (DAP) a mean value of 3.2 g nitrogen was produced per 100 g of organic matter fermented in the rumen. Considerable variation was found in the results and if the data of Miller (1973) are included together with those of Walker and Nader (1970), a somewhat lower value is obtained. In the context of discussing the principles here, the absolute value is not particularly important, but a value of 2.5 g microbial nitrogen per 100 g of organic matter fermented is probably close to the mean based on literature estimates and is preferred here since it was established with diets similar to those that will be discussed in more detail later, (Ørskov, Fraser and McDonald, 1972). If it is further assumed that 70 per cent of the organic matter digested is fermented in the rumen, the microbial nitrogen in relation to total organic matter digested will be in the region of 1.8 g nitrogen/100 g DOM. If digested organic matter contains 4.5 kcal/g (19 KJ/g) then microbial nitrogen amounts to about 4.0 g/Mcal digested (1.0 g/MJ). Again, assuming that 82 per cent of the DOM is metabolisable energy, the value will be about 5.0 g nitrogen/Mcal of ME (1.2 g/MJ).

When a purified diet in which urea provided 95 per cent of the nitrogen was given to young early weaned lambs, we have recorded a nitrogen retention of about 0.74 g nitrogen/100 g of DOM when these diets were given at near maximal intake. If this is taken to represent the maximum nitrogen retention that can be achieved from microbial protein alone, then in relation to the value of 1.8 quoted above, it amounts to a retention of about 40 per cent of the microbial protein produced. In view of the fact that microbial nitrogen contains 15 to 20 per cent of nucleic acid nitrogen (Smith, 1969), and that some nitrogen is used for maintenance and some bacterial nitrogen excreted in the faeces, then this is not an unreasonable estimate. In *Figure 26.1* the value of 0.74 g nitrogen/100 g DOM has been superimposed on the three curves relating feeding level to nitrogen utilisation. It should be emphasised that the value was determined with lambs which had the capacity to retain nitrogen at a much greater rate than 0.74 g nitrogen/ 100 g DOM.

From the principle illustrated in *Figure 26.1* it is possible to estimate the adequacy to the animal of microbial nitrogen, but before doing so it is as well to point out that diets are seldom if ever devoid of protein and their protein is not completely fermented in the rumen. As a result nitrogen retention is generally higher when basal diets are supplemented with nitrogen to secure an adequate supply of NPN. For example, with barley-based diets supplemented with urea, a retention of 0.94 ± 0.19 g nitrogen/100 g DOM (Ørskov *et al.,* 1972) was obtained This value is also included in *Figure 26.1* as a dotted horizontal line. It would have been interesting and desirable to have included here data from roughage diets varying in protein content and also other cereal diets, but this is not yet available.

It is possible to estimate from *Figure 26.1* the extent to which a reasonably constant supply of microbial nitrogen in relation to energy input can meet the changing demand of the host animal. In the early stages of growth, microbial protein is insufficient, also, when barley is given as the energy source, its undegraded protein together with microbial protein is insufficient to ensure maximum nitrogen retention. The insufficiency relative to requirement will largely depend on the level of feeding, as illustrated.

When the feeding level is restricted, either because the physical nature of the roughage in the diet imposes a limitation on voluntary intake, or because the amount of a concentrated food given is restricted; then, depending on the extent of the restriction and the type and breed of the animal, the requirement may never exceed that which is available from the microbial protein. When a certain stage of maturity is reached, then even at maximal intake (*Figure 26.1*), the requirement can be met and exceeded by microbial protein. It should be emphasised that for the vast majority of ruminant production systems, the requirement is met and in many instances exceeded by the microbial protein and the protein from the diet which escapes degradation.

Figure 26.1 does pose a number of important questions which will now be discussed. How do we assess the requirement when the maximum attainable rate of protein deposition exceeds the microbial supply of protein. At which stage of development or weight does the animal require no protein supplementation? and how do we assess the microbial requirement for nitrogen?

Assessment of Requirement when the Potential Nitrogen Retention of the Animal is not met by Basal and Microbial Protein

Ideally the maximum attainable rate of protein deposition should be assessed for each genotype in question and its relationship to age and weight described as in *Figure 26.1*, for the diet to which it is to be applied. It is clear, however, that the work involved in doing so would be too vast, although a description of a response curve for a genotype would be valuable data on, for instance, a new breed or strain of animal. Information obtained from several experiments by Mr. C. Fraser and myself has been plotted in *Figure 26.2a* with barley-based diets for Suffolk × (Finnish Landrace × Dorset Horn) male lambs. In addition, data obtained when barley-based diets were given to Friesian steers by Dr. M. Kay, Mr. N.A. McLeod and Mr. A. Macdearmid (also from this Institute) has been used. The data is included in *Figure 26.2b*. In *Figure 26.3* the relationship of the maximum attainable rate of protein deposition and live weight for these two species and breeds has been given. This includes, for each species, over 80 nitrogen balance trials with *ad libitum* intake of barley based diets and with protein provided in excess of the animal's attainable rate of protein deposition. To describe these curves it is essential that protein is given in excess, and that the diets with which the protein status is to be related are offered at the relevant level of energy input.

For the male Suffolk × (Finnish Landrace × Dorset Horn) lambs the maximum attainable rate of protein deposition from 10 to 35 kg was described by a linear regression of the form:

$$Y = 2.30 - 0.037X \quad (n = 44, RSD = 0.21)$$

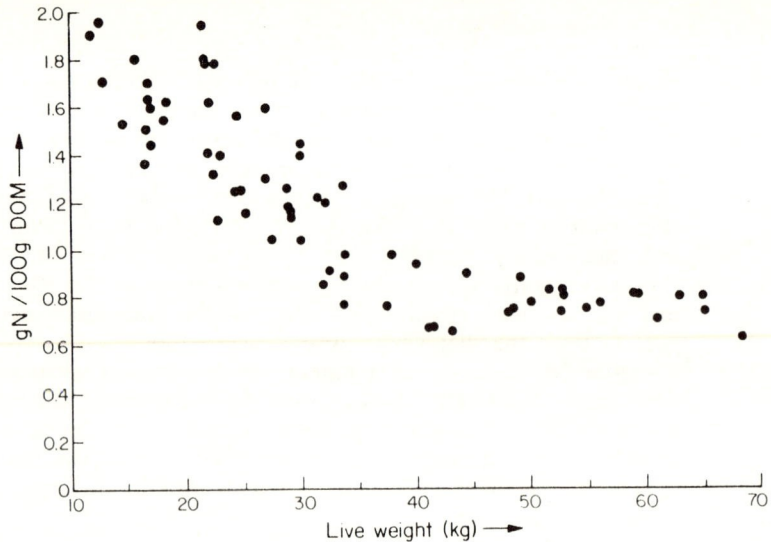

Figure 26.2a Plot of individual values of nitrogen retention per 100 g of DOM when barley based diets with excess of protein were given to Suffolk cross male lambs

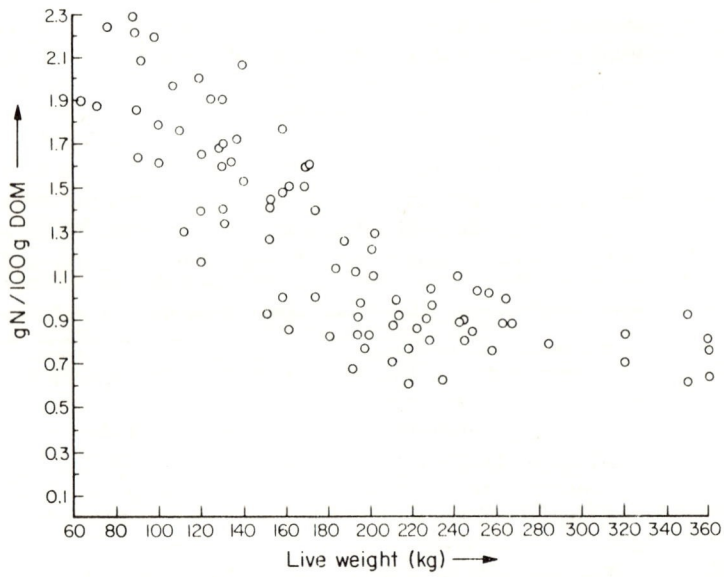

Figure 26.2b Plot of individual values of nitrogen retention per 100 g of DOM when barley based diets with excess of protein were given to Friesian steers

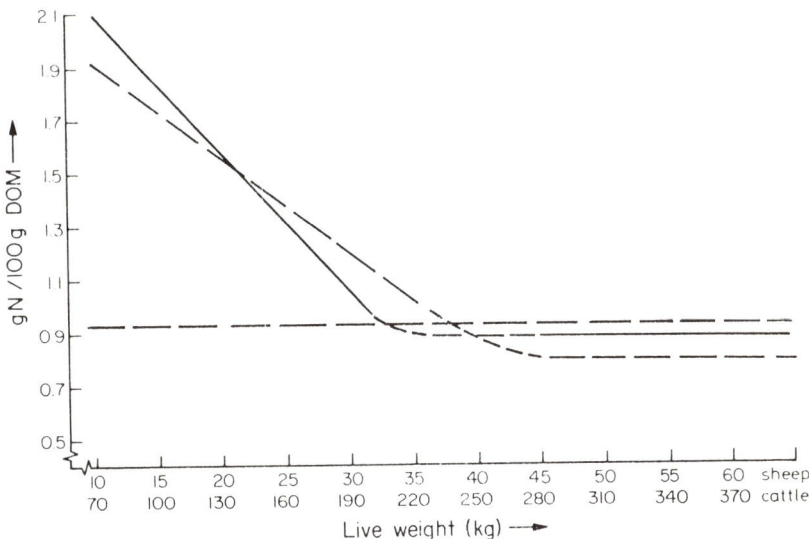

Figure 26.3 The effect of weight of Friesian cattle (———) and Suffolk cross males lambs (— — —) on nitrogen retention with ad libitum *feeding of barley based diets with excess of protein supplements. The horizontal line indicates potential rate of retention when barley supplemented with NPN is given*

where Y is again nitrogen retained/100 g DOM. After 200 kg weight there was no relationship and the mean value was 0.88 ± 0.15 g nitrogen/100 g DOM (n = 26).

While the weights of the two species are obviously different, it is interesting to note the similarity in response, a linear decrease in the attainable rate of protein deposition per unit of energy input with increasing weight being followed by an almost constant one. The equations would imply that when the quantity, g nitrogen/100 g DOM, had fallen to less than 0.94 then the basal diet supplemented if necessary with NPN would be sufficient to meet the animals potential capacity for nitrogen retention of both species. For the Suffolk cross lambs this would occur at about 37 kg live weight and for the Friesian cattle at about 200 kg live weight.

Having described the maximum attainable rate of protein deposition it is now possible to obtain some estimate of the amount of protein required when this potential is not met by microbial and undegraded dietary nitrogen. The sophistication of the parameters used for practical purposes must bear some relation to the variation encountered within the animal. The individual values plotted in *Figure 26.2a* and *Figure 26.2b* are a good example of the variability within sex and genotype.

Having obtained the amounts retained when only the basal diet is given, the only information required in addition, is that of the utilisation of the protein supplement. Such information was obtained for instance by Ørskov, Fraser and Corse (1970) with fish meal, soya bean meal and casein, both when each was given so that it was exposed to degradation in the rumen and when the rumen was by-passed by using the closure of the oesophageal groove to conduct food directly to the abomasum. The utilisation is dependent on the value of the protein source as a complementary protein to that already available from rumen micro-organisms and on the extent to which the particular protein source is degraded in the rumen. If the protein is given so that the rumen is by-passed, the utilisation will be more efficient. This may be a simplified statement, it has not been tested whether the determined utilisation will depend on the level included, i.e. whether the utilisation of the supplementary protein is a constant one.

The utilisation of fish meal determined by Ørskov, Fraser and Corse (1970) to be 49.4 per cent when given conventionally so that the protein entered the rumen, has been used here to calculate the amount of supplement required for lambs and cattle.

In *Table 26.1* the maximum attainable rate of protein deposition at 5 kg intervals for the lambs and in *Table 26.2* at 20 kg intervals for cattle has been obtained from the data given in *Figure 26.3*. Some tentative estimates of supplementary protein allowances can now be made. In order to arrive at the amount of supplementary protein required in addition to the basal diet, the value of 0.94 g nitrogen has been subtracted from the potential at each weight. The efficiency of utilisation of fish meal mentioned earlier was subsequently used to express the amount of nitrogen required as fish meal, which finally has been expressed as g crude protein per 100 g dry matter assuming that the barley diets contained 81 per cent DOM, and again assuming that the value of 49.4 per cent applies at each level. The supplementation of NPN will be discussed in more detail later.

It would of course be impracticable to alter the protein concentration at each 5 kg interval in the case of lambs and at 20 kg intervals in the case of calves, especially if the food is compounded into pellets. With lambs given fish meal, a uniform concentration of about 16 per cent crude protein in DM has been found to give results not significantly different from those resulting from a stepwise reduction in protein concentration during fattening (Ørskov and Fraser, unpublished results). If soya protein had been used instead of fish meal, a higher concentration would be required because of its poorer utilisation (Ørskov, Fraser and Corse, 1970). On the other hand, if the supplements were given so that the rumen was by-passed, the same nitrogen could be retained with about 30 per cent less of the supplement (Ørskov, Fraser and Corse, 1970).

There are several other methods that have been developed in recent years to protect protein from rumen degradation, by chemical treatments (Ferguson, Hemsley and Reis, 1967; Leroy and Zelter, 1970). It should be emphasised here that in order to assess the effect of protection of protein from rumen degradation, it is essential to know the animals attainable rate of protein deposition since protection will only be advantageous if their potential for nitrogen deposition is greater than the quantity supplied by microbial and basal protein. Many experiments

Table 26.1 *The effect of weight of male lambs on potential nitrogen retention in relation to digestible organic matter (DOM) and the amount of fish meal required as supplementary nitrogen. The conversion of DOM to live weight has also been calculated together with conversion of dry matter with barley based diets*

Live weight kg	Maximal attainable rate of nitrogen deposition g/100 g DOM	Attainable rate above basal contribution g nitrogen/100 g DOM[1]	Nitrogen required as fish meal nitrogen/ 100 g DOM[2]	NPN required g/100 g DOM	Fish meal required g crude protein/ 100 g DM	NPN required g crude protein/100 g DM	Food conversion	
							Kg DOM/kg gain[3]	Kg DM/kg gain[4]
10	1.93	0.99	2.00	-	9.9	-	1.45	1.83
15	1.75	0.81	1.64	-	8.1	-	1.60	2.03
20	1.56	0.62	1.26	-	6.2	-	1.80	2.27
25	1.37	0.43	0.87	-	4.3	-	2.07	2.59
30	1.19	0.25	0.51	0.17	2.4	0.9	2.35	2.98
35	1.01	0.06	0.12	0.37	0.5	1.9	2.64	3.37
40-70	0.79	-	-	0.43	-	2.2	3.55	4.49

[1] Assuming microbial and undegraded protein to result in a retention of 0.94 g nitrogen/100 g DOM
[2] Assuming the utilisation of fish meal to be 49.4 per cent
[3] Assuming 2.8 g nitrogen/100 g live weight gain
[4] Assuming 81 per cent of DOM in barley dry matter and 10 per cent crude protein

Table 26.2 *The effect of weight of Friesian male cattle on potential nitrogen retention in relation to digestible organic matter (DOM) and the amount of fish meal required as supplementary nitrogen. The conversion of DOM to live weight gain has also been calculated together with conversion of dry matter with barley based diets*

Live weight kg	Maximal attainable rate of nitrogen deposition g/100 g DOM	Attainable rate above basal contribution g nitrogen/100 g DOM[1]	Fish meal required nitrogen/100 g DOM[2]	NPN required g/100 g DOM	Fish meal required g crude protein/100 g DM	NPN required g crude protein/100 g DM	Food conversion	
							Kg DOM/kg gain[3]	Kg DM/kg gain[4]
70	2.11	1.17	2.37	-	11.7	-	1.33	1.68
90	1.93	0.99	2.00	-	9.9	-	1.45	1.84
110	1.76	0.82	1.66	-	8.2	-	1.59	2.01
130	1.58	0.64	1.29	-	6.4	-	1.77	2.24
150	1.40	0.46	0.93	-	4.6	-	2.00	2.53
170	1.22	0.26	0.53	0.16	2.6	0.8	2.30	2.91
190	1.05	0.11	0.22	0.33	1.1	1.7	2.67	3.38
210 and over	0.88	-	-	0.43	-	2.2	3.18	4.03

[1] Assuming microbial and undegraded protein to result in a retention of 0.94 g nitrogen/100 g DOM
[2] Assuming the utilisation of fish meal to be 49.4 per cent
[3] Assuming 2.8 g nitrogen/100 g live weight gain
[4] Assuming 81 per cent of DOM in barley dry matter and 10 per cent crude protein

have shown no benefit from protection of protein for growing animals, (Faichney and Lloyd Davies, 1972). In many cases this is not surprising since an effect of protection can only be accurately assessed when the potential rate of nitrogen deposition by the animal is known.

The stage or weight at which protein supplementation is no longer required with the barley-based diet used here would be when the maximum attainable rate of protein deposition became equal to the value of 0.94 g nitrogen/100 g DOM or less. This indeed has been supported by experimental data. There may however be many practical conditions in which it would be desired to assess whether protein supplementation is required but where no facilities were available for measuring the nitrogen metabolism of the animal. In this respect, it is of interest to note that the expression of nitrogen retained per unit of organic matter intake is also an expression of the efficiency of conversion of organic matter to live weight gain, because nitrogen retention is closely correlated with live weight gain.

In *Tables 26.1* and *26.2* an estimated conversion ratio of DOM to live weight has been included. This estimate is based on an equation, incorporating live weight gains for lambs varying from 100 to 400 g/d, and 84 nitrogen balance trials. It was found to be 2.8 g nitrogen/100 g live weight gain (Ørskov, unpublished results). It is possible also then to calculate from the feed efficiency data, the point at which protein supplementation is not required. If only microbial protein was available, no protein supplementation would be required when the amount of DOM required/kg gain exceeded 3.8 kg. With barley-based diets the value would be 3.0 kg DOM or converted to dry matter (81 per cent DOM) when more than 3.7 kg was required per kg live weight gain. It should be emphasised that the values quoted are rate of food conversion and not accumulated food conversion.

The concept illustrated in *Figures 26.1* and *26.3* implies that no response to protein supplementation will be found when the animal has reached the stage at which sufficient protein to meet its maximal attainable rate is available from microbial and undegraded dietary protein, provided that the diet contains sufficient nitrogen for microbial growth. Unpublished results from this laboratory showed in support of this that lambs weighing about 55 kg live weight and fed *ad libitum* on barley-based diets did not alter their nitrogen retention when 7 g of nitrogen in the form of fish meal was given to them in such a manner that the rumen was by-passed. On the other hand, Broster *et al.* (1969) showed that dairy heifers increased their nitrogen retention when a low protein diet, which would be expected to supply adequate nitrogen for the microbes, was supplemented with ground-nut meal. More recent work, however, (Broster and Smith, personal communication) has shown that steers retaining 0.8 g nitrogen/100 g DOM did not increase their nitrogen retention if the protein supplement was protected by formalin treatment. This would be in agreement with the principles illustrated in *Figures 26.1, 26.2* and *26.3*.

Assessment of Microbial Requirement

In the area shown in *Figures 26.1* and *26.3* where the maximal attainable rate of protein deposition is met or exceeded by microbial and undegraded basal protein, the only parameter that it is necessary to determine is the microbial requirement for nitrogen. The microbial requirement must be met in order to achieve maximal digestibility and intake. It will be an attribute of the diet and relatively independent of the animal to which it is given.

In order to estimate the microbial requirement it will be necessary again to return to the value of 1.8 g nitrogen per 100 g of DOM referred to earlier as being the approximate amount of microbial nitrogen produced. This would also imply that a similar amount of nitrogen is required per unit of DOM. While some recycling of nitrogen does take place and may be greater with mature than young animals, (Ørskov, Fraser and McDonald, 1971), the importance of the recycling is uncertain and it is probably of little quantitative importance in the present calculations (Nolan, Norton and Leng, 1973).

The amount of nitrogen required is in the main determined by the digestibility or fermentability of the dietary organic matter and by the amount of the dietary nitrogen which is degraded in the rumen. The value of 1.8 g nitrogen/100 g DOM was arrived at by assuming that 70 per cent of the organic matter digested was fermented in the rumen. While this is an acceptable average estimate, it would not apply to diets such as ground maize where less than that would be fermented in the rumen (Ørskov, Fraser and Kay, 1969; Beever, Coelho da Silva and Armstrong, 1970), but it would certainly apply to barley-based diets.

In *Figure 26.4* the relationship of organic matter digestibility and the required concentration of degradable nitrogen has been given. Since, as mentioned earlier, the nitrogen in the feed is not generally completely degraded, a second relationship has been included, assuming that 70 per cent of the nitrogen from the diet was degraded in the rumen. This value is close to our estimate of the extent of degradation of barley protein. There are extremely few estimates available in the literature of this parameter, while there seem to be many more estimates of the extent of degradation of protein supplements (Ørskov, Fraser and McDonald, 1971; Schoeman, De Wet and Burger, 1973). If the approximate digestibility of organic matter is known, and if the extent of degradation of nitrogen is assumed to be about 70 per cent, then it is possible to estimate whether or not the nitrogen content of the basal diet is sufficient to satisfy the microbial requirement. If the nitrogen content is lower than the amount required for microbial protein synthesis, supplementation of nitrogen is necessary to achieve maximum intake and digestibility. The difference between the amount in the diet and that required according to *Figure 26.4* should be reduced by 30% if the supplement consists of NPN provided it is completely degradable and utilised. In fact this is a situation in which less nitrogen is required in the form of urea than in the form of protein supplements which are only partly degraded. Experimental evidence for this has recently been

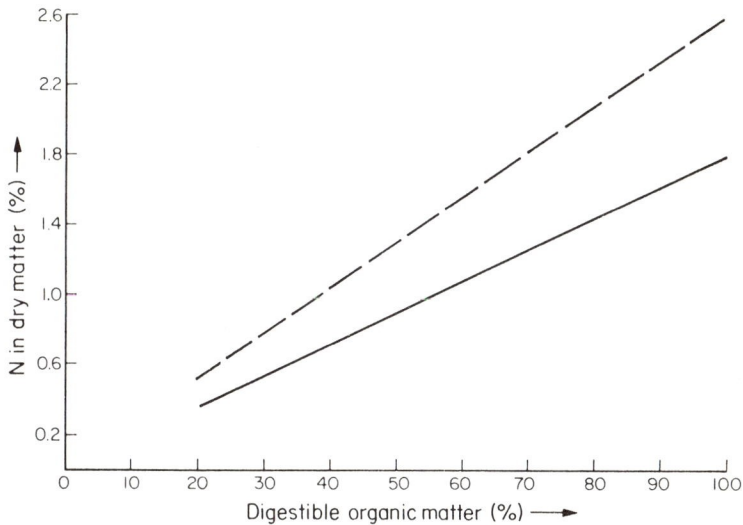

Figure 26.4 The nitrogen concentration required in basal diets to meet the microbial needs in relation to digestible organic matter content. It is assumed that 100 per cent (———) or 70 per cent (— — —) of the nitrogen in the diet is degraded in the rumen.

reported from our laboratory (Ørskov, *et al.*, 1974) and is shown in *Figure 26.5*. Here the maximum digestibility of a barley diet was attained at a lower nitrogen content when the supplement was urea rather than when it was fish meal. In *Table 26.3* estimates of the critical criteria of some basal feeds have been given. When the basal feeds provided insufficient nitrogen to satisfy the microbial requirement, the supplementary nitrogen required has been calculated both as crude protein and as urea, assuming it is completely degraded and utilised.

Protein and Non-protein Nitrogen Supplementation and the Effect of Protein Quality

As mentioned in the previous section when the stage has been reached where the potential protein utilisation by the animal is met by microbial nitrogen, then NPN can be used as the only supplement, if supplementation is required. Since the quantity of even the most limiting amino acid is in excess, the amino acid requirement is for all practical purposes irrelevant. It should again be re-emphasised that this occurs in the vast majority of feeding systems for growing ruminants. Only with early weaning and feeding of concentrated foods is the requirement not met, as shown in *Figures 26.1* and *26.3*.

Table 26.3 *Calculation of adequacy or inadequacy of nitrogen for microbial fermentation in different feedstuffs with differing digestibility of organic matter and differing nitrogen contents*

Feedstuff	Organic matter digestibility (%)	Crude protein (CP) (% of dry matter)	Required for rumen microbes		Excess or deficit of CP (%)	CP to be supplied as NPN (%)	NPN as urea (% of dry matter)
			100 % availability	70% availability			
Maize	84	10.4	9.5	13.6	−3.2	2.2	0.8
Wheat	83	11.4	9.3	13.3	−1.9	1.3	0.5
Oats	70	12.4	7.9	11.3	+1.1	-	-
Barley (low protein)	81	10.0	9.1	13.0	−3.0	2.1	0.8
Barley (high protein)	86	13.6	9.7	13.9	−0.3	0.2	0.1
Barley straw	44	4.1	5.0	7.1	−3.0	2.1	0.8
Red clover hay	64	21.4	7.2	10.3	+11.1	-	-
Fescue hay (early cut)	67	18.1	7.5	10.7	+7.4	-	-
Fescue hay (late cut)	56	6.0	6.3	9.0	−3.0	2.1	0.8
Turnips	90	14.0	10.1	14.4	−0.4	0.3	0.1
Potatoes	85	9.0	9.6	13.7	−4.7	3.3	1.2

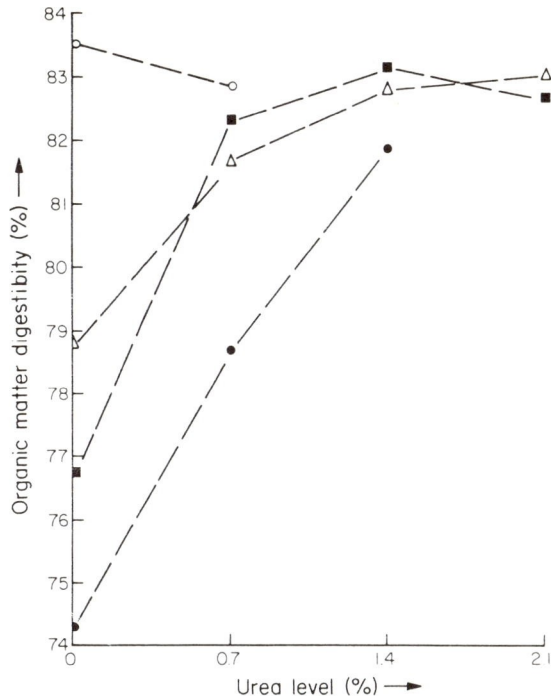

Figure 26.5 Effect on organic matter digestibility in lambs when barley diets containing four increasing levels of fish meal (g/kg O (●) 30 (■), 60 (▲) and 90 (○) are supplemented with four levels of urea, where each increment of urea is isonitrogenous with each increment of fish meal. From Ørskov et al., 1974. Reproduced by courtesy of the British Journal of Nutrition.

The amino acid composition of the supplementary protein when supplied for early growth is undoubtedly very important and likely to be mainly responsible for differences in protein utilisation when protein supplements are given post-ruminally (Ørskov, Fraser and Corse, 1970). Many experiments have shown that sulphur-containing amino acids, methionine and cysteine, appear to be limiting in microbial protein (Armstrong and Annison, 1973; Nimrick et al., 1970a,b), so that protein supplements with a relatively high content of sulphur-containing amino acids are most efficiently utilised. The other important factor determining the efficiency of utilisation of the protein supplements is the extent of degradation if the supplements are given so that they enter the rumen. Better utilisation can be achieved if the supplements are protected by chemical treatments, as discussed earlier, or given so that the rumen is by-passed. It should be emphasised that in most instances sufficient protein to meet the animal's potential for protein deposition can reach the small intestine (Ørskov, Fraser and McDonald, 1971), with conventional feeding of the protein supplement since most protein supplements are only partly degraded. It is therefore only possible to spare protein by protection; it is not possible to alter the maximum attainable rate of protein deposition by the animal. Benefits from protection of protein should therefore not be expected if it is compared iso-nitrogenously with levels of protein given conventionally in amounts adequate to meet the animal's requirement for tissue synthesis.

The question of whether both NPN, to satisfy the microbial requirement and protein, to satisfy any additional requirement, should be used, needs further clarification. In the case where the protein supplement is protected from ruminal degradation, then NPN should be used to satisfy the microbial requirement at all times. However, if the protein is given in the conventional form, which is most often the case, then the inevitable degradation, albeit only partial, is often sufficient to supply the needs of the microbes additional to the basal protein. This is illustrated further in *Tables 26.1* and *26.2*, where the amount of NPN required is calculated on the assumption that 50 per cent of the protein supplement was degraded to yield ammonia in the rumen. It was demonstrated experimentally from data by Ørskov *et al.* (1974) and illustrated in *Figure 26.6*. The maximum efficiency of food utilisation was not achieved with urea supplementations for the young early weaned lamb; at the low levels of fish meal inclusion there is an additional response to urea but at the highest level of fish meal, which was close to the requirement by the young lambs, there was no response in feed efficiency to supplements of urea.

The information required for the model of estimating nitrogen requirement is in most cases not available. There is a great deal of information required on the maximum attainable rate of protein deposition (*Figures 26.1, 26.2* and *26.3*) under different dietary regimes. Knowledge of this parameter is extremely important since this in itself can lead to an accurate prediction on matters of protein or NPN utilisation.

More precise information is needed on the efficiency of utilisation of different protein supplements by the young ruminants, and whether their efficiency is dependent on their level of inclusion. An accurate determination of this however pre-supposes a knowledge of the attainable rate of protein deposition by the animal.

Finally, but not least, more accurate information is required on the degradability of nitrogen from basal diets, both cereal and roughages, since this would lead to a reasonably accurate prediction of the need for NPN supplementation provided the digestibility of the diet is known.

Voluntary Intake in Relation to Protein Utilisation

Many years ago Osborne and Mendel (1918) showed that rats increased their intake when protein insufficiencies were corrected. The young ruminant tends to react in a slightly different manner owing to the microbial fermentation in the forestomach. If the microbial requirement for nitrogen is not met, voluntary intake will be low. When this deficiency is corrected, intake will increase owing to increases in the rate of rumen fermentation. When the animal has a requirement for

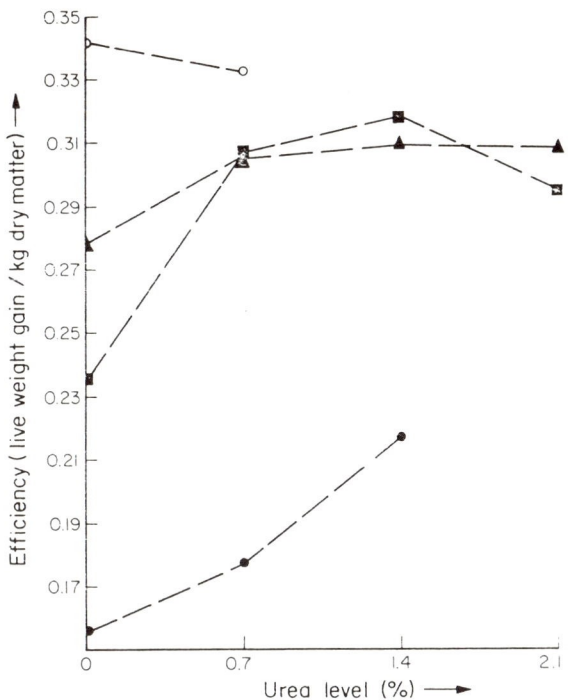

Figure 26.6 The effect on efficiency of food utilisation with male lambs of barley diets containing four increasing levels of fish meal (g/kg 0 (•) 30 (■) 60 (▲) and 90 (○) are supplemented with urea, where each increment of urea is isonitrogenous with each increment of fish meal. From Ørskov et al., 1974. Reproduced by courtesy of the British Journal of Nutrition

nitrogen which is in excess of the microbial protein production, intake of the basal diet will increase further by a protein supplement whether this is given so that it enters the rumen, (Ørskov *et al.,* 1972), or whether the rumen is bypassed (Ørskov, Fraser and Pirie, 1973). On the other hand, if the basal diet does not contain sufficient nitrogen for rumen fermentation we have not been able to increase intake substantially by giving protein supplements post-ruminally (Mehrez and Ørskov, unpublished results).

The consequence of this, as referred to earlier, is that even if the microbial and the undegraded dietary protein is in excess of the need of the animal, the microbial requirement must be met to ensure maximum intake, which in many systems of ruminant production is the most limiting factor (Blaxter, 1962). The concentration of protein necessary to ensure maximum intake and production will therefore remain relatively constant after the stage has been reached, (*Figures 26.1* and *26.3*) at which the potential protein utilisation is met from microbial and basal protein.

Effect of Protein Nutrition on Body Composition

As with monogastric animals, if the diets are insufficient in protein more fat will be laid down (Andrews and Ørskov, 1970b). However, with ruminants it is not possible to alter the protein:energy ratio to the same extent since if the microbes are deficient in nitrogen, intake of food will decrease. With the pre-ruminant lamb it has been shown that it is possible to create large changes in body composition by altering the protein:energy ratio in the milk substitute (Jagusch, Norton and Walker, 1970). Even so, it is possible to cause some changes in the nitrogen content of gains by altering the protein:energy ratio (Andrews and Ørskov, 1970b), but again it is important to consider that if and when the requirement for nitrogen is met by microbial and undegraded basal protein, then changes in body composition are difficult to produce because it is almost impossible to create a protein insufficiency and maintain high levels of intake. This would occur with late weaned lambs and is probably responsible for the conclusions of Reid *et al.* (1968) that altering the protein:energy ratio had no effect on body composition.

Conclusions

(1) The rate of protein deposition by the young ruminant is most appropriately expressed as nitrogen retained per unit of energy digested, since the microbial protein formed tends to be proportional to the amount of organic matter fermented.
(2) The attainable rate of protein deposition per unit of energy digested increases with level of feeding and decreases with stage of maturity.
(3) With systems involving early weaning and high intake of cereals, the maximum attainable rate of protein deposition cannot be met by microbial protein alone and protein supplements must be added to the diet. The supplementary protein can be most efficiently utilised if the rumen is by-passed and if the amino acid composition forms an optimum combination with the microbial protein. Protein sources high in sulphur-containing amino acids have been shown to be the most efficiently utilised.
(4) For the vast majority of feeding systems for growing ruminants, the requirement can be met by microbial protein and some of the basal protein which may escape the rumen undegraded. Food intake will be restricted if the microbial requirement is not met, hence the latter must be met even if it provides protein in excess of the host animal's requirements. For each basal feed, the microbial need for nitrogen will depend on the digestibility of the feed's organic matter and the degradability of its protein, and will be relatively independent of the animal to which it is given. If the basal diet has to be supplemented to meet the microbial needs, non-protein nitrogen may be a preferred source.
(5) A system for estimating the requirement has been suggested involving firstly, an assessment of the maximal attainable rate of

protein deposition of the animals concerned with the basal diet to be fed, and secondly, an estimation of the utilisation of the protein supplement. If nitrogen balance trials cannot be conducted, some of this information can be obtained from the food conversion ratio.

References

ANDREWS, R.P. and ØRSKOV, E.R. (1970a). *J. agric. Sci., Camb.*, **75**, 11
ANDREWS, R.P. and ØRSKOV, E.R. (1970b). *J. agric. Sci., Camb.*, **75**, 19
ARMSTRONG, D.G. and ANNISON, E.F. (1973). *Proc. Nutr. Soc.*, **32**, 107
BALCH, C.C. (1967). *Wld. Rev. Anim. Prod.*, **3**, 84
BEEVER, D.E., COEHLO DA SILVA, J.F. and ARMSTRONG, D.G. (1970). *Proc. Nutr. Soc.*, **29**, 43A
BLAXTER, K.L. (1962). *Energy metabolism of ruminants.* Hutchinson, Scientific and Technical, London
BROSTER, W.H., TUCK, V.J., SMITH, T. and JOHNSON, V.W. (1969). *J. agric. Sci., Camb.*, **72**, 13
FAICHNEY, G.J. and LLOYD DAVIES, H. (1972). *Aust. J. agric. Res.*, **23**, 167
FERGUSON, K.A., HEMSLEY, J.A. and REIS, P.J. (1967). *Aust. J. Sci.*, **30**, 215
HOBSON, P.N. (1971). *Progr. Indust. Microbiol.*, **9**, 141
HUME, I.D. (1970). *Aust. J. agric. Res.*, **21**, 297
HUNGATE, R.E. (1966). *The rumen and its microbes.* Academic Press, London and New York
ISHAQUE, M., THOMAS, P.C. and ROOK, J.A.F. (1971). *Nature, New Biology*, **231**, 253
JAGUSCH, K.T., NORTON, B.W. and WALKER, D.M. (1970). *J. agric. Sci., Camb.*, **75**, 279
LEROY, F. and ZELTER, S.Z. (1970). *Annls Biol. anim. Biochim. Biophys.*, **10**, 401
MILLER, E.L. (1973). *Proc. Nutr. Soc.*, **32**, 79
McINTYRE, K.H. and WILLIAMS, V.J. (1970). *Aust. J. agric. Res.*, **21**, 95
NIMRICK, K., HATFIELD, E.E., KAMINSKI, J. and OWENS, F.N. (1970a). *J. Nutr.*, **100**, 1293
NIMRICK, K., HATFIELD, E.E., KAMINSKI, J. and OWENS, F.N. (1970b). *J. Nutr.*, **100**, 1301
NOLAN, J.V., NORTON, B.W. and LENG, R.A. (1973). *Proc. Nutr. Soc.*, **32**, 93
ØRSKOV, E.R. (1970). in *University of Nottingham 4th Nutrition Conference of Feed Manufacturers.* Eds. H. Swan and D. Lewis. p.3. Churchill, London
ØRSKOV, E.R. and BENZIE, D. (1970). *Br. J. Nutr.*, **23**, 415
ØRSKOV, E.R. and BOYNE, A.W. (1973). *Proc. Nutr. Soc.*, **32**, 88A

ØRSKOV, E.R. FRASER, C. and CORSE, E.L. (1970). *Br. J. Nutr.*, **24**, 803
ØRSKOV, E.R., FRASER, C. and KAY, R.N.B. (1969). *Br. J. Nutr.*, **23**, 217
ØRSKOV, E.R. FRASER, C. and McDONALD, I. (1971). *Br. J. Nutr.*, **25**, 243
ØRSKOV, E.R., FRASER, C. and McDONALD, I. (1972). *Br. J. Nutr.*, **27**, 491
ØRSKOV, E.R., FRASER, C.; McDONALD, I. and SMART, R.E. (1974). *Br. J. Nutr.*, **31**, 89
ØRSKOV, E.R., FRASER, C. and PIRIE, R. (1973). *Br. J. Nutr.*, **30**, 361
ØRSKOV, E.R., McDONALD, I., FRASER, C. and CORSE, E.L. (1972). *J. agric. Sci., Camb.*, **77**, 351
OSBORNE, T.B. and MENDEL, L.B. (1918). *J. biol. Chem.*, **35**, 19
REID, J.T., BENSADOUN, A., BULL, L.S., BURTON, J.H., GLEESON, P.A., HAN, L.K., JOO, Y.D., JOHNSON, D.E., McMANUS, W.R., PALADINES, O.L., STROUD, J.W., TYRRELL, H.F., VAN NIEKERK, B.D.H., WELLINGTON, G.H. and WOOD, J.D. (1968). *Proc. Cornell Nutrition Conference*, p.18
SCHOEMAN, E.A., DE WET, P.J. and BURGER, W.J. (1973). *Agroanimalia*, **4**, 35
SMITH, R.H. (1969). *J. Dairy Res.*, **36**, 313
THOMAS, P.C. (1973). *Proc. Nutr. Soc.*, **32**, 85
WALKER, D.J. and NADER, C.J. (1970). *Aust. J. agric. Res.*, **21**, 747

27

FUTURE ROLE OF COMPUTER SIMULATION IN RESEARCH AND ITS APPLICATION TO RUMINANT PROTEIN NUTRITION

J.L. BLACK, G.J. FAICHNEY and N. McC. GRAHAM
C.S.I.R.O., Division of Animal Physiology, Prospect, N.S.W., Australia

Introduction

Special features of protein nutrition in ruminants result from the capacity of rumen micro-organisms to degrade dietary proteins and to incorporate the released ammonia into microbial protein. Breakdown of dietary proteins can lead to a wastage of nitrogen when more ammonia is produced than is needed for microbial growth and the excess is absorbed and excreted as urea in the urine. Hence it is possible for ruminants to suffer from amino acid deficiencies despite high intakes of protein. Such deficiencies can be overcome by the chemical or physical treatment (protection) of proteins or amino acids to reduce their degradation within the rumen. On the other hand, the capacity of rumen micro-organisms to utilise ammonia means that ruminants can be fed non-protein nitrogen (NPN) as a substitute for more expensive proteins. Although the potential of such dietary manipulations for improving the efficiency of ruminant production has long been recognised, they have not yet had widespread successful application in practice.

NPN has extensive commercial use (Chalupa, 1973; Leng *et al.*, 1973), yet the efficacy of its use with many high concentrate diets and in many pastoral situations where animals are grazing on low quality roughages has been seriously questioned (Loosli and McDonald, 1968; Leng *et al.*, 1973). In contrast, protected proteins and amino acids have not been widely adopted, partly because they have been only recently developed but also because of conflicting results from investigations of their use, such as those for calves (Broster, Smith and Broster, 1972; Faichney and Davies, 1973; Davies and Faichney, 1973).

Nevertheless, we see future trends in ruminant protein nutrition to be the refinement and successful application of techniques which make use of the special features of the ruminant digestive system. However, the productive situations in which their use will be beneficial must first be identified and the likely advantages predicted, so that managerial decisions can be soundly based. Because of the complexity of the interactions between an animal, its diet, its physiological state and the environment, it is our view that conventional feeding trials are not the most appropriate way of providing this information. The results obtained from such trials often apply only to the conditions of a particular experiment and to cover all possible productive situations would be extremely time consuming and costly. The purpose of this paper is to illustrate the shortcomings of the conventional approach by considering, as an example, the results of investigations into the effects on wool growth of increasing the absorption of amino acids. We suggest that, as an alternative, research should be orientated toward providing the fundamental information needed for the development of computer models which can realistically simulate animal performance and which can therefore be used to assess the merit of alternative management strategies.

Wool Growth Responses to Increased Absorption of Amino Acids

More experiments have been concerned with the effects of increasing the intestinal absorption of amino acids on wool growth than its effects on any other form of ruminant production. Nevertheless, even after a great deal of experimentation, it remains difficult to predict the magnitude of the response to be expected.

All experiments with non-breeding adult sheep indicate that wool growth is stimulated by the intestinal digestion of casein (Reis and Schinckel, 1961, 1964; Ferguson, Hemsley and Reis, 1967; Egan, 1970; Langlands, 1971; and Barry, 1972). However, increases in wool growth have ranged from 17 to 181 per cent, and the casein intake needed for maximum response has varied from 25 to 120 g/day. The results of Reis (1969) suggest that there would be no response in sheep grazing on high quality pasture. The magnitude of the response appears to be influenced by both the composition and intake of the basal diet. This is supported by the findings of Black, Robards and Thomas (1973) who infused the entire diet of wethers into the abomasum and found an optimum ratio of protein to energy absorption for wool growth. Clearly, wool growth is influenced by net energy intake and the pattern of amino acids absorbed but, as yet, the optimum has only been determined in this one situation.

The physiological state of the animal also affects wool growth responses. For example, with growing lambs, Faichney (1971) found that formaldehyde treatment of a casein supplement increased the intestinal absorption of amino acids, stimulated live weight gain, but had no effect on wool growth. Wright (1971), on the other hand, reported that treated casein increased both wool growth and live weight gain of lambs. With breeding sheep, Barry (1969) reported that, in contrast to ewes fed high intakes of a basal diet, ewes fed low intakes did not grow more wool when supplemented with treated casein, but the birth weights of their lambs increased. Similarly, Henderson et al. (1970) showed that wool growth in ewes responded to treated casein only during early pregnancy and late lactation.

The amino acid composition of proteins digested in the small intestine also influences wool growth. The experiments of Reis and Colebrook (1972) suggest that proteins with a high content of all essential amino acids (casein, egg protein) can produce a substantial increase in the wool growth of non-breeding adult sheep, whereas proteins which are either low in sulphur-amino acids (soyabean protein) or deficient in other essential amino acids (wheat gluten, zein, gelatin) produce a smaller increase or even reduce wool growth. Since the nitrogen: sulphur ratio of microbial protein is high relative to that of wool (Bird, 1973) and since much of the protein digested by sheep is of microbial origin, it is not surprising that increased absorption of sulphur-amino acids alone can produce a marked increase in wool growth (Reis, 1967; Wickham, 1970; Reis, Tunks and Downes, 1973). However, the magnitude of the response is variable and Robards (1971) suggested that it is influenced by the nature of the basal diet. For methionine, but not cystine, the level of absorption also appears to be critical in sheep fed roughage diets, with the response declining as methionine supplementation exceeds 6 g/day. Moreover, Reis and Tunks (1974) have recently found that the wool growth of sheep fed only wheat grain was unaffected by the abomasal administration of cystine and was reduced by the administration of methionine; this situation has not been studied with other grains. Despite extensive research, the precise mixture of amino acids required in the diet for maximum wool growth is not known for any situation.

In addition, it has been found that low temperatures (Downes and Hutchinson, 1969) and photoperiodic changes (Hutchinson, 1965) can affect wool growth. Furthermore, during heat stress, many physiological functions are altered and wool growth can be depressed (Thwaites, 1968). However, the impact of such factors on the response to increased amino acid absorption has not been investigated.

It is apparent that we can recognise a number of situations where an increase in wool growth would probably result from an increased absorption of amino acids, but there remain several areas of uncertainty. This applies particularly to growing, pregnant and lactating animals and to those diets with which sulphur-amino acid supplementation does not produce a response. Even when a response is expected, its magnitude cannot be predicted because, firstly, a significant quantity of vital information is missing and, secondly, even if all relevant information was available, the situation is too involved for the necessary calculations to be done by hand. Nevertheless, success in the use of NPN and widespread acceptance of protected amino acids in any form of ruminant production is unlikely until these problems can be solved.

Computer Simulation of Ruminant Production

Because of the complexity of the interactions between an animal, its diet and the environment, we suggest that computer programmes which simulate the animal provide a practical way of predicting performance under any given conditions. If models are based on an understanding of the mechanisms involved in these interactions, they can be used to predict the outcome of variations in any of the determinants of animal performance and thereby provide the background necessary for sound managerial decisions. In addition, such models can be useful in providing guidelines for determining the priorities of future research.

Computer programmes which simulate animal systems can be constructed at many levels. To be of use in solving the problems outlined above a model would need to include: (1) quantitative prediction of the major nutrients absorbed by the animal; (2) quantitative assessment of the utilisation of each absorbed nutrient as it is affected by the physiological state of the animal and the environment. Given starting conditions to represent the state of an animal and its environment, such a model can calculate, for a unit of time, protein and energy deposition in body tissues, wool, foetus and milk and so provide a new set of starting conditions for the next unit of time. An important feature of a computer model is the ease with which a long sequence of such calculations may be performed to simulate the lifetime performance of an animal. Transient changes in factors affecting production can be allowed for in these calculations. For example, pre-determined patterns of change in diet and environment can be introduced and, in addition, account can be taken of variation in basic mechanisms between and within animals. Furthermore, allowance can be made for events of uncertain cause but known probability of occurrence.

Quantitative prediction of nutrient absorption

To predict the quantity of nutrients absorbed by ruminants it would be necessary to develop sub-models for the stomach, small intestine and large intestine. This requires, for each section, consideration of the inflow of material, the reactions occurring, the resulting absorption and the outflow of material.

For the stomach, the factors to be considered include food intake, the chemical and physical nature of the diet, the pattern and frequency of feeding and the entry of nitrogen, sulphur and other materials from the body. The products of fermentation, and the yield of microbial cells, can be assessed from the stoichiometry of the chemical reactions as determined by the substrates fermented, and the strains and growth rates of micro-organisms present. Factors to be taken into account here include the supply of essential nutrients such as nitrogen, sulphur and branched chain volatile fatty acids (VFA) as well as rumen pH, the rate of passage of material from the stomach and the physical and chemical nature of the diet. These calculations allow the determination of both the absorption of VFA and ammonia from the stomach, and the quantity and composition of dietary and microbial material flowing to the small intestine. At this point the potential response to dietary supplementation with NPN or sulphur can be assessed.

Much of the quantitative information needed for these sub-models is already available (Nolan and Leng, 1972; Waldo, Smith and Cox, 1972; Hobson, 1972; Miller, 1973; Thomas, 1973). Indeed, Baldwin,

Lucas and Cabrera (1970) have shown that the construction of models at this level of complexity for the rumen is quite feasible. They have used such a model to predict the effects of dietary manipulation on the production of VFA, methane, heat loss associated with fermentation and the yield of microbial cells. The predictions for several diets agree reasonably well with experimental observations. The model predicted that grinding lucerne hay reduced the production of VFA and the yield of microbial cells by about 15 per cent because of the increased passage of unfermented holocellulose from the rumen. This prediction is consistent with the results of Hinders and Owen (1968) which showed that grinding lucerne reduced both the ruminal digestion of the fibrous portion of the diet and VFA concentration in cattle. Again, in agreement with experimental observations, the model predicted that feeding the same total quantity of lucerne hay once daily instead of twice daily reduced the digestion of holocellulose immediately post-feeding because of a decrease in the population of cellulolytic microorganisms. Lower yields of VFA and microbial cells were predicted for a straw plus urea diet than for lucerne, and the effect of removing urea on these parameters was assessed. The model was also able to predict that a reduction in the rate of conversion of urea to ammonia would stimulate the production of microbial cells when a straw plus urea diet was given.

These early attempts at modelling the stomach are encouraging. Inevitably several areas are still obscure, as indicated by Baldwin, Lucas and Cabrera (1970), but the problems are readily amenable to appropriate research. One problem not discussed by Baldwin concerns the changes, apparently spontaneous, which sometimes occur in microbial populations and may be responsible for marked fluctuations which have been observed in the flow of nitrogen to the intestines (Ishaque, Thomas and Rook, 1971; Faichney, 1974). Thus, in studies of the process of digestion, divergent results are often obtained in successive periods of observation with one animal and diet, or in simultaneous observations on two animals treated identically. If the probability of such changes occurring can be stated it is relatively easy to programme a computer to select appropriate predictive equations and print out either the most likely response to a diet or a list of possible responses. Evaluation of such a time series by hand would be so laborious as never to be attempted. Another problem is the estimation of the extent of degradation of feedstuffs within the rumen and the digestion of that dietary material passing to the intestines. These are influenced by many factors and can often be extremely variable for the one type of diet. As suggested by Miller (1973), chemical or *in vitro* methods may be required to adequately assess this situation.

Quantitative assessment of the utilisation of absorbed nutrients

It is necessary to know the major pathways by which absorbed nutrients are utilised in order to predict the balances of fat and protein within the body and the quantity and composition of products formed. Fluxes through these pathways are influenced by the amount of individual nutrients absorbed in relation to the needs of an animal. When nutrients are limiting, the competition between tissues for these nutrients becomes important. For example, in pregnant ewes it appears that the demand of the foetus for amino acids takes precedence over that of the wool follicle (Barry, 1969). Similarly, during early lactation, body tissues are often mobilised to support milk production (Broster, 1973). Furthermore, some nutrients are often preferentially utilised by tissues to fulfil particular needs. During lipogenesis, absorbed fats are more likely to be deposited in adipose tissue than to be oxidised to supply ATP requirements (Chudy and Schiemann, 1969). Glucose, in ruminants, is unlikely to be used for lipogenesis because of the very low levels of the enzyme ATP citrate lyase (Ballard, Hanson and Kronfeld, 1969). The physiological state of an animal and the environmental conditions to which it is exposed have major influences on the utilisation of individual nutrients.

A large number of studies with radioactive tracers have been used to determine the fluxes of individual nutrients through various metabolic pathways (e.g. Ford and Reilly, 1969; Annison and Armstrong, 1970; Wolff and Bergman, 1972). However, these results refer only to the particular experimental conditions used. Unfortunately, our knowledge of the mechanisms which control the fluxes of absorbed nutrients between various tissue pools in ruminants is limited and further research is needed. Hence there is not yet sufficient information available for the construction of a comprehensive model of a whole animal, although limited models of parts of the system have been attempted (Baldwin and Smith, 1971). Nevertheless, there is at present sufficient information to construct useful models which are based on superficial correlations.

An animal's utilisation of absorbed nutrients can be defined in terms of, firstly, the energy needed for maintenance, activity, regulation of body temperature and net synthesis of body tissues and secondly, the substrates, such as amino acids, needed in the synthesis of tissues and products. There is at present considerable information in a number of these areas for sheep. Graham, Searle and Griffiths (1974) have studied in detail the basal energy expenditure of sheep and shown it to be a function of body weight, age, growth rate and digestible energy intake immediately prior to its measurement. Energy expenditure during feeding, rumination and locomotion has been documented (Arnold, 1962; Blaxter, 1962; Graham, 1966). Similarly, the energy expenditure of sheep in the cold can be calculated from considerations of the amount and type of diet eaten, length of fleece, wind speed and ambient temperature (Graham, 1966). However, there is not sufficient information to reliably calculate energy expenditure associated with high ambient temperatures and high radiation load, and little is known of the energy cost of protein synthesis. Energy balance can be predicted from energy intake and energy expenditure as described by Blaxter (1962).

Information is available which permits the prediction of nitrogen balance and hence net protein synthesis in penned lambs in thermo-neutral conditions (Black and Griffiths, 1975). Equations for the prediction of nitrogen balance were established for liquid-fed lambs, ranging in weight from 3 to 38 kg, but with an appropriate adjustment for the loss of endogenous nitrogen in faeces (Black, Pearce and Tribe, 1973), they can also be applied to ruminant lambs. The greatest nitrogen balance (NB_N) that could result from the nitrogen absorbed (NA) if energy intake was not limiting the response is given by the equation:

$$NB_N = BV.NA - EN \qquad (1)$$

where BV is biological value of absorbed protein and EN is the total endogenous loss of nitrogen in urine and faeces. If energy intake is limiting the response, nitrogen balance (NB_E) will be less and given by the equation:

$$NB_E = a\,NE - bW^{0.75} - c\,NE.W^{0.75} + d\,(W^{0.75})^2 \qquad (2)$$

where NE is estimated net energy intake, W is body weight, and a, b, c, and d are constants. There remains a lack of knowledge of the requirements for individual amino acids and the variation which may result from changes in live weight and energy intake. This knowledge is essential for the realistic assessment of the biological value of absorbed nitrogen. In addition, more quantitative information is required about the effects on protein synthesis of high energy expenditure during either cold exposure or increased physical activity.

Although the effects of protein and energy absorption on wool growth in adult sheep have been investigated (Black, Robards and Thomas, 1973), there is not sufficient information to predict the wool growth of sheep in various physiological states and environmental conditions. When protein supply is inadequate, amino acids appear to be preferentially used for body tissue gain (Black, Robards and Thomas, 1973), foetal growth (Barry, 1969) and milk formation (Henderson et al., 1970) rather than for wool growth. Wool growth does not stop in these situations, but the manner in which the absorbed amino acids are partitioned between the various possible sites of deposition needs elucidation. The relative effects of protein and energy absorption in pregnant and lactating animals and the extent to which body protein reserves are used to sustain foetal growth and milk production also need further research.

Application of an Animal Model

Computer models which are sufficiently comprehensive to fully answer questions regarding the productive situations where either NPN or 'protected' amino acids may have a role and the magnitude of response resulting from their use can not yet be developed. Nevertheless, with the information currently available, Graham, Black and Faichney (unpublished) have constructed a model to simulate growth in lambs. At present the model does not deal separately with reactions within the rumen nor with the utilisation of individual nutrients in the calculation of protein and energy balances. The absorption and utilisation of nutrients is calculated either from correlations such as those mentioned above or from relevant experimental results. This model is now used to illustrate the way in which computer models may assist in improving the efficiency of ruminant protein nutrition. The examples given are concerned with growth in lambs rather than wool growth, because there is not yet enough quantitative data to obtain satisfactory predictive relationships for the wool growth responses to absorbed nutrients.

Faichney (1971) grew two groups of lambs from 20 to 40 kg with a diet containing 54 per cent roughage, 36 per cent high starch concentrates and 10 per cent casein. The daily dry matter intake was 3.78 per cent of live weight. The casein in the diet fed to one group was treated with formaldehyde and this group grew significantly faster than the other group which was fed untreated casein. In the lambs fed the untreated diet, 148 g of non-ammonia crude protein was digested in the small intestine per kg of digested organic matter (DOM), whereas the value was 239 g/kg DOM for lambs given the treated diet (Faichney and Weston, 1971). It was assumed that amino acids account for 80 per cent of digested crude protein (20 per cent nucleic acids; Smith and McAllan, 1970) and that the biological value of absorbed true protein is 0.8 (Hungate, 1966); the model was then used to predict the body weight up to which lambs given the untreated diet were deficient in protein and the increased growth rate that would result from protein treatment. However, some of the dietary protein (lucerne and maize) would have escaped fermentation in the rumen (Hungate, 1966) and the absorption of protein by lambs given the untreated diet would have been greater than would occur if all the digested protein was of microbial origin. For lambs dependent solely on microbial protein, the body weight up to which they would be deficient in protein was also predicted. The absorption of crude protein in these lambs was estimated at 114 g/kg DOM by assuming that 90 per cent of dietary organic matter is digested in the rumen (Hume, Moir and Somers, 1970), that 27 g microbial nitrogen was synthesised per kg organic matter fermented (Thomas, 1973) and that 75 per cent of microbial nitrogen was digested in the small intestine (Coelho da Silva *et al.*, 1972)

In *Figure 27.1* the continuous line represents the nitrogen balance (NB_E) determined by net energy intake and live weight (i.e. if nitrogen absorption was in excess of requirement) and broken lines represent the nitrogen balances (NB_N) that would result if nitrogen absorption was inadequate at the three levels of protein absorption. Thus, for any one situation, the actual nitrogen balance is given by the lower of the continuous and appropriate broken line, and the difference between them divided by biological value represents the deficiency or excess of absorbed nitrogen. When given treated casein, the lambs would not have been deficient in protein at any body weight, but when given the untreated casein diet, they would have been deficient in protein up to a body weight of 25 kg. Animals digesting only microbial protein would have been deficient in protein up to a body weight of 38 kg. The predicted live weight gains of the lambs given

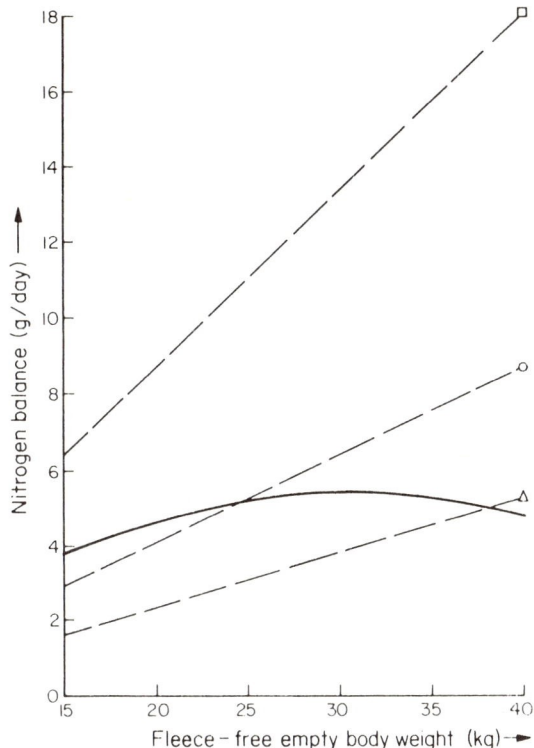

Figure 27.1 Effect of body weight on the predicted nitrogen balance of lambs when either net energy intake (———) or nitrogen absorption (— — —) is limiting the response for the levels of absorption resulting from a treated casein diet (□), an untreated casein diet (○) and microbial protein (△). Nitrogen balance for each level of absorption is given by the lower of the continuous or broken lines

the treated and untreated diets where, respectively, 165 and 155 g/day over the weight range from 15 to 40 kg. These are almost identical to the values of 166 and 154 g/day observed by Faichney (1971). The model provides considerably more information on which managerial decisions can be based than could simply be obtained from the original experimental data. The prediction indicates that, at 15 kg, lambs receiving the untreated diet required an additional absorption of 7.0 g/day of casein and that this would have resulted in an increased growth rate of 27 g/day. Similarly, if the protein digested by the lambs had been solely of microbial origin, they would have grown at 112 g/day, and at 15 kg, could have used an extra 17.5 g/day of absorbed casein to increase growth rate by 70 g/day.

Faichney (1971) fed lambs at only one level of intake but, using the model, it is possible to predict the likely effect of altering the plane of nutrition on the body weight at which lambs given the untreated casein diet would cease to respond to additional protein absorption. Assuming that 148 g crude protein was absorbed per kg

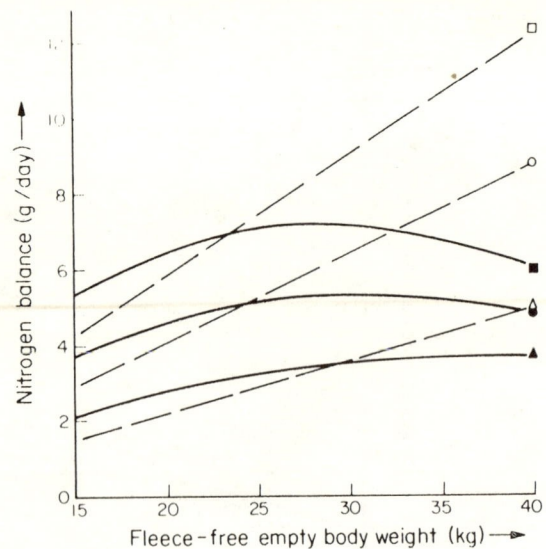

Figure 27.2 Effect of body weight on the predicted nitrogen balance of lambs when either net energy intake (———) or nitrogen absorption (– – –) is limiting the response for three levels of dry matter intake [5 per cent (□), 3.78 per cent (○) and 2.5 per cent (△) of live weight per day] of an untreated casein diet. Nitrogen balance for each level of intake is given by the lower of the continuous or broken lines

DOM irrespective of energy intake, the predicted effects of three levels of feeding (2.5, 3.78 and 5.0 per cent of live weight per day) on nitrogen balance are shown in *Figure 27.2*. The body weight below which lambs would have been deficient in protein, decreased from 29 kg with an intake of 2.5 per cent of live weight, to 23 kg when intake was 5 per cent of live weight. This example illustrates the errors inherent in extrapolating the results beyond the conditions under which they were originally obtained. To discover the magnitude of these differences experimentally would require considerable effort. The predicted effect of increasing intake on the weight of body fat in lambs growing from 15 to 40 kg is illustrated in *Figure 27.3*. This suggests that raising the plane of nutrition would increase the amount of fat in the body of lambs at any weight, and that the major effect would occur as feed intake is raised from 2.5 to 3.78 per cent of body weight. Similar predictions of the effects of energy intake on body composition of lambs given adequate or near adequate levels of protein have been presented elsewhere (Black, 1974).

Nevertheless, the predicted increase in body fat with increasing level of feeding is inconsistent with the results of Andrews and Ørskov (1970), who found that increasing the intake of a barley-soyabean diet decreased the amount of fat in lambs of similar weight. However, Ørskov and Fraser (1973) showed recently that the amount of crude

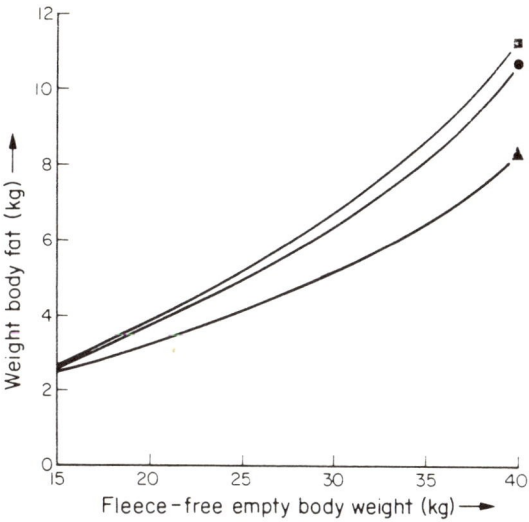

Figure 27.3 The predicted effect of body weight on the weight of body fat in lambs fed three levels of dry matter intake [5 per cent (■), 3.78 per cent (●) and 2.5 per cent (▲) of live weight per day] of an untreated casein diet

protein absorbed in the small intestine per kg DOM increased as the intake of a similar barley-soyabean diet was raised, presumably because more dietary protein escapes degradation. It can be calculated from Ørskov and Fraser (1973) that the amount of crude protein digested in the small intestine per kg DOM probably increased from less than 105 g to more than 170 g as the intake of a high protein diet was raised from 280 to 1110 g/day. If it is assumed that in the experiment of Andrews and Ørskov (1970) the amount of crude protein absorbed from the small intestine in lambs given the three levels of intake were 105, 140 and 170 g/kg DOM, the predicted effects on nitrogen balance and body composition are shown respectively in *Figures 27.4* and *27.5*. The predictions suggest that lambs on the lowest level of intake would have been deficient in protein up to body weights of at least 40 kg, whereas those on the medium level of intake would have been deficient up to 28 kg and lambs on the highest intake would not have been deficient at any body weight (*Figure 27.4*). It is also predicted that the lambs given the lowest level of energy intake would have been the fattest (*Figure 27.5*) as was in fact found (Andrews and Ørskov, 1970). In general, the model suggests that, at any body weight, animals given protein deficient diets will contain more body fat than animals given adequate protein intakes (Black, 1974), but energy intake is a confounding factor and the outcome of any particular situation is difficult to predict without the aid of a computer. Although an increase in protein digestion in the intestines offers a plausible explanation for the experimental observations of Andrews and Ørskov (1970), this result cannot necessarily be extended to other situations. Miller (1973) found that doubling the

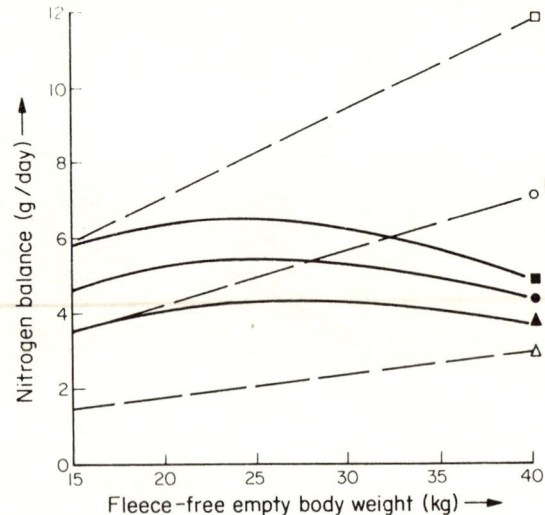

Figure 27.4 Effect of body weight on the predicted nitrogen balance of lambs when either net energy intake (———) or nitrogen absorption (— — —) is limiting the response for the high (□), medium (0) and low (△) levels of intake fed to lambs by Andrews and Ørskov (1970) when it is assumed that the crude protein digested in the small intestine per kg organic matter digested was, respectively, 170 g, 140 g and 105 g. Nitrogen balance for any one intake is given by the lower of the continuous or broken lines

intake of a diet containing sunflower seed meal did not increase the amount of meal escaping degradation in the rumen. This observation illustrates the importance of accurate prediction of the net result of reactions within the rumen.

These few examples show how computer models could be used to improve the efficiency of animal production. With the development of more comprehensive models the effects on animal performance of altering dietary or environmental conditions could be rapidly assessed and the most suitable managerial strategy readily determined. Economic factors can be introduced into models and the financial implications of any management decision could therefore be readily assessed. Hence in the future it is possible that computer models can be used to provide information on such diverse questions as the likely economic advantage of providing an NPN supplement to cattle grazing dry pastures in Northern Australia, or the probability that a certain proportion of a sheep flock will die following a May shearing in Wales.

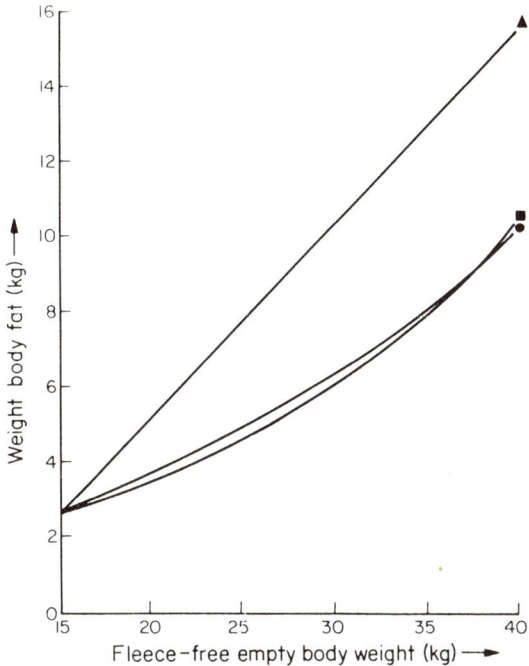

Figure 27.5 The predicted effect of body weight on the weight of body fat in lambs fed three levels of intake. Symbols are the same as in Figure 27.4

Conclusions

Because of the complexity of the interactions between an animal, its diet, its physiological state and the environment, we believe comprehensive computer models to be the only way in which animal requirements and performance can be satisfactorily predicted in any given circumstance. Future research in ruminant protein nutrition should be designed to provide information needed for the construction of comprehensive computer models. These models could then be used to define those situations in which particular dietary manipulations are applicable and to assess the magnitude of the likely benefit, thereby providing a sound basis for managerial decisions. With the development of national computer networks, the availability of a comprehensive model at a central location would facilitate, for individual producers scattered throughout a country, the formulation of rations from the feedstuffs they have available, lead to improvements in management and optimise the efficiency of resource utilisation.

References

ANDREWS, R.P. and ØRSKOV, E.R. (1970). *J. agric. Sci., Camb.*, **75**, 19
ANNISON, E.F. and ARMSTRONG, D.G. (1970). in. *Physiology of Digestion and Metabolism in the Ruminant*. p.422. Ed. A.T. Phillipson. Oriel Press, Newcastle-upon-Tyne
ARNOLD, G.W. (1962). *J. Br. Grassl. Soc.*, **17**, 41
BALDWIN, R.L., LUCAS, H.L. and CABRERA, R. (1970). in *Physiology of Digestion and Metabolism in the Ruminant*. p.319. Ed. A.T. Phillipson. Oriel Press, Newcastle-upon-Tyne
BALDWIN, R.L. and SMITH, N.E. (1971). *J. Dairy Sci.*, **54**, 583
BALLARD, F.J., HANSON, R.W. and KRONFELD, D.S. (1969). *Fedn Proc. Fedn Am. Socs exp. Biol.*, **28**, 218
BARRY, T.N. (1969). *Proc. N.Z. Soc. Anim. Prod.*, **29**, 218
BARRY, T.N. (1972). *N.Z. J. agric. Res.*, **15**, 107
BIRD, P.R. (1973). *Aust. J. biol. Sci.*, **26**, 1429
BLACK, J.L. (1974). *Proc. Aust. Soc. Anim. Prod.*, **10**, 211
BLACK, J.L., PEARCE, G.E. and TRIBE, D.E. (1973). *Br. J. Nutr.*, **30**, 45
BLACK, J.L., ROBARDS, G.E. and THOMAS, R. (1973). *Aust. J. agric. Res.*, **24**, 399
BLACK, J.L. and GRIFFITHS, D.A. (1975). *Br. J. Nutr.*, **33**, 399
BLAXTER, K.L. (1962). *The Energy Metabolism of Ruminants*. Hutchinson, London
BROSTER, W.H. (1973). *Proc. Nutr. Soc.*, **32**, 115
BROSTER, W.H. SMITH, T. and BROSTER, V.J. (1972). *Bienn. Rep. natn. Inst. Res. Dairying*, 69
CHALUPA, W. (1973). *Proc. Nutr. Soc.*, **32**, 99
CHUDY, A. and SCHIEMANN, R. (1969). *Publ. Eur. Ass. Anim. Prod.*, **12**, 161
COELHO DA SILVA, J.F., SEELEY, R.C., BEEVER, D.E., PRESCOTT, J.H.D. and ARMSTRONG, D.G. (1972). *Br. J. Nutr.*, **28**, 357
DAVIES, H.L. and FAICHNEY, G.J. (1973). *Aust. J. exp. Agric. Anim. Husb.*, **13**, 142
DOWNES, A.M. and HUTCHINSON, J.C.D. (1969). *J. agric. Sci.*, **72**, 155
EGAN, A.R. (1970). *Aust. J. agric. Res.*, **21**, 85
FAICHNEY, G.J. (1971). *Aust. J. agric. Res.*, **22**, 453
FAICHNEY, G.J. (1974). *Proc. Aust. Soc. Anim. Prod.*, **10**, 398
FAICHNEY, G.J. and DAVIES, H.L. (1973). *Aust. J. agric. Res.*, **24**, 613
FAICHNEY, G.J. and WESTON, R.H. (1971). *Aust. J. agric. Res.*, **22**, 461
FERGUSON, K.A., HEMSLEY, J.A. and REIS, P.J. (1967). *Aust. J. Sci.*, **30**, 215
FORD, E.J.H. and REILLY, P.E.B. (1969). *Res. Vet. Sci.*, **10**, 409
GRAHAM, N. McC. (1966). *Proc. Aust. Soc. Anim. Prod.*, **6**, 364
GRAHAM, N. McC., SEARLE, T.W. and GRIFFITHS, D.A. (1974). *Aust. J. agric. Res.*, **25**, 957
HENDERSON, A.E., BRYDEN, J.McG., MAZZITELLI, F.E. and MILLIGAN, K.E. (1970). *Proc. N.Z. Soc. Anim. Prod.*, **30**, 186
HINDERS, R.G. and OWEN, F.G. (1968). *J. Dairy Sci.*, **51**, 1253
HOBSON, P.N. (1972). *Proc. Nutr. Soc.*, **31**, 135

HUME, I.D., MOIR, R.J. and SOMERS, M. (1970). *Aust. J. agric. Res.*, **21**, 283

HUNGATE, R.E. (1966). *The Rumen and Its Microbes.* Academic Press, New York

HUTCHINSON, J.C.D. (1965). in *Biology of the Skin and Hair Growth.* p.565. Eds. A.G. Lyne and B.F. Short. Angus and Robertson, Sydney

ISHAQUE, M., THOMAS, P.C. and ROOK, J.A.F. (1971). *Nature, New Biol.*, **231**, 253

LANGLANDS, J.P. (1971). *Aust. J. exp. Agric. Anim. Husb.*, **11**, 9

LENG, R.A., MURRAY, R.M., NOLAN, J.V. and NORTON, B.W. (1973). *Aust. Meat Res. Committee Rev.*, No.15, 1

LOOSLI, J.K. and McDONALD, I.W. (1968). *F.A.O. Agric. Studies*, No.75

MILLER, E.L. (1973). *Proc. Nutr. Soc.*, **32**, 79

NOLAN, J.V. and LENG, R.A. (1972). *Br. J. Nutr.*, **27**, 177

ØRSKOV, E.R. and FRASER, C. (1973). *Proc. Nutr. Soc.*, **32**, 68A

REIS, P.J. (1967). *Aust. J. biol. Sci.*, **20**, 809

REIS, P.J. (1969). *Aust. J. biol. Sci.*, **22**, 745

REIS, P.J. and COLEBROOK, W.F. (1972). *Aust. J. biol. Sci.*, **25**, 1057

REIS, P.J. and SCHINCKEL, P.G. (1961). *Aust. J. agric. Res.*, **12**, 335

REIS, P.J. and SCHINCKEL, P.G. (1964). *Aust. J. biol. Sci.*, **17**, 532

REIS, P.J. and TUNKS, D.A. (1974). *Aust. J. agric. Res.*, **25**, 919

REIS, P.J., TUNKS, D.A. and DOWNES, A.M. (1973). *Aust. J. biol. Sci.*, **26**, 249

ROBARDS, G.E. (1971). *Aust. J. agric. Res.*, **22**, 261

SMITH, R.H. and McALLAN, A.B. (1970). *Br. J. Nutr.*, **24**, 545

THOMAS, P.C. (1973). *Proc. Nutr. Soc.*, **32**, 85

THWAITES, C.J. (1968). *Proc. Aust. Soc. Anim. Prod.*, **7**, 259

WALDO, D.R., SMITH, L.W. and COX, E.L. (1972). *J. Dairy Sci.*, **55**, 125

WICKHAM, G.A. (1970). *Proc. N.Z. Soc. Anim. Prod.*, **30**, 209

WOLFF, J.E. and BERGMAN, E.N. (1972). *Am. J. Physiol.*, **223**, 447

WRIGHT, P.L. (1971). *J. Anim. Sci.*, **33**, 137

LIST OF DELEGATES

Adamson, A.H.	A.D.A.S. (M.A.F.F.), Government Buildings, Lawnswood, Leeds
Aelten, G. van	Aan-en Verkoopvennootschap van de Belgische Boerenbond N.V., Merksem, Belgium
Annison, Dr E.F.	Unilever Research Laboratory, Colworth House, Sharnbrook, Bedford
Armstrong, Prof. D.	Department of Agricultural Biochemistry, University of Newcastle-upon-Tyne
Arnal, M.	Laboratoire d'Étude du Metabolisme Azoté, I.N.R.A., Beaumont, France
Baardseth, I.P.	Norwegian Food Research Institute, P.O.50. 1432. AS-NLH, Norway
Banton, C.L.	Trouw-G.B.-Ltd., Harston, Cambridge
Basson, W.D.	Research Institute for Animal and Dairy Science, Irene, South Africa
Beever, Dr D.E.	Grassland Research Institute, Hurley, Maidenhead, Berks.
Bellingham, Dr F.	Calor Agriculture Limited, Slough, Bucks.
Benedictus, N.	Centraal Veevoederbureau, Bornsesteeg 45, Wageningen, Netherlands
Ben-Ghedalia, Dr D.	Department of Agricultural Biochemistry, University of Newcastle-upon-Tyne
Berende, P.	ILOB, Wageningen, Haarweg 8, Netherlands
Bertram, H.L.	Degussa, Hanau, Frankfurt, Germany
Bican, Dr P.	Swiss Federal Research Station for Animal Production, Liebefeld, Berne, Switzerland

List of delegates

Bielorai, Dr Rachael	Division of Animal Nutrition, Agricultural Research Organisation, Volcani Centre, Beit Dagan, Israel
Bjørnstad, J.	Department of Poultry Science, Box 17, Ås-NLH, Norway
Black, Dr J.L.	C.S.I.R.O., Division of Animal Physiology, Prospect, Blacktown, Australia
Blanc, B.	Station Fédérale de Recherches Laitières, Liebefeld, Berne, Switzerland
Boekholt, H.A.	Agricultural University, Wageningen, Haarweg 10, Netherlands
Boer, F. de	Institute on Animal Feeding and Nutrition Research, Keern 33, Hoorn, Netherlands
Boeve, J.	Provimi, P.O. Box 5062, Rotterdam, Netherlands
Bond, K.I.	J. Bibby Agriculture Ltd., Farm Products Division, Richmond House, Rumford Place, Liverpool
Boorman, Dr K.N.	University of Nottingham School of Agriculture. Sutton Bonington, Loughborough, Leicestershire
Boxall, Dr R.C.	Spillers Ltd., Kennett Nutritional Centre, Kennett, Newmarket, Suffolk
Braude, Dr R.	National Institute for Research in Dairying, Shinfield, Reading, Berkshire
Brenninkmeijer, Dr C.	Hendrix' Voeders B.V., Wim de Körverstraat 39, Boxmeer, Netherlands
Broome, Dr A.W.V.	I.C.I. Ltd., Pharmaceuticals Division, Mereside, Alderley Park, Macclesfield, Cheshire
Bruckental, Dr I.	Institute of Animal Science, Agricultural Research Organisation, Volcani Centre, Beit Dagan, Israel
Bruel, A. van den	D.S.M. Central Laboratory, P.O. Box 18, Geleen, Netherlands

List of delegates 495

Bujard, Eliane	Research Laboratory, Nestle's Products, La Tour de Peilz, Switzerland
Burger, Dr I.H.	Animal Studies Centre, Pedigree Petfoods Ltd., Melton Mowbray, Leicestershire
Burns, R.A.	Department of Agriculture and Food Chemistry, Queen's University, Belfast
Buttery, Dr P.J.	University of Nottingham School of Agriculture, Sutton Bonington, Loughborough, Leicestershire
Caffrey, Dr P.	University College, Lyons, Newcastle P.O., Co. Dublin, Eire
Caine, S.	Department of Biochemistry, Royal Infirmary, Glasgow
Chaudhry, S.A.	F.A.O. Veterinary Faculty, Royal Veterinary and Agricultural University, Copenhagen, Denmark
Calet, Dr C.	Station de Recherches Avicoles INRA, Nouzilly, France
Carlyle, E.	Department of Biochemistry, Royal Infirmary, Glasgow
Carpenter, Dr K.J.	Department of Applied Biology, Cambridge University, Downing Street, Cambridge
Carr, J.R.	Pig Research Centre, Massey University, Palmerston North, New Zealand
Chalmers, Dr Margaret I.	Rowett Research Institute, Bucksburn, Aberdeen
Chamberlain, Dr A.G.	University College of North Wales, Bangor, Gwynedd, Wales
Chamberlain, D.G.	The Hannah Research Institue, Ayr, Scotland
Champredon, C.	Centre de Recherches I.N.R.A., Beaumont, France
Clark, R.D.	Beecham Agricultural Products, Great West Road, Brentford, Middlesex

List of delegates

Clemens, Dr M.J.	National Institute for Medical Research, Mill Hill, London
Cole, Dr D.J.A.	University of Nottingham School of Agriculture. Sutton Bonington, Loughborough, Leicestershire
Cooke, Dr B.C.	Crosfields House, The Promenade, Bristol
Court, R.D.	Division of Tropical Agronomy, Cunningham Laboratories, C.S.I.R.O., Lucia, Queensland, Australia
Cuperlović, Dr M.	Institute for the Application of Nuclear Energy in Agriculture, Veterinary Medicine and Forestry, Zemun, Baranjska 15, Yugoslavia
Cuthbertson, Sir David	Department of Pathological Biochemistry, University of Glasgow and Glasgow Royal Infirmary
Dammers, Dr J.	Koninklijke Wessanen N.V., Wormerveer, Netherlands
Davies, A.	I.C.I. Ltd., Pharmaceuticals Division, Alderley Park, Macclesfield, Cheshire
Dekker, J.	Koopmansmeelfabrieken B.V., Leeuwarden, Netherlands
Delort-Laval, J.	INRA, Laboratoire de Technologie des Aliments des Animaux, Jouy en Josas, France
Donaldson, I.A.	A.R.C. Institute for Research on Animal Diseases, Compton, Berkshire
Dumont, Rollande	Chare des Productions Animales, E.N.S.S.A.A., Dijon, France
Eggum, Dr B.O.	Department of Animal Physiology, Biochemistry and Analytical Chemistry, Rolighedsvej, Copenhagen, Denmark
Eikelenboom, Dr G.	Research Institute for Animal Husbandry, Driebergseweg, Zeist, Netherlands

List of delegates 497

Elsley, Prof. F.W.H.	The Edinburgh School of Agriculture, West Mains Road, Edinburgh
Erbersdobler, Prof. H.	Institut für Tierphysiologie der Universität München, München, West Germany
Es, Dr A.J.H. van	Institute for Feeding and Nutrition Research, Postbus 67, Hoorn, Netherlands
Evans, Dr P.J.	Unilever Research, Colworth House, Sharnbrook, Bedford
Fahnenstich, Dr R.	Degussa, Postfach 602, Hanau, Frankfurt, Germany
Finot, Dr P.A.	Research Laboratory, Nestle's Products, La tour de Peilz, Switzerland
Fisher, Dr C.	Poultry Research Centre, Roslin, Midlothian
Fleck, Dr A.	Department of Biochemistry, Royal Infirmary, Glasgow
Ford, Dr J.E.	National Institute for Research in Dairying, Shinfield, Reading, Berkshire
Frape, Dr D.L.	Spillers Ltd., Kennett Nutrition Centre, Kennett, Newmarket, Suffolk
Fuller, Dr M.F.	Rowett Research Institute, Bucksburn, Aberdeen
Gergely, Dr Adrienne	Phylaxia, Szallas.u.5-7, Budapest, Hungary
Gericke, Dr H.	Bayer AG, Veterinär-Entwicklung, 56 Wuppertal 1, Postfach 13, West Germany
Glyn-Jones, J.	Calor Agriculture Ltd., Calor House, Windsor Road, Slough, Buckinghamshire
Gotthardsson, S.I.	SHS, Eskilstuna, Sweden
Granborg, M.	F.D.B.s Centrallaboratorium, Albertslund, Denmark
Grankuist, H.B.	Lantbrukshogskolan Kungsängens gård, Uppsala, Sweden

498 *List of delegates*

Griffiths, Dr T.W.	Agricultural Institute, Dunsinea, Castleknock, Dublin, Eire
de Groote, G. Jnr.	Ryksstation voor Kleinveeteelt, Laan-9220, Merelbeke, Belgium
Hagemeister, Dr H.	Bundesanstalt für Milchforschung, Hermann-Weigmann-Strasse, Kiel, West Germany
Hall, Dr T.C.	University of Wisconsin, Madison, U.S.A.
Hansen, N.G.	National Research Institute of Animal Nutrition, Roskildevej, Hilleroed, Denmark
Harris, Dr C.I.	Rowett Research Institute, Bucksburn, Aberdeen
Harrison, D.G.	Grassland Research Institute, Hurley, Berkshire
Henderickx, Prof. H.K.	Faculty of Agricultural Science, University of Ghent, Bosstraat 1, Melle, Belgium
Hillcoat, J.B.	Unilever Research, Colworth House, Sharnbrook, Bedford
Hiller, Dr Günter	Henkel and Cie. GmbH, Düsseldorf, Henkelstrasse 67, West Germany
Hof, G.	Agricultural University, Haagsteeg no.4, Wageningen, Netherlands
Hogan, Dr J.P.	C.S.I.R.O., Ian Clunies Ross Animal Research Laboratories, Blacktown, New South Wales, Australia
Holm, Dr H.	Institute for Nutrition Research, School of Medicine, University of Oslo, Blindern, Oslo, Norway
Holmes, J.J.	E.B. Bradshaw and Sons Ltd., Bell Mills, Driffield, Humberside
Homb, Prof. Thor	Agricultural University of Norway, Vollebekk, Norway
Hunt, M.	Feed Services (Livestock) Ltd., Corsham, Wiltshire

List of delegates 499

Jensen, K.	Royal Agricultural and Veterinary University, Bulowsvej, Copenhagen, Denmark
Juhasz, Maria	Phylaxia, Szallas.u.5-7, Budapest, Hungary
Kan, Dr C.A.	Spelderholt Institute for Poultry Research, Beekbergen, Netherlands
Karue, Dr C.N.	Animal Production Division, E.A.A.F.R.D., Nairobi, Kenya
Kaufmann, Prof. Dr W.	Bundesanstalt für Milchforschung, Hermann-Weigmann-Strasse, Kiel, West Germany
Kempen, Dr G.J.M. van	C.L.O. Institute for Animal Nutrition, Hoogland, Netherlands
Kielanowski, Prof. J.	Institute of Animal Physiology and Nutrition, Jablonna, Warsaw, Poland
Kihlberg, Dr R.	Nutrition Unit, Karolinska Institute, Stockholm, Sweden
Kitchen, D.I.	Pauls and Whites (Foods) Ltd., Ipswich, Suffolk
Krause, Dr J.	Biotest-Serum-Institut GmbH, 6072 Dreieichenhain, Postfach 21 - West Germany
Kubasek, F.O.T.	I.C.I. Ltd., Jealott's Hill Research Station, Bracknell, Reading
Laksesvela, Dr B.	Veterinary College of Norway, Oslo, Norway
Landis, Prof. J.	Institut für Tierernährung Eidg. Techn. Hochschule, Universität Strasse, Zurich, Switzerland
Langlands, Dr J.P.	C.S.I.R.O. Pastoral Research Laboratory, Armidale, New South Wales, Australia
Leitgeb, Dr R.	Institut für Tierproduktion, Vienna, Austria
Lettner, Dr F.	Institut für Tierproduktion, Vienna, Austria
Levin, G.	Nutrition Unit, Karolinska Institute, Stockholm, Sweden

List of delegates

Lewis, Prof. D.	University of Nottingham School of Agriculture, Sutton Bonington, Loughborough, Leicestershire
Limborgh, C.L. van	Trouw and Co. N.V. Keizersgracht 702, Amsterdam, Netherlands
Lindsay, Dr D.B.	A.R.C. Institute of Animal Physiology, Babraham, Cambridge
Ljungquist, B.G.	Department of Medical Biochemistry, Gothenburg, Sweden
Low, Dr A.G.	National Institute for Research in Dairying, Shinfield, Reading, Berkshire
Lykkeaa, J.	National Institute of Animal Science, Copenhagen, Denmark
McCracken, Dr K.J.	Agriculture and Food Chemistry Research Division, Newforge Lane, Belfast
McDonald, A.D.	I.C.I., Agricultural Division, Billingham, Teeside
McFarquhar, Dr A.M.	Bernard Matthews Ltd., Norwich, Norfolk
McMeniman, N.P.	Department of Agricultural Biochemistry, University of Newcastle-upon-Tyne
Mackie, I.L.	Beecham Research Laboratories, Nutritional Research Centre, Tadworth, Surrey
Mackie, R.I.	Veterinary Research Institute, Onderstepoort, South Africa
Macrae, Dr J.C.	Hill Farming Research Organisation, Penicuik, Midlothian
Madsen, Dr A.	Forsøgslaboratoriet Afd. for forsøg med svin og hests, Copenhagen, Denmark
Madsen, Dr J.	Royal Veterinary and Agricultural University, Institute of Animal Science, Copenhagen, Denmark

List of delegates

Manchester, Dr K.L.	Department of Biochemistry, University College, Gower Street, London W.C.1
Martens, Dr H.	School of Veterinary Medicine, Hanover, West Germany
Mendes Pereira, E.	Laboratoire d'Études du Metabolisme Azoté, Beaumont, France
Mepham, Dr B.	University of Nottingham School of Agriculture, Sutton Bonington, Loughborough, Leicestershire
Mercer, Dr J.R.	Unilever Research Laboratory, Sharnbrook, Bedford
Miller, W.S.C.	Spillers Ltd., Kennett Nutritional Centre, Kennett, Newmarket, Suffolk
Milligan, Dr L.P.	University of Alberta, Edmonton, Alberta, Canada
Mills, Dr E.	Department of Biochemistry, Royal Infirmary, Glasgow
Millward, Dr D.J.	London School of Hygiene and Tropical Medicine, Keppel Street, London
Milner, C.K.	Shell Research Ltd., Sittingbourne Research Centre, Sittingbourne, Kent
Mitchell, R.M.	Isaac Spencer and Co. (Aberdeen) Ltd., Wellheads Road, Dyce, Aberdeen
Møller, P.E.H.	Aarhus Oliefabrik a/s, Aarhus, Denmark
Mortensen, H.P.	Forsøgslaboratoriet Afd. for forsøg med svin og hests, Copenhagen, Denmark
Muftic, Dr A.	University of Sarajevo, Sarajevo, Yugoslavia
Munck, Dr B.G.	Institute of Medical Physiology A, Juliane Maries Vej 28, Copenhagen, Denmark
Munro, Dr H.N.	Massachusetts Institute of Technology, Cambridge, Massachusetts, U.S.A.
Narang, Dr M.P.	Department of Animal Physiology and Chemistry, 25, Rolighedsvej, Copenhagen, Denmark

List of delegates

Neale, Dr R.J.	University of Nottingham School of Agriculture, Sutton Bonington, Loughborough, Leicestershire
Nicholas, Dr G.A.	Rowett Research Institute, Bucksburn, Aberdeen
Nielsen, Dr H.E.	Forsøgslaboratoriet, Afd. for forsøg med svin og hests, Copenhagen, Denmark
Nikolic, Dr Judith A.	Institute for Application of Nuclear Energy in Agriculture, Veterinary Medicine and Forestry, Baranjska, Yugoslavia
Njaa, Dr L.R.	Government Vitamin Laboratory, Bergen, Norway
Obracević, Prof. C.	Faculty of Agriculture, Nemanjina 6, Beograd-Zemun, Yugoslavia
O'Connell, Dr W.	University College, Lyons, Newcastle, Co. Dublin, Eire
Olieman, F.J.	Mengvoeder UT-Delfia B.V., Maarssen, Netherlands
Olsson, G.I.	Agricultural College of Sweden, Kungsängen Research Station, Uppsala, Sweden
Omstedt, Dr P.T.	Nutrition Unit, Karolinska Institute, Stockholm, Sweden
O'Neill, Dr S.J.B.	Devenish Feed Supplements, Belfast
Oostendorp, D.	Agricultural Bureau, Dutch Nitrogen Fertilizer Industry, Wageningen, Netherlands
Opstvedt, J.	Norwegian Herring Oil and Meal Industry Research Institute, Fyllingsdalen, Norway
Ørskov, Dr E.R.	Rowett Research Institute, Bucksburn, Aberdeen
Osborne, Dr D.F.	The Grassland Research Institute, Maidenhead, Berkshire
Owers, Dr M.J.	Pauls and Whites Foods, Research and Advisory Department, New Cut West, Ipswich

List of delegates 503

Pain, Dr Virginia M.	London School of Hygiene and Tropical Medicine, St. Pancras Way, London N.W.1
Perry, Dr B.N.	A.R.C. Meat Research Institute, Langford, Bristol
Perry, F.G.	Trouw G.B. Ltd., Marston, Cambridge
Pfeffer, E.	Institut für Tierphysiologie und Tierernährung der Universität Göttingen, Oskar-Kellner-Weg 6, Göttingen, West Germany
Pimblett, I.J.	Pfizer, Ltd., Sandwich, Kent
Pion, R.	INRA, Laboratoire d'Étude du Metabolisme Azoté, Beaumont, France
Playne, Dr M.J.	Division of Tropical Agronomy, CSIRO Davies Laboratory, Townsville, Queensland, Australia
Płonka, Dr S.	Institute of Animal Production, Sarego Street, Krakow, Poland
Poppe, Prof. S.	Sektion Tierproduktion Universität, Rostock, East Germany
Porter, P.	RHM Research Ltd., Lincoln Road, High Wycombe, Buckinghamshire
Porter, Dr J.W.G.	National Institute for Research in Dairying, Shinfield, Reading
Radcliffe, J.D.	Rowett Research Institute, Bucksburn, Aberdeen
Rerat, Dr A.A.	INRA, Centre National de Recherches Zootechniques. Jouy en Josas, France
Robb, Dr. J.	Unilever Research Laboratories, Colworth House, Sharnbrook, Bedford
Robert, J.C.	Société de Chimie, Organique et Biologique, Commentry-Alliei, France
Roberts, Dr R.	J. Bibby and Sons Ltd., Rumford Place, Liverpool

List of delegates

Robinson, Dr A.	I.C.I. Pharmaceuticals Division, P.O. Box 25, Alderley Park, Macclesfield, Cheshire
Roets, E.	Veterinary Faculty of the University of Ghent, Casinoplein 24, Ghent, Belgium
Rogers, Prof. Q.R.	School of Veterinary Medicine, University of California, Davis, California, U.S.A.
Rook, Prof. J.A.F.	Hannah Research Institute, Ayr, Scotland
Rowlinson, A.	Feed Services (Livestock) Ltd., Corsham, Wiltshire
Saelzer, Dr V.	Bundesanstalt für Milchforschung, Hermann-Weigmann-Strasse, Kiel, West Germany
Sambeth, Dr W.	Farbwerke Hoechst AG, Abteilung Veterinarmedizinische Forschung, Frankfurt/Main 80, West Germany
Schürch, Prof. A.	Institut für Tierernährung, Zurich, Switzerland
Scott, L.J.	Colborn Vitafeeds Ltd., Barton Mills, Canterbury, Kent
Sender, Dr P.M.	London School of Hygiene and Tropical Medicine, Keppel Street, London W.C.1
Sibbald, Dr. I.R.	Animal Research Institute, Central Experimental Farm, Ottawa, Ontario, Canada
Simonsson, Dr A.A.A.	Lantbrukshogskolan Lövsta, Hammarby, Uppsala 1, Sweden
Sinnott, M.L.	Department of Agriculture and Fisheries, Dublin, Eire
Slater, Dr J.S.	Animal Diseases Research Association, Moredun Institute, Edinburgh, Scotland
Smekalov, Dr N.A.	HU Union Research Institute of Animal Husbandry, Podolsk, Dubrovici, Moscow, U.S.S.R.
Smith, Diana C.	I.C.I. Research Station, Jealott's Hill, Bracknell, Berkshire

Smith, Dr R.H.	National Institute for Research in Dairying, Shinfield, Reading
Smith, T.L.	Cooper Nutrition Products Ltd., Stepfield, Witham, Essex
Smithers, Dr M.J.	Imperial Chemical Industries Ltd., Pharmaceuticals Division, Mereside, Alderley Park, Macclesfield, Cheshire
Spaendonck, R. van	Laboratory of Animal Nutrition, University of Ghent, Merelbeke, Belgium
Srivastava, K.N.	Department of Animal Physiology and Chemistry, Rolighedsvej 25, Copenhagen, Denmarl
Stobo, Dr I.J.F.	National Institute for Research in Dairying, Shinfield, Reading
Stothers, Prof. S.C.	University of Manitoba, Winnipeg, Manitoba, Canada
Stranks, M.H.	A.D.A.S. (M.A.F.F.), Wye, Ashford, Kent
Svanberg, U.S.O.	Department of Medical Biochemistry, University of Gothenburg, Gothenburg, Sweden
Symons, Dr L.E.A.	Division of Animal Health, C.S.I.R.O. McMaster Laboratory, Glebe, New South Wales, Australia
Syrjala, Dr L.	University of Helsinki, Helsinki, Finland
Terpstra, Dr K.	Spelderholt Institute for Poultry Research, Beekbergen, Netherlands
Terry, R.A.	Grassland Research Institute, Hurley, Maidenhead, Berkshire
Thomas, Dr P.C.	The Hannah Research Institute, Ayr, Scotland
Thomke, Dr S.	Department of Animal Husbandry, Agricultural College, Uppsala, Sweden
Thompson, F.	Rumenco Group, Derby Road, Burton-on-Trent

List of delegates

Thompson, Dr J.R.	Department of Animal Science, University of Alberta, Edmonton, Alberta, Canada
Treacher, Dr R.J.	Institute for Research on Animal Diseases, Compton, Newbury, Berkshire
Treullé, J.G.	Kemovit A/S, Rygards Allé 131, 2900 Hellerup, Denmark
Vantcheva, Dr Z.	Research Institute of Animal Breeding, Kostinbrod, Sofia, Bulgaria
Vaz Portugal, Prof. A.	Vale de Santarem, Portugal
Wakeling, Dr A.E.	I.C.I. Pharmaceuticals Division, Alderley Park, Macclesfield, Cheshire
Waldo, D.R.	U.S. Department of Agriculture, Barc-East, Beltsville, U.S.A.
Walker, Dr D.M.	University of Sydney, Sydney, New South Wales, Australia
Walstra, Dr P.	Research Institute for Animal Husbandry, Schoonoord, Zeist, Netherlands
Waterworth, D.G.	I.C.I., Jealotts Hill Research Station, Bracknell, Berkshire
Webb, N.W.C.	I.C.I. Chemical Industries Ltd., Pharmaceuticals Division, Mereside, Alderley Park, Macclesfield, Cheshire
Webster, Dr A.J.F.	Rowett Research Institute, Bucksburn, Aberdeen, Scotland
Weerden, Dr E.J. van	ILOB, Haarweg 8, Wageningen, Netherlands
White, Dr F.	Rowett Research Institute, Bucksburn, Aberdeen
Whittemore, Dr C.T.	University of Edinburgh, Edinburgh, Scotland
Wilby, D.T.	W.F. Tuck and Sons Ltd., The Mills, Burston, Diss, Norfolk

Wilde, Dr R. de	Laboratory of Animal Nutrition, University of Ghent, Merelbeke, Belgium
Wilson, Dr B.J.	A.R.C. Poultry Research Centre, King's Building, West Mains Road, Edinburgh, Scotland
Wilson, Dr P.N.	BOCM Silcock Ltd., Basing View, Basingstoke, Hampshire
Witt, G.T.	Seemeel Ltd., Carpenters Road, Stratford, London E.15

INDEX

Alanine, effect on lysine transport, 78
Albumin catabolism, 110
Amino acid (*see also* under individual names)
  absorption, 112-114, 378
  and wool growth, 478
  anabolism – Michaelis constants, 282
  availability, 119, 139-158, 264, 378-381, 385
  biological value, 404
  blood, 259-262
  carcass composition, and, 212, 262-264, 311, 384
  catabolism – Michaelis constants, 183, 282
  digestibility of, 142, 152, 241, 379, 381, 386, 434
  energy source, as, 183-195
  faecal, 241
  food, in, 404
  food intake and, 310
  gluconeogenesis, and, 185-191
  imbalances, nutritional and metabolic effects of, 279
  inter-relationships, 313, 314
  intestinal mucosa, 263
  liver, 262
  oxidation, 193
  pattern, in dairy cow intestine, 434
  plasma, 117-128, 259-262, 269, 286, 419
  potassium, and, 315
  renal absorption of, 316
  requirement,
    age, effect of, 309, 387-389
    calf, pre-ruminant, 265
    cow (lactation), 441-455

Amino acid *continued*
  requirement *continued*
    duck, 318
    estimation technique, 369, 372, 383-385
    factorial estimates of, 421
    game birds, 319
    genetics, effect of, 308
    goose, 319
    guinea fowl, 319
    lamb, pre-ruminant, 269
    laying hen, 328
    maintenance (pig), 383
    meat chicken, 265, 305, 316
    non-protein nitrogen and, 469
    pig (boar), 374-377
    pig (growing), 265, 369-382, 383-394
    pig (lactation), 362
    pig (pregnancy), 361
    pullet, 326
    quail, 318
    rat, 265
    ruminants, 417-424
    sex, effects of, 387-389
    sperm production, 374-377
    turkey, 318
  rumen bacteria, 397
  small intestine, digesta, composition of, 107, 124
  supply,
    blood, to the, 447
    dietary protein and, 401
    intestine of dairy cow, to, 425-438
    protein synthesis, and, 19-36, 56-64
  simulation of, 442

Amino acid *continued*
 synthesis
  rumen bacteria, by, 397, 398
  supply of, and, 56-64
 transport, 73-95
  basic, 77-80
  basolateral membrane, across, 85, 86
  effect of hormones, 36, 43
  K_t values, 80
  neutral, 75-76
  specificity of, 75-78
 uptake by rumen bacteria, 397
Ammonia, 168-177
 absorption from rumen, 169-171, 177
 concentration, in blood, 172, 173
 microbial protein synthesis and, 225, 401
 detoxification by liver, 172
 production from urea, 173
 toxicity, 169, 171, 174
 transport, pH dependence, 171-173
Animal model, application of, 483
Antinutritional factors, 244
Appetite, *see* Food intake
Arginase, 308, 314
Arginine, lysine interaction, 314
Arginine requirement (*see also* Amino acid requirement)
 chick, 308, 314
 pig (growing), 393
Available lysine, 141-144

Basolateral membrane, amino acid transport across, 85, 86
Biological value (BV), 240, 417, 418
Blood metabolites, as measures of protein adequacy, 250, 359
Branched chain fatty acids, 219
Brush border membranes, 75, 88

Choline, 314
Compensatory growth, 372, 390
Competitive exchange diffusion, 83
Computer simulation and ruminant protein nutrition, 477-491

Cysteine, 152
Cystine (*see also* Methionine and cystine; Sulphur amino acids), 152

Deoxyribonucleic acid,
 hormones, and, 37
 transcription, 3
Dwarfing gene (poultry), 309

Egg production,
 efficiency of protein utilisation, 345
 energy cost of, 210, 211
 requirements for, *see* Amino acids; Protein
Endogenous nitrogen, 240
 absorption, 121
 composition, 108-110
 digestibility, 111
Enzymes (*see also* under individual names)
 adaptation of, 105-107
 digestive, 102-107
 gastrectomy, effect of, 103
Energy,
 amino acid as source of, 183-195
 cost of egg production, 210-211
 cost of fat deposition, 202-204
 cost of milk production, 210-213
 cost of protein deposition, 202-204, 207-215
 cost of protein synthesis, 199, 400
 cost of urea synthesis, 197-198
 cost of uric acid synthesis, 198, 201
 nitrogen interactions in rumen, 217
 requirements,
  maintenance, 209, 212
  rumen micro-organisms, 220
 value of high-energy phosphate bonds, 209
Epithelial cells,
 crypt, 73
 isolated, 74, 77, 87

Fat deposition, energy cost of, 202-204
Fluorodinitrobenzene, available lysine and, 144
Flux, uni-directional, 74
Food,
  choice, imbalance, and, 284
  intake,
    effect of sex of pigs, 375
    imbalance, and, 284
    protein absorption in ruminants, and, 485
    protein utilisation in ruminants, and, 472
    utilisation, methionine deficiency, and, 201
Furosine, 145-147

Glucocorticoids, 41
Gluconeogenesis, 185-191
  glucagon, 191
  role of insulin, 192
Glutamic acid, proline interaction, 316
Glycine, serine, methionine, interaction, 314
Growth hormone, 21

High energy phosphate bonds, energy value of, 209
Histidine requirement (see also Amino acid requirement)
  pig (growing), 393
Hormones,
  amino acid transport, effect on, 36, 43
  deoxyribonucleic acid, effect on, 35
  glucocorticoids, 39
  growth, 21
  insulin, 21-28, 41, 61
  mechanism of action, 6
  messenger ribonucleic acid, effect on, 37
  oestradiol, 39
  oestrogen, 39
  progesterone, 39
  protein catabolism, and, 44

Hormones *continued*
  RNA synthesis, effect on, 5
  steroid, 19, 20, 28, 37-38
  translation rate, 37

Imino acid transport, 80-82
Indirect measures of protein adequacy (see also Protein assessment), 249, 359, 385
Insulin, 21-28, 41, 61
Isoleucine,
  inhibition of lysine transport, 79
  requirement (see also Amino acid requirement),
    pig (growing), 392-393

Kwashiorkor, plasma amino acids, 250

Lactation,
  amino acid requirements of cow, 441-455
  milk production, energy cost of, 210-213
  nitrogen metabolism in sow, 358, 359
  protein and amino acid requirements of sow, 362, 364
Leucine,
  effect on lysine transport, 78
  requirement (see also Amino acid requirement)
    pig (growing), 393
Liver perfusion, 197
Lysine,
  arginine interaction, 314
  availability, 141-144
  carcass composition, pigs, 384
  digestibility, 379
  ϵ-fructose, 148-151
  ketoglutarate reductase, 308
  sugar reactions, with, 151
  transport,
    alanine, effect of, 78
    inhibition by other amino acids, 79
    leucine, effect of, 78
    transstimulation by methionine, 86

512 *Index*

Lysine *continued*
  requirement (*see also* Amino acid requirement)
    laying hen, 336
    meat chicken, 310
    pig (boar), 375-377
    pig (growing), 370, 381, 383-385, 389-391
    pig (lactation), 364
    pullet, 326

Maillard reaction, 141, 145, 146
Messenger ribonucleic acid,
  hormones, and, 35
  synthesis, 4
  translation, 4
Methionine, 152, 153
  absorption, 124-126
  carcass composition, pigs, 384
  deficiency and food utilisation, 201
  plasma concentration and rate of passage, 420
  transstimulation of lysine transport, 86
Methionine and cystine,
  digestibility, 379
  requirement (*see also* Amino acid requirement)
    feathering and, 313
    laying hen, 329, 334
    meat chicken, 309
    pig (boar), 375-377
    pig (growing), 371, 381, 383-385, 391
    pullet, 326
3-Methylhistidine, 9, 64
Microbial markers, 222
Milk production (*see also* Lactation)
  energy cost of, 210-213
Model, use of in animal nutrition, 483

Net protein ratio (NPR), 234
Net protein utilisation (NPU), 240
Niacin, tryptophan interaction, 379

Nitrogen (*see also* Amino acid; Protein; Non-protein nitrogen)
  balance, *see* Nitrogen retention
  balance index, 240
  digestibility, 279
  endogenous,
    absorption, 121
    composition, 108-110
    digestibility, 111
  energy, interactions in rumen, 217
  gas, rumen micro-organisms and, 219
  metabolism,
    pigs (lactation), 358, 359
    ruminants, 407, 411
  passage through wall of digestive tract, 159
  requirements,
    rumen bacteria, 218, 221, 468
    rumen protozoa, 219, 224, 225, 468
  retention,
    growth in pigs and, 384
    pigs (sows), 356
    sheep, 461, 484
  supply,
    intestine of dairy cow, to, 425-439
    tissues, from, 447
  utilisation, ruminants, 397-415
Non-protein nitrogen,
  laying hen, 340
  meat chicken, 316
  ruminants, 405
    amino acid requirement and, 469
    ammonia, 404
    commercial use and, 477
    dietary, utilisation of (ruminants), 405
    factors affecting utilisation, 457-476
    metabolism, 407
    plasma urea, and, 406
    protein supply, and, 435
    substrate, and, 406
Nucleolus, 4
Nutrient absorption, quantitative prediction of, 480

Index   513

Oestradiol, 39
Oestrogen, 39

Pancreatic enzymes, 103-106
Pepsins, 102, 105
Peptidases, 75, 90
Peptide transport, 89-91
 across brush border, 90
Phenylalanine plus tyrosine requirement (*see also* Amino acid requirement)
 pig (growing), 393
Polysomes, 10, 21-24, 30
 imbalance and, 289
Pregnancy, protein requirements of sow, 356, 357, 361, 362
Progesterone, 39
Proline, glutamic acid interaction, 316
Proteases, 140, 143
Protein,
 absorption, 97
  food intake, and, 485
 assessment, 353, 478
  biological value (BV), 240
  energy, effect of, 236
  enzyme activities, and, 252
  factorial methods, 353, 355, 383
  growth studies, 384
  indirect methods, 249, 359, 385
  muscle fibres, and, 254
  net protein ratio (NPR), 234
  net protein utilisation (NPU), 240
  nitrogen balance, 359
  nitrogen balance index, 240
  nitrogen growth index, 233
  optimum protein ratio, 234
  partial carcass measure, and, 254
  plasma amino acids, and, 250, 359
  plasma proteins, and, 249
  plasma urea, and, 251
  protein efficiency ratio (PER), 234
  repletion methods, 237
  separate feeding, effect of, 236
  urinary allantoin, and, 242
  urinary nitrogen, and, 242
  urinary urea, and, 242

Protein *continued*
 breakdown, 401
 catabolism (*see also* Protein turnover)
  effect of diet, 64
  effect of insulin, 61
  hormones, 46
  kinetics, 62
  measurement, 62
 deposition,
  eggs, 210
  energy cost of, 202-204, 207-215
  milk, 210-213
  ruminant, in, 459, 461
  unitary cost of protein (ECPD), 207
 digestion, 97
  kinetics of, 111, 112
  ruminant, 434, 487
  type of protein, effect of, 107-128
 fermentation, 402
 microbial, 430, 434
  synthesis, 400, 457, 472
  nitrogen, dietary and, 450
  milk nitrogen and, 450
 nutrition,
  computer simulation, 477-489
  pig (breeding sow), 353-367, 374-377
  pig (growing), 265, 369-382, 383-394
  ruminant and body composition, 474
  sheep, 461
 protection, 464
 requirement (*see also* Amino acid; Nitrogen requirement)
  dairy cow (lactating), 441-455
  duck, 318
  factors affecting (laying hens), 342
  game bird, 319
  goose, 319
  guinea fowl, 319
  laying hen, 324
  pig (boar), 374

Protein *continued*
  requirement *continued*
    pig (growing), 265, 369-382, 383-394
    pig (lactation), 358, 362, 374-377
    pig (pregnancy), 356, 357, 361, 362
    pullet, 328
    quail, 318
    ruminant young, 457-476
    turkey, 318
  supply,
    measurement in dairy cow, 426
    ruminant, 401
    tissues from, 447
  synthesis,
    absorbed nutrients, and, 481
    adaptation, 16
    amino acid supply, 8-14, 19-33, 56-64
    anabolic hormones, 19, 21
    energy cost of, 199, 400
    hormones, effect of, 37-48
    imbalance, and, 288
    mechanism, 3-18
    microbial, 400, 457, 472
  turnover,
    brain, 65
    energy expenditure, 54
    heart, 65
    kidney, 65
    liver, 65
    measurement, 49-53, 62
  utilisation,
    in ruminant, 401, 457-466, 472
    meat chicken, 200
    muscle, 65
    sheep, 200
    tissue organelles, 66
Protein efficiency ratio (PER), 234

Requirement (*see also* Amino acid; Protein etc.)
  response curves, theory of, 324
Ribosomes, 8, 10

RNA,
  synthesis, effect of hormones, 5
  transfer, imbalance and, 291
Rumen micro-organisms,
  amino acid synthesis by, 397, 398, 399
  carbon skeleton requirements, 219
  composition, 217
  diet, effect on, 218, 225
  fermentation, 431, 460
  pH, effect on, 218
  protein production, efficiency, 224
  rumen ammonia, and, 225
  uptake of amino acids, 398

Serine, glycine, methionine interaction, 314
Short circuit current, 87
Sodium gradient hypothesis, 86-88
Steroid hormones, 19, 21, 28, 39, 40, 41
Stomach emptying,
  carbohydrates and fats, effects of, 100
  heated and unheated proteins, and, 147-148
  proteins, effect of, 98-100
Streptococcus zymogenes, 144
Sugar transport, interaction between amino acids, 89
Sulphur amino acids (*see also* under individual names)
  availability, 152-154
  glycine, serine interaction, 314
  requirements,
    feathering, and, 313
    laying hen, 329, 334
    meat chicken, 309
    pullet, 326

Threonine requirement (*see also* Amino acid requirement)
  pig (growing), 387
Transconcentration effects, 82-84
Translation rate, hormones and, 37

Tryptophan,
  inhibition of lysine transport, 79
  requirement (*see also* Amino acid requirement)
    laying hen, 337
    pig (growing), 391-392
    pig (lactation), 362
  transport, effect of lysine, 77

Urea,
  blood concentration in ruminal vein, 164
  concentration,
    gradient across rumen, 165
    in body fluids, 166, 167
  conversion to rumen ammonia, 163
  degradation, 162, 163, 168
  metabolism, 406, 407
  passage through rumen wall, 160-163
  re-cycling, 160
  synthesis, 197, 198

Urease activity, 165, 168
Uric acid, 407
  excretion, 202
  synthesis, energy cost of, 198, 201
Unitary cost of protein deposition (ECPD), 207

Valine,
  inhibition of lysine transport, 79
  requirement (*see also* Amino acid requirement)
    pig (growing), 393

Wool growth, amino acid absorption, and, 478

Y_{ATP} in rumen, 220, 400